Lecture Notes in Biomathematics

Managing Editor: S. Levin

19

Measuring Selection in Natural Populations

Edited by
F. B. Christiansen and T. M. Fenchel

Springer-Verlag
Berlin Heidelberg New York 1977

Editors

Freddy B. Christiansen
Tom M. Fenchel
Department of Ecology and Genetics
Ny Munkegade, Aarhus Universitet
DK-8000 Aarhus C/Denmark

Library of Congress Cataloging in Publication Data
Main entry under title:

Measuring selection in natural populations.

(Lecture notes in biomathematics ; 19)
Proceedings of the symposium "Measuring selection in
natural populations," held in Sandbjerg Manor House in
southern Jutland May 10-14, 1976.
Bibliography: p.
Includes index.
1. Natural selection--Congresses. 2. Evolution--
Congresses. I. Christiansen, Freddy B., 1946-
II. Fenchel, Tom. III. Series.
QH375.M4   575.1   77-11040

AMS Subject Classifications (1970): 92-02, 92 A 05, 92 A 10, 92 A 15

ISBN-13: 978-3-540-08435-8   e-ISBN-13: 978-3-642-93071-3

DOI: 10.1007/978-3-642-93071-3

PREFACE

The present volume constitutes the proceedings of the Symposium: "Measuring Selection in Natural Populations", held in memory of late professor Ove Frydenberg. The Symposium took place in Sandbjerg Manor House in Southern Jutland May 10-14, 1976. The purpose of the symposium was to reflect contemporary research on the mechanisms of biological evolution.

The arrangement of the symposium was possible only with the assistance and work of many other people. Professor Ole Barndorff-Nielsen took an active part in the planning and arrangement of the symposium. Our gratitude is also due to Ellen Christensen, Jens Ole Frier, Arno Jensen, Preben Jensen and Anne Nissen, all of whom took care of the practical arrangements at the symposium. The publication was prepared with the assistance of Arno Jensen, Preben Jensen, Verner Blak Nielsen and Kirsten Svendsen.

The Symposium was financed by grants from The Danish Natural Science Research Council, The Danish Atomic Energy Commission and The University of Aarhus.

Aarhus, June 1977

Freddy B. Christiansen   Tom M. Fenchel

LIST OF PARTICIPANTS

AND CONTRIBUTORS

ALLARD, R.W.: Department of Genetics, University of California, Davis, California 95616, USA.

ANDERSEN, A. HOLST: Department of Theoretical Statistics, Institute of Mathematics, University of Aarhus, Ny Munkegade, DK-8000 Aarhus C, Denmark.

AYALA, F.J.: Department of Genetics, University of California, Davis, California 95616, USA.

BARKER, J.S.F.: Department of Animal Husbandry, University of Sydney, Sydney 2006, New South Wales, Australia.

BARNDORFF-NIELSEN, O.: Department of Theoretical Statistics, Institute of Mathematics, University of Aarhus, Ny Munkegade, DK-8000 Aarhus C, Denmark.

BEARDMORE, J.: Department of Genetics, University College of Swansea, Singleton Park, Swansea SA2 8PP, Great Britain.

BENGTSSON, B.O.: Institute of Genetics, University of Lund, S-223 62 Lund, Sweden.

BERG, UFFE: Department of Zoology, University of Copenhagen, Universitetsparken 15, DK-2100 København Ø, Denmark.

BODMER, WALTER F.: Genetics Laboratory, Department of Biochemistry, University of Oxford, South Parks Road, Oxford, OX1 3QU, Great Britain.

BUNDGAARD, J.: Department of Ecology and Genetics, University of Aarhus, Ny Munkegade, DK-8000 Aarhus C, Denmark.

CAVALLI-SFORZA, LUIGI L.: Department of Genetics, Stanford University School of Medicine, Stanford, California 94305, USA.

CHARLESWORTH, B.: School of Biological Sciences, University of Sussex, Falmer, Brighton BN1 9QG, Great Britain.

CHARLESWORTH, D.: School of Biological Sciences, University of Sussex, Falmer, Brighton BN1 9QG, Great Britain.

CHRISTENSEN, BENT: Department of Zoology, University of Copenhagen, Universitetsparken 15, DK-2100 København Ø, Denmark.

CHRISTENSEN, ELLEN: Department of Ecology and Genetics, University of Aarhus, Ny Munkegade, DK-8000 Aarhus C, Denmark.

CHRISTIANSEN, FREDDY B.: Department of Ecology and Genetics, University of Aarhus, Ny Munkegade, DK-8000 Aarhus C, Denmark.

CLARHOLM, MARIANNE: Department of Ecology and Genetics, University of Aarhus, Ny Munkegade, DK-8000 Aarhus C, Denmark.

CLARKE, B.*: Genetics Department, University of Nottingham, University Park, Nottingham NG7 2RD, Great Britain.

CLEGG, M.T.: Department of Botany, University of Georgia, Athens, Georgia 30602, USA.

DIJKEN, F.R. van: Department of Population and Evolutionary Biology, Rijksuniversiteit, Utrecht, The Netherlands.

DOLL, HANS: Agricultural Research Department, Danish AEC, Risø, DK-4000 Roskilde, Denmark.

DOYLE, R.: Department of Biology, Dalhousie University, Halifax, Nova Scotia, Halifax B3H 4H1, Canada.

EANES, WALTER F.: Department of Ecology and Evolution, State University of New York, Stony Brook, N.Y. 11794, USA.

EDWARDS, A.W.F.: Gonville and Caius College, Cambridge, Great Britain.

EWENS, W.J.: Department of Biology, University of Pennsylvania, Philadelphia, PA 19104, USA.

FELDMAN, M.: Department of Biological Sciences, Stanford University, Stanford, California 94305, USA.

FENCHEL, T.: Department of Ecology and Genetics, University of Aarhus, Ny Munkegade, DK-8000 Aarhus C, Denmark.

FRIER, J.O.: Department of Ecology and Genetics, University of Aarhus, Ny Munkegade, DK-8000 Aarhus C, Denmark.

GAFFNEY, PATRICK M.: Department of Ecology and Evolution, State University of New York, Stony Brook, N.Y. 11794, USA.

---

*Due to ill health the contribution by Professor Clarke could not be included in this volume. It will be published elsewhere.

GILLESPIE, J.H.: Department of Biology, University of Pennsylvania, Philadelphia, PA 19104, USA.

HALKKA, O.: Department of Genetics, University of Helsinki, P. Rautatiekatu 13, SF-00100, Helsinki 10, Finland.

HARTL, D.: Department of Biological Sciences, Purdue University, Lilly Hall of Life Sciences, West Lafayette, Indiana 47907, USA.

HJORTH, J.P.: Department of Ecology and Genetics, University of Aarhus, Ny Munkegade, DK-8000 Aarhus C, Denmark.

HOFFMANN-JØRGENSEN, J.: Institute of Mathematics, University of Aarhus, Ny Munkegade, DK-8000 Aarhus C, Denmark.

HOORN, A.J.W.: Department of Population and Evolutionary Biology, Rijksuniversiteit, Utrecht, The Netherlands.

JENSEN, S. KOLDING: Department of Ecology and Genetics, University of Aarhus, Ny Munkegade, DK-8000 Aarhus C, Denmark.

JELNES, JENS: Department of Zoology, University of Copenhagen, Universitetsparken 15, DK-2100 København Ø, Denmark.

JOHNSON, G.B.: Department of Biology, Washington University, St. Louis, Missouri 63130, USA.

JONES, J.S.: Royal Free Hospital, School of Medicine, University of London, 8 Hunter Street, London WC1 1BP, Great Britain.

JONG, G. de: Department of Population and Evolutionary Biology, Rijksuniversiteit, Utrecht, The Netherlands.

KAHLER, A.L.: Department of Genetics, University of California, Davis, California 95616, USA.

KEIDING, N.: Institute of Mathematical Statistics, University of Copenhagen, Universitetsparken 15, DK-2100 København Ø, Denmark.

KOEHN, R.K.: Department of Ecology and Evolution, State University of New York, Stony Brook, N.Y. 11794, USA.

LAKOVAARA, S.: Department of Genetics, University of Oulu, Postilokero 191, SF-90101 Oulu 10, Finland.

LANGLEY, C.: National Institute of Environmental and Health Sciences, Research Triangle Park, North Carolina, USA.

LANKINEN, PEKKA: Department of Genetics, University of Oulu, Postilokero 191, SF-90101 Oulu 10, Finland.

LIGNY, WILHELMINA de: Netherlands Institute for Fisheries Investigation, Ijmuiden, The Netherlands.

LOKKI, J.: Department of Genetics, University of Helsinki, P. Rautatie-katu 13, SF-00100 Helsinki 10, Finland.

MAYNARD SMITH, J.: School of Biological Sciences, University of Sussex, Falmer, Brighton, Sussex, BN1 9QG, Great Britain.

MICHOD, R.: Department of Zoology, University of Georgia, Athens, Georgia 30602, USA.

MIKKOLA, ERKKI: Department of Mathematics, University of Helsinki, P. Rautatiekatu 13, SF-00100, Helsinki 10, Finland.

MUKAI, T.: Department of Biol. Faculty, Kyushu University, Hakazaki, Fukuokashi 812, Japan.

NIELSEN, B.V.: Department of Ecology and Genetics, University of Aarhus, Ny Munkegade, DK-8000 Aarhus C, Denmark.

NIELSEN, J. TØNNES: Department of Ecology and Genetics, University of Aarhus, Ny Munkegade, DK-8000 Aarhus C, Denmark.

O'DONALD, P.: Department of Genetics, University of Cambridge, Cambridge CB4 1XH, Great Britain.

PARKIN, D.T.: Department of Genetics, University of Nottingham, Nottingham NG7 2RD, Great Britain.

POULSEN, E.T.: Institute of Mathematics, University of Aarhus, Ny Munke-gade, DK-8000 Aarhus C, Denmark.

RICHARDS, LAURA J.: Department of Biology, Dalhousie University, Halifax, Nova Scotia, Halifax B3H 4J1, Canada.

ROUGHGARDEN, J.: Department of Biological Science. Stanford University, Stanford, California 94305, USA.

SANDFAER, J.: Agricultural Research Department, Danish AEC, Risø, DK-4000 Roskilde, Denmark.

SAURA, A.: Department of Genetics, University of Helsinki, P. Rauta-tiekatu 13, SF-00100 Helsinki 10, Finland.

SCHARLOO, W.: Department of Population and Evolutionary Biology, Rijks-universiteit, Utrecht, The Netherlands.

SIMON, CHRISTINE M.: Department of Ecology and Evolution, State University of New York, Stony Brook, N.Y. 11794, USA.

SIMONSEN, V.: Department of Ecology and Genetics, University of Aarhus, Ny Munkegade, DK-8000 Aarhus C, Denmark.

SLOTH, L.: Department of Ecology and Genetics, University of Aarhus, Ny Munkegade, DK-8000 Aarhus C, Denmark.

SUOMALAINEN, E.: Department of Genetics, University of Helsinki, P. Rautatiekatu 13, SF-00100 Helsinki 10, Finland.

THOMPSON, E.A.: Statistical Laboratory, Department of Pure Mathematics and Mathematical Statistics, Cambridge University, 16 Mill Lane, Cambridge CB2 1SB, Great Britain.

THOMSON, G.J.: Genetics Laboratory, Department of Biochemistry, University of Oxford, South Parks Road, Oxford, OX1 3QU, Great Britain.

THÖRIG, G.E.W.: Department of Population and Evolutionary Biology, Rijksuniversiteit, Utrecht, The Netherlands.

VAETH, M.: Department of Theoretical Statistics, Institute of Mathemathics, University of Aarhus, Ny Munkegade, DK-8000 Aarhus C, Denmark.

WARD, R.D.: Department of Genetics, University College of Swansea, Singleton Park, Swansea SA2 8PP, Great Britain.

ØSTERGAARD, H.: Department of Ecology and Genetics, University of Aarhus, Ny Munkegade, DK-8000 Aarhus C, Denmark.

OVE FRYDENBERG 1929-1975

The symposium "Measuring Selection in Natural Populations" was held
in the memory of the late Ove Frydenberg, who died at the age of only
45 a little more than a year ago; the event was a tragic loss to his
friends and family and to Danish science. His international contribution to genetics will appear from the following bibliography and is
probably, at least in part, known to most population geneticists. Also
the paper given by Freddy Christiansen will account for a research project Ove Frydenberg initiated and worked on during the last years of
his life. I would therefore mainly here say a few words about his contributions to the Danish scientific community and about Ove Frydenberg as a person.

  Ove Frydenberg originally studied veterinary science at the School
of Agriculture and Veterinary Science in Copenhagen but after a few
years he switched to the University of Copenhagen where he studied zoology and genetics. Although he eventually graduated in genetics, his
first scientific contribution was a classical study of the distribution and taxonomy of Danish fruitflies. It remained characteristic
for him to have a broad scientific outlook and to take interest in
all kinds of biology, as long as the work was characterized by quality and yielded real insight.

  After his graduation, he spent one year in Brazil together with

Dobzhansky followed by a year in Madison with Sewall Wright. In 1957
he became lecturer in genetics at the University of Copenhagen. After
another year in Brazil, as a visiting professor, he became professor
of genetics and founder of the Department of Genetics at the University of Aarhus in 1967.

Fairly recently it has become unsuitable for scientists to live
in an ivory tower and many of them now talk much about their own social relevance and usefulness. Ove Frydenberg always felt a true obli-
gation to be useful to the surrounding society, something which in
no way hindered him in doing good science at the same time. At various
times during his career he served as member of the Research Council,
as member of the Atomic Energy Commission, as lecturer in genetics
for dentistry students, as consultant and participant in projects concerned with the improvement of barley strains, and as member of a
commission for initiating and administrating scientific studies on
Greenland. He was very active in the scientific education of the public
at large. For a period he was a newspaper science journalist; he wrote
popular articles on various biological topics, lectured for a variety
of audiences and edited a highschool biology textbook. This activity
was in particular centered around how scientific thought and insight
may contribute to rational decisions of the society. He was, for example very active in showing how genetical facts and insight could
defy irrational - but still widely held - superstitions about race
and eugenics; topics on which he co-authored a book in Danish. He
also spent much time in explaining the genetic effects of radiation
and recently played a role in bringing rationality and sense of proportions into an otherwise heated public debate on nuclear reactors.

His preoccupation with the popularization of science in order to
distribute rationality and insight in the general public was in a way
paradoxical; Ove Frydenberg could have a quite arrogant attitude and
remained a pessimist with respect to man and society. He did certainly
not join the bandwagon of pollution hysteria or the doomsday missio-
naries, this would have been contrary to his nature; rather he was
fundamentally sceptical with respect to human nature and human motives.

To anyone who ever met Ove Frydenberg two characteristics were
evident; his mastering of the Danish language and his unusually well
developed sense of humor. His sense of humor was a very important
and strong component of his nature and to him a vital quality of life.
He could quite consciously evaluate the substance of a scientific
lecture by the humoristic sense of the lecturer and he had a keen eye

for pompousness, bores of all persuasions, and pathos.

To Ove Frydenberg, language was above all the communication of
clear thoughts; this perhaps is related to the fact that he had only
little sense of music and arts and was completely alien to philosophy,
religion and metaphysics. In science he was wary of false analogies
and the extension of science into the field of speculation. He was
eager to demonstrate how population genetical theory shows that it is
irrational to perform eugenics against recessive genetical disorders
but he would very reluctantly take part in a "nature-nurture" debate
where science can contribute very little and is mainly used as a cover for one or another political or social attitude.

His sense of humor, personal charm and his precision in verbal
and written language made him a brilliant teacher - and also a successful diplomat. During his time in Århus he was a kind of unoffi-
cial dean for all the biology departments. He could rapidly analyse
a situation and find - often untraditional - solutions to faculty
problems. He favored rapid solutions and although he was perfection-
istic and valued quality highly, he resented slowness disguised as
perfection and prudence just as he resented narrowmindedness in science
administration and in other aspects of life. Although he was himself
a few times victim to the jealousy and narrowmindedness of lesser
spirits, he remained a central figure to whom everyone came for ad-
vice and solutions; for the biologists in Århus and in fact for the
whole Danish scientific community. This role was also natural to Ove
Frydenberg, due to his general and broad insight in a variety of biological and other natural sciences.

The broad biological outlook of Ove Frydenberg meant that we could
have chosen between several different topics for a symposium held in
the honor of his memory. Still the theory of evolution and problems
of natural selection was especially close to him. This is evident
from much of his scientific activity and popular writings, and also
from the fact that he revised and republished the original Danish
translation from 1872 of "The Origin of Species". Also, evolutionary
biology is if anything the unifying principle which gives coherence
and perspective to biological sciences.

                                        Tom Fenchel

BIBLIOGRAPHY OF OVE FRYDENBERG

1, Frydenberg, O. 1955. Survey of Danish *Drosophila* species.
   Dros. Inf. Serv. 29: 119.

2. Frydenberg, O. 1955. On the types of *Drosophila picta* Zett. and
   S. spurca Zett. (Drosophilidae, Dipt.) and a new description of
   the former.
   Ent. Medd. (Copenhagen) 27: 104-112.

3. Frydenberg, O. 1956. The Danish species of *Drosophila* (Dipt.).
   Ibid. 27: 249-294.

4. Frydenberg, O. 1956. The synonymy, relationship, and distribution of *Drosophila confusa* Stæger (Dipt.).
   Ibid. 27: 295-308.

5. Frydenberg, O. 1956. *Drosophila pallidipennis* from Peru.
   Dros. Inf. Serv. 30: 115.

6. Frydenberg, O. 1956. *Drosophila pallidipennis* from Peru (Diptera,
   Drosophilidae).
   Rev. Brasil. Biol. 16: 287-294.

7. Frydenberg, O. 1956. Two new species of *Drosophila* from Peru
   (Drosophilidae, Diptera).
   Rev. Brasil. Ent. 6: 57-64.

8. Frydenberg, O. 1956. On the observation of heterosis in natural
   populations of *Drosophila*. Proceedings from "*Terceira Semana de
   Genética no Brasil*": 25-26.

9. Frydenberg, O. 1957. Equilibrium of a lethal mutant.
   Dros. Inf. Serv. 31: 120

10. Frydenberg, O. 1958. The Bennet population cages.
    Dros. Inf. Serv. 32: 167-168.

11. Frydenberg, O. 1958. Den Biologiske evolutionsteori. (The biolo-
    gical theory of evolution). *Udviklingsproblemer*, Nordisk Sommer-
    universitet 1958: 24-34.

12. Frydenberg, O and Zeuthen, E. 1960. Oxygen uptake and carbon
    dioxide output related to the mitotic rhythm in the cleaving eggs
    of *Dendraster excentricus* and *Urechis caupo*.
    C. R. Trav. Carlsberg 31: 423-455.

13. Frydenberg, O. and Sick, K. 1960. Selection of *st* and *cn*.
    Dros. Inf. Serv. 34: 78-79.

14. Frydenberg, O., Sick, K. and Hennningsen, K. 1960. Lack of back-
     mutations in the *cn*, *bw* and *e* loci.
     Dros. Inf. Serv. 34: 78.

15. Frydenberg, O. and Sick, K. 1960. Desemination by cold shocks.
     Dros. Inf. Serv. 34: 79.

16. Frydenberg, O. and Sick, K. 1960. The selection of *st* and *cn*
     alleles on different genetical backgrounds in *Drosophila melano-*
     gaster.
     Hereditas 46: 601-621.

17. Pedersen, P.O. and Frydenberg, O. 1962. The genetics program
     in Danish dental education. In *Genetics and Dental Health*, Edit.
     C.J. Witkop, jun. McGraw Hill Co. p. 275-280.

18. Sick, K., Westergaard, M., and Frydenberg, O. 1962. Hæmoglobin
     pattern and chromosome number of American, European, and Japanese
     eels (*Anguilla*).
     Nature 193: 1001-1002.

19. Frydenberg, O. 1962. Estimation of some genetical and vital
     statistics parameters of Bennet populations of *Drosophila mela-*
     nogaster.
     Hereditas 48: 83-104.

20. Frydenberg, O. 1962. Om genetiske skader af ioniserende strå-
     ling. (On genetic hasards of ionizing radiation).
     Meddelelser fra Sundhedsstyrelsen (Copenhagen) 116: 861-867.

21. Frydenberg, O and Sick, K. 1962. The selective value of a cinna-
     bar mutant of *Drosophila melanogaster* in populations homozygous
     for an unsuppressible vermillion mutant.
     Hereditas 48: 313-323.

22. Tryde, G., Frydenberg, O., and Brill, N. 1962. An assessment of
     the tactile sensibility in human teeth. An evaluation of a quan-
     titative method.
     Acta Odont. Scand. 20: 233-256.

23. Frydenberg, O. 1962. The modification of polymorphisms in some
     artificial populations of *Drosophila melanogaster*.
     Hereditas 48: 423-441.

24. Frydenberg, O. and Spärck, J.V. 1963. *Arv og race hos mennesket.*
     (Heredity and human races). Berlinske Forlag, Copenhagen. 160 pp.

25. Frydenberg, O. 1963. Some theoretical aspects of the scoring of
     mutation frequencies after mutagenic treatment of barley seeds.
     Radiation Botany 3: 135-143.

26. Frydenberg, O. 1963. Population studies of a lethal mutant in
     Drosophila melanogaster I. Behaviour in populations with discrete
     generations.
     Hereditas 50: 89-116.

27. Sick, K., Frydenberg, O., and Nielsen, J.T. 1963. Hæmoglobin
     patterns of plaice, flounder, and their natural and artificial
     hybrids.
     Nature 198: 411-412.

28. Frydenberg, O., Doll, H., and Sandfaer, J. 1964. The mutation
     frequency in different spike categories in barley.
     Radiation Botany 4: 13-25.

29. Frydenberg, O. 1964. Population studies of a lethal mutant in
     Drosophila melanogaster II. Behaviour in populations with over-
     lapping generations.
     Hereditas 51: 31-66.

30. Frydenberg, O. 1964. Long-term instability of an *ebony* polymor-
     phism in artificial populations of *Drosophila melanogaster*.
     ibid. 51: 198-206.

31. Frydenberg, O. and Sandfaer, J. 1965. The vitality, productivity
     and radiosensitivity of recurrently irradiated barley populations.
     The Use of Induced Mutations in Plant Breeding, FAO/IAEA Tech-
     nical Meeting on the Use of Induced Mutations in Plant Breeding,
     Rome May 1964: 175-183.

32. Frydenberg, O., Møller, D., Nævdal, G. and Sick, K. 1965. Hae-
     moglobin polymorphism in Norwegian cod populations.
     Hereditas 53: 257-271.

33. Frydenberg, O. and Nielsen, G. 1965. Amylase isozymes in germi-
     nating barley seeds.
     Ibid. 54: 123-139.

34. Philip, J., Frydenberg, O., and Sele, V. 1965. Enlarged chromo-
     some no. 1 in a patient with primary amenorrhoea.
     Cytogenetics 4: 329-339.

35. Frydenberg, O. and Jacobsen, P. 1966. The mutation and the se-
     gregation frequencies in different spike categories after chemi-
     cal treatment of barley seeds.
     Hereditas 55: 227-248.

36. Frydenberg, O. and Høegh-Guldberg, O. 1966. The genetic differ-
     ences between Southern English *Aricia agestis* Schiff. and Scot-
     tish *A. artaxerxes* F.
     Ibid. 56: 145-158.

37. Sick, K., Bahn, E., Frydenberg, O., Nielsen, J.T., and von Wett-
     stein, D. 1967. Haemoglobin polymorphism of the American fresh
     water eel *Anguilla*.
     Nature 214: 1141-1142.

38. Frydenberg, O., Nielsen, J.T., and Sick, K. 1967. The population
     dynamics of the haemoglobin polymorphism of the cod.
     Ciência e Cultura 19: 111-117.

39. Frydenberg, O., Nielsen, J.T., and Simonsen, V. 1969. The main-
     tenance of the haemoglobin polymorphism of the cod.
     Jap. J. Genet. 44, Suppl. 1: 160-165.

40. Frydenberg, O., Nielsen, G., and Sandfaer, J. 1969. The inheri-
     tance and distribution of α-amylase types and DDT responses in
     barley.
     Z. Pflanzenzüchtg. 61: 201-215.

41. Nielsen, G. and Frydenberg, O. 1969. Research note.
     Isozyme Bulletin 2: 37-38.

42. Nielsen, G. and Frydenberg, O. 1971. The inheritance and dis-
     tribution of esterase isozymes in barley.
     Barley Genetics II: 14-22.

43. Nielsen, G. and Frydenberg, O. 1971. Chromosome localization
     of the esterase loci *Est-1* and *Est-2* in barley by means of triso-
     mics.
     Hereditas 67: 152-154.

44. Nielsen, G and Frydenberg, O. 1972. Distribution of esterase
     isozymes, α-amylase types, and DDT reactions in some European
     Barleys.
     Z. Pflanzenzüchtg. 68: 213-224.

45. Simonsen, V. and Frydenberg, O. 1972. Genetics of *Zoarces* pop-
     ulations II. Three loci determining esterase isozymes in eye and
     brain tissue.
     Hereditas 70: 235-242.

46. Frydenberg, O., Gyldenholm, A.O., Hjorth, J.P., and Simonsen, V.
     1973. Genetics of *Zoarces* populations III. Geographic variations
     in the esterase polymorphism *EstIII*.
     Ibid. 73: 233-238.

47. Sick, K., Frydenberg, O., and Nielsen, J.T. 1973. Haemoglobin
     patterns of second-generation hybrids between plaice and floun-
     der.
     Heredity 30: 244-245.

48. Christiansen, F.B., Frydenberg, O., and Simonsen, V. 1973. Gene-
    tics of *Zoarces* populations IV. Selection component analysis of
    an esterase polymorphism using population samples including moth-
    er-offspring combinations.
    Hereditas 73: 291-304.

49. Frydenberg, O. and Simonsen, V. 1973. Genetics of *Zoarces* pop-
    ulations V. Amount of protein polymorphism and degree of genic
    heterozygosity.
    Ibid. 75: 221-232.

50. Frydenberg, O. 1973. Genernes evolutionære levetid. (The evo-
    lutionary lifetime of genes).
    Naturens Verden 1973, No. 9-10: 320-326.

51. Christiansen, F.B. and Frydenberg, O. 1973. Selection component
    analysis of natural polymorphisms using population samples in-
    cluding mother-offspring combinations.
    Theoret. Popul. Biol. 4: 425-445.

52. Christiansen, F.B., Frydenberg, O., Gyldenholm, A.O., and Simon-
    sen, V. 1974. Genetics of *Zoarces* populations VI. Further evi-
    dence, based on age group samples, of a heterozygote deficit in
    the *EstIII* polymorphism.
    Hereditas 77: 225-236.

53. Christiansen, F.B. and Frydenberg, O. 1974. Geographical pat-
    terns of four polymorphisms in *Zoarces viviparus* as evidence of
    selection.
    Genetics 77: 765-770.

54. Frydenberg, O. 1974. På jagt efter ældgamle gener. (In pursuit
    of ancient genes).
    Naturens Verden 1974, No. 8-9: 269-274.

55. Frydenberg, O. 1974. Hiroshima and Nagasaki atombombernes ge-
    netiske følger. (The genetic consequences of the Hiroshima and
    Nagasaki atomic bombs).
    Ibid., No. 10: 305-315 & 324-328.

56. Christiansen, F.B. and Frydenberg, O. 1975. Selection compo-
    nent analysis of natural polymorphisms using mother-offspring
    samples of successive cohorts.
    In *Population Genetics and Ecology*. (S. Karlin and E. Nevo). *Aca-
    demic Press, New York*: 277-301.

57. Frydenberg, O. 1975. The expression of mutants induced by a
    transitory rise of the mutation rate in a human population with
    some inbred mariages. Proceedings of "The 1'st Scientific Confer-
    ence of The Iraqi Atomic Energy Commission", Iraq, April 1975.

58. Christiansen, F.B., Frydenberg, O., Hjorth, J.P., and Simonsen,
    V. 1976 (*post humous*). Genetics of *Zoarces* populations IX.
    Geographic variation at the three phosphoglucomutase loci.
    Hereditas 83: 245-256.

59. Christiansen, F.B. and Frydenberg, O. 1977 (*post humous*).
    Selection mutation balance for two nonallelic recessives produc-
    ing an. inferior double homozygote.
    Amer. J. Hum. Genet. 29: 195-207.

60. Christiansen, F.B., Frydenberg, O., and Simonsen, V. 1977 (*post
    humous*). Genetics of *Zoarces* populations X. Selection component
    analysis of the *EstIII* polymorphism using samples of successive
    cohorts.
    Hereditas in preparation.

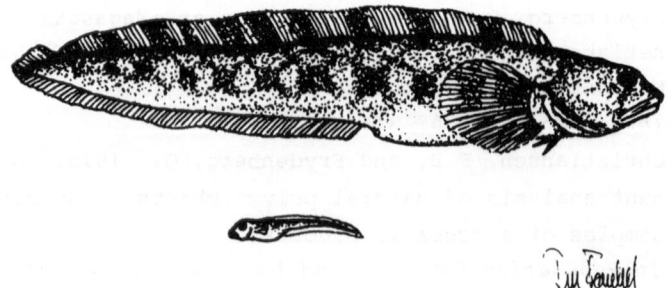

Zoarces viviparus (L.)

INTRODUCTION

The concept of natural selection is central to the theory of evolution
which again is the unifying principle of all biological sciences. Yet
many people - and among them even some biologists working in other
fields- have vague and incomplete ideas about natural selection. Thus,
it is a widespread conviction that natural selection is modelled by say
a bacterial culture to which an antibiotic is added so that out of a
billion cells only one mutant survives to proliferate. Evolutionary
biologists hold somewhat more sophisticated ideas about natural selection, but many aspects are still incompletely understood or even controversial.

The problems of natural selection are attacked from a diversity of
viewpoints corresponding to the diversity of the scientists preoccupied
with the concept. The basis of the concept is formed by the naturalist's
observations on variation and adaptation but has also inspired other
students of biological and non-biological fields. Today, the background
of an evolutionary biologist may be that of a mathematically orientated
theoretician, a biochemically orientated laboratory experimentalist, or
a field orientated naturalist.

Evolution is concerned with the explanation of the adaptation of
phenotypes to the environment through changes in the genotype of individuals. This short formulation actually covers a universe of mechanisms
which are far from being understood. In this volume, however, we will
mainly address the problems from the viewpoint of population genetics.
Population genetics centers around the main observation that lasting
changes in the genotypic composition of a species can only originate
through alterations in the genotypic composition at individual gene loci.
Therefore, a common approach in population genetics is to study the population dynamics of individual gene loci or some other unity of genetics.
The forces acting on the variation at the gene level has proven difficult
to study, if of no other reasons, because of the often formidable difficulties in obtaining reliable estimates of the basic descriptive parameters in natural populations such as population sizes, migration rates
and gene frequencies.

While we structured the volume into its five main chapters, difficulties arose since many of the contributions could equally well have
been placed under two headings. Thus several of the papers found under

the heading *Study of Selection* also deal in general with the description of polymorphisms, and similarly several of the papers under the heading *Study of Polymorphisms* address the problem of direct measurement of selection. The heading *Ecology and Evolution* accentuates the problem as all studies on natural selection approach an ecological problem, that is, fitness is always a measure related to some biotic or abiotic environment of the organism. In the present *Ecology and Evolution* chapter we have placed only the contributions where the observational or theoretical approach is mainly that of an ecologist. To make a special section on *Human Evolution* is in a sense an arbitrary taxonomical idiosyncrasy only justified by the general interest in that particular species.

In Chapter 1, *Study of Selection*, two contributions, *viz.* those of Christiansen and Allard *et al.*, discuss the measurement of selection in natural populations in terms of selection components. Both demonstrate the necessity of this approach for the analysis of polymorphisms in terms of natural selection and that the study of the genetic structure of a single life cycle stage may be very misleading. Allard *et al.* proceed to discuss the components of selection involved in the reproduction in plants and this is accompanied by a detailed analysis of these components in several plant species. The study of reproduction fitness components in animal populations is undertaken by Eanes *et al.* and by O'Donald. The former contribution concentrates on the formal treatment of changes in genotypic distribution between adults and adults involved in the breeding. O'Donald takes a more descriptive approach to the problem of reproductive selection in his studies of evolution of reproduction characteristics in birds. Here he discusses the evolutionary consequences of territoriality in the arctic skua.

The traditionally more widely studied selection component of zygotic selection or viability selection is discussed by Mukai and by Scharloo *et al.* Mukai gives a comprehensive description of the variance in viability of local *Drosophila* populations in terms of genome structure. The large and substantial experimental evidence on genetic load and linkage disequilibrium (in general) in natural populations is contrasted with the inability to explain these as the results of variations at the level of isozyme loci. Scharloo *et al.* study the variation at enzyme loci in *Drosophila* in relation to variations in the environment which supposedly influence the enzyme substrates. They are able to show dependence in viability selection of the environment for varia-

tion at the amylase locus. They proceed to discuss further possibilities for the direct expression of enzyme variation in the phenotypic
variation. The contribution by Hartl discusses the selective forces
associated with the segregation distorter chromosomes in natural populations of *Drosophila*. This polymorphism, albeit it may be considered
pathological, provides an inspiring example of interacting selection
components and of interacting loci which induce a secondary modification in the genome. The work by Jones and Parkins attempts to follow
selection in transferred populations. Although their experiments gave
negative results, the approach may in future contribute significantly
to our understanding of selection. The chapter concludes by a methodo-
logical discussion of inference based on the comparison of genotypic
proportions with the Hardy-Weinberg distribution. Barndorff-Nielsen
discusses the use of conditional methods with arguments from plausibility theory.

During the last couple of decades, and in particular after the
method of enzyme electrophoresis introgressed into population genetics,
it has become clear that most natural populations are highly polymorphic. Evidence for selection on electrophoretically defined polymorphisms is still sparce and in many cases inconclusive. Also appreciable
selection forces on each of the polymorphic loci seem to require an unacceptable large phenotypic variation with respect to fitness. It has,
therefore, been questioned whether the high level of polymorphism is a
result of selection. Two schools have appeared, the "neutralists" and
the "selectionists". In their extreme versions, the former claims that
the majority of polymorphisms are selectively neutral and the latter
that the polymorphisms are maintained by selection. This controversy
is reviewed in the contributions by Ayala and by Ewens in Chaper 2.
Since the controversy is formulated as a general characteristic of
biochemical polymorphisms, the natural approach is to seek general patterns in the polymorphisms which necessitate the assumption of selection for explanation. The theoretical problems of this approach is reviewed in the paper by Ewens. The chapter, *Study of Polymorphism*, also
contains six contributions illustrating experimental approaches. In
spite of much effort this central problem of evolutionary biology has
not yet been settled with a clear and unambigous answer. One approach
would be to see whether the pattern of differentiation between populations is consistent with neutralism as a null hypothesis. This approach
is taken by Ayala, by Christiansen, and by Lakovara *et al*. The impres-

sive and thorough study of Ayala on *Drosophila* species yields data which
are inconsistent with a hypothesis that random genetic drift is the
sole determinant of the differentiation between species. He shows that
the gene frequency structure of the different species clusters around
certain discrete patterns which he interprets as adaptive foci. Lakovara *et al*. study the pattern of the α-*Gpdh* locus through the family
of Drosophilidae and show that it is generally monomorphic within each
species, having one out of eleven possible alleles and with no apparent
correlation between the allele of the species and its predominant environment. Christiansen describes the geographic variation in the polymorphisms of *Zoarces viviparus*. He argues that a distinct dichotomy in
pattern of variation exists which is difficult to understand assuming
neutral loci. A different approach is represented by Beardmore and Ward
and by Langley, who search for evidence of correlations within single
populations in the distribution between loci. Beardmore and Ward look
for correlations in genotypic distributions through the initial life
stages in the plaice, and they find some interesting patterns in the
multi locus heterozygosity. Langley reviews a large body of data on linkage disequilibrium in *Drosophila*. He shows that there is no strong,
consistent non-random association among the biochemically defined loci.
      The study of polymorphism at the single gene loci is dependent on
a methodology to disclose the variation. It has long been appreciated
that electrophoresis is unable to detect all variation. Two contributions, those of Johnson and Nielsen, persue the possibilities for the
study of "cryptic" variation. Johnson refines the electrophoretical
technique by using a gel sieving analysis and he reviews investigations
using this technique for several loci in several organisms. All of these
investigations disclose an appreciable amount of genetic variation
within ordinary electrophoretic alleles. The technique also allows a
more sophisticated interpretation of the biochemical background for
electrophoretic variation. Nielsen studies a new dimension in the possibility of variation at biochemically defined "loci". His work on the salivary amylase locus in the bank vole demonstrates natural populations
with polymorphism in the number of genes coding for the enzyme in addition to ordinary electrophoretic variation. Through the distribution of
the vole there is variation in the polymorphism with respect to the different electrophoretic forms of the enzyme, whereas the chromosomal polymorphism seems rather constant. The chapter concludes with two theo-
retically defined contributions. Edwards reformulates the theory of se-

lection at a locus with three alleles using geometrical arguments. Gillespie attempts to establish a unifying theory which by assumptions on the function of enzyme loci will predict properties of quantitative characters. The attempt is successful in that by reasonable assumptions on single loci most global characterstics are produced.

In an outbreeding species which is switched to parthenogenesis, the individuals could be expected to increase their average fitness by a factor of two. Why then are parthogenetic species so relatively rare or why do they seem to show a short life span in an evolutionary time scale. The question can also be phrased by asking how the widespread occurrence of sex can be explained. Maynard Smith discusses this and related problems offering a number of models to account for the maintenance of sex in eukaryotic species. The papers of Christensen *et al.* and Saura *et al.* approach some of these problems by studying the distribution pattern and the genetic variation of species which occur in sexual as well as parthenogenetic forms. Such systems will undoubtedly play an important role in answering the questions posed by Maynard Smith. Closely related to the problem of the evolution and maintenance of sexual dimorphism is the problem of stable sex ratios. Charlesworth studies the evolution of the sex ratio in a population with overlapping generations and argues that deviant sex ratios can be stable. Bengtson provides a model of a genetic system in which natural selection will lead to sex ratios deviating strongly from 1:1. The model involves a breakdown of the chromosomal determination of sex and is applied to certain microtine rodents from which such deviating sex ratios are known.

Natural selection will only be completely understood in an ecological context as evolution is a function of the environment. Chapter 4 contains contributions which in different ways approach the study of selection from the ecologist's point of view. Barker in his paper first discusses the ideal system to study natural selection and describes one which in many ways conforms with these ideas: Fruitflies living in rots of *Opuntia* cactus in Australia. The habitats occur as isolated islands with small *Drosophila* populations in which the effect of local environment can be followed, and the organism is amenable to genetic analysis in the laboratory. Halkka and Mikkola have studied similarly structured populations of spittle bugs on Finnish skerries, and analyze the selection maintaining a color polymorphism. Doyle and Richards studied the interaction of selection and migration in a coarse grained habitat for a quantitative genetic trait in the marine polychaete *Spirorbis*. Host-

parasite interactions have been suggested to be a major agent of evolution in either of the interaction species (Clarke, at the symposium),
and the work of Sandfaer provides a nice example of this; it demonstrates
that strong selection forces may be induced by an epidemic in a susceptible type when grown in competition with a type that carries the parasite endemically. The two last contributions in the ecology chapter are
concerned with the biotic environment on the evolution of species. In
Fenchel and Christiansen's paper a model for the evolution of Mendelian
species involved in exploitative competition is described. Evolution of
the niche positions and the development of the niche widths by polymorphism are studied, and the results are discussed in the light of observational data. Roughgarden uses a much broader approach, *i.e.* the coevolution within whole communities. By the simplifying assumption that
the interaction coefficients are unchanged by the evolution of the interacting species, he formulates general principles for community coevolution.

Traditional use of evolutionary arguments on human populations has
been seriously questioned by the works of Feldman and Cavalli-Sforza.
In their contribution they study the interaction between Mendelian inheritance and the inheritance of aquired characters, the latter being
of different types, *viz.* inheritance by learning from parents or other
family members or inheritance by infection from parents as is the case
in the system studied by Sandfaer. They show that even a moderate phenotypic transmission will have quite significant evolutionary consequences. Thompson studies the potential possibilities of using ideomorphic alleles for inference on population structure, and she applies the
theory to tribal populations. Thomson and Bodmer persue an interesting
aspect of one of the most significant contributions of human genetics
to evolutionary studies, *viz.* the histocompability systems. They provide
interesting speculations and data concerning associations between HLA
alleles and various diseases.

The last decade has in many ways been fruitful and exiting for evolutionary biologists; we hope this volume demonstrates that this also
applies to the problem of measuring selection in natural populations.
We have attempted to give a broad spectrum of ideas and approaches relevant to the topic. This diverse representation may give a caleidoscopic
impression, but science is very much an evolutionary process itself fed
by this diversity. The diversity of a science is itself subject to selection and competition and as in real evolution of the biological world

most lines will eventually go extinct. However, given a preliminary selection on quality, it is very difficult to make predictive models of
the future development of our science. Nevertheless, we hope that this
collection of papers show that a fruitful development depends on a broad
outlook, that it must draw from a great variety of other biological di-
sciplines, and that it must combine empirical observations with theoretical thought in order to reach a synthesis of and insight in the evolutionary process.

TABLE OF CONTENTS

1. STUDY OF SELECTION

2. STUDY OF POLYMORPHISM

3. SEX AND EVOLUTION

4. ECOLOGY AND EVOLUTION

5. HUMAN EVOLUTION

CHAPTER 1. STUDY OF SELECTION

ESTIMATION OF MATING CYCLE COMPONENTS OF SELECTION IN PLANTS

R.W. Allard, A.L. Kahler and M.T. Clegg

At any instant the gene pool of a population is expressed as an array
of genotypes and one of the main goals of evolutionary genetics is to
determine how the genotypic frequency distributions of populations are
transformed over generations. In considering the factors that cause
the array of genotypes in a population to change in an evolutionary
sense it is convenient to identify two main components of selection,
acting in two distinct parts of the life cycle. The first is viability.
Viability selection is concerned with changes which occur in the array
of genotypes within any generation, n, and it thus reflects the probability of survival from fertilization to reproductive maturity. To
estimate *net* viability selection, we have to know the composition of
the array of genotypes in the population at the instant of fertilization and we have to know the array of genotypes at the time of reproduction. It is also informative to know the composition of the array
of genotypes at various intermediate stages, e.g. infancy, childhood,
youth, because selection may not operate with the same intensity, or
even in the same direction, at different stages from fertilization to
reproductive maturity.

The second component of selection is fertility. This includes all
aspects of selection associated with reproduction. To determine the
net effect of this component, we require estimates of the composition
of the array of genotypes of the reproducing adults and the composi-

tion of the array of new zygotes at fertilization. However, to understand how the array of new zygotes comes about we need to know: (1)
the mating system of the adults because it determines the patterns in
which the gametes are brought together to form the array of zygotes
which initiate a generation; (2) the genetic composition of the array
of male and female gametes produced by the adults because differential
fertility among parental genotypes may result in different numbers and
kinds of gametes; (3) the viability of genetically different gametes;
and (4) the ability of gametes of different genotype to effect fertilization. Fertility selection is thus concerned with changes in the array of genotypes which occur in the transition from generation n adults
to generation n+1 zygotes.

ESTIMATION OF SELECTION COMPONENTS

The estimation of selection components is a laborious procedure and
estimation will often, if not always, be incomplete in that potentially important components may be ignored. When we consider the number of
components and the small size and ephemeral nature of some of the structures involved (e.g. the male gametophytes), it is evident that problems of estimation are especially troublesome in the reproductive cycle.
Estimation of reproductive cycle parameters has indeed lagged, and this
is particularly unfortunate because it is precisely this stage of the
life cycle that can generate complex modes of selection as a result
of the confounding of the effects of differential mating behavior with
the effects of differential fertility of adults and differential survival and functioning of gametes.

Annual plants have a number of technical advantages for the study
of components of selection in nature. One advantage is lack of complex
age structure in a population so that demographic changes are not confounded with the effects of selection. Other advantages are associated
with sedentary habit which makes it easy to collect experimental ma-
terials and also makes it possible to identify population units precisely. Consequently it is not surprising that annual plants have often been chosen as experimental organisms in studies of components of
selection (review of earlier literature in Allard, Jain and Workman,
1968; more recent studies are reviewed in Clegg and Allard, 1973). This
paper describes a sample of these studies and considers briefly their

implications regarding population structure and evolutionary change.

In discussing the estimation of selection components, it is informative to start with some data on the frequency of peroxidase allozyme variants in adult plants of a natural population of ryegrass, *Lolium multiflorum*. This is an outbreeding species with a well-developed self-incompatibility system. It can be seen from Table 1 that the adult frequencies fit Hardy-Weinberg expectations very closely. This same population was also sampled in the seedling stage in the next generation. The results, given in Table 2, show that the fit to Hardy-Weinberg expectations is now extremely poor due to a deficiency of heterozygotes and a substantial excess of both homozygotes. This is a common result in ryegrass, i.e. for the seedling array to be deficient in heterozygotes and for the proportion of heterozygotes to increase sharply from seedling to adult stages, until in many cases there is even an excess of heterozygotes over Hardy-Weinberg expectations at maturity. This sort of result makes two things clear: first, that something occurs during the reproductive cycle that causes a deficiency of heterozygotes at the time of fertilization; and second, that there is strong viability selection favoring heterozygotes at one or more developmental stages after fertilization and prior to adulthood. These ryegrass data illustrate the point that, if we are to understand how selection works, it is nec-

Table 1. "Goodness of Fit" of *Lolium multiflorum* adults to Hardy-Weinberg expectations (Peroxidase). N = 100 p = 0.8100 q = 0.1900

	Genotype		
	11	*12*	*22*
Observed	64	34	2
Expected	p^2	$2pq$	q^2
Expected No.	65.61	30.78	3.61
O - E	-1.61	+3.22	-1.61

$\chi^2_{[1]}$ = 1.095   0.20 < P < 0.50

Table 2.   "Goodness of Fit" of *Lolium multiflorum* seedlings to Hardy-
Weinberg expectations (Peroxidase). N = 972   p = 0.8179
q = 0.1821

	Genotype		
	11	*12*	*22*
Observed	675	240	57
Expected	p^2	$2pq$	q^2
Expected No.	650.23	289.54	32.23
O - E	24.77	-49.54	24.77

$\chi^2_{[1]}$ = 28.457   P < < 0.001

essary to separate the life cycle, particularly the reproductive cycle,
into as many stages as possible. This procedure has been advocated re-
peatedly by Prout on theoretical grounds (Prout, 1971a and earlier) and
Prout (1971b) has shown in laboratory experiments with *Drosophila* that
very strong selection does in fact occur during one or more stages in
the reproductive cycle. Others, e.g. Bundgaard and Christiansen (1972)
and Anderson and Watanabe (1974), also have laboratory results showing
that selection during the reproductive cycle can be no less important
than viability selection and that it is sometimes the dominant evolutionary force. Christiansen and Frydenberg have (1973) presented a
thorough investigation of experimental designs for the detection of selection at various life cycle stages but in applying their methods to
data on an esterase polymorphism in a natural population of the eelpout, they obtained evidence for zygotic selection only (Christiansen
and Frydenberg, 1976).

In plants the technique of electrophoresis has provided a convenient approach to the problem of breaking the reproductive cycle into
parts, primarily because it allows numerous genetic markers to be detected in the seedling stage very shortly after fertilization has occurred. Given markers at this stage of development it is possible to
design simple experiments which allow estimation of parameters that

bear on the main components of the reproductive cycle, and incidentally also on parameters of viability selection. The experimental procedure
is a straightforward two-step process: first, seeds are collected from
a large random sample of individual mature plants in nature and these
seeds are germinated in the laboratory and the seedlings are assayed
electrophoretically for several loci to determine the genetic composition of progeny arrays from single maternal individuals; second, a
random sample of seeds collected in bulk from the adult population is
germinated in the laboratory and assayed electrophoretically to determine genotypic frequencies in generation n+1. The genotypic composition
of each progeny array depends on the maternal genotype, gene frequencies in the pollen pool, and the mating system itself: consequently,
given data on the genetic composition of the array of progeny from a
random sample of maternal individuals, adult gene frequencies and mating system parameters can be estimated according to the model shown in
Table 3 (Brown and Allard, 1970). It can be seen from this table that
the observation of one of each kind of homozygote in a progeny identifies positively the maternal individual as a heterozygote. Failure to
observe an A_2A_2 homozygote in a progeny array identifies A_1A_1 maternal

Table 3. Single progeny conditional probabilities for one locus with
two alleles (From Brown and Allard, 1970)

Maternal Parent	Progeny		
	A_1A_1	A_1A_2	A_2A_2
A_1A_1	1-qt	qt	0
A_1A_2	(s + 2pt)/4	1/2	(s + 2qt)/4
A_2A_2	0	pt	1-pt

p = frequency of allele A_1 in the pollen pool

q = frequency of allele A_2 in the pollen pool

s = proportion of ovules self fertilized

t = 1-s = proportion of ovules fertilized by random outcrossing

individuals with increasing certainty as N (progeny size) increases.
Similarly, failure to observe an A_1A_1 individual in a progeny identifies A_2A_2 maternal individuals. The probability of correct identification is a function of gene frequency and the outcrossing rate. So
long as p or q is not nearly 0 or 1, three or four progeny are adequate
for correct identificaction (P > .99) in heavily inbreeding populations
(t < .10) and in outcrossing populations (t near 1) eight or nine progeny are adequate (P > .99). The system is a practical one because the
cost and labor of assaying the numbers of progeny indicated above are
not great.

The data from the progeny arrays allow us to make direct estimates
of the genetic composition of the array of genotypes among reproducing
adults in generation n, gene frequencies (p and q) in the pollen pool,
and also the mating system parameters s and t. The data from the bulk
seed harvests allow direct estimates of the genetic composition of the
array of new zygotes in generation n and n+1 shortly after the reproductive cycle. Further, since 99% or more of seeds give assayable seedlings under laboratory conditions, the data provide estimates of zygotic frequencies before selection has had opportunity to modify the initial zygotic array in any significant way. Once the genotypes of the
reproducing adults in generation n, and the unselected seedling array
in generation n+1, are known, estimates can be made of parameters which
specify the expected composition of the male and female gametic pools
and the actual composition of the two pools. The difference between the
expected and the actual compositions allows estimates of the combined
effects of fertility and selection in the gametophytic stages of the
life cycle. In short, while not all components of selection in the reproductive cycle can be estimated, those which are most likely to be
of interest can be.

This system also allows the partitioning of *viability* selection into components associated with specific stages of the life cycle. Thus,
to determine the amount of selection which occurs between any two developmental stages, say from germination to age three weeks, samples
are taken at each stage and grown to maturity under conditions where
survival is high. Two procedures have been used to insure high survival:
transplanting into greenhouse pots or by reducing competition in the
field by thinning. Seeds are then collected at maturity and assayed to
infer maternal frequencies and thus the composition of the population
at each stage; any difference in composition can be attributed to selection operating in the interval between the two stages.

MATING SYSTEMS AND SELECTION ON MATING SYSTEMS

Let us now return to the *Lolium* data and determine the effect of the mating system itself on the progeny array of generation n+1. Direct estimates show that t = 0.83 and s = 0.17, indicating that the total inbreeding which occurred is equivalent to that which would result if the only inbreeding were due to selfing in the amount of 17%. At inbreeding equilibrium this mating system is expected to lead to a theoretical Inbreeding Coefficient of F = (1-t)/(1+t) = 0.093. When this numerical value of F is substituted in Wright's Equilibrium Equation (Table 4) the fit of expected to observed numbers is much improved over that obtained assuming F = 0, as in Table 2. It is therefore clear that a modest amount of inbreeding is responsible for nearly all of the deficiency of heterozygotes at the seedling stage. (Controlled mating studies show that nearly all ryegrass plants will set some seeds on selfing and that some plants are moderately self compatible. This leakage in the self-incompatibility system is probably responsible for most of the breeding observed. However, some of it may be due to local seed drop or other causes of population subdivision: local seed drop increases the likelihood that neighbors will be relatives and neighbors have a tendency to mate together.) However, there is still a signifi-

Table 4. "Goodness of Fit" of *Lolium multiflorum* seedlings to Wright's Equilibrium Law (peroxidase). N = 972   p = 0.8179   q = 0.1821   t = 0.83 ± 0.056   F = (1-t)/(1+t) = 0.093

	Genotype		
	11	*12*	*22*
Observed	675	240	57
Expected	$p^2 + pqF$	$2pq(1-F)$	$q^2 + pqF$
Expected No.	663.78	262.43	45.78
O - E	11.22	-22.43	11.22

$\chi^2_{[1]} = 4.7189$, $\hat{F} = 0.17$ $0.02 < P < 0.05$

cant departure from expectations according to Wright's equilibrium law, which indicates that some factor or factors in addition to the inbreeding measured at the peroxidase locus operate during the reproductive cycle.

These results raise three main questions about the effect of the mating system itself on the initial zygotic frequency distribution in any generation: Is the mating system the same for all genotypes in a single population? Do all populations of a species have the same mating system? Is the mating system itself under selection and subject to evolutionary change? Let us take up each of these questions in turn. Is the mating system the same for all genotypes in a single population? Estimates of outcrossing in ryegrass made using five electrophoretically detectable loci provide evidence on this point. The results given in Table 5 show the amount of outcrossing differs from locus to locus, varying from 76% with locus PGM to 99% for Locus PGI. The reason for this difference is not known, but it can be speculated that it may be associated with linkage between the electrophoretically detectable loci and incompatibility loci that differ in frequency of self-compatibility versus self-incompatibility alleles. Whatever the basis for the difference it is clear that different genotypes do not outcross to the same

Table 5. Estimates of outcrossing from five enzyme loci in a population of *Lolium multiflorum*

Locus	Outcrossing (t in %)
PGM	76* ± 2.1
Phosphatase	81* ± 2.4
Peroxidase	92* ± 2.6
GOT	94* ± 2.0
PGI	99* ± 2.9

*t significantly smaller than 1.00 (random mating)

extent and hence that there is genetic variability for mating system
within this population. Second, do all populations of a species have the
same mating system? Estimates of the amount of outcrossing made in different populations of ryegrass indicate they do not, and that the mating system in fact often changes significantly over short distances. As
an example, in a study of four sites located approximately 25 meters
apart along a transect up a west-facing slope, t varied from 0.81 to
1.02, which is highly significant statistically. This transect also
provides evidence concerning the third question: Is the mating system
itself under selection and subject to evolutionary change? The bottom
site and the top site on the transect top are both flat, well-watered
areas whereas the two central sites on the transect occupy arid areas
on a steep, westfacing part of the slope. Interestingly, the outcrossing rate was high at the bottom and the top of the slope and low in
the two intermediate arid sites. This is a common observation in plants,
i.e. for the amount of outcrossing to be lower under arid conditions
than under mesic conditions. The data in Table 6 give further evidence
concerning differences in the mating system among different populations of six representative species. In *Avena barbata*, for example,
the range in amount of outcrossing has varied from 0.1% or less in
some populations to more than 7.5% in other populations, a 75-fold difference in the amount of outcrossing. In this species the amount of
outcrossing tends to be 1% or less in the more arid sites, and above
5% in the most mesic sites. *Collinsia sparsiflora* shows a particularly
wide range in amount of outcrossing, varying from much less than 1% in
some populations to more than 84% in other populations. Again the amount
of outcrossing is correlated with environment, the lower amounts of
outcrossing occurring in populations which occupy the most arid sites.
Evidently these differences reflect genetic adaptations, at least in
part, because populations which occupy arid sites maintain their low
outcrossing rates in wet years and populations which occupy mesic sites
outcross at high rates in dry years. Thus the main conclusion to be
reached from Table 6 is that the amount of outcrossing differs from
population to population in many of the species and that the amount
of outcrossing is a genetic adaptation to the environment. This implies that the mating system is itself under selection and that it is
subject to evolutionary change. However, it would be reassuring to
have direct and unequivocal evidence on this point and Composite Cross
V, an experimental population of barley, has provided an opportunity
to measure change in mating system over time. Since most of our re-

Table 6. Outcrossing rates (t) determined from enzyme data in
            several plant species.

Species	Range in $\bar{t}$ (%)	Number of populations
Avena barbata	0.1 - 7.5	21
Avena fatua	0.1 - 1.6	9
Bromus mollis	7.0 - 14.0	2
Lollium multiflorum	68.0 - 104.3	11
Collinsia sparsiflora	0.0 - 84.0	11
Hordeum vulgare	0.0 - 8.5	3

maining examples are from Composite Cross V we will now describe this
population briefly.

Composite Cross V (CCV) was developed by the late H.V. Harlan from
intercrosses from 30 barley varieties representing the major barley
growing areas of the world (Kahler and Allard, 1970). In 1937 the 30
parents were crossed in pairs and during the next three years the F_1
hybrids of each cycle were again paircrossed to produce ultimately a
single hybrid stock involving all 30 parents. This grand hybrid stock
was then allowed to reproduce by its natural mating system, which in
the barley species is usually one of about 99% of self fertilization
and about 1% of outcrossing. The initial selfed generation, designated
F_2, was grown in 1941 and the F_3 and all subsequent generations were
grown from random samples taken from the harvest of the previous generation. The plot was managed according to normal agricultural practice and no conscious selection was practiced at any time. Viable seeds
of the earlier generations have been maintained by keeping part of the
harvest of each generation in storage and growing these seeds at 5-
to 10-generation intervals.

The experimental procedure followed in estimating the amount of
selfing vs. outcrossing in this population, and in estimating other
reproductive cycle parameters to be discussed shortly, involved growing two plots of each of generation 8, 19 and 28 in a uniform field

nursery (Kahler, Clegg and Allard, 1975). A census was conducted on the first plot of each generation to estimate adult genotypic frequencies. The census involved taking a single spike from each of 1,100 randomly-chosen adult plants of each of the three generations. Nine seeds from each spike were germinated in a laboratory germinator and the 9 (occasionally only 8) seedlings surviving to age 7 days were assayed electrophoretically. In total, 28,950 progeny from 3,309 adults were studied. The second plot of each generation was harvested en masse at maturity to provide bulk seed of generations 9, 20 and 29. The first plot could not be used for this census because the sampling of the adult plants damages the population. Eleven hundred random seeds of each of generations 8 and 9, 19 and 20, and 28 and 29 were germinated and the resulting seedlings (6,570 in total) were assayed electrophoretically for three esterase loci, EA, EB and EC. It should be noted that more than 99% of seeds germinated and produced assayable seedlings in the germination chamber; hence the assays very closely reflect initial zygotic frequencies. The three generations were grown in the same year and the same nursery at Davis, California, thus allowing comparisons of mating system parameters and the parameters of selection at three stages in the evolutionary history of the population, unconfounded by environmental differences.

Estimates of the amount of outcrossing in the three generations of CCV are given in Table 7. It can be seen that the amount of outcrossing increased by about 85 per cent in the transition from generation 8 to generation 19 and by another 8 per cent in the transition from generation 19 to generation 28, thus, the outcrossing rate doubled in the 20-generation span studied. Theory by Fisher (1941) and by Karlin and McGregor (1974) indicates that, other things being equal, selfing is expected to increase in populations because modifier genes increasing selfing will be transmitted to a greater proportion of an individual's progeny. Other things are evidently not equal in CCV because selfing decreased and outcrossing increased, presumably leading to increased recombination. This result supports the results shown in Table 6 in indicating that the mating system is itself under selection and that the mating system is adjusted by selection to be in proper balance with the total recombinational system. The adjustment can be over very wide ranges, even within a single species, as illustrated by *Collinsia* in which different populations vary from heavily inbred to very nearly random mating.

Table 7.   Outcrossing rates (averaged over three enzyme loci) in
            three generations of Composite Cross V

Generation	$\bar{t}$ (%)[a]
8	0.52 ± 0.11
19	0.96 ± 0.10
28	1.04 ± 0.10

a: From Kahler, Clegg and Allard (1975). Additional data provide these revised
outcrossing estimates.

REPRODUCTIVE CYCLE COMPONENTS OF SELECTION

Let us now turn from the question of selection as it affects the mating system itself to the effects of selection on other components of
the reproductive cycle. In a study of viability and fertility components of selection in a natural population of the slender wild oat, *A-
vena barbata*, it was found that heterozygotes increased markedly in
frequency from the time of fertilization to adulthood, i.e. that strong
net viability selection took place and that it took the form of heterozygote advantage (Table 8).

Strong net fertility selection also took place; however, in this
case heterozygotes were intermediate between the homozygotes. Net selective values over the entire life cycle, given approximately by the
product of the viability and fertility estimates, indicate overall advantage of heterozygotes. To look at viability vs. fertility selection in more detail it is convenient to return to Composite Cross V,
in which more stages of the life cycle have been investigated. Table 9
shows a typical result in this population for a single-locus case. Note
the frequency of the A_1A_1 homozygote increases from zygote to adult
stage indicating that the linkage block marked by the A_1 allele is
favored in viability selection. However, the frequency falls off from
the adult stage to the zygote stage indicating that the A_1A_1 homozygote is at a disadvantage in fertility selection. The A_2A_2 and A_3A_3
homozygotes follow the opposite pattern: they decrease in frequency

Table 8. Estimates of viability and fertility components of selection (average for three enzyme loci) in *Avena barbata* (From Clegg and Allard, 1973)

	Genotype		
	11	*12*	*22*
Viability	1.00	1.82	1.03
Fertility	1.00	0.87	0.83
Total	1.00	1.59	0.84

Table 9. Single-locus genotypic frequencies in CCV. generation transition 28-29 (From Clegg, Kahler and Allard, in preparation)

Genotype	Zygote	Adult	Zygote
A_1A_1	.58	.65	.60
A_2A_2	.35	.30	.33
A_3A_3	.05	.03	.06
A_1A_2, A_1A_3 A_2A_3	.02	.02	.01
N	1102	1052	1139

S.E. < 0.02

from zygote to adult stage and increase in frequency from adult to zygote stage, indicating that they are at a disadvantage in viability selection but favored in fertility selection. Heterozygotes are infrequent and hence their frequencies are characterized by erratic fluctuations due to small sample sizes; however, heterozygotes often increased from adult to zygote stages indicating that they are highly
favored in fertility components of selection. Table 10 expresses the
change shown in Table 9 in terms of selective values relative to the
A_1A_1 homozygote taken as a standard. Note that the differences between
the A_1A_1 and A_2A_2 homozygotes are about 20% and also that these differences are statistically significant. The comparison between the A_1A_1
and the A_3A_3 homozygote is also large and statistically significant.
Numerous significant results of the type shown in this example were
found among the large number of comparisons which can be made in single
generation transitions from generation n to generation n+1. Table 11
shows average results for homozygotes and for heterozygotes over the
entire data set. Some homozygotes are superior and some are inferior
to the standard genotype (fitness 1) and the main selective value of
homozygotes is near unity for both viability and for fertility selection. However, heterozygotes, on the average, are highly favored in
both viability and fertility selection.

Table 10. Single-locus viability and fertility values in CCV, generation transition 28-29 (From Clegg, Kahler and Allard, in
preparation)

Genotype	Viability	Fertility
A_1A_1	1.00	1.00
A_2A_2	0.77* (.06)	1.17 (.11)
A_3A_3	0.56* (.13)	1.99*(.45)

*Significantly larger or smaller than unity, P < 0.05

Table 11. Single locus selective values (grand averages) over four
enzyme loci and three generation transitions (8-9, 19-20,
28-29) in CCV (From Clegg, Kahler and Allard, in preparation)

	Viability	Fertility
Homozygotes	1.03	1.04
Heterozygotes	1.63	2.14

Let us now break the reproductive cycle down still further and ask
this question -- "What are the joint effects of fertility and gametophytic selection during the reproductive cycle?" Two examples are giv-
en in Table 12. For locus EA, allele 2 has an advantage over allele 1
in both pollen and ovules (note from Table 9 that the A_1A_1 homozygote
is favored in viability selection). For locus EB allele 1 is inferior
in the pollen but superior in the ovules and the reverse is the case
for allele 2: it is superior in the pollen but inferior in the ovules.
Numerous other cases of strong selection associated with fertility and/
or gametophytic selection were found in the total data set.

So much for single-locus analysis; let us now turn to the multilocus situation. Earlier studies of Composite Cross V have shown that,
in the initial generation, the population genotype is unorganized in
the sense that alleles at loci EA, EB and EC occur nearly at random
with respect to each other, i.e., the population is close to linkage
equilibrium, D = 0 (Weir, Allard and Kahler, 1974). However, these
three loci are tightly linked; also the inbreeding which occurs in the
population further restricts recombination among these loci, as well
as among loci located on different chromosomes. Thus it is expected
that favorable combinations of alleles at different loci, including unlinked loci, will be protected from break up due to segregation and
that selection will increase their frequency in the population (Allard,
1975). In CCV the linkage disequilibrium parameter rapidly reached very
high values (D > 50 per cent of the theoretical maxima) indicating that
the population genotype has become organized into coadapted multilocus

Table 12. Pollen and ovule "fitness" values for Loci EA and EB, generation transition 28-29 in CCV (From Clegg, Kahler and Allard,
in preparation)

	Allele *1*	Allele *2*
Locus EA		
Pollen	0.68* (0.13)	1.54 (0.28)
Ovule	0.92* (0.03)	1.08 (0.07)
Locus B		
Pollen	0.90 (0.12)	1.22 (0.35)
Ovule	1.02 (0.03)	0.86 (0.07)

*Significantly smaller than unity, $P < 0.05$

blocks of concordant alleles at different loci (Clegg, Allard and Kahler, 1972). The development of coadapted blocks of genes in this population leads to this question: "Do the multilocus genotypes differ in
components of selection associated with viability and fertility?" Table
13 gives a sample of data for 3-locus homozygotes which bear on this
question. It can be seen that some of these homozygotes, for example
the *111/111* homozygote, decrease in frequency from zygote to adult stage
but increase from adult to zygote stage, indicating that they are at a
disadvantage in viability selection but favored in fertility selection.
The pattern is the opposite for other homozygotes. The selective values,
relative to the *111/111* homozygote as a standard, show that the differentials are frequently quite large and that viability and fertility
selection frequently operate in opposite directions. Multilocus heterozygotes were infrequent and estimates of components of selection for
single genotypes were characterized by large and erratic fluctuations
over generations. However, the viabilities of heterozygotes were nearly
always larger than unity and they tended to increase with increasing
heterozygosity.

Another question we can ask is -- "Can the multilocus selective effects be predicted from the single locus selective values?" One way of

Table 13. Three-locus genotypic frequencies (Loci EA, EB, EC) and selective values for the five most frequent (among eight) homozygotes for generation transition 28-29 in CCV (From Clegg,
Kahler and Allard, in preparation)

	Frequency in generation			Selective value		
	28	28	29			
Genotype	Zygote	Adult	Zygote	Viability	Fertility	Total
111/111	.191	.170	.180	1.000	1.000	1.000
211/211	.310	.283	.295	1.030	.990	1.020
112/112	.142	.186	.142	1.472	.722	1.063
122/122	.230	.259	.226	1.272	.841	1.070
212/212	.085	.055	.083	.727	1.420	1.032

examining the distribution of selective effects at the three-locus level
is to compare marginal single-locus fitness estimates with joint three-
locus estimates. A large number of comparisons are possible, far too
many to consider here. However, when they are made it is clear that the
marginal effects, i.e. the single-locus effects, do not accurately predict the three-locus effects (Clegg, Kahler and Allard, in preparation).

In this experiment three generation transitions, generations 8 to
9, 18 to 19, 28 to 29, were examined in a single environment. Consequently we can ask a still further question: "Have the selective effects changed over time at the multilocus level?" Again documentation
concerning this point requires a large number of comparisons and we
will consequently give only the result obtained when the appropriate
comparisons are made. The result is that components of selection, both
those of viability selection and fertility selection, did in fact change
as the population advanced from generation 8 to generation 19 and from
generation 19 to generation 29. Thus, as selection restructured the population, the mating system changed and also the patterns of viability
and fertility selection changed over generations concomitant with the

rapidly growing correlations in allelic state over loci observed in the
distribution of gametic types by Clegg et al. (1972).

Summary.

   Results from studies of natural and field populations of plants show that selection is likely to operate in many different stages in the life cycle. It is important to recognize how little information is provided by studies of only a single
stage in the life cycle. The adult plant data on *Lolium* which appear in Table 1 illustrate this point well. These adult stage data indicate no departure from Hardy-
Weinberg expectations. But when data were obtained from an additional stage in the
life cycle, the real situation turned out to be one involving substantial departure
from random mating, strong viability selection, and there were also indications of
other departures from Hardy-Weinberg assumptions. The conclusion is obvious: without detailed information about patterns of selection in the critical stages of the
life cycle we have to be careful about specific conclusions and we also have to be
careful about broad generalizations concerning selective neutrality, genetic load,
maintenance of genetic variability and similar questions. The likelihood of oppos-
ing types of selection operating at different stages in the life cycle may open new
avenues for the maintenance of genetic variability in populations.

   The multilocus data bring out an additional point: multilocus frequency distributions cannot be predicted within a single generation, nor over any sequence of
generations, from knowledge of single-locus genotypic frequencies. The impact of selection on any specific locus is a function of the cumulative selective effects o-
ver many loci, indeed perhaps on the entire genome. It is important to keep this in
mind when we interpret experiments in which loci are studied one at a time and it
should also be kept in mind as well when we develop models of evolutionary change.

Acknowledgements.

   This work was supported in part by National Science Foundation Grant BMS 73-
01113-A01.

REFERENCES

Allard, R.W.   1975.   The mating system and microevolution.
      Genetics 79: 115-126.

Allard, R.W-, Jain, S.K., and Workman, P.L.   1968.   The genetics of inbreeding pop-
      ulations.
      Adv. in Genet. 14: 55-131.

Anderson, W.W., and Watanabe, T.K.   1974.   Selection by fertility in *Drosophila pseu-
      doobscura*.
      Genetics 77: 559-564.

Brown, A.H.D., and Allard, R.W.   1974.   Estimation of the mating system in open-
      pollinated maize populations using isozyme polymorphisms.
      Genetics 66: 133-145.

Bundgaard, J., and Christiansen, F.B.   1972.   Dynamics of polymorphisms. I. Selection
      components in an experimental population of *Drosophila melanogaster*.
      Genetics 71: 439-460.

Christiansen, F.B., and Frydenberg, O. 1973. Selection component analysis of natural polymorphisms using population samples including mother-offspring combinations. *Theoret. Popul. Biol.* **4**: 425-445.

Christiansen, F.B., and Frydenberg, O. 1976. Selection component analysis of natural polymorphisms using mother-offspring samples of successive cohorts. In Population Genetics and Ecology, Editors, Samuel Karlin and Eviatar Nevo. pp. 277-301.

Clegg, M.T., and Allard, R.W. 1973. Viability versus fecundity selection in the Slender Wild Oat, *Avena barbata* L. *Science* **181**: 667-668.

Clegg, M.T., Allard, R.W., and Kahler, A.L. 1972. Is the gene the unit of selection? Evidence from two experimental plant populations. *Proc. Nat. Acad. Sci.* (U.S.) **69**: 2474-2478.

Clegg, M.T., Kahler, A.L., and Allard, R.W. Estimation of life cycle components of selection in an experimental plant population. (in preparation).

Fisher, R.A. 1941. Average excess and average effect of a gene substitution. *Ann. Eugen.* **11**: 53-63.

Kahler, A.L., Clegg, M.T., and Allard, R.W. 1975. Evolutionary changes in the mating system of an experimental population of barley (*Hordeum vulgare* L.) *Proc. Nat. Acad. Sci.* (U.S.) **72**: 943-946.

Karlin, S., and McGregor, J. 1974. Towards a theory of the evolution of modifier genes. *Theor. Pop. Biol.* **5**: 59-103.

Prout, T. 1971a. The relationship between fitness components and population prediction in *Drosophila*. I. The estimation of fitness component. *Genetics* **68**: 127-149.

Prout, T. 1971b. The relationship between fitness components and population prediction in *Drosophila*. II. Population prediction. *Genetics* **68**: 151-167.

Weir, B.S., Allard, R.W., and Kahler, A.L. 1974. Further analysis of complex allozyme polymorphisms in a barley population. *Genetics* **78**: 911-919.

1. STUDY OF SELECTION

POPULATION GENETICS OF *ZOARCES VIVIPARUS* (L.): A REVIEW

Freddy Bugge Christiansen

The now almost classical controversy as to whether the vast amount of
biochemically defined genetic variation in natural populations needs
to be explained by adaptive differences between genotypes is an over-
whelming argument for detailed studies which have the potential of
revealing and describing the action of natural selection on these
polymorphisms. However, the mere existence of the controversy empha-
sizes the difficulty involved in designing such studies so that a
reasonably clear answer emerges. Among several possible approaches to
the problem, we will discuss the possibility of measuring selection
solely by direct observation of the genetic composition of a natural
population.

For this purpose it is preferable to choose an organism where extensive sampling may be performed with relative ease without pro-
foundly disturbing the population, and where the breeding structure
of populations is so simple that it does not require attention in the
sampling procedures. A commercially exploited marine organism fulfils
the first requirement and for the second a fish with one discrete
breeding event per season would be most useful. Among the most successful investigations of such organisms is that of the North Atlantic cod, *Gadus morrhua* L. (Frydenberg, Møller, Naevdal & Sick 1965;
Frydenberg, Nielsen & Sick 1967; Frydenberg, Nielsen & Simonsen 1969;
Schmidt 1930; Sick 1961, 1965a,b). However, marine fishes are not well

suited for breeding in the laboratory, so observations in nature cannot be supported by formal genetic analyses. This problem has been overcome by turning to the eelpout, *Zoarces viviparus* (L.), which, as the
name indicates, is viviparous. As analyses on fetuses are possible,
the population of pregnant females constitutes a population of breeding "experiments", which provide formal genetic information about the
variation of the population (Schmidt 1917, 1918, 1921, 1922). Ove Frydenberg also perceived that samples of these pregnant females provide
the observer with detailed knowledge about the population that can
be used in attempts to measure selection. We therefore launched a
large scale project on the genetics of natural populations using the
eelpout as the model organism.

We will return to the measurement of selection using pregnant females, but first we will review our present knowledge of eelpout populations.

EELPOUT POPULATIONS

The eelpout is a marine teleost, which is very common in shallow waters around Denmark with its distribution reaching the English Channel, the east coast of Britain, the White Sea, and the inner Baltic
and Gulf of Bothnia. Individuals are sexually mature at the age
of one and a half years. The fish mate in late summer, and the females carry the young for some five months until they are delivered
in midwinter at a size of about 5 cm. The size of the adult fish varies with age and location. In Kalø Cove (location 18, Figure 1) females aged 2, 3 and 4 years have an average length of 25 cm, 30 cm and
32 cm. Males are slightly smaller. A few fish as old as 10 years have
been recorded and lengths up to 45 cm seen. Fecundity of the females
at the age of 2 years is around 50 offspring and at age 3 around 150
offspring, and offspring counts exceeding 300 are not uncommon. Fish
can be aged using the otoliths (Gyldenholm & Jensen 1976).

The biology of the eelpout allows extensive formal genetic analyses of polymorphism by assaying mother-offspring combinations when
the polymorphs may be scored in the fetuses. Among the known electrophoretically detectable polymorphic loci (Frydenberg & Simonsen 1973),
four loci *EstIII*, *HbI*, *PgmI* and *PgmII*, have been analysed thoroughly
(Hjorth 1971, 1974, 1975; Simonsen & Frydenberg 1972). Others, e.g.

Ada and *Got*, are now under investigation. The major variation in all
of these loci is contributed by two codominant alleles. The *EstIII*
locus recombines freely with *HbI*, *PgmI* and *PgmII* and the two phosphoglucomutase loci also show no linkage. Analyses of segregations within
litters show that they may stem from more than one mating. The analyses are thus limited to investigations of the inheritance of the female alleles and of the segregation ratio from heterozygous females.

The amount of protein polymorphism and the degree of genic heterozygosity of the eelpout was investigated by Frydenberg and Simonsen
(1973) in a study of 24 proteins and functionally different enzymes
determined by 32 gene loci. Approximately 30 per cent of the loci
are polymorphic (two or more alleles occuring at a frequency of at
least .01) and about 10 per cent of the loci in a random individual
are heterozygous. (All the loci have one or two common alleles, with
rare variants comprising less than .01 of the samples). Samples from
two localities were investigated: one, Kalø Cove, is situated just
North of Aarhus in the Danish Belt Sea (location 18, Figure 1), and
the other, Bornholm, is an island in the Baltic (location 43, Figure
1).

Table 1. Symbols for the genes referred to in the text

Symbol	Name of enzyme/ protein coded for
Ada	Adenosine deaminase
EstIII	Esterase
Got	Glutamate oxaloacetate transaminase
LdhI	Lactate dehydrogenase
Me	Malate dehydrogenase (NADP), malic enzyme
PgmI *PgmII* *PgmIII*	Phosphoglucomutase
To	Tetrazolium oxidase
HbI	Haemoglobin

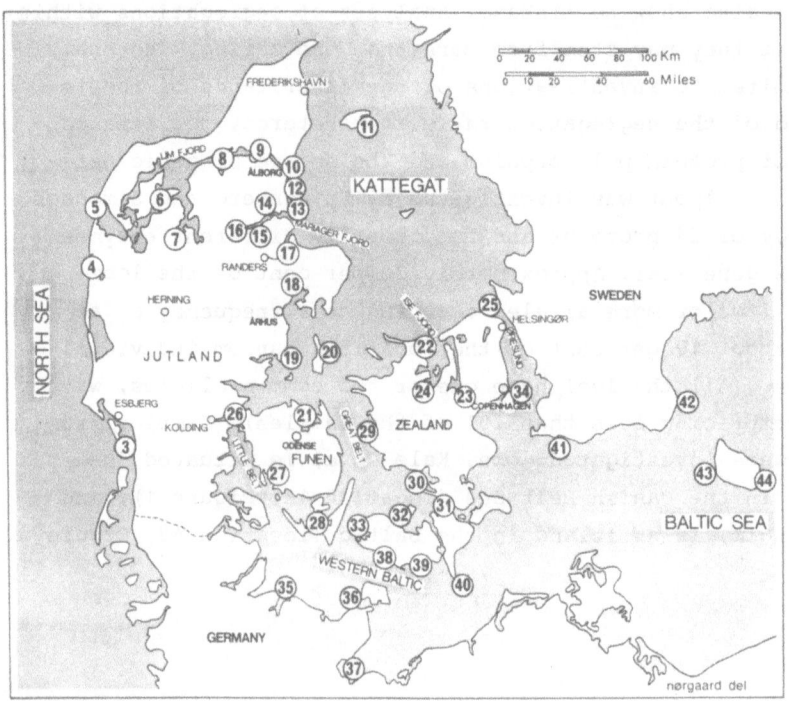

Figure 1.   Location of the sampling stations.

    The data from these two populations may give a preliminary im-
pression on the amount of differentiation at different loci within
the Danish waters. Figure 2 shows the difference in gene frequency of
the most common allele between the Kalø Cove and Bornholm populations
for the nine loci where the rarest variant is at a frequency of at
least .05 in one of the populations. The loci fall into two groups,
in one which the gene frequency difference between localities is below .05, i.e. *Ada*, *Got*, *LdhI*, *PgmI*, *PgmII* and *To*, and one where the
gene frequency difference is considerably greater, i.e. *EstIII*, *Me*
and *HbI*. In the eelpout, neither the heterozygosity nor the amount of
geographical variation at a locus seems to be correlated with the presumed function of the enzyme as suggested by Gillespie & Kojima (1968).

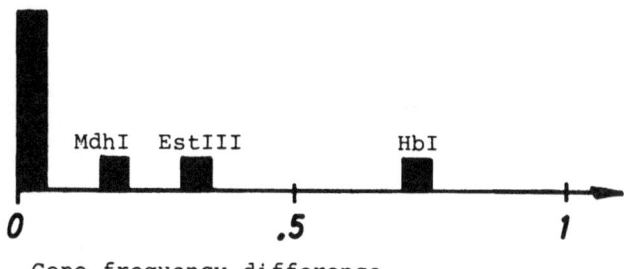

Figure 2. Difference in gene frequency between Kalø Cove and Bornholm at nine loci where the rarest allele has a frequency of at least .05 at one of the two locations. The six loci with a gene frequency difference below .05 is *Ada*, *Got*, *LdhI*, *PgmI*, *PgmII* and *To*. (Data from Frydenberg & Simonsen 1973).

Geographic structure of the population.

Two of the geographically constant loci, *PgmI* and *PgmII*, and two of the varying loci, *EstIII* and *HbI*, were subjected to more extensive study of the geographical variation during the years 1969-71 (Frydenberg, Gyldenholm, Hjorth & Simonsen 1973; Hjorth & Simonsen 1975; Christiansen, Frydenberg, Hjorth & Simonsen 1976). For the esterase and haemoglobin loci, 46 and 43 localities were investigated, most of which are situated in Denmark and along the adjacent coasts of the Baltic (Figure 1). At 30 of the localities, the populations were assayed for the phosphoglucomutase polymorphisms as well. Most samples comprised about 200 adult individuals.

The dichotomy in geographical distribution pattern of the common alleles at these four loci (as indicated in Figure 2) was confirmed by this larger study (Figure 3). The *Pgm* loci show very little variation whereas the *EstIII* and *HbI* loci show marked clinal variation, from the Kattegat through the Danish Belts to the Baltic. In addition *EstIII* shows a cline along the west coast of Jutland. Clines through the Belts are remarkably parallel (Figure 4). This agrees with an interpretation that both clines are maintained by limited diffusion like migration between the Kattegat population and the population of the Baltic. The differentiation between Kattegat and the Baltic Sea

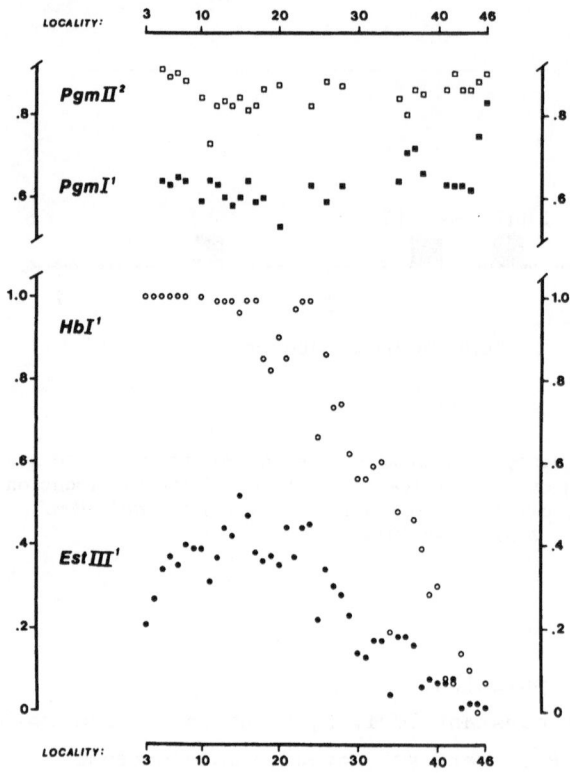

Figure 3.   Gene frequencies at the various sampling locations which appear arranged in the geographical order of Figure 1. (Data from Frydenberg *et al.* 1973, Hjorth & Simonsen 1975, Christiansen *et al.* 1976, after Christiansen & Frydenberg 1974).

of the *EstIII* and *HbI* loci and the concomitant similarity at the two *Pgm* loci led Christiansen and Frydenberg (1974) to the conclusion that selective forces have to be incorporated in the explanation of the geographical distribution in either or both of the two groups of loci. However, if one is free to postulate a suitable population structure in the past, then each pattern may be explained by current theory as a result of what may be termed purely demographic processes such as random genetic drift and migration, but the same processes are unlikely to account for both the observed patterns. In general, the geographic distribution of gene frequencies is inadequate for the

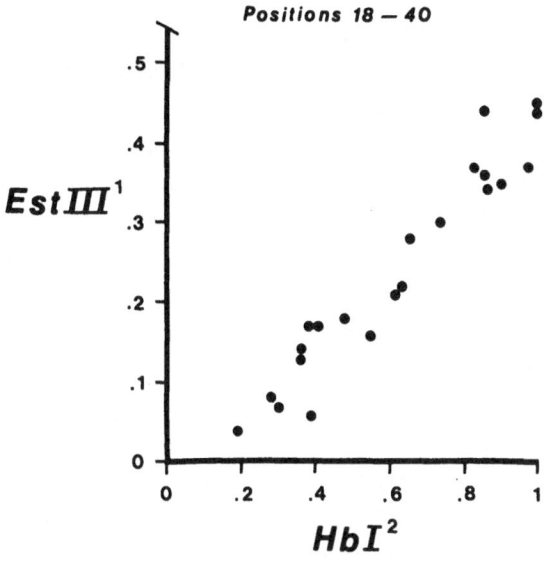

Figure 4. Relationship between gene frequencies at the *EstIII* and *HbI* loci at locations 18-40. (Data from Frydenberg *et al*. 1973, and Hjorth & Simonsen 1975).

inference of selection (Ewens & Feldman 1976; Ewens 1977). However, the geographic structure of the present data allows a conservative reduction of the data so the test for neutrality of the polymorphisms may be performed through a comparison of gene frequencies between two virtually isolated populations, a situation where theory is well developed (Wright 1940, 1951; Kimura & Maruyama 1971; Kimura & Ohta 1971).

Accepting that selection is a determinant of the geographical distribution of either the constant or the variable loci (or both), we may, at least for the varying loci, try to identify possible environmental agents of selection through a comparison between the variation of environmental variables and the variation in the gene frequencies. The differentiation in *EstIII* and *HbI* may immediately be interpreted as a function of the environmental differences between the marine and warm Kattegat and the brackish and cold Baltic. However, alternative environmental variables has to be mobilized to interpret the North Sea Cline of *EstIII*.

Additional information may be gained by considering the microgeographical variation in the complex of estuaries and fjords that comprises the coastline of Denmark. For esterase, Frydenberg *et al.* (1973)
found a tendency of the *EstIII*[1] allele to be more common inside the
two large Kattegat fjords: Isefjord (location 22-24) and Mariager Fjord
(location 13-16), a tendency which is supported by later investigations (Christiansen, Frydenberg and Simonsen *in preparation*), but
which was not, however, found in the German fjord Slien. These indications of microgeographical variation are in the opposite direction
to that expected if the Belt cline is a result of the salinity differences between Kattegat and the Baltic Sea, but they might suggest
that temperature is the relevant environmental variable.

Thus, there is far from conclusive evidence that the variable
loci are those subject to selection, and the differentiation in the
gene frequencies of *EstIII* and *HbI* between the Kattegat and the Baltic may just reflect that these two populations have been separated
for a sufficiently long time to show a major divergence due to random
genetic drift. This divergence must have occurred during the 10,000
years after the latest glacial period, so it requires a period of very
low population sizes, far lower than that of the present day populations. However, during the foundation of the populations or during
the period where the Baltic Sea was cut off from the Kattegat some
years ago, the population sizes may have allowed ample divergence
(Christiansen and Frydenberg 1974).

The geographical constancy of the *Pgm* loci may equally well be
explained as due to either balancing selection, to large population
sizes, or to large migration between the Kattegat and the Baltic. The
general picture of constancy was, however, modified by a more careful
analysis of the data, and subtle, but significant, microgeographical
and regional variations was revealed in these loci (Christiansen, Frydenberg, Hjorth and Simonsen 1976). For example, differences between
the Limfjord (location 5-9) and the Kattegat (location 12-17) were demonstrated and the island Læsø (location 11) showed a unique gene frequency in *PgmII*. These small deviations from the constancy for the *Pgm*
loci should be compared to deviations from the common variation in the
EstIII and *HbI* loci (Figure 4). However, due to the much lower power
of such an analysis in the varying loci, variations of the order seen
in the *Pgm* loci would not be expected to show, and no pattern in the
deviations have emerged. These considerations on the variation between
local populations were supported by considering the distribution of

rare variants at the *HbI*, *PgmI*, *PgmII* and *PgmIII* loci. Significant differentiation was detected down to a distance of about 50 km along a
coast, a magnitude similar to that observed by Schmidt (1917) for morphological variation.

The generalization of the conclusion of Christiansen and Frydenberg (1974) to cover the importance of selective forces as determinants of observed geographical patterns rests heavily on the four
chosen loci as representative of four average polymorphic loci. From
Figure 2 we see that a larger sample of loci has not altered the ratio
between varying and constant loci to any great extent, and in fact the
variation of the *Me* locus is of the same nature as the previously considered variation, i.e. clinal and parallel to the *EstIII* variation
(Figure 5; Christiansen, Frydenberg & Simonsen *in preparation*).

Thus from the geographical investigation we see that selective
forces are acting as determinants of the geographical distribution
pattern at a sizable proportion of the loci investigated. We cannot,
however, say on which loci, and we cannot even argue that selection is

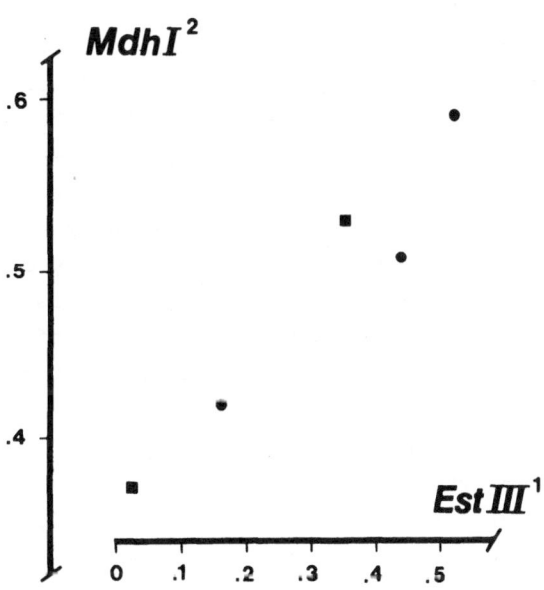

Figure 5. Relationship between gene frequencies at the *EstIII* and *Me* loci.
● : Data from Frydenberg and Simonsen (1973). ■ :Data from Christiansen, Frydenberg and Simonsen (in prep.).

acting through adaptive differences on the loci considered, as gene
frequency differentiations may have occurred by "hitch-hiking" on local
or global adaptive substitutions (Maynard Smith & Haig 1974; Thomson
1976). This is a common conclusion from this kind of study, and the
only advantage contributed in this case by the choice of the organism
is that the formal genetics of the markers in question are reasonably
well investigated. This might, however, be more easily achieved through
the choice of an organism which is adapted to human care in the laboratory. The advantage of the eelpout stems as mentioned from the possibility of a refined analysis of possible selective forces through the
collection of incomplete family data.

ANALYSIS OF SELECTION COMPONENTS

In the preceeding section we have argued exclusively from the observed gene frequencies. This may seem odd, as indications of selection in
the samples might be expected to reveal themselves through a comparison of observed genotypic proportions with the corresponding Hardy-
Weinberg expectations. However, deviations from Hardy-Weinberg proportions may emerge from a multitude of causes, and even if we could
be certain that selection was the only possible cause of deviations,
comparisons with Hardy-Weinberg proportions cannot reveal the structure of selection (Frydenberg 1956; Wallace 1958, Lewontin and Cockerham 1959). Further, the magnitude of selection cannot be determined,
as the only selective force which may distort genotypic proportions,
used in Hardy-Weinberg comparisons, is, in principle, differential
survival of the zygotes to the time of observation. Thus, from genotypic proportions we may gain at best only a limited and incomplete
knowledge of selection acting during only a part of the life cycle.
These remarks are still valid even if the method is extended to following the adult population through a parent-offspring transition
(Prout 1965, 1969; Christiansen, Bundgaard & Barker 1977).

    Our aim, of course, is to measure the total selection. However,
any observation in a population separates the life cycle of the organism into two parts, that from zygotes to the time of observation,
and that from the time of observation to the production of the gametes which unite to produce the next zygote population. Thus the total
selection acting on a polymorphism is by the observation separated in-

to two *selection components*, one pre-observational and one post-observational.

This pure *ad hoc* concept of selection components as it emerges
from an experimental procedure, has a dual theoretical concept. In
theoretical considerations it must be realized that selection may
work on different organizational levels of the genetic material, i.e.
gametes, zygotes and breeding pairs (Bundgaard & Christiansen 1972).
In a description of the total selection working on a polymorphic locus, these selection components have to be specified individually to
have any predictive and descriptive value. In addition we have to
specify the manner in which gametes are combined to produce zygotes.
This can often be achieved by specifying the mating patterns of the
breeding individuals.

These specifications to a large degree may be met through the
analysis of population data which includes mother-offspring combinations.

Mother-offspring data.

In eelpout populations mother-offspring data for *EstIII* can be
collected from late October to January. A sample of the adult population is assayed for genotype after separation according to sex, and
sorting females into pregnant and non-pregnant classes. For each
pregnant female one random fetus is assayed for genotype (only one
fetus is assayed as a second would not add information because of the
possibility of multiple matings). The data with this subdivision takes
the form shown in Table 2.

Mother-offspring data for *EstIII* polymorphism has been collected
in six consecutive years, 1969-74, with an additional adult sample in
the early spring of 1973 referred to as "1972" (Christiansen, Frydenberg & Simonsen 1973; Christiansen, Frydenberg, Gyldenholm & Simonsen
1974; Christiansen, Frydenberg & Simonsen 1977). From 1971 and on,
the adults have been classified by age, so that we can follow the different ages of the population of fish born in a specific year, i.e. a
cohort (Table 3). Age classes are separated only to five years old as
the total size of the older classes (6+) is too small to contribute to
the analysis. The total sample over all years is shown in Table 4. For
the samples collected in 1969-71, the number of fetuses in each pregnant female was counted. This provided information on fecundity.

Table 2. Structure of mother-offspring samples

Adult genotype	Genotype of offspring			Mothers	Non-pregnant females	Adult females	Adult males	all adults
	11	12	22					
11	C_{11}	C_{12}		F_1	S_1	$A_{\female 1}$	$A_{\male 1}$	A_{01}
12	C_{21}	C_{22}	C_{23}	F_2	S_2	$A_{\female 2}$	$A_{\male 2}$	A_{02}
22		C_{32}	C_{33}	F_3	S_3	$A_{\female 3}$	$A_{\male 3}$	A_{03}
Sum	C_{01}	C_{02}	C_{03}	F_0	S_0	$A_{\female 0}$	$A_{\male 0}$	A_{00}
$\male$ gametes	M_1	M_2	M_0					

$$M_1 = C_{11} + C_{21} + C_{32},$$
$$M_2 = C_{12} + C_{23} + C_{33},$$
$$M_0 = F_0 - C_{22}.$$

Table 3. Cohorts represented in the samples
Numbering is according to year of birth

Age class	Year of sample					
	1969	1970	1971	1972	1973	1974
0	69	70	71	72	73	74
1			70	–	–	73
2			69	70	71	72
3			68	69	70	71
4			67	68	69	70
5			66	67	68	69
6+						

Besides the obvious information about the composition and development of the adult population, the mother-offspring combination and
the fecundity counts contain information about the process of reproduction. The mothers are representative of the females that contribute gametes to the zygote population, and as each offspring "carries"
a male gamete the male gamete pool within each female genotype is
also   reproduced in the data. From this information and the fecundity counts the expected population of newborn zygotes can be predicted.
We thus have a fairly complete picture of the genetic composition of
the population through the whole of its life-cycle. The only stage
which is not represented is that of the males taking part in the reproduction; only the gametes produced by them are known.

No difference in fecundity between the different mother-offspring
types have been recorded, neither in the data from 1969 and 1970
(Christiansen *et al.* 1973) nor in the data from 1971 which because of
the age determination allows a more refined analysis. Therefore, the
zygote population is simply represented by the marginal sum of the
maternal genotypes in the table of mother-offspring combinations.

The possibilities for using mother-offspring data to measure selection have been analysed by Christiansen and Frydenberg (1973,
1976). They devised a statistical procedure for extracting information
from the data about selection, mating and stability of the population
composition through time. This procedure is constructed in such a way

Table 4. Pooled data on the *EstIII* polymorphism in Kalø Cove 1969-74

Adult genotype	Genotype of offspring 11	12	22	Mothers	Non-pregnant females	Adult females[a]	Adult males	All adults	Adolescents
11	305	516		821	43	1008	693	1701	35
12	459	1360	877	2696	161	3344	2332	5676	125
22		877	1541	2418	160	3029	2201	5230	93
Sum	764	2753	2418	5935	364	7381	5226	12607	253
-gam.	1641	2934	4575						

a: Including adult females from "1972", which is not sorted into pregnant and non-pregnant females.
Data from Christiansen, Frydenberg & Simonsen 1977).

that each of several selection components can be simply analysed on
the basis of assumptions previously tested in the data.

In the analysis it is natural to distinguish the selection components shown in Table 5, as each of them corresponds to a simple comparison of the observed classes. Table 5 shows the χ^2-tests for the
hypothesis of absence of the different effects of interest in the main
body of data. Below those, under the heading of "additional fitting"
is the sum of the tests applied to introduce the irregular observational classes into the analysis, i.e. the samples 1969, 1970, "1972".
the age classes 6+ and adolescent fish from the samples of 1971 and
1974 (Table 3). These "fittings" may be interpreted as parts of the
main analysis (Christiansen *et al.* 1977), but to simplify the discussion they will be neglected here. The sum of all the tests in the Table
is a global test for random mating and the absence of selection. The
fit of the data to this hypothesis is satisfactory. The distribution of
the individual tests is also satisfactory, and only one significant result is found in the analysis. This is juvenile zygotic selection (component 8b). The χ^2-tests, however, do not summarize all the information contained in the data as they are independent of the observed
pattern of variation and only measure the magnitude of variation, e.g.,
the χ^2-test for time homogeneity (5 of Table 5) do not change its value by a shuffle of the time sequence. The procedure for the analysis
of data is discussed in more detail in the following and the complete
analysis is given by Christiansen *et al.* (1977). Initially we will discuss the hypothesis of the absence of female gametic selection as this
occupies a central position in supporting the genetic hypothesis that
the considered esterase variation is determined by two codominant alleles at an autosomal locus.

(1) *Female gametic selection*. If the heterozygous females produce their
alleles in a ratio of 1:1 in their gametes then we expect that half
of their offspring will be heterozygotes. Table 5 shows the test value for this hypothesis when tested within each female age class of
each sample. Neither this nor a more detailed analysis of the variation in segregation ratio with time and age provides us with any reason to reject the hypothesis of Mendelian segregation of the alleles
in heterozygous females. The observed incidence of heterozygous offspring from heterozygous mothers in the mother-offspring data is .504
(±.010), and with the similar data of Frydenberg and Simonsen (1972)
the observed segregation ratio becomes .496 (±.006).

A well-known problem in electrophoretic variation is the existence

Table 5.   Analysis of the mother-offspring data on *EstIII*

(From Christiansen, Frydenberg & Simonsen 1976)

Effect tested	χ^2	d.f.	P
1) *Female gametic selection*	18	20	·59
2) Random mating			
a) in each female age group	16	26	.94
b) between ages	22	16	·14
3) *Female sexual selection*	4	6	·72
4) *Adult zygotic selection*			
a) in females	22	18	·25
b) in males	18	16	·30
5) Time homogeneity			
a) in adult females	9	10	·49
b) in adult males	9	10	·53
c) in male gametes	10	5	·07
6) *Male reproductive selection*	0	1	·73
7) *Fecundity selection*	-	-	-
8) *Juvenile zygotic selection*			
a) equality of the sexes	1	2	·54
b) pooled over sex	6	1	·01
Additional fitting	40	44	.31
Random mating and no *selection*	175	175	

of rare recessive alleles known as null alleles. These will manifest themselves in the mother-offspring combinations as a decrease in the number of heterozygous offspring from heterozygous mothers, but more important as the appearance of the exceptional mother-offspring combinations *11,22* and *22,11*. These will each occur in a frequency of *pqr* where *p* and *q* are the gene frequency of thr normal *1* and *2* alleles and *r* is the frequency of the recessive null allele. In total we have observed 4 exceptional cokbinations out of 5939 giving an estimate of *r* as .0015. The existence of rare recessive nulls of course introduces a bias towards observing an excess of individuals which are phenotypically homozygotes. However, at the *Es$III* locus, their frequency is so low that they may be neglected (as they have been in Table 4).

Female gametic selection, whenever the gene frequency among the fertilizing males gametes deviates from 1/2, will produce a deviation from the expectation of half heterozygous offspring from heterozygous mothers. Thus the results may be summarized as follows: The data supports our genetic interpretation of the variation in esterase and the data provides us with no reason to believe that the allelic constitution at the *EstIII* locus has any influence on the success of female eggs.

(2) *Random mating*. If the population mates at random, we expect the frequency among fertilizing male gametes to be independent of the female type. In each female age group this may be tested by comparing the male gamete frequencies among mother genotypes (2a in Table 5) and as there are no indications for deviations from random mating in this part of the data, we can estimate the transmitted male gamete frequency in each of the female age groups. Random mating in addition implies that the male gamete frequency in each of the females age groups is homogenous. This is tested by (2b) of Table 5, and as before there are no indications of deviations, so we may proceed with the analysis on the assumption that mating is random between females and male gametes.

(3) *Female sexual selection*. Female sexual selection is the differential "survival" of mature females into the stage of breeding females. For the age group 2 this selection component is simple and well defined, but as mature females have the possibility of repeated breeding, the survival between breeding events becomes an aspect of sexual selection. In the data, however, it is natural to distinguish these two aspects of sexual selection, so we will consider as sexual selection only the differential breeding of mature females present in

the population at any time. The residual survival component between
breeding events will be referred to as *adult zygotic selection,* in
that we recognize its physiological similarities to zygotic selection
proper, i.e. differential survival of zygotes to the time of maturity. To avoid confusion we will refer to this latter component as juvenile zygotic selection.

By this definition, female sexual selection will appear in the
data as difference in genotypic composition between pregnant and non-
pregnant females within each age class. The test for absence of female sexual selection in Table 5 is shown in (3). This certainly does
not point to any significant female sexual selection.

(4) *Adult zygotic selection.* If there are differences in viability
between genotypes through adult life, these will cause heterogeneity
between the age classes within the cohorts. This hypothesis is tested
by (4a) and (4b) in Table 5, and the test values give no reason to
reject the hypothesis of age homogeneity within the cohorts. However,
these tests are sums of homogeneity tests within each cohort, and inspection of the individual tests show that for the males in cohort 67
(Table 3) there is significant heterogeneity between 4 year old fish
in 1971 and 5 year old fish in 1972. This is supported by the trend
shown in Fig. 6, in that both homozygotes were inferior to the heterozygote in the survival through the year 1971-1972. This effect
seems to disappear in the following years. The females did not show
a similar pattern.

5) *Temporal homogeneity of the population.* Homogeneity of the population over time is most reasonably tested with reference to the population of zygotes. At present we know that the zygotes are produced by
random mating between adult females and male gametes, so the zygote
population will be the same in all years if the adult females are homogeneous between cohorts and the male gametes are homogeneous through
time. This is tested as (5a) and (5c) in Table 5, where we have also
included a similar comparison of males between cohorts, (5b). Neither
of these tests are significant. Figure 7 shows the variation of the
male gamete frequency through time, and it is seen that the high test
value is due to an extremely low frequency of the allele *EstIII* in the
sample of 1973. For the adults, some temporal heterogeneity is indicated in the gene frequency as both the oldest and the youngest cohort
show extreme values (Fig. 8.).

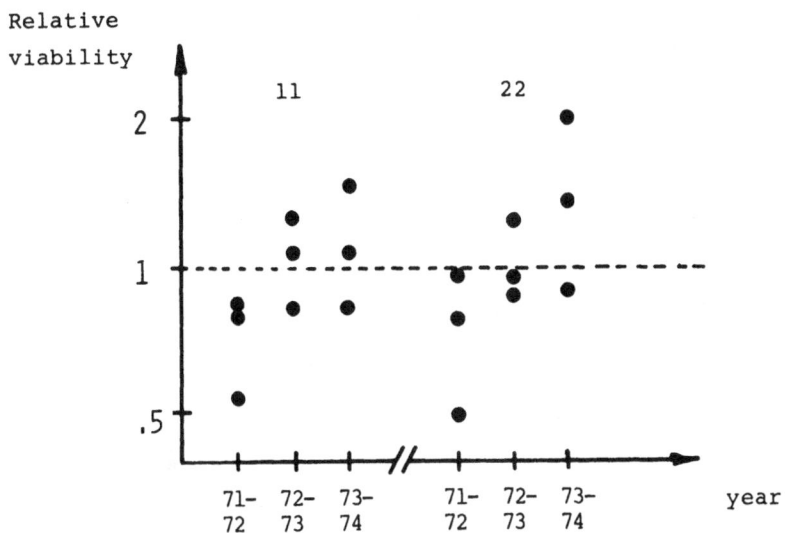

Figure 6. Year to year adult viability of the *EstIII* genotypes among male homozygotes
relative to that of the heterozygote (From Christiansen, Frydenberg & Simonsen 1977)

6) *Male reproductive selection*. Male reproductive selection confounds
male gamete selection and male sexual selection (referred to here in
the same way as female sexual selection), it is analysed by a comparison between the transmitted male gametes pooled over years and adult
males pooled over cohorts and ages. The test value gives no reason to
believe that there are differences between these two observational
classes. The effect of male sexual selection is only represented by
its genic effect, so a detailed description of this component cannot
be obtained from mother-offspring data. In particular, if male sexual
selection is the only force working on the polymorphism, and if the
population is at a gene frequency equilibrium, then no deviation from the
hypothesis of absence of male sexual selection is expected.
7) *Juvenile zygotic selection*. The estimation of the zygote population
is now made by random union of gametes from the adult females and males. The accuracy of this estimation may be enhanced by first pooling
the sexes as tested by (8a) of Table 5. The question of difference be-

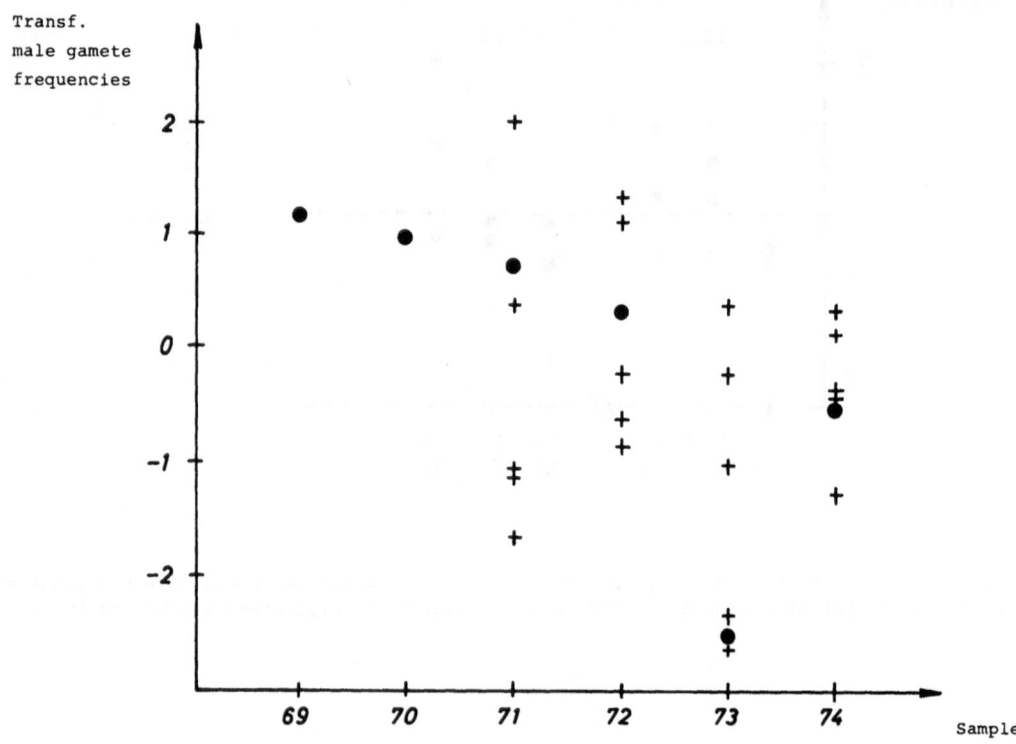

Fig. 7.   Variation of gene frequency at *EstIII* among transmitted male gametes in
in successive samples. Gene frequencies are shown as deviations from the mean over
.all samples, deviations which are approximately normal distributed with zero mean
and unit variance. ● : deviation of sample mean. + : deviation of age class of a
sample. (From Christiansen, Frydenberg & Simonsen 1977).

tween the sexes among adults corresponds to the hypothesis that juvenile zygotic selection, if present, is equal in the two sexes. The
test value is less than expected so the pooling can legitimately be
carried out and we can infer that the random union of gametes from the
adults will produce Hardy-Weinberg proportions in the zygotes. This of

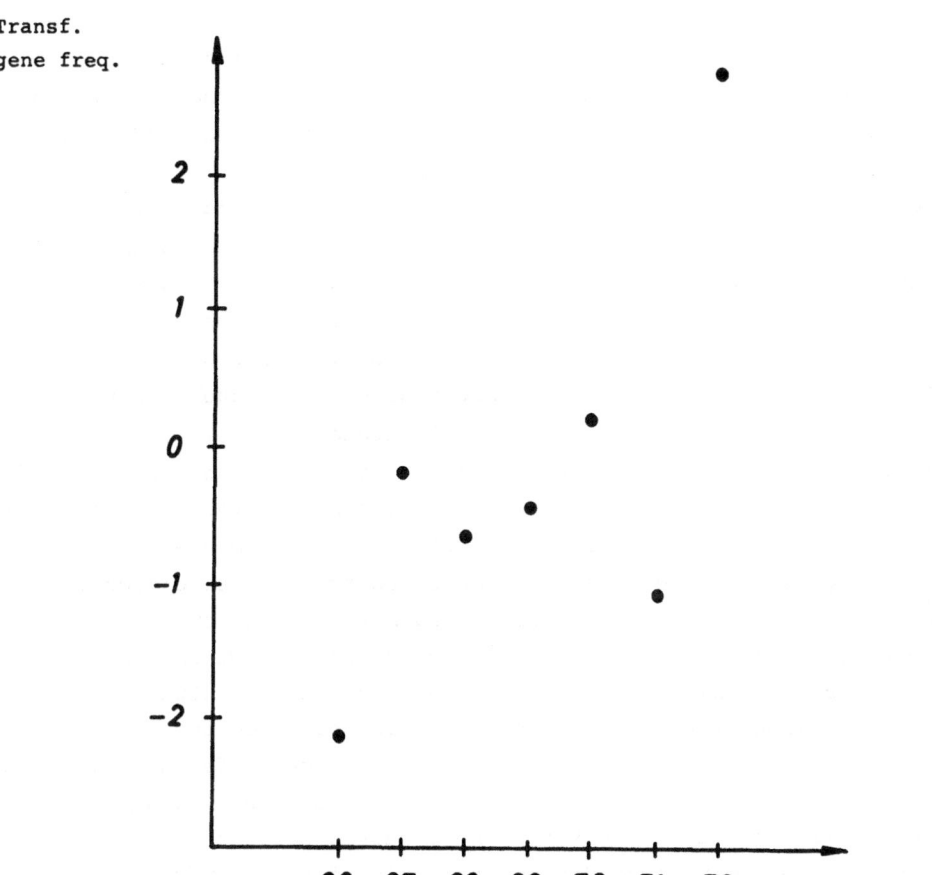

Fig. 8.   Variation of gene frequency at *EstIII* among adults in successive cohorts.
Gene frequencies are shown as deviations from the mean over all cohorts similar to
Fig. 7. (From Christiansen, Frydenberg & Simonsen 1977).

course assumes that fecundity selection is absent.

The absence of juvenile zygotic selection may now be tested by
comparing the distribution in adults to the estimated Hardy-Weinberg
proportions of the observed gene frequency of the whole data, i.e.
.360 (±003). This test (8b in Table 5) shows significance at the one
per cent level. Thus the genotypic proportions among adults deviate
from that expected from our knowledge about the breeding of the population. Inspection of the genotypic proportions shows that the reason
for the significance is a deficit of heterozygotes corresponding to
viabilities of 1.07 and 1.04 for the homozygotes *11* and *22* relative to
the heterozygote. This tendency recurs in both sexes and in the largest cohorts, but cannot be demonstrated in all age classes and all
cohorts. This, however, is not to be expected as the estimates of viabilities are subject to a large random error, in most observational
classes far exceeding the magnitude of the above effects.

The deficit of heterozygotes among adults may be due to several
reasons other than differential survival through the juvenile stage
(Christiansen *et al.* 1974). One already mentioned is the existence of
rare recessive null alleles in the *EstIII* locus. However, the observed
effect would require a gene frequency of the recessive $r = .011$, an
order of magnitude higher than that observed from the exceptional
mother-offspring combinations. Inbreeding may be ruled out as there is
no tendency to genotypic assortative mating and in fact the population
of zygotes shows a slight excess of heterozygotes. A Wahlund effect of
population mixing (Wahlund 1928) can be ruled out from our data on the
geographical variation: Significant gene frequency differentiation can
be demonstrated over distances of about 50 km but gene frequency differences sufficient to account for the observed deficit of heterozygotes under extreme mixing conditions can only be found over much greater distances. Thus we must conclude that the deficit of heterozygotes
is due to a difference between the genotypes within the local population, and the easiest explanation is then juvenile zygotic selection.

CONCLUSIONS

We have demonstrated that it is indeed possible to make extensive investigations in natural populations that allow estimation of most of
the components of the total selection working on a polymorphism. How-

ever, we may recall the point made by Christiansen and Frydenberg (1973) that even with this powerful and discriminate procedure, selection forces of considerable magnitude may still be hidden by the expected random variation in the samples. This is probably best illustrated by referring to the results of Table 5, where the test for juvenile zygotic selection is significant, but not overwhelmingly so, even though the estimated viability differences are of the order of magnitude of 5 per cent. This comment is further illustrated by Lewontin (1974).

In the mother-offspring data we have met important problems of consistency. As already mentioned in the review of the results of the analysis, we have the rather trivial problem of lack of consistency due to lack of statistical power in the genotypic distributions among adults in all ages in all cohorts. However, the suggestion in Fig. 6 of changes through time in adult zygotic selection in males raises the more fundamental question of whether we should even expect consistency. A similar remark pertains to the suggestions of heterogeneity in Fig. 7 and 8. The expectation of consistent trends is based on the assumption that possible selective forces on the polymorphism are constant through time. But this assumption has no foundation in the data whatsoever. Besides the suggestions of heterogeneity in the genetic composition of the population, the data show marked time heterogeneities of another kind, the most impressive being the observation that the dominating age class in all samples except that of 1974 is the 1969 cohort, and in 1974 only the 1972 cohort is larger (Christiansen *et al.* 1977). Thus the environment of the population is likely to be changing considerably from year to year. One suggestive environmental correlation has emerged: In the summer of 1972 reports from local fisherman show that oxygen concentration in the cove was low. This event coincides with the loss of adult male homozygotes (Fig. 6), which in turn is inconsistent with the general observation of a deficit of heterozygotes. Thus the present data probably asks more questions than it answers, but we are reasonably well convinced that the data demonstrate differences in fitness between the *EstIII* genotypes in the Kalø Cove population.

If for the moment, we accept that the only selection component acting on the *EstIII* polymorphism is underdominant juvenile zygotic selection, then we may ask whether that is consistent with the existence of the present polymorphism. If the Kalø Cove were an isolated population, then polymorphism would certainly not be expected. However, the

Kalø Cove population is situated in the *EstIII* cline between Kattegat
and the Baltic Sea, so the genetic composition in that particular population need not bear any immediate relationship to the local selective forces (Karlin & Richter-Dyn 1976). On the other hand, it is hard
to imagine that a migration rate sufficient to stabilize a polymorphism
under divergent selection could leave the Kalø population and the Hov
population (location 18 and 19, Figure 1) different in the frequency of
ideomorphic alleles at the *PgmII* locus (Chirstiansen *et al* 1976). However, these arguments are founded on the assumption that the population is at an equilibrium gene frequency with respect to selection.
The available data does not provide convincing information on this
(ref. Figure 8). Comparisons over a longer time span are probably needed. At the present day the environment is in a historical context quite
exceptional because of the eelgrass (*Zostera*) catastrophe in the 1930's.
This altered the habitat of the eelpouts considerably, and presumably
also had a profound influence on the physical and chemical characteristics of the water in the Danish Belt Sea (Nordenberg 1960; Rasmussen
1973).

   The investigation of the esterase polymorphism in the Kalø Cove
has now been terminated, but since 1975 we have undertaken a similar
study of four polymorphic loci, including *EstIII*, at Hov. This provides
us with the possibility of studying interactions between loci, but more
important, it provides us with more single locus data of the kind
discussed above with comparatively little extra effort. This extension
is especially important when we consider the possibility of selective
values varying through time, as it increases the possibility of pick-
ing up the effect of extreme environmental conditions with respect to
a locus. An alternative extension of the work would be to obtain additional data on the geographical variation. Here, the west coast of
Sweden is an obvious area for study as it would provide a test for
consistency in the *EstIII* clinal variation. Similarly, the turn of esterase cline in the Limfjord and the peculiar characteristics of this
area at the *Pgm* loci render it an interesting subject for investigation.
The geographical variation at additional loci (especially the *Me* locus)
also calls for attention.

   In conclusion, the study of the genetics of *Zoarces* populations
has proved inspiring and rewarding for those that have participated,
especially through the dynamic leadership of Ove Frydenberg. At present,
the project above all has developed a suitable model organism and a
coherent methodology for asking sensible questions about the dyna-

mics of natural polymorphisms.

Acknowledgments.

The eelpout studies have been supported by two research grants from Carlsberg Foundation of Copenhagen. I am indebted to Vibeke Simonsen for permission to use the unpublished data referred to in this review. Jørgen Bundgaard called my attention to the implications of *Zostera* catastrophe, and Tom Fenchel contributed by pointing out the temperature variation in the Danish waters in connection with the *EstIII* variation. The manuscript was commented upon by Stuart Barker, Steve Jones and Vibeke Simonsen.

REFERENCES

Bundgaard, J., and Christiansen, F.B. 1972. Dynamics of polymorphisms: I. Selection components in an experimental population of *Drosophila melanogaster*. *Genetics* 71: 439-460.

Christiansen, F.B., Bundgaard, J., and Barker, J.S.F. 1977. On the structure of fitness estimates under post-observational selection. *Evolution*. (in press).

Christiansen, F.B., and Frydenberg, O. 1973. Selection component analysis of natural polymorphisms using population samples including mother-offspring combinations. *Theoret. Popul. Biol.* 4: 425-445.

Christiansen, F.B., and Frydenberg, O. 1974. Geographical patterns of four polymorphisms in *Zoarces viviparus* as evidence of selection. *Genetics* 77: 765-770.

Christiansen, F.B., and Frydenberg, O. 1976. Selection component analysis of natural polymorphisms using mother-offspring samples of successive cohorts. *In* Karlin and Nevo (1976) pp. 207-301.

Christiansen, F.B., Frydenberg, O., Gyldenholm, A., and Simonsen, V. 1974. Genetics of *Zoarces* populations VI. Further evidence, based on age groups samples, of a heterozygote deficit in the *EstIII* polymorphism. *Hereditas* 77: 225-236.

Christiansen, F.B., Frydenberg, O., Hjorth, J.P., and Simonsen, V. 1976. Genetics of *Zoarces* populations. IX. Geographic variation at the three phosphoglucomutase loci. *Hereditas* 83: 245-256.

Christiansen, F.B., Frydenberg, O., and Simonsen, V. 1973. Genetics of *Zoarces* populations IV. Selection component analysis of an esterase polymorphism using population samples including mother-offspring combinations. *Hereditas* 73: 291-304.

Christiansen, F.B., Frydenberg, O., and Simonsen, V. 1977. Genetics of *Zoarces* populations X. Selection component analysis of *EstIII* polymorphism using samples of successive cohorts. *Hereditas*. (in prep.).

Ewens, W.J. 1977. Selection and neutrality.
   This symposium.

Ewens, W.J., and Feldman, M.W. 1976. The theoretical assessment of selective neutrality.
   In Karlin and Nevo (1976) pp. 303-337.

Frydenberg, O. 1956. On the observation of heterosis in natural populations, *in*
   Terceira Semana de Genetica no Brazil, pp 25-26. Escola Superior de Agricultura
   "Luiz de Queivoz". Piraciaba, Brazil.

Frydenberg, O., Gyldenholm, A., Hjorth, J.P., and Simonsen, V. 1973. Genetics of
   Zoarces population III. Geographic variations in the esterase polymorphism
   EstIII.
   Hereditas: 233-238.

Frydenberg, O., Møller, D., Naevdal, G., and Sick, K. 1965. Haemoglobin polymorphism in Norwegian cod population.
   Hereditas **53**: 257-271.

Frydenberg, O., Nielsen, J.T., and Sick, K. 1967. The population dynamics of the
   haemoglobin polymorphism of the cod.
   Ciencia e Cultura **19**: 111-117.

Frydenberg, O., Nielsen, J.T., and Simonsen, V. 1969. The maintenance of the haemoglobin polymorphism of the cod.
   Japanese J. Genet. **44**, S1: 160-165.

Frydenberg, O., and Simonsen, V. 1973. Genetics of *Zoarces* populations V. Amount
   of protein polymorphism and degree of genic heterozygosity.
   Hereditas **75**: 221-232.

Gillespie, J.M., and Kojima, K. 1968. The degree of polymorphisms in enzymes in-
   volved in energy production compared to that in non specific enzymes in two
   Drosophila ananassae populations.
   Proc. Nat. Acad. Sci. **61**: 582-585.

Gyldenholm, A.O., and Jensen, A. 1975. Otolith growth zones, age and size distributions in the eelpout, *Zoarces viviparus* L.
   J. Fish. Biol. (in prep.).

Hjorth, J.P. 1971. Genetics of *Zoarces* populations. I. Three loci determining the
   phosphoglucomutase isozymes in brain tissue.
   Hereditas **69**: 233-242.

Hjorth, J.P. 1974. Genetics of *Zoarces* populations. VII. Fetal and adult hemoglobins
   and a polymorphism common to both.
   Hereditas **78**: 69-72.

Hjorth. J.P. 1975. Molecular and genetic structure of multiple hemoglobins in the
   eelpout, *Zoarces viviparus* L.
   Biochem. Genet. **13**: 379-391.

Hjorth, J.P., and Simonsen, V. 1975. Genetics of *Zoarces* populations VIII. Geographic variation common to the polymorphic loci HbI and EstIII.
   Hereditas **81**: 173-184.

Karlin, S., and Nevo, E. 1976. Editors of *Population Genetics and Ecology*.
   Academic Press, New York.

Karlin, S., and Richer-Dyn, N. 1976. Some theoretical analysis of migration selection interaction in a cline: a generalized two range environment.
   In Karlin and Nevo (1976) pp. 659-706.

Kimura, M., and Maruyama, T. 1971. Pattern of neutral polymorphism in a geographically structured population.
   Genet. Res. **18**: 125-131.

Kimura, M., and Ohta, T. 1971. *Theoretical Aspects of Population Genetics.*
    Princeton University Press, Princeton.

Lewontin, R.C. 1974. *The Genetic Basis of Evolutionary Change.*
    Columbia Univ. Press, New York.

Lewontin, R.C., and Cockerham, C.C. 1959. The goodness-of-fit test for detecting
    natural selection in random mating populations.
    Evolution 13: 561-564.

Maynard Smith, J., and Haigh, J. 1974. The hitch-hiking effect of a favourable gene.
    Genet. Res. 23: 23-35.

Nordenberg, C.-B. 1960. Zur Kenntnis der Verbreitung der grössere Seenadel (Syng-
    natus acus L.) im Oresund und in der südwestlichen Ostsee.
    Kungl. Fysiografiska Sällskapets i Lund Förhandlingar 30: 107-113.

Prout, T. 1965. The estimation of fitnesses from genotypic frequencies.
    Evolution 19: 546-551.

Prout, T. 1969. The estimation of fitnesses from population data.
    Genetics 63: 949-967.

Rasmussen, E. 1973. Systematics and ecology of the Isefjord marine fauna (Denmark).
    Ophelia 11: 1-507.

Schmidt, J. 1917. *Zoarces viviparus* L. and local races of the same.
    C.R. Trav. Lab. Carlsberg. 13 (3): 277-397.

Schmidt, J. 1918. Racial Studies in Fishes I. Statistical Investigations with *Zo-
    arces viviparus* L.
    J. Genet. 7: 105-118.

Schmidt, J. 1930. The Atlantic cod (*Gadus callarias* L.) and local races of the same.
    C.R. Trav. Lab. Carlsberg. 18: 1-72.

Sick, K. 1961. Haemoglobin polymorphism in Fishes.
    Nature 192: 894-896.

Sick, K. 1965a. Haemoglobin polymorphism in cod in the Baltic and the Danish Sea.
    Hereditas 54: 19-48.

Sick, K. 1965b. Haemoglobin polymorphism of cod in the North Atlantic Ocean.
    Hereditas 54: 49-69.

Simonsen, V., and Frydenberg, O. 1972. Genetics of *Zoarces* populations II. Three
    loci determining esterase isozymes in eye and brain tissue.
    Hereditas 70: 235-242.

Thomson, G. 1976. The effect of a selected locus on linked neutral loci.
    Genetics in press.

Wahlund, S. 1928. Zusammensetzung von Populationen und Korrelationserscheinungen
    vom Standpunkt der Vererbungslehre aus betrachtet.
    Hereditas 11: 65-106.

Wallace, B. 1958. The comparison of observed and calculated zygotic distributions.
    Evolution 12: 113-115.

Wright, S. 1940. Breeding structure of populations in relation to speciation.
    Amer. Natur. 75: 232-248.

Wright, S. 1951. The genetical structure of populations.
    Ann. Eugenics. 15: 323-354.

1. STUDY OF SELECTION

A STUDY OF SEXUAL SELECTION IN NATURAL POPULATIONS OF THE
MILKWEED BEETLE, *TETRAOPES TETRAOPHTHALMUS*

Walter F. Eanes, Patrick M. Gaffney, Richard K. Koehn and
Christine M. Simon*

Panmixia in natural populations is one of the fundamental prerequisites
for the conformity of observed phenotypic frequency distributions to
the polynomial expectations of Hardy-Weinberg equilibrium. When mating
is non-random, the observed and expected phenotypic distributions may
be incongruous and thereby provide a means to discover various aspects
of the mating structure within populations. The contribution of certain
population members to the mating population can be unequal and studies
of sexual selection are concerned with elucidating this differential
contribution by various population subgroups as well as factors that
influence the contribution. Terms such as "mating selection" and "sexual selection" have been used to describe a diversity of phenomena of
this type. In this paper, sexual selection is the differential contribution to the mating population by certain population members. When
the magnitude of the differential contribution varies as a function
of the relative frequency of the contributing group, the sexual selection is "frequency-dependent".

Sexual selection has received much theoretical treatment (Wright
1921, *et seq.*, 1934; Crow and Felsenstein 1968; Karlin 1968a; 1968b;

---

* Authors are listed alphabetically

and others). Most empirical studies of sexual selection have been conducted on laboratory populations of Drosophila (Petit, 1958; Ehrman
et al. 1965; Ehrman 1966;1972; Ehrman and Petit 1968; Spiess and Spiess
1969; Bundgaard and Christiansen 1972; reviewed by Petit and Ehrman
1969 and Ehrman and Parsons 1976).

Though laboratory studies have also been done with Tribolium (Sinnock 1970), the housefly (Childress and McDonald 1973) and the wasp
Mormoniella (Grant *et al.* 1974) frequency-dependent sexual selection
was observed in many of the Drosophila studies that is stronger in
unique males and hence has been termed the "rare male effect" (see
Ehrman and Parsons 1976, for extensive review).

A mating system wherein certain phenotypes, particularly rare phenotypes, have a mating advantage, could be of major evolutionary significance, if it could be shown to exist in natural populations; so
far, this has not been done. Unfortunately, mating patterns are very
difficult to investigate in nature and the few known examples are restricted to systems of conspicuous phenotypic variation such as the
Blue-Snow Goose complex (Cooch and Beardmore 1959), song patterns in
Mockingbirds (Howard 1974), melanic patterns in birds (O'Donald 1976,
this symposium), and differential mating of Tiger Swallowtail morphs
(Levin 1973). Only one attempt has been made to estimate mating fitness in a population by the use of enzyme markers (Christiansen *et al.*
1973) and sexual selection could not be demonstrated.

There are several reasons why few investigations have been made of
sexual selection in natural populations. Difficulties arise in three
areas: the rarity of breeding characteristics of organisms in nature
that lend themselves to the study of sexual selection, (e.g. mechanics
of breeding, natural phenotypic markers, and so forth), the numerically
large samples necessary for estimating mating fitness by means other
than direct observation, and the statistical evaluation of genetic information obtained from various subdivisions of the population. In this
paper we will enumerate these difficulties in detail as they have been
encountered in a study of sexual selection in the Four Spotted Milkweed Beetle, *Tetraopes tetraophthalmus* (Forrester), a cerambycid beetle
distributed throughout eastern North America. We will point out the important characteristics of organisms that must be considered in designing similar studies in the future, the kind of data that can reasonably be expected and the statistical problems that arise when the
data is analyzed. We will emphasize certain difficulties as they arise
with regard to the use of enzyme loci as genetic markers, since a com-

plete understanding of the importance of sexual selection in nature
should not ultimately depend upon those few organisms exhibiting phenotypic diversity of more conspicuous characters (e.g., plummage color,
behaviorial traits, etc.).

Lastly, we will present certain data that suggest frequency-dependent sexual selection in Tetraopes, which in view of the difficulties
enumerated below, must be interpreted with caution.

POPULATION CHARACTERISTICS

An inability to distinguish the mating and non-mating components of a
natural population excludes most organisms from study. Clearly, to
study mating selection, one must be able to observe and sample the
mating population and preferably, to understand the dynamics of this
activity (i.e., single versus multiple mating, time spent *in copulo*,
and so forth). During this study, substantial proportions of the total
population of Tetraopes were mating at any one time (about 40 per cent),
but we do not know how these proportions may have varied over the breeding season of the summer months. Mating pairs remain *in copulo* for extended periods (days), yet estimates of mating structure made at one
point in time cannot be informative of temporal variation in mating
structure. Mating pairs and solitary beetles can be collected separately, but evidence of mating selection from such collections must be interpreted in a context of total ignorance of potential temporal changes
in mating patterns. In Tetraopes, a single collection so devastates
the population that subsequent sampling is either impossible or would
be unrepresentative of a natural system. With the use of enzyme markers, field observations are not possible, so that studied organisms
must be abundant enough for multiple sampling with minimal disturbance
of the population, or have such a protracted breeding season that a
single sample adequately estimates the breeding structure. The advantage of Tetraopes for easy identification of the mating population at
any one time, is lost to the unavailability of multiple samples.

Our collections of Tetraopes ranged in sample size from 199 to
435. It became apparent during the data analyses that even the largest
of these was less than optimal for some of our purposes. For example,
a single sample was subdivided by sex, mating participation, and geno-
type(s) which substantially reduces the sample sizes of various sub-

groups treated in the statistical analyses. Very large samples are required to detect numerically small, but biologically significant, among-
group differences and samples an order of magnitude larger than we
have taken would not be unreasonable.

While extraordinarily large samples and an identifiable mating
population are necessary prerequisites for a study of mating selection in nature, both can be accomodated by the choice of an appropriate study organism. When both conditions can be met, estimates of mating fitness can be obtained. There is, however, no *a priori* basis for
designing the study of frequency-dependent sexual selection, since
this will additionally depend on the existence of substantial interpopulation variation of genetic composition. While laboratory populations are amenable to manipulation of their base frequencies (e.g.
sex ratio phenotypic frequencies, etc.), in nature, multiple populations must be found that (1) inhabit relatively comparable environments so as not to drastically alter the among-population ecological
influences on mating and (2) exhibit some real variance in gene frequencies. Thus, frequency-dependent sexual selection can be detected
in nature only when sexual selection occurs within populations and natural frequency variations exist with which both the magnitude and pattern of sexual selection can be compared. It is at this point that
sampling problems become acute. Samples of constant size will not
estimate individual variables with equal accuracy. More importantly,
if the "rare effect" is to be studied in nature, genotypes of principal interest will be those composed of least common alleles, that
vary close to fixation among populations. These will be inadequately
represented in samples of any size. Parameters estimated from them
will have very large confidence intervals and these will be hetero-
scadastic over the sampled frequency range. We will return to this
problem later.

Natural populations of the Four Spotted Milkweed Beetle, *Tetraopes
tetraophthalmus*, fulfill the basic requirements (except where otherwise noted) for an investigation of sexual selection. On Long Island,
the species is distributed as discrete semi-isolated populations, as
it is restricted to patches of the Common Milkweed (*Asclepias syriaca*).
Large patches of *A. syriaca* are uncommon on Long Island and because of
the low vagility of this beetle species, sampled populations (the only
populations discovered with sufficient abundance of animals for study)
probably represent relatively isolated units. Males and females are
easily distinguishable from one another on the bases of size and ex-

ternal genitalia. Geographic variation and mating fitness of certain
morphological characters have been previously studied (Mason 1964;
1972) and implicated female preference manifested as stabilizing selection on males.

METHODS

Collections of *T. tetraophthalmus* were made at two localities in July,
1974, and seven localities in July, 1975, on Long Island, New York
(Figure 1). The largest distance between any two localities is about
50 miles, but one set of localities was separated by only 400 meters
of woodland. Additional large populations were sought on Long Island,
but not found. Adults emerge in early July and are found on milkweed
plants until mid-August. Collections were made by hand, moving from
plant to plant in one direction through the population. All beetles

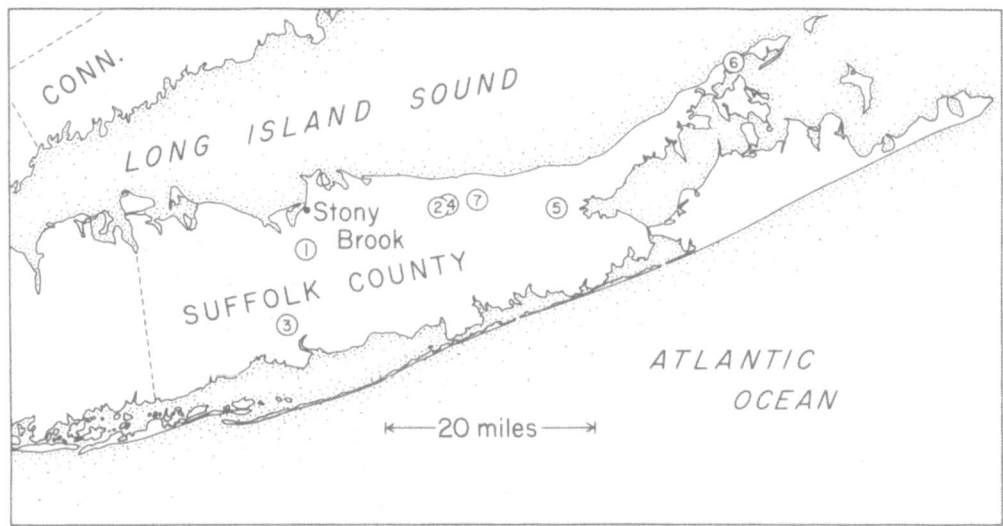

Figure 1.   Geographic distribution of sample localities.

encountered were sorted into two groups representing solitary individuals and copulating pairs. Total sample size averaged 300 individuals per locality, and the proportion of mating individuals was estimated by a parallel census of the population.

Specimens were stored at $-60^{\circ}C$ until electrophoretic analyses could
be done. Storage had no detected effects on electrophoretic patterns.
The thorax of individual beetles was mechanically homogenated in 0.2 M
sucrose containing an equal volume of two percent phenoxyethanol and
electrophoresis was done on extracts. Homogenates were centrifuged for
20 minutes at about 48,000 g. Horizontal starch gel electrophoresis
was performed in a 12 percent gel (Electrostarch, Otto Hiller, Inc.,
Madison, Wisconsin). While we initially screened for large numbers of
enzyme loci in this species only two polymorphic enzymes were discovered: phosphoglucomutase (*PGM*) and leucine naphthylamidase (*LN*). PGM
was resolved on a tris-maleate buffer system consisting of: 0.10 M
tris - 0.10 M maleic acid - 0.10 M ETDA - 0.10 M magnesium chloride,
pH 7.4. The gel buffer was a 1:9 dilution of this electrode buffer.
The electrophoretic run was for approximately five hours at 1.5 mA
per centimeter width. Leucine naphthylamidase was resolved on a phosphate buffer consisting of: 0.138 potassium phosphate - 0.061 M so-
dium hydroxide, pH 6.7. The gel buffer consisted of a 1:19 dilution of
this electrode buffer and running conditions were the same as for PGM
above. "Leucine naphthylamidase" is detected with L-leucyl-β-naphthy-
lamide as substrate.

A variety of statistical tests were performed on these data. Zygotic frequency distributions of relevant sub-groups were compared to
Hardy-Weinberg expectations by the commonly used chi-square test. Genotypic compositions of various distributions such as sexes, mating versus non-mating samples and so forth, were compared by a RxC contingency chi-square on allele counts as well as by a RxC test of genotypic
frequencies. The two tests are not wholly independent, but with multiple alleles and departures from Hardy-Weinberg distributions, they
can estimate different sources of heterogeneity.

Kendall rank correlations were determined between genotype frequencies and mating fitness over localities in order to test for the
presence of frequency-dependent sexual selection. Mating fitness is
computed as:

$$W_i = \frac{P_i'}{P_i} \times \frac{P_j}{P_j'}$$

where P_i is the frequency of the ith genotype in the total population, P'_i is the frequency of the ith genotype in the mating portion of the population, P_j is the frequency of the most common genotype in the total population, and P'_j is the frequency of the same genotype in the mating proportion of the population (Bundgaard and Christiansen 1972). This estimate of mating fitness has both advantages and disadvantages, (discussed later), but generally measures the *relative* success of individual genotypes to contribute to the mating population. Mating fitnesses were computed separately for each sex.

No estimates of the proportion of the population mating were made in the two 1974 collections. The base frequencies used for computing mating fitness were estimated by weighting 1974 samples by the proportion mating in the 1975 collections. The mating proportions among populations did not vary significantly in 1975 ($\chi^2_6 = 6.04$).

RESULTS

Genotypic distributions and allele frequencies estimated at the *PGM* and *LN* loci for total population samples and various subdivisions are given in Tables 1 and 2. Samples were divided into mating males, mating females, non-mating males, non-mating females.

Interlocality variation.

Significant differentiation was observed among localities for all alleles at both loci. The level of differentiation is marked considering the relatively small geographic distances that separate the sampled populations. Three alleles were segregating at the *PGM* locus in all populations. Significant interpopulation differences occur in counts of all three alleles ($\chi^2_{16} = 384$; P << .001) and genotypic frequencies (*PGM*, $\chi^2_{40} = 416$; P << .001). Two alleles were observed at the *LN* locus and their frequencies were highly heterogeneous among localities ($\chi^2_{16} = 177$; P << .001). Among locality variation occurred in both sexes.

Table 1. Summary of genetic composition at the *PGM* locus in nine samples of *T. tetraophthalmus* subdivided by sex, mating (M), and non-mating (N) groups. Sampled in 1975, except where noted. χ^2_3 is for a test of Hardy-Weinberg proportions

Locality	Sex	$PGM^{A/A}$	$PGM^{A/B}$	$PGM^{B/B}$	$PGM^{A/C}$	$PGM^{B/C}$	$PGM^{C/C}$	N	$\chi^2_{(3)}$	f(A)	f(B)
	M ♂	1	13	17	6	28	12	77	1.53	.136	.487
	N ♂	1	13	26	10	27	6	83	1.61	.150	.554
1a (1974)	M ♀	1	12	25	5	21	10	74	2.50	.128	.560
	N ♀	1	3	16	4	12	4	40	2.80	.112	.587
	Total	4	41	84	25	88	32	274	1.48	.135	.542
	M ♂	1	12	11	7	25	9	65	1.76	.161	.453
	N ♂	3	33	15	9	29	13	102	12.09**	.235	.457
1b	M ♀	2	13	9	1	28	9	62	10.86*	.145	.475
	N ♀	1	16	22	12	28	5	84	3.08	.178	.523
	Total	7	74	57	29	110	36	313	13.84**	.186	.476
	M ♂	0	3	19	0	18	1	41	3.10	.036	.719
	N ♂	1	6	33	2	25	4	71	1.45	.070	.683
2	M ♀	0	5	13	1	22	2	43	4.67	.069	.616
	N ♀	1	6	25	0	22	2	56	5.02	.071	.696
	Total	2	20	90	3	87	9	211	8.62	.064	.680
	M ♂	1	13	20	5	16	3	58	.45	.172	.594
	N ♂	5	30	34	7	26	7	109	2.17	.215	.568
3	M ♀	2	26	21	2	10	0	61	4.98	.262	.639
	N ♀	1	16	24	2	14	2	59	1.66	.169	.661
	Total	9	85	99	16	66	12	287	5.74	.207	.608
	M ♂	0	22	72	6	46	12	158	3.26	.088	.670
	N ♂	0	3	24	2	7	3	39	4.92	.064	.743
4a (1974)	M ♀	0	14	75	1	60	10	160	3.92	.046	.700
	N ♀	0	5	8	0	8	1	22	2.87	.113	.659
	Total	0	44	179	9	121	26	379	5.42	.069	.690
	M ♂	0	10	35	0	21	5	71	4.55	.070	.711
	N ♂	0	23	51	5	38	10	127	4.28	.110	.641
4b	M ♀	4	11	28	4	20	8	75	6.06	.153	.580
	N ♀	0	12	35	2	18	2	69	1.19	.101	.724
	Total	4	56	149	11	97	25	342	6.17	.109	.659
	M ♂	0	1	41	3	26	6	77	5.27	.026	.707
	N ♂	0	1	43	0	30	14	88	4.90	.005	.664
5	M ♀	0	0	25	2	42	6	75	7.31	.013	.613
	N ♀	0	1	22	0	26	9	58	.71	.008	.612
	Total	0	3	131	5	124	35	298	3.52	.013	.652
	M ♂	0	6	30	1	48	14	99	2.78	.035	.575
	N ♂	0	4	22	0	57	18	101	6.77	.019	.519
6	M ♀	0	6	21	2	43	19	91	2.10	.044	.500
	N ♀	0	7	39	0	73	25	144	6.62	.024	.548
	Total	0	23	112	3	221	76	435	15.84**	.029	.537
	M ♂	0	6	21	2	24	2	55	2.73	.072	.654
	N ♂	0	4	32	1	32	7	76	.50	.032	.651
7	M ♀	2	12	22	1	17	2	56	2.74	.151	.657
	N ♀	0	4	16	1	20	7	48	.97	.052	.583
	Total	2	26	91	5	93	18	235	4.39	.074	.640

* $P < .05$;   ** $P < .01$

Table 2. Summary of genetic composition at the *LN* locus.
See Table 1 caption

Locality	Sex	$LN^{A/A}$	$LN^{A/B}$	$LN^{B/B}$	N	$\chi^2_{(1)}$	$f(A)$
	M ♂	26	41	29	96	2.01	.484
	N ♂	19	34	31	84	2.53	.428
1a	M ♀	21	46	23	90	.04	.488
(1974)	N ♀	12	16	11	39	1.24	.512
	Total	78	137	94	309	3.79	.474
	M ♂	16	32	17	65	.01	.492
	N ♂	20	52	31	103	.04	.446
1b	M ♀	15	26	23	64	1.95	.437
	N ♀	13	45	26	84	.80	.422
	Total	64	155	97	316	.02	.447
	M ♂	1	23	17	41	4.29	.304
	N ♂	8	32	29	69	.03	.347
2	M ♀	9	20	10	39	.02	.487
	N ♀	4	19	27	50	.06	.270
	Total	22	94	83	199	.36	.346
	M ♂	14	30	15	59	.01	.491
	N ♂	26	58	24	108	.59	.509
3	M ♀	10	25	26	61	.87	.368
	N ♀	17	33	9	59	1.14	.567
	Total	67	146	74	287	.09	.487
	M ♂	13	62	79	154	.02	.285
	N ♂	2	18	21	41	.57	.268
4a	M ♀	17	62	73	152	.47	.315
(1974)	N ♀	1	8	13	22	.02	.227
	Total	33	150	186	369	.12	.292
	M ♂	10	24	36	70	2.92	.314
	N ♂	11	54	63	128	.01	.296
4b	M ♀	13	17	44	74	14.50**	.290
	N ♀	9	21	19	49	.54	.398
	Total	43	116	162	321	8.43**	.314
	M ♂	11	34	33	78	.21	.359
	N ♂	13	42	44	99	.34	.343
5	M ♀	11	33	30	74	.15	.371
	N ♀	6	27	25	58	.04	.334
	Total	41	136	132	309	.40	.352
	M ♂	33	48	14	95	.26	.600
	N ♂	28	51	22	101	.01	.529
6	M ♀	28	39	25	92	2.10	.516
	N ♀	46	65	31	142	.78	.552
	Total	135	203	92	430	.92	.550
	M ♂	16	22	17	55	2.19	.490
	N ♂	12	45	19	76	2.87	.453
7	M ♀	9	23	22	54	.49	.379
	N ♀	18	30	26	74	2.38	.445
	Total	55	120	84	259	.98	.444

** P < .01

Intralocality variations.

The differences in genetic composition that occur between the mating and non-mating populations may be attributed to sexual selection.
Of the 36 possible test comparisons of the mating and non-mating populations (18 on allele counts and 18 on zygotic frequencies, one of
each for each sex in each sample), nine comparisons indicate significant heterogeneity between them. The significant test results were
obtained only in comparisons of females, but at both loci and in five
of the nine samples (Table 3).

When performing a large number of statistical tests, some significant test results are always expected by chance. However, in view of
the significance levels we have observed relative to the total number
of tests performed, the test results must be considered to reflect
some real differences in genetic composition between mating and non-
mating female groups. The sources of heterogeneity are, however, variable from one comparison to another. For example, allele counts at
the *PGM* locus are different between mating and non-mating females in
samples 4b and 7 (Tables 1 and 3), where the frequencies of PGM^b and
PGM^c differ by .07 and .14. However, the frequency of PGM^b increases
(PGM^c decreases) in the mating group in one sample (7), but decreases
in the other (4b). There is a reduction of PGM^a in both samples in
the mating population. Differences are stronger at the *LN* locus, but
are also variable in pattern. The mating female group has a .10 and
.20 lower frequency of LN^a in samples 4b and 3, respectively (Tables
2 and 3), but LN^a is .21 higher in mating females in sample 2. Comparisons of zygotic frequencies of mating and non-mating females (Table
3) indicate a difference between the two groups also exists in population sample 1b.

Patterns of sexual selection.

Since significant heterogeneity has been detected both among and
within population samples, it was desirable to investigate the possibility that there is an among-population frequency-dependent pattern
in estimates of sexual selection. We were particularly interested in
examining a potential relationship between relative mating fitness and
frequency of particular rare genotypes. At the *LN* locus, the three
genotypes resulting from two alleles are not of equivalent frequencies in population samples. The frequency of the $LN^{a/a}$ homozygote is

Table 3. Summary of the significant test results of statistical comparisons of mating and non-mating female populations within nine locality samples. Sexes were treated separately, resulting in eighteen total tests of each type on each locus (nine for each sex). All comparisons of mating versus non-mating males gave non-significant test results.

Chi-Square test	Locus	d.f.	Test result χ_2	P	Locality
R × C test on allele counts					
	PGM	2	6.60	< .05	4b
		2	10.58	< .01	7
	LN	1	8.92	< .005	2
		1	9.54	< .005	3
		1	9.21	< .005	4b
R × C test on zygotic frequencies					
	PGM	5	13.54	< .025	1b
	LN	2	8.53	< .025	2
		2	11.15	< .005	3
		2	6.24	< .05	4b

usually less common than either of the other two genotypes. The six genotypes resulting from the assortment of three alleles at the *PGM* locus are also of unequal frequency in population samples ($PGM^{a/a}$ << $PGM^{a/c}$ < $PGM^{c/c}$ ≃ $PGM^{a/b}$ < $PGM^{b/c}$ ≃ $PGM^{b/b}$). Because of the allele frequency variation among localities, there is sometimes a change in the order of adjacent genotypes in their relative frequencies, though $PGM^{a/a}$ and $PGM^{a/c}$ are without exception the least common of the six genotypes. $PGM^{a/a}$ is so rare in our samples that in cannot be analyzed with regard to frequency-dependence (only 28 individuals of genotype $PGM^{a/a}$ were observed in samples totalling 2,774 individuals).

Mating fitnesses (W_i) of the ith genotypes were computed separate-

ly for each sex and each locality. Kendall rank correlations were computed separately for each sex between W_i and P_i. The computation of W_i
for the ith genotype of frequency P_i, is relative to the frequency of
a "standard" genotype (P_j). For genotypes at the *LN* locus W was computed relative to the heterozygote, $LN^{a/b}$, resulting in four individual tests, two homozygous genotypes in each sex. At the *PGM* locus,
$PGM^{b/c}$ (most common) was used as the standard in the relative computation of W for $PGM^{a/b}$, $PGM^{a/c}$, $PGM^{b/b}$, and $PGM^{b/c}$ ($PGM^{a/a}$ was too rare
to be tested), giving in all, eight tests of frequency-dependent sexual
selection at this locus.

Of the twelve total tests of a rank correlation between relative
mating fitness and relative genotypic frequency, a significant relationship is suggested for $PGM^{a/b}$ in males, $PGM^{a/c}$ in females, and $LN^{a/a}$
in females (Figure 2). In all three cases, the correlation is negative,
which reflects an increase in values of W with a relative decrease in
genotype frequency.

DISCUSSION

Our results provide evidence that sexual selection in females is occurring with respect to both loci in some of the Tetraopes populations.
There were substantial differences in genetic composition between mating and non-mating samples at the time of collection, and these differences are generally of the same magnitude as the among locality differences in genetic composition. Two genotypes at the *PGM* locus (one
in each sex) exhibit patterns of mating fitness consistent with frequency-dependent sexual selection, however, apparent patterns of sexual selection as illustrated in Figure 2 should be interpreted with
caution. The sampling variance of mating fitness is itself frequency-
dependent and the statistical analysis of such patterns has many difficulties. As the relevant genotypes become rare in each collection,
the fitness estimates associated with each carry exceedingly large
sampling variances. Their magnitudes are unknown, but it is reasonable to assume that none of the W_i values in Figure 2 are significantly
different from unity. These difficulties are again suggested in Figure
2 for the relationship between mating fitness and the frequency of $LN^{a/a}$
genotype in females. There is no significant frequency-dependent pattern as tested by the rank correlation, yet all the fitnesses above

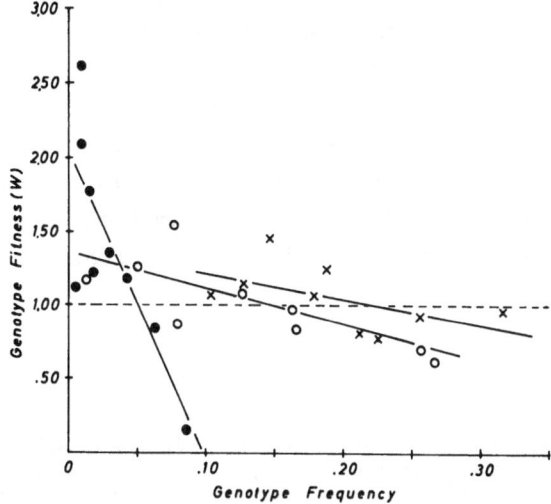

Figure 2. Relationships between genotype-specific mating fitnesses and genotype frequencies among populations. x: *LN-AA* in females. O: *PGM-AB* in males. ●: *PGM-AC* in females.

the mean frequency are greater than 1.0 while all the fitnesses are less than one for the samples below mean genotype frequency.

The computation of W_i as a standardized fitness removes the inherent frequency-dependent properties of computing fitness as a simple ratio of selected to preselected frequencies (Bundgaard and Christiansen 1972). The ith genotype can decrease in representation due to mating selection, yet if the standardizing genotype is also underrepresented, the fitness of the ith genotype can still be greater than one.

It should also be emphasized that our collections were made at single points in time and may not reflect total mating structure over an entire season. It is possible that all of the non-mating indivi-

duals had already mated or would ultimately mate. Repeated sampling
over time would provide insight concerning total mating structure, but
is impossible in Tetraopes since the single collections approach the
total population.

The results here provide an example of how sexual selection in
natural populations could be studied, and provide preliminary estimates of both expected outcomes and the types of difficulties of interpretation that may be anticipated. It seems likely these problems will
serve to discourage future studies of this type. The favorable characteristics possessed by an organism such as Tetraopes are not typical of the vast majority of other animal species, and even in suitable
organisms the statistical affirmation of the patterns of sexual selection suggested here will require prohibitively large sample sizes.

Summary.

Two enzyme loci were used to study sexual selection in nine natural populations
of the Milkweed Beetle, *Tetraopes tetraophthalmus*. Of thirty-six test comparisons,
nine indicated significant differences between mating and non-mating females. Allele
frequencies between mating and non-mating females differed by up to .21, but no differences occurred between mating and non-mating males.

The genetic composition of population samples varied significantly among the
nine collections. Among-population frequency-dependent estimates of sexual selection
were investigated by rank correlations of mating fitnesses with genotypic frequencies. Significant negative correlations were observed for a total of three of the
seven genotypes analyzed at both loci. These results suggest that frequency-dependent sexual selection occurs in natural populations of Tetraopes. However, we enumerate the many difficulties that arise in the collection and analyses of data of
this type, and as a consequence, our results should be interpreted with caution.

Acknowledgement.

We are grateful to Freddy Christiansen for substantial statistical guidance.
Lee Ehrman commented on an earlier version of the manuscript. I. Kornfield assisted
our collecting. This work was supported by USPHS Research Grant GM-21133, NSF Research Grant BMS 74 02522 and USPHS Career Award GM-28963. This is contribution
number 177 from the Program in Ecology and Evolution, State University of New York,
Stony Brook, N.Y. 11794.

REFERENCES

Bundgaard. J.̄, and Christiansen, F.B. 1972. Dynamics of polymorphism: I. Selection components in an experimental population of *Drosophila melanogaster*. *Genetics* 71: 439-460.

Childress, D., and McDonald, I.C. 1973. Tests for frequency-dependent mating success in the house fly. *Behav. Genet.* 3: 217-223.

Christiansen, F.B., Frydenberg, O., and Simonsen, V. 1973. Genetics of *Zoarces* populations. IV. Selection component analysis of an esterase polymorphism using population samples including mother-offspring combinations. *Hereditas* 73: 291-304.

Cooch, F.G., and Beardmore, J.A. 1959. Assortative mating and reciprocal differences in the Blue-Snow Goose complex. *Nature* 183: 1833-1834.

Crow, J.F., and Felsenstein, J. 1968. The effect of assortative mating on the genetic composition of a population. *Eugen. Quart.* 15: 85-97.

Ehrman, L. 1966. Mating success and genotype frequency in *Drosophila*. *Anim. Behav.* 14: 332-339.

Ehrman, L. 1972. A factor influencing the rare male mating advantage in *Drosophila*. *Behav. Genet.* 2: 69-78.

Ehrman, L., and Parsons, P.A. 1976. *The Genetics of Behavior*. Sinauer Associates, Inc., Sunderland, Mass. 390 pp.

Ehrman, L., and Petit, C. 1968. Genotype frequency and mating success in the *willistoni* species group of *Drosophila*. *Evolution* 22: 649-658.

Ehrman, L., Spassky, B., Pavlovsky, O., and Dobzhansky, Th. 1965. Sexual selection, geotaxis, and chromosomal polymorphism in experimental populations of *Drosophila pseudoobscura*. *Evolution* 19: 337-346.

Grant, B., Snyder, G.A., and Glessner, S.F. 1974. Frequency-dependent mate selection in *Mormoniella vitripennis*. *Evolution* 28: 259-264.

Howard, R.D. 1974. The influence of sexual selection and interspecific competition on Mockingbird song (*Mimus polyglottos*). *Evolution* 28: 428-438.

Karlin, S. 1968a. Equilibrium behavior of population genetic models with non-random mating. Part I. Preliminaries and special mating systems. *J. Appl. Prob.* 5: 231-313.

Karlin, S. 1968b. Equilibrium behavior of population genetic models with non-random mating. Part II. Pedigrees, homozygosity, and stochastic models. *J. Appl. Prob.* 5: 487-566.

Levin, M.P. 1973. Preferential mating and the maintenance of the sex-limited dimorphism in *Papilio glaucus*: Evidence from laboratory matings. *Evolution* 27: 257-264.

Mason, L.G. 1964. Stabilizing selection for mating fitness in natural populations of *Tetraopes*. *Evolution* 18: 492-497.

Mason, L. 1972. Natural insect populations and assortative mating.
Am. Midl. Nat. 88: 150-157.

O'Donald, P. 1976. Mating preferences and their genetic effects in models of sexual selection for colour phases of the Artic Skua, pp. 411-430.
In : Population Genetics and Ecology (S. Karlin and E. Nevo, eds.), Academic Press, N.Y. 832 pp.

Petit, C. 1958. Le determinisme genetique et psycho-physiologique de la competition sexuelle chez *Drosophila melanogaster*.
Bull. Biol. France Belg. 92: 248-329.

Petit, C., and Ehrman, L. 1969. Sexual selection in *Drosophila*, pp. 177-223.
In: Evolutionary Biology, vol. 3, (Th. Dobzhansky, M.K. Hecht and W.C. Steere, eds.), Appleton-Century-Crofts, N.Y.

Sinnock, P. 1970. Frequency dependence and mating behavior in *Tribolium castaneum*.
Amer. Natur. 104: 469-476.

Spiess, L.D., and Spiess, E.B. 1969. Minority advantage in interpopulational matings of *Drosophila persimilis*.
Amer. Natur. 103: 155-172.

Wright, S. 1921. Systems of mating. I. The biometric relations between parent and offspring.
Genetics 6: 111-123.

Wright, S. 1921. Systems of mating. II. The effects of inbreeding on the genetic composition of a population.
Genetics 6: 124-143.

Wright, S. 1921. Systems of mating. III. Assortative mating based on somatic resemblance.
Genetics 6: 144-161.

Wright, S. 1921. Systems of mating. IV. The effects of selection.
Genetics 6: 162-166.

Wright, S. 1921. Systems of mating. V. General considerations.
Genetics 6: 167-178.

Wright, S. 1934. The method of path coefficients.
Anns. Math. Stat. 5: 161-215.

1. STUDY OF SELECTION

HOW DOES THE GENOME CONGEAL?

Daniel L. Hartl

A question originally inspired by R.A. Fisher - Why does the genome
not congeal? - has been a stimulus for theoretical study of linkage
modification (Kojima and Schaffer 1964; Turner 1967a, 1967b; Nei 1967,
1969; Lewontin 1971; Feldman 1972; Karlin and McGregor 1974; Strobeck,
Maynard Smith and Charlesworth 1976). In this paper I wish to discuss
some experimental evidence that bears on the reverse question -How
does the genome congeal? - because in some species the genome has
congealed, at least parts of it. Among plants the most famous case is
that of translocation complexes in *Oenothera* (Stebbins 1950). Among
animals we have the celebrated paracentric inversions of *Drosophila
pseudoobscura* (Dobzhansky 1970). Many other instances of congealment
of parts of the genome are known (Stebbins 1950, White 1973).

I believe that the evolution of coadapted genetic systems with
restricted recombination occurs in a sequence of three rather distinct
steps: (1) initiation by epistasis, (2) consolidation by congealment,
and (3) improvement by coadaptation. Step (1) envisions the chance
appearance of strongly epistatic mutations at two distinct loci and
the subsequent maintenance of linkage disequilibrium between them.
Step (2) occurs because genetic systems with permanent linkage disequilibrium will tend to evolve reduced recombination (Turner 1967b).
And step (3) occurs because reduction in recombination along a chromosome tends to select for a chain of epistatically favorable alleles

within or near the congealed segment (Kojima and Schaffer 1964).

One experimental approach to determining the validity of these
ideas is to take such a system apart genetically, piece by piece, and
then try to reconstruct its evolution under experimental conditions.
For this purpose I have been studying *segregation distortion* in *Drosophila melanogaster*. In this case, almost the whole of the right arm
of the second chromosome has become congealed by means of inversions.
Yet the system is amenable to genetic dissection in the laboratory,
and the information obtained can be used to reconstruct or infer the
evolution of the supergene.

Natural populations of *D. melanogaster* are polymorphic for a supergene called *SD* (stands for *segregation distorter*). The principal
selective forces that maintain the polymorphism are easily identified:
heterozygous SD/SD^+ males produce a gross excess of *SD* bearing offspring due to the dysfunction of the SD^+ bearing sperm, the ratio of
SD to SD^+ among functional sperm ranging from 6:1 to 99:1 (Sandler,
Hiraizumi and Sandler 1959; review in Hartl and Hiraizumi 1976). However, the meiotic drive is counterbalanced by zygotic selection: homozygous *SD/SD* males are sterile or nearly sterile due to the dysfunction of almost all their sperm (Hartl 1969). (In nature, *SD/SD* males
and females are inviable because most naturally-occurring *SD* chromosomes carry recessive lethals). It might be argued that this system
is pathological because one of the selective forces is a high degree
of meiotic drive, a phenomenon which is relatively uncommon despite the
occurrence of examples in a wide variety of organisms (Zimmering, Sandler and Nicoletti 1970). However, it seems to me that the most important aspect of the system is not the precise nature of the selective
mechanisms. What is most important is that the alleles within the
supergene have strong epistatic interactions. Indeed, I suspect that
any genetic system with sufficiently strong epistasis to initiate the
evolution of a supergene would appear pathological merely because of
the unusual strength of the epistasis.

The *SD* supergene consists of two mutations straddling the centro-
mere of chromosome 2 and tied together by a pericentric inversion
which includes about the middle 1/8 of the chromosome. (Genetic analysis of the system has been made possible by the discovery of an *SD*
chromosome which lacks the inversion (Sandler *et al.* 1959; Hartl 1974).)
The greater part of the inversion involves centromeric heterochromatin;
since crossing over in the centromeric heterochromatin rarely occurs,
zygotic loss due to aneuploid gametes produced by crossing over in

females is minimal, and crossing over does not occur in males of *D. melanogaster*.

One of the mutations in the *SD* supergene is known as *Sd* (for *segregation distorter*); *Sd* occupies a locus in the left arm of the chromosome, and its wild-type allele is denoted Sd^+. The other mutation is known as *Rsp* (for *responder*); *Rsp* occupies a locus in the right arm of the chromosome, and its wild-type allele is designated Rsp^+. Both loci are close to the centromeric heterochromatin; *Rsp* may be in it (McClanahan 1974). In any case, an *SD* chromosome is genotypically *Sd Rsp*.

The loci *Sd* and *Rsp* interact as follows: *Rsp* is the allele driven by the meiotic drive, and Sd^+ behaves as a recessive suppressor of drive. That is to say, Rsp/Rsp^+ males will produce an excess of *Rsp*-bearing sperm except when the males are homozygous Sd^+/Sd^+. Segregation in *Rsp/Rsp* and Rsp^+/Rsp^+ homozygotes is normal, but *Sd Rsp/Sd Rsp* homozygotes are male sterile.

The loci *Sd*, *Rsp* and the pericentric inversion constitute only part of the rather complex system as it has evolved in nature. Most naturally-occurring *SD* chromosomes carry, near the tip of the right arm, an enhancer allele called *St* (*stabilizer*) which increases the amount of meiotic drive (Sandler and Hiraizumi 1960a); moreover, the enhancer is tied to the *Sd Rsp* complex by one or more paracentric inversions which contain a set of polygenic enhancers (Miklos and Smith-White 1971) and usually a recessive lethal. (Review in Hartl 1975a.)

In this paper I will consider only the initial step in the evolution of the *Sd Rsp* supergene - the establishment of a two-locus polymorphism maintained because of epistatis and marked by permanent linkage disequilibrium. This provides the precondition for selection of reduced recombination, which is followed, in turn, by the selection of favorably interacting alleles within or near the congealed block of chromosome. I presume that the part of the *SD* complex in the right arm of the chromosome involving paracentric inversions and polygenic enhancers arises by a comparable sequence of events, the strong epistasis required for initiation in this case being provided by the *St* allele.

The experiments reported here provide experimental support for the following points about the population dynamics of the *Sd Rsp* polymorphism. First, populations that are polymorphic with a low frequency of *Sd Rsp* tend to evolve a remarkably high frequency of Sd^+ *Rsp*. Second, the Sd^+ *Rsp* chromosomes in a population derive originally by recom-

bination and not by mutation. Third, populations containing Sd and
Rsp but no pericentric inversion evolve a stable, two-locus polymorphism marked by linkage disequilibrium. And fourth, the Sd Rsp polymorphism is maintained because of epistasis; that is, the Rsp/Rsp^+
polymorphism is not established in the absence of the Sd allele.

FORMAL GENETICS OF SEGREGATION DISTORTION

The extensive, complex, often contradictory literature on the formal
genetics and developmental biology of segregation distortion has been
reviewed by Hartl and Hiraizumi (1976). For present purposes only
the broadest outline is necessary. The segregation ratios conform to
the rule mentioned earlier: meiotic drive in favor of Rsp occurs in
Rsp/Rsp^+ heterozygous males unless the males are homozygous Sd^+/Sd^+;
that is, Sd^+ behaves as a recessive suppressor of the meiotic drive
of Rsp.

   The mechanism of meiotic drive in Rsp/Rsp^+ males is the induced
dysfunction of a large proportion of the Rsp^+ bearing sperm. Although
the details of the molecular mechanism have not been fully worked out,
recent evidence suggests that the developmental lesion in the dysfunctional sperm may involve a defective transition from lysine-rich histones to arginine-rich histones, a transition which normally occurs
during spermatogenesis in $D.$ $melanogaster$ (Das, Kaufmann and Gay 1964;
Kettaneh and Hartl 1976). In any case, segregation distorter males
produce dysfunctional sperm, and consequently they suffer a reduction
in fertility. Half the sperm from Rsp/Rsp^+ males showing meiotic drive
are dysfunctional; in population cages this results in about a 20% reduction in fertility (Hartl 1970a). Sperm dysfunction also occurs in
other genotypes, however. Virtually all the sperm are dysfunctional
in males of genotype Sd Rsp/Sd Rsp, thus the males are sterile or nearly sterile (Hartl 1973). Of equal evolutionary importance is the fact
that males of genotype Sd Rsp/Sd^+ Rsp produce only half as many functional sperm as do wild-type males, although the segregation ratio
from such males is normal (Hartl 1969; Hihara 1974). Indeed, abnormalities in spermatogenesis have been observed in all Rsp/Rsp genotypes
so far examined (Hauschteck-Jungen and Hartl 1977).

ANALYSIS OF POPULATIONS

Results from studies of three populations will be discussed. One
is a natural population in Raleigh, North Carolina; some of these results have appeared elsewhere (Hartl and Hartung 1975) and will be
mentioned only briefly. The second population is a long-term experimental population established eight years ago by Professor Y. Hiraizumi and generously provided to me for analysis (Hartl 1977); my
comments here will be largely confined to recent unpublished results.
The third population is a natural population from southern Texas;
this population is of interest because it appears to be free of Sd
Rsp.

Analysis of natural and artificial populations first involves extracting a large number of second chromosomes. Males from the population of interest are mated individually to laboratory females from
a standard homozygous strain bearing the second-chromosomal mutations
cn and bw. (cn = $cinnabar$ eyes; bw = $brown$ eyes; the cn bw/cn bw
double homozygote has $white$ eyes; with respect to the loci involved
in segregation distortion, the cn bw chromosome is Sd^+ Rsp^+.) From each
such mating a $single$ +/cn bw son is selected and backcrossed to cn bw.
(The + symbol is here used to designate a cn^+ bw^+ chromosome.) Each
strain so produced carries a single +-chromosome isolated from the
population, and the chromosome is maintained intact by repeated back-
crossing of +/cn bw males to cn bw females. Since recombination does
not occur in normal males of $D.$ $melanogaster$, the genetic integrity
of the isolated second chromosome is maintained, and the repeated back-
crossing gradually replaces the genetic background with that from the
cn bw strain. (The wild Y chromosome is usually not replaced as the Y
seems to be free of modifiers of segregation distortion (Hartl 1970b).)

Each of the extracted + chromosomes is then made heterozygous in
males with each of four genetically-marked tester chromosomes and the
segregation ratios of the males determined. The tests are:

(A) +/Sd^+ Rsp^+. An $excess$ of + bearing offspring in this test
indicates that the + chromosome is genetically Sd Rsp. If the segregation ratio is normal, the + chromosome is said to be a $non-distorter$.

(B) +/Sd Rsp. An absence of meiotic drive indicates that the tested chromosome is $insensitive$ to segregation distortion.

(C) +/Sd Rsp^+. A $deficiency$ of Sd Rsp^+ bearing offspring indicates
that the + chromosome carries Rsp. If the + chromosome is also an insensitive nondistorter (tests B and A), then its genotype is diagnosed

as Sd^+ Rsp.

(D) $+/Sd^+$ Rsp. An *excess* of Sd^+ Rsp bearing offspring indicates
that the + chromosome is genetically Sd Rsp^+.

Chromosomes that are sentitive to distortion (test B) and which
segregate normally in tests (A), (C) and (D) are classified as Sd^+ Rsp^+;
chromosomes that are insensitive to distortion (test B) and which segregate normally in tests (A), (C) and (D) are considered to be Sd^+ Rsp^+,
but carrying a dominant suppressor (possibly polygenic) of distortion.

From each of the four tests we try to obtain 500 offspring to
estimate the segregation ratio, or 2000 offspring altogether for each
tested chromosome. Analysis of 50 chromosomes from a population thus
entails examination of 100,000 offspring, a not unreasonably large
number.

POPULATION DYNAMICS OF Sd Rsp

At this point it is appropriate to return to the propositions posed
in the introduction.

(1) *Natural populations containing Sd Rsp also contain a high frequency of* Sd^+ *Rsp*. It has been known for several years that natural
populations maintain a high frequency of insensitive second chromosomes (Kataoka 1967; Hartl 1970b). However, only recently have appropriate tester chromosomes been derived which allow an ascertainment
of whether the insensitive chromosomes are Sd^+ Rsp. Hihara (1974)
noted a frequency of insensitive chromosomes of 19% in the Odate, Japan, population. She studied one of the insensitive chromosomes in
detail, and finding that males bearing Sd Rsp along with this insensitive homologue had a reduction in fertility characteristic of Sd $Rsp/$
Sd^+ Rsp, she inferred that the insensitive chromosome was Sd^+ Rsp. On
this basis she suggested that all of the insensitive chromosomes in
the population might be of this genotype. Hartl and Hartung (1975)
examined 75 second chromosomes from the Raleigh population and found
the frequency of insensitive second chromosomes to be 0.48 ± 0.06.
Of these, the vast majority (16/17) were shown to be Sd^+ Rsp.

The Sd^+ Rsp chromosomes found in natural populations have a peculiar property first noted by Hihara (1974): they are partially sensitive to some Sd Rsp chromosomes. This is shown in Table 1 for ten
Sd^+ Rsp chromosomes from the Raleigh population. The Table shows the

recovery of the $R-1$ chromosome from $R-1/Sd^+$ Rsp males and the recovery
of the $SD(NH)-2$ chromosome from $SD(NH)-2/Sd^+$ Rsp males. (Both $R-1$ and
$SD(NH)-2$ are Sd Rsp; $R-1$ is a derivative of an SD from Madison, Wisconsin, and the segregation ratio of $R-1/Sd^+$ Rsp^+ males is normally
about 0.85; $SD(NH)-2$ is from Odate, and the segregation ratio of
$SD(NH)-2/Sd^+$ Rsp^+ males is normally about 0.99.) As the data clearly
show, the Sd^+ Rsp chromosomes from Raleigh are completely insensitive
to $R-1$ but only partially insensitive to $SD(NH)-2$. This result will
be of use later (section 3) when the origin of Sd^+ Rsp chromosomes in
artificial populations is considered.

(2) *Populations containing Sd and Rsp evolve a stable two-locus
polymorphism with permanent linkage disequilibrium.* Experimental populations initially containing only Sd Rsp and Sd^+ Rsp^+ rapidly accumulate insensitive second chromosomes but establish an equilibrium
frequency of Sd Rsp of about 10%. Hiraizumi, Sandler and Crow (1960)
found 20% insensitive second chromosomes after only 20 generations;
Watanabe (1967) found 30% insensitive second chromosomes after 66
generations. Determination of the genotypes of such insensitive chromosomes has only recently become possible.

I have examined an artificial population established by Hiraizumi
eight years ago. Initially containing 42% Sd Rsp and 58% Sd^+ Rsp^+,
the population now, after 208 generations, contains 12% Sd Rsp, 9%
Sd^+ Rsp^+ and, remarkably, 79% Sd^+ Rsp (Hartl 1977). The population appears to be in equilibrium because it has maintained the same frequencies for over a year. Because of the high frequency of Sd^+ Rsp
in the population, the frequency of Sd Rsp^+ chromosomes is kept exceedingly low (Sd^+ Rsp/Sd Rsp^+ males produce a deficiency of Sd Rsp^+
bearing offspring). Thus, Sd Rsp^+ chromosomes are virtually nonexis-
tent, and the equilibrium value of the linkage disequilibrium parameter
D is 0.011. This amount of linkage disequilibrium is equal to its
maximum possible value, given the allele frequencies.

(3) *The Sd$^+$ Rsp chromosomes selected in populations initially
derive from recombination and not from mutation.* The issue here is not
whether Sd^+ Rsp chromosomes are selected: certainly they are, once
they have arisen. The issue is, where do they come from in the first
place, mutation or recombination? At first sight this question would
seem almost impossible to answer, but because of a lucky happenstance
the issue can be decided using the Hiraizumi population described a-
bove. The Sd Rsp chromosome used in setting up the cage was $R(cn)-14$,
a derivative of $R-1$ (Hiraizumi and Nakazima 1967). The $R-1$ chromosome

has been the object of extensive recombinational analysis (Hartl 1974), and one of the characteristic features of Sd^+ Rsp recombinants from $R-1$ is that they are insensitive to distortion by $SD(NH)-2$. This property distinguishes them from Sd^+ Rsp chromosomes in natural populations because the latter are only partially insensitive to distortion by $SD(NH)-2$ (Table 1). Thus, if the Sd^+ Rsp chromosomes in the Hiraizumi population derive originally by recombination in Sd^+ Rsp^+/Sd Rsp females, then the insensitive chromosomes in the population should be insensitive to both $R-1$ and $SD(NH)-2$. Results of tests with twelve Sd^+ Rsp chromosomes from the cage show this unquestionably to be the case (Table 2).

Both $R-1$ and $R(cn)-14$ lack the pericentric inversion usually associated with Sd Rsp in natural populations. One might therefore argue that mutation is the source of Sd^+ Rsp in nature, even though recombination is the source in the Hiraizumi population. However, Sd^+ Rsp recombinants can arise even from Sd Rsp's which carry the pericentric inversion because the Sd locus is outside of the inversion, though close to it. Although recombination between Sd and Rsp is reduced, the frequency of recombination is still much higher than the mutation rate (Sandler and Hiraizumi 1960b). (There is a limit to how large the pericentric inversion may be and still be selected, because crossing over within the inversion loop leads to aneuploid gametes and zygotic loss. In fact, the main source of reduction in recombination due to the naturally-occurring inversion may be asynapsis in the region of the breakpoints.)

(4) *Natural populations free of Sd Rsp chromosomes are not polymorphic for Rsp*. Since the evolution of a supergene is most likely in systems in which epistasis is essential to the maintenance of a two-locus polymorphism (Turner 1976b), this is an important point. The proposition has been tested by examining a natural population discovered by Y. Hiraizumi in southern Texas. Among over 400 chromosomes extracted from this population, none are Sd Rsp. Thus the frequency of Sd Rsp is either zero or very close to zero in this population.

Figure 1 presents histograms of the segregation ratios obtained in tests of second chromosomes from the Texas population. Panel A presents the proportion of + bearing offspring in matings of $+/Sd^+$ Rsp^+ males (test A described earlier), where + represents a chromosome from the Texas population. Each segregation ratio is based on the pooled results of crosses of 7-10 males of each tested genotype. The harmonic mean number of offspring per test in this experiment was 640; the mean

Table 1. Segregation ratios of two *Sd Rsp* chromosomes, *R-1* and
SD(NH)-2, when heterozygous with *Sd$^+$ Rsp* chromosomes
from the Raleigh population, showing that the chromosomes from Raleigh are completely insensitive to distortion by *R-1* but only partially insensitive to distortion by *SD(NH)-2*. The ratios given in the table
are the proportions of *R-1* bearing (or *SD(NH)-2* bearing) offspring from males of the indicated genotypes

| | Segregation ratio ± s.e. and (N) from | |
chromosome number	*R-1/Sd$^+$ Rsp*	*SD(NH)-2/Sd$^+$Rsp*
5	0.50 ± 0.02 (641)	0.83 ± 0.01 (799)
10	0.52 ± 0.02 (491)	0.72 ± 0.02 (325)
18	0.49 ± 0.02 (506)	0.84 ± 0.01 (1708)
22	0.56 ± 0.02 (532)	0.82 ± 0.01 (931)
24	0.50 ± 0.02 (446)	0.82 ± 0.01 (1437)
25	0.50 ± 0.02 (516)	0.76 ± 0.01 (1288)
28	0.46 ± 0.03 (326)	0.90 ± 0.01 (965)
33	0.52 ± 0.02 (482)	0.72 ± 0.01 (1486)
59	0.52 ± 0.02 (754)	0.82 ± 0.01 (1796)
65	0.45 ± 0.02 (608)	0.89 ± 0.01 (967)

segregation ratio was 0.49. As can be seen in panel A, there were no
Sd Rsp chromosomes in this sample of 54 chromosomes. Several males
produced segregation ratios significantly *less* than 1/2. These are so-
called *Mr* (*male recombination*) chromosomes (Hiraizumi 1971, Slatko
and Hiraizumi 1973). *Mr* chromosomes are associated with a recovery
from heterozygotes of less than 50%, induced recombination in males,
and an enhanced mutation rate. *Mr* is polymorphic in the Texas population, and apparently in other populations as well (Voelker 1973).

Panel B provides the proportion of *Sd Rsp* bearing offspring in
matings of *+/Sd Rsp* males (test B). Here the harmonic mean number of

Table 2.   Segregation ratios of two *Sd Rsp* chromosomes, *R-1* and
            SD(NH)-2, when heterozygous with *Sd*$^+$ *Rsp* chromosomes
            from Hiraizumi's long-term artificial population, show-
            ing that the chromosomes from the population are com-
            pletely insensitive to distortion by both *R-1* and
            SD(NH)-2. The ratios given in the table are the pro-
            portions of *R-1* bearing (or *SD(NH)-2* bearing) off-
            spring from males of the indicated genotypes

| | Segregation ratio ± s.e. and (N) from | |
chromosome number	*R-1/Sd*$^+$ *Rsp*	*SD(NH)-2/Sd*$^+$ *Rsp*
5	0.50 ± 0.02 (1098)	0.49 ± 0.03 (241)
11	0.51 ± 0.03 (309)	0.50 ± 0.02 (510)
24	0.53 ± 0.02 (709)	0.51 ± 0.02 (458)
27	0.49 ± 0.03 (365)	0.49 ± 0.02 (483)
31	0.53 ± 0.02 (520)	0.54 ± 0.02 (410)
39	0.50 ± 0.02 (561)	0.51 ± 0.02 (416)
53	0.52 ± 0.02 (592)	0.48 ± 0.03 (248)
68	0.53 ± 0.02 (541)	0.54 ± 0.04 (175)
78	0.50 ± 0.02 (490)	0.50 ± 0.03 (253)
92	0.53 ± 0.03 (321)	0.52 ± 0.02 (440)
102	0.46 ± 0.02 (597)	0.53 ± 0.03 (257)
108	0.53 ± 0.03 (314)	0.56 ± 0.03 (279)

offspring was 593, and the mean segregation ratio was 0.71. As can be
seen, most, if not all, of the 51 tested chromosomes are sensitive to
distortion. The mean segregation ratio is somewhat smaller than that
typically observed for *Sd Rsp/Sd*$^+$ *Rsp*$^+$ males (which is about 0.85), but
this may be due to polygenic modifiers. One chromosome *reverses* distortion: the *Sd Rsp* chromosome is recovered in only 28% of the offspring. This chromosome is currently being examined in greater detail.

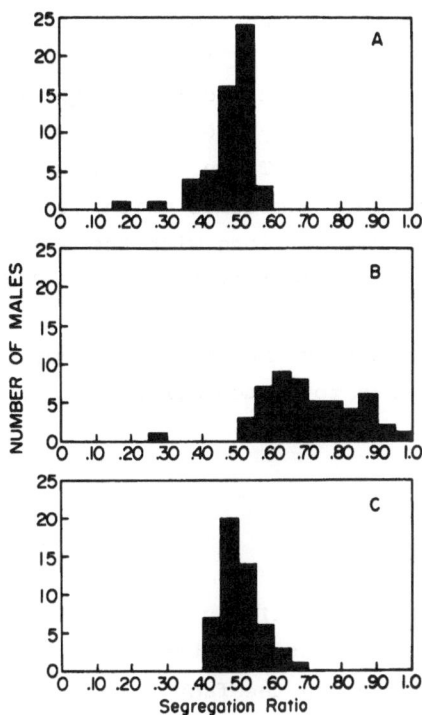

Figure 1.   Profiles of segregational properties of about 50 chromosomes from the southern Texas population. Panel A is the proportion of + bearing offspring from +/Sd⁺ Rsp⁺ males; a segregation ratio significantly in excess of 0.50 would indi‐cate that the tested chromosome is *Sd Rsp* (none are). The chromosomes having se‐gregation ratios significantly less than 0.50 are *Mr* chromosomes (see text). Panel B is the proportion of *Sd Rsp* bearing offspring from +/Sd Rsp males; segregation ratios in excess of 0.50 indicate that the tested chromosomes are sensitive to distortion. Panel C gives the proportion of *Sd Rsp⁺* bearing offspring from +/Sd Rsp⁺ males; segregation ratios significantly less than 0.50 would indicate that the tested chromosomes carry *Rsp* (none do).

Panel C is the test for Sd^+ Rsp. Here are plotted the proportions
of Sd Rsp^+ bearing offspring in matings of $+/Sd$ Rsp^+ males (test C).
The harmonic mean number of offspring in this experiment was 281; the
mean segregation ratio was 0.51. As seen in panel C, there is no evidence for a skewing of the distribution to the left; if anything, the
left side looks truncated. It appears from this test that none of the
chromosomes in the Texas population is Sd^+ Rsp.

To amplify the above point the results of some individual tests
are shown in Table 3. Here are presented the segregation ratios of
certain chromosomes which in test B of Figure 1 showed little or no
distortion, along with the segregation ratios of these same chromosomes as ascertained in test C of Figure 1. (The peculiar chromosome
that reverses distortion is also shown.) Again, there is no evidence
that any of the tested chromosomes is Sd^+ Rsp; the Texas population
appears to be monomorphic for Rsp^+.

DISCUSSION

While not entirely conclusive, the above studies bear out the propositions stated in the introduction. The strong epistasis between Sd
and Rsp and the fact that Sd^+ Rsp chromosomes are insensitive to distortion leads to a mutual adjustment of allele frequencies in populations. A two-locus polymorphism is generated, and permanent linkage
disequilibrium is sustained.

What happens next? Presumably, selection for all the other genetic
changes mentioned earlier. A pericentric inversion linking Sd more
tightly with Rsp is selected (we are in the process of looking for such
inversions in Hiraizumi's population cage), epistatic enhancers like
St on the second chromosome are selected, and these become linked more
tightly to each other and to the Sd Rsp supergene by paracentric inversions, and so on. In short, the second chromosome begins genetically to "congeal", all as a consequence of the strong epistasis initially between Sd and Rsp. [Theoretical aspects of the progression have
been studied by Prout, Bundgaard and Bryant (1973), Thomson and Feldman (1974), Hartl (1975b), Feldman and Krakauer (1976), and Thomson
and Feldman (1976)].

If this sort of evolutionary progression occurs in other organisms,
then it accounts for some otherwise very mysterious facts. For example,

Table 3. Segregation ratios of *Sd Rsp* (*R-1*) and *Sd Rsp*[+] (*R(cn)-10*) when heterozygous with certain chromosomes from the southern Texas population which are insensitive or partially insensitive to distortion by *R-1*. (Chromosome 104 *reverses* the meiotic drive by *R-1*.) The lack of distortion in *R(cn)-10/+* males (*i.e.* the lack of a deficiency of *R(cn)-10* bearing offspring) shows that none of the tested chromosomes is *Sd*[+] *Rsp*. The ratios given in the table are the proportion of *R-1* bearing (or *R(cn)-10* bearing) offspring from males of the indicated genotypes.

	Segregation ratio ± s.e. and (N) from	
chromosome number	*R-1/+*	*R(cn)-10/+*
1	0.60 ± 0.02 (1004)	0.61 ± 0.04 (175)
2	0.56 ± 0.02 (407)	0.53 ± 0.07 (55)
39	0.56 ± 0.02 (901)	0.44 ± 0.02 (895)
55	0.51 ± 0.02 (778)	0.52 ± 0.02 (403)
63	0.54 ± 0.02 (593)	0.48 ± 0.02 (808)
113	0.57 ± 0.02 (970)	0.48 ± 0.02 (934)
145	0.60 ± 0.01 (1557)	0.50 ± 0.02 (433)
104	0.28 ± 0.02 (568)	0.49 ± 0.03 (345)

inversion polymorphisms are widespread on the X chromosome of *Drosophila pseudoobscura* and on the third chromosome; the second chromosome is, however, essentially monokaryotypic (Dobzhansky 1970). Why? It surely does not have to do with the external environment, for the external environment has been the same for the second chromosome as for the third throughout the history of the species. Another example: balanced translocation complexes are widespread in *Oenothera* species, but are not found in *O. hookeri* (Stebbins 1950). Why?

The argument that inversions in *D. pseudoobscura* and translocations in *Oenothera* are intrisically advantageous is untenable; if the

argument were correct, then the second chromosome of *D. pseudoobscura*
ought to carry polymorphic inversions and *O. hookeri* ought to have
translocations. Indeed, it seems much more probable that such chromosome aberrations are intrinsically disadvantageous and are selected
in spite of it.

A plausible force for initiating supergene evolution is to be
found in strong epistatic interactions between alleles on the same
chromosome or, in the case of *Oenothera*, on different chromosomes. If
epsitatis is sufficiently strong, and just how strong it must be in
finite populations has not been calculated, then the epistatic selection will generate significant linkage disequilibrium. Once that has
occurred, selection will favor modifiers that have favorable epistasis
with the alleles that already interact beneficially. This creates an
even greater amount of linkage disequilibrium. At one point an inversion (or, in the case of *Oenothera*, a translocation) occurs which con-
geals the evolving supergene, and subsequently more epistatic modifiers will be selected (Kojima and Schaffer 1964, Turner 1976b).

Now, I believe that the amount of epistasis required to produce
the preconditions for reduction of recombination and initiation of supergene evolution is very large. I also believe that sufficiently
large amounts of epistasis are rather rare, even though small amounts
of epistasis may be common. I therefore would account for the absence
of inversions on *D. pseudoobscura*'s second chromosome by the argument
that, by chance, no alleles on this chromosome interact strongly e-
nough relative to their distance apart to get the process started
(Dobzhansky 1970). The same sort of explanation goes for *O. hookeri*,
and the fact that *O. hookeri* is genetically essentially homogeneous
(Levy and Levin 1975) fits quite nicely: one can hardly have linkage
disequilibrium in the absence of genetic variation.

In sum, then, the strong epistasis between *Sd* and *Rsp*, and the
manifold evolutionary consequences this has caused is, I think, a model
for the evolution of tightly linked, highly epistatic, multilocus
complexes in general.

Summary.

Segregation distorter chromosomes which are polymorphic in most natural populations of *Drosophila melanogaster* carry polygenic complexes of epistatic modifiers
and have become genetically "congealed" by means of pericentric and paracentric
inversions. This genetic system is proposed as an experimental model for the evolution of supergenes. The evolutionary progression is envisioned as consisting of
a sequence of three distinct stages: first, the appearance of strongly epistatic
alleles which generate a two-locus polymorphism marked by linkage disequilibrium;
second, the congealment of the two-locus system by means of genetic modifiers of
linkage or chromosome aberrations; and third, the evolutionary improvement of the
supergene by selection of coadapted, epistatic modifiers within it. The two strongly epistatic alleles of the segregation distorter system are known as *Sd* and *Rsp*.
Evidence is here presented that populations containing *Sd* and *Rsp* and no inversions
do evolve a stable two-locus polymorphism marked by linkage disequilibrium, even
though such populations accumulate a high frequency of Sd^+ *Rsp* chromosomes derived
originally by recombination. Epistasis is apparently essentail for the maintenance
of the *Rsp* polymorphism, because a natural population lacking *Sd Rsp* has been found
to be monomorphic for Rsp^+.

Acknowledgements.

This work was done while the author was supported by Research Career Award
GM0002301. The research was supported by N.I.H. grant GM21732 and N.S.F. grant
GB43209.

REFERENCES

Das, C.C., Kaufmann, B.P., and Gay, H. 1964. Histone-protein transition in *Drosophila melanogaster*. I. Changes during spermatogenesis.
Exp. Cell Res. 35: 507-514.

Dobzhansky, Th. 1970. *Genetics of the Evolutionary Process*. Columbia University
Press, New York.

Feldman, M.W. 1972. Selection for linkage modification: 1. Random mating populations.
Theoret. Popul. Biol. 3: 324-346.

Feldman, M.W., and Krakauer, J. 1976. Genetic modification and modifier polymor-
phisms.
In Population Genetics and Ecology, S. Karlin and E. Nevo eds. Academic Press,
New York.

Hartl, D.L. 1969. Dysfunctional sperm production in *Drosophila melanogaster* males
homozygous for the segregation distorter elements.
Proc. Nat. Acad. Sci. USA 63: 782-789.

Hartl, D.L. 1970a. A mathematical model for recessive lethal segregation distor-
ters with differential viabilities in the sexes.
Genetics 66: 147-163.

Hartl, D.L. 1970b. Meiotic drive in natural populations of *Drosophila melanogaster*.
IX. Suppressors of *segregation distorter* in wild populations.
Can. J. Genet. Cytol. 12: 594-600.

Hartl, D.L. 1973. Complementation analysis of male fertility among the segregation distorter chromosomes of *Drosophila melanogaster*.
Genetics 73: 613-629.

Hartl, D.L. 1974. Genetic dissection of segregation distortion. I. Suicide combinations of *SD* genes.
Genetics 76: 477-486.

Hartl, D.L. 1975a. Segregation distortion in natural and artificial populations of
Drosophila melanogaster.
In D.L. Mulcahy (ed.) *Gamete Competition in Plants and Animals*. North-Holland,
Amsterdam.

Hartl, D.L. 1975b. Modifier theory and meiotic drive.
Theoret. Popul. Biol. 7: 168-174.

Hartl, D.L. 1977. Mechanism of a case of genetic coadaptation in populations of
Drosophila melanogaster.
Proc. Nat. Acad. Sci. USA (in press).

Hartl, D.L., and Hartung, N. 1975. High frequency of one element of *segregation
distorter* in natural populations of *Drosophila melanogaster*.
Evolution 29: 512-518.

Hartl, D.L., and Hiraizumi, Y. 1976. Segregation distortion.
In M. Ashburner and E. Novitski (eds.) *The Genetics and Biology of Drosophila*.
Vol. 1b, Academic Press, New York.

Hauschteck-Jungen, E., and Hartl, D.L. 1977. Spermiogenesis in normal and *segregation distorter* males of *Drosophila melanogaster*. Submitted to Genetics.

Hihara, Y.K. 1974. Genetic analysis of modifying system of segregation distortion
in *Drosophila melanogaster*. II. Two modifiers for *SD* system on the second chromosome of *D. melanogaster*.
Japan. J. Genetics 49: 209-222.

Hiraizumi, Y., Sandler, L., and Crow, J.F. 1960. Meiotic drive in natural populations of *Drosophila melanogaster*. III. Populational implications of the segregation distorter locus.
Evolution 14: 433-444.

Hiraizumi, Y., and Nakazima, K. 1967. Deviant sex ratio associated with segregation distortion in *Drosophila melanogaster*.
Genetics 55: 681-697.

Hiraizumi, Y. 1971. Spontaneous recombination in *Drosophila melanogaster* males.
Proc. Nat. Acad. Sci. USA 68: 268-270.

Karlin, S., and McGregor, J. 1974. Towards a theory of the evolution of modifier
genes.
Theoret. Popul. Biol. 5: 59-103.

Kataoka, Y. 1967. A genetic system modifying segregation-distortion in a natural
population of *Drosophila melanogaster* in Japan.
Japan. J. Genetics 42: 327-337.

Kettaneh, N.P., and Hartl, D.L. 1976. Histone transition during spermiogenesis
is absent in *segregation distorter* males of *Drosophila melanogaster*.
Science 193: 1020-1021.

Kojima, K., and Schaffer, H.E. 1964. Accumulation of epistatic gene complexes.
Evolution 18: 127-129.

Levy, M., and Levin, D.A. 1975. Genic heterozygosity and variation in permanent
translocation heterozygotes of the *Oenothera biennis* complex.
Genetics 79: 493-512.

Lewontin, R.C. 1971. The effect of genetic linkage on the mean fitness of a popul-
    ation.
    Proc. Nat. Acad. Sci. USA 68: 984-986.

McClanahan, M.L. 1974. Irradiation-induced mutations in the segregation-distorter
    region in *Drosophila melanogaster*. M.S. thesis, University of Washington, Seattle.

Miklos, G.L.G., and Smith-White, S. 1971. An analysis of the instability of segre-
    gation-distorter in *Drosophila melanogaster*.
    Genetics 67: 305-317.

Nei, M. 1967. Modification of linkage intensity by natural selection.
    Genetics 57: 625-641.

Nei, M. 1969. Linkage modification and sex difference in recombination.
    Genetics 63: 681-689.

Prout, T., Bundgaard, J., and Bryant, S. 1973. Population genetics of modifiers
    of meiotic drive. I. The solution of a special case and some general implica-
    tions.
    Theoret. Popul. Biol. 4: 446-465.

Sandler, L., Hiraizumi, Y., and Sandler, I. 1959. Meiotic drive in natural popul-
    ations of *Drosophila melanogaster*. I. The cytogenetic basis of segregation-dis-
    tortion.
    Genetics 44: 233-250.

Sandler, L., and Hiraizumi, Y. 1960a. Meiotic drive in natural populations of
    Drosophila melanogaster. IV. Instability at the segregation-distorter locus.
    Genetics 45: 1269-1287.

Sandler, L., and Hiraizumi, Y. 1960b. Meiotic drive in natural populations of
    Drosophila melanogaster. V. On the nature of the *SD* region.
    Genetics 45: 1671-1689.

Slatko, B.E., and Hiraizumi, Y. 1973. Mutation induction in the male recombina-
    tion strains of *Drosophila melanogaster*.
    Genetics 75: 643-649.

Stebbins, G.L. 1950. *Variation and Evolution in Plants*. Columbia University Press,
    New York.

Strobeck, C., Maynard Smith, J., and Charlesworth, B. 1976. The effects of hitch-
    hiking on a gene for recombination.
    Genetics 82: 547-558.

Thomson, G.J., and Feldman, M.W. 1974. Population genetics of modifiers of meio-
    tic drive. II. Linkage modification in the segregation distortion system.
    Theoret. Popul. Biol. 5: 155-162.

Thomson, G.J., and Feldman, M.W. 1976. Population genetics of modifiers of meio-
    tic drive. III. Equilibrium analysis of a general model for the genetic control
    of segregation distortion.
    Theoret. Popul. Biol. 10: 10-25.

Turner, J,R.G. 1967a. Why does the genotype not congeal?
    Evolution 21: 645-656.

Turner, J.R.G. 1967b. On supergenes: I. The evolution of supergenes.
    Amer. Natur. 101: 195-221.

Voelker, R.A. 1973. An analysis of the genetic control and recombinational pro-
    ducts of male crossing over in *Drosophila melanogaster*.
    Genetics 74: s286 (abstract).

Watanabe, T.K. 1967. Persistence of lethal genes associated with *SD* in natural
    populations of *D. melanogaster*.
    Japan. J. Genetics 42: 375-386.

White, M.J.D. 1973. *Animal Cytology and Evolution*, 3rd ed. Cambridge University
    Press, Cambridge.

Zimmering, S., Sandler, L., and Nicoletti, B. 1970. Mechanisms of meiotic drive.
    Ann. Rev. Genet. 4: 409-436.

1. STUDY OF SELECTION

ATTEMPTS TO MEASURE NATURAL SELECTION BY ALTERING GENE FREQUENCIES IN NATURAL POPULATIONS

J.S. Jones and D.T. Parkins

Experiments designed to demonstrate the action of natural selection on a polymorphic locus can provide convincing results only if they relate to a polymorphism whose genetic and ecological environment reflects that found in nature. Two important practical problems arise from this. First, laboratory experiments using only a small sample of alleles from a natural population may produce a misleading appearance of selection on the locus under investigation as a result of sampling disequilibria between this locus and others which are more responsive to selection. This effect can be avoided by using adequate samples from nature (Jones and Yamazaki 1974; Powell 1973). Secondly, investigations of polymorphism in, for example, a *Drosophila* population cage are open to the criticism that the laboratory environment lacks important components of selection which may be present in nature. A failure to demonstrate selection in such conditions may therefore not be a real indication of selective neutrality. Thus, attempts to simulate the natural environment (such as adding ethanol to the medium when studying the alcohol dehydrogenase polymorphism; see Clarke, this symposium) sometimes reveal previously undetected selective differences. Experiments on populations in their natural environment avoids this problem.

In view of the hundreds of experiments which have been carried out on laboratory populations of various animals (in particular of *Drosophila*), remarkably few attemps have been made to measure selection by ex-

perimenting with them in the field. The few experiments which have been
made have often succeeded in demonstrating strong natural selection.
Comparison of the survival of melanic and non-melanic *Biston betula-*
ria released into polluted and unpolluted woodlands showed clearly
that visual selection by predators affected the fitness of the morphs
(Kettlewell 1961). Similar results were obtained for mimetic Lepidoptera released into a natural environment (Benson 1972; Brower, Cook
and Croze 1967). All these experiments have involved either transient
polymorphisms, introduced species, or native species whose appearance
has been experimentally altered. We have attempted to measure natural
selection by perturbing gene frequencies in populations of land snails
and of *Drosophila* in their own habitats. This work has not yet succeeded in demonstrating measurable selective differences between alleles, but has at least provided some interesting incidental ecological and genetic information on the experimental organisms.

EXPERIMENTAL MANIPULATION OF *CEPAEA VINDOBONENSIS* POPULATIONS

Polymorphic *Cepaea* snails have been widely used in research on the genetic structure of natural populations, and there is now good evidence
that morph distribution is affected by components of the environment
such as predation and climate (Cain and Currey 1963; Jones 1973), although random processes play a part in at least some populations (Goodhart 1963). The eastern European species *C. vindobonensis* shows an
association between gene frequency and microclimate which is particularly suited to experimental investigation, as this snail has a relatively simple visible polymorphism. In the Lika region of northern Yugoslavia, faint banded snails (which are a pale yellow colour) are
found almost exclusively on hillsides, while dark and fully pigmented
snails occur primarily in the basins which form a prominent topogra-
phical feature of this limestone area. Morph frequencies are influenced by climatic selection. Dark coloured snails heat up and hence become active more rapidly than do faint in the early morning after the
inversions of temperature which frequently reduce the night minimum
temperature of the basins to several degrees below that found on the
hillsides (Jones 1973, 1974). We have attempted to measure the intensity of climatic selection by examining the survival of marked snails
transferred between hillsides and basins.

A sample of 6500 adult and near-adult snails were collected from
hillside colonies containing both faint and dark individuals and were
marked by drilling holes in specific bands near the lip of the shell.
In each of the 31 separate transfers (during July and August, 1971)
half the marked snails were transferred to a basin, and half to a
different hillside as a control. In 10 experiments, snails from more
than one donor population were released at a single site and these
were coded and analysed separately. All releases were made in the evening into ten-metre lengths of hedgerow (mainly *Prunus spinosa* and
Crataegus spp.) chosen to be as similar to each other as possible.
Whenever possible, an equivalent number of native snails were removed
from the release site before introducing experimental animals.

One thousand snails were recaptured during July and August, 1972
(Table 1). The proportion recaptured was higher on hillsides than in
basins, and there seemed in general to be a higher recovery rate in
localities where the native snail population was sparse. Marked snails
were captured up to 100 metres from a release site, but it is uncertain
whether this is a meaningful measure of natural migration because of
the tendency of introduced animals to move more actively than do na-
tives (Krebs and Myers 1974; Lomnicki 1969).

In our discussion, release localities are referred to as "sites",
and the hillside-basin pairs of sites as "experiments". In the group of
experiments in which transferred animals were from the same colony,
live marked snails were found at both hillside and basin sites in 18
cases. The data are essentially in the form of an 18 x 2 x 2 x 2 contingency table, and have been analysed accordingly (Fienberg 1970).
The number recaptured at each release site is regarded as a binomial
variate, and expected values for the contingency table are calculated
using the General Linear Interactive Modelling program of Nelder (1974).
This computes a goodness of fit χ^2 for a "basic" model using only the
simple marginal totals, and then for models made successively more complex by incorporating interaction components between the marginal totals
(Table 2).

Model I depends on marginal totals alone, and shows that the results
as a whole display significant heterogeneity. Model II incorporates the
interaction in survival between sites and the 18 experiments. This model
gives a significantly better fit to the observed data, indicating that
the ratio of survival on hillsides to survival in basins varies between
experiments. Neither of these results is particularly surprising. They
may stem from differences in the genetic background of the populations

Table 1. Experimental transfers of *C. vindobonensis*. Grid references are to the 55/49 square of the Universal Transverse Mercator grid. Rel. = No. of snails released; Rec. = No. of snails recaptured

Expt. No.	Locality of Donor Population	Morphs in Donor Population		Grid.Ref.	Hillside				Grid.Ref.	Basin			
		Faint	Dark		Faint Rel.	Faint Rec.	Dark Rel.	Dark Rec.		Faint Rel.	Faint Rec.	Dark Rel.	Dark Rec.
1	ZVTA LOKVA	86	103	061802	43	10	52	20	113754	43	7	51	9
2	ZVTA LOKVA	80	156	145829	40	6	78	6	124833	40	5	78	9
3	ZVTA LOKVA	140	144	101668	70	25	72	27	226656	70	12	72	10
4	ZVTA LOKVA	34	54	249773	17	3	27	9	240784	17	0	27	0
5	LONCARI	41	71	058803	21	3	36	6	134733	20	5	35	6
6	JEZERANE	131	32	097668	65	17	16	6	216667	66	6	16	3
8	PROKIKE	147	59	144883	74	25	29	10	128862	73	9	30	2
9	VRZICI	63	78	056804	32	4	39	5	115749	31	7	39	3
10	LALICI	22	173	057803	11	3	87	23	110731	11	3	86	15
13	JEZERANE	122	81	291675	60	5	40	6	264665	62	12	41	12
19	LONCARI	37	166	165894	18	3	83	19	127858	19	0	83	3
21	KRIZPOLJE	58	174	192751	28	2	88	12	227717	30	3	86	14
22	ZVTA/LOKVA	103	96	166897	51	7	49	3	125851	52	6	47	4
23	LONCARI	43	154	056814	21	11	77	3	092832	22	0	77	5
24	ZVTA LOKVA	80	47	175760	40	11	39	10	237722	40	4	38	5
25	PROKIKE	48	75	040769	24	0	37	0	113736	24	2	38	4
26	KRIZPOLJE	48	183	055804	25	3	91	18	085836	23	7	92	12
27	STAJNICE	49	123	163890	25	10	62	10	129860	24	5	62	10
28	STAJNICE	63	92	056810	31	4	46	18	111836	32	6	46	10
29	STAJNICE	49	150	180755	25	3	75	12	206711	24	3	75	26
31	MISC	80	84	175761	40	7	42	4	130783	41	0	42	0

| | | Morphs in Donor Population | | Hillside | | | | | Basin | | | | |
Expt. No.	Locality of Donor Population	Faint	Dark	Grid.Ref.	Faint Rel.	Faint Rec.	Dark Rel.	Dark Rec.	Grid.Ref.	Faint Rel.	Faint Rec.	Dark Rel.	Dark Rec.
7	SVICA	0	72	047815	43	3	36	8	112836	43	2	36	3
	KUTEREVO	86	0										
12	LALIC	8	164	061801	32	3	83	16	122743	32	6	82	20
	ZHALVENICA	56	1										
15	SVICA	0	121	042766	132	4	70	2	111777	131	13	71	2
	JEZERANE	263	20										
16	SVICA	0	125	060802	107	6	75	5	136736	106	46	75	18
	JEZERANE	213	25										
17	SVICA	0	85	093668	42	10	46	15	223657	45	0	47	0
	JEZERANE	87	8										
18	SVICA	0	140	055820	26	8	73	23	114836	26	1	74	6
	JEZERANE	52	7										
20	GLAVACE	21	43	042766	24	2	37	0	117779	24	8	38	17
	JEZERANE	27	32										
32	SVICA	0	175	098840	100	6	96	2	056806	101	4	96	3
	JEZERANE	201	17										
35	SVICA	0	129	961672	106	43	70	12	224656	106	5	72	3
	JEZERANE	212	11										
30	MISCELLANEOUS	47	142	101831	24	5	71	13	057816	23	2	71	4

Table 2. Analysis of survival of transferred *Cepaea vindobonensis.*
The models include successive interactions between sites,
experiments and morphs (see text for further explanation)

Model	Component	Goodness-of-fit χ^2	d.f.	Improvement χ^2	d.f.	P
I	'Basic'	135.9	52			
				71.9	17	< .001
II	+ Expt. x site	64.0	35			
				36.6	17	< .01
III	+ Expt. x morph	27.4	18			
				0.1	1	>0.7
IV	+ Site x morph	27.3	17			
				27.4	17	>0.05
V	+ Expt. x site x morph	0.01	0			

from which the snails originate, or from some general differences in
the environments into which they were introduced. These variables are
notoriously difficult to control under field conditions.

Model IV is central to our analysis, as it incorporates the interaction between the survival of the two morphs and the site into which
they were introduced. It does not improve the fit to the data beyond
that provided by Model III. We may therefore conclude that there were
no significant differences in the survival of faint and dark *C. vindobonensis* between hillsides and basins during the period of our experiment. Finally, Model V, which includes the three-way interaction between sites, experiments and morphs does not give a significant improvement in fit, indicating that the linear model which we have used is
appropriate.

This sequential analysis seems to be biologically reasonable. The
least ecologically relevant effects are eliminated first, and the most
important left until last. Changing the sequence in which the model is
analysed does not significantly alter our conclusions. The ten less
satisfactory experiments in which the transferred snails originated
from different hillside populations also provide no evidence for differences in survival between faint and dark snails.

Although information on the distribution of genes, on microclimate,
and on behaviour combines to suggest that morph frequencies in *C. vin-*

dobonensis are affected by climatic selection, our experiments have
shown no differences in survival over one year between faint and dark
snails transferred to a foreign habitat. It may be that "selective
crises" involving very low temperatures in frost hollows or very high
temperatures on hillsides occur only occasionally, and that these "cri-
sis" levels of climatic selection did not operate during the course
of this experiment. However, the closely related species *C. nemoralis*
very frequently finds itself at a temperature approaching its lethal
limit (Richardson 1974), and the existence of a temperature difference
of 1°C between faint and dark *C. vindobonensis* in sunshine (Jones 1973)
is probably of physiological importance on many occasions during a
single year. Our experiments are only extensive enough to identify
rather large selective differentials. It may be that the average intensity of climatic selection is low, and that the accumulation of modifying genes has led to sharp differentiations of population structure
even under weak environmental selection (Clarke 1966; Endler 1973;
Slatkin 1973). Surveys of allozyme variation in *C. nemoralis* populations have not yet provided unequivocal information on whether population differentiations for visible markers are indeed accompanied by
parallel differentiation at large numbers of loci (Johnson 1976; Jones
and Selander 1976).

Perhaps the most likely explanation for our failure to detect climatic selection is that climate acts primarily on some component of
fitness other than the adult viability measured in these experiments.
Frydenberg and his collaborators have shown very clearly that this component of fitness, although it may be the easiest to measure, is by no
means the most important in the few organisms in which fitness components have been properly analysed (Bundgaard and Christiansen 1972;
Christiansen, this symposium). Is is possible to suggest mechanisms
whereby different behavioural responses to climate might cause, for
example, differences in mating activity between faint and dark *C. vindobonensis*. This component of fitness is known to be very important in
Drosophila , and its action in *C. vindobonensis* would not be detected
in the experiments described here. Differences in survival of young
snails (which are subject to much higher mortality than are adults;
Wolda and Kreulen 1973), rate of development of juveniles or fecundity
of adults would also remain undetected in our experiments. Experimental
manipulation of a population in the hope of detecting natural selection
must therefore encompass more than one complete generation transition
if it is to succeed in demonstrating any selective differences which

may exist between alleles (Prout 1971). Such experiments are scarcely
likely to be feasible in *Capaea*, which has a generation interval of a-
bout five years, but may be possible in *Drosophila*.

ATTEMPTS TO MANIPULATE GENE FREQUENCY IN *DROSOPHILA PSEUDOOBSCURA*

The importance of natural selection in controlling the distribution of
allozymes in natural populations remains obscure. Fitness differences
between alleles at the esterase-5 locus in *D. pseudoobscura* have been
investigated by Yamazaki (1971). A careful laboratory examination of
all components of fitness revealed no differences between the Est-5$^{1.00}$
and Est-5$^{1.12}$ alleles. In collaboration with S.H. Bryant, D.W. Crumpacker, R.C. Lewontin, J.A. Moore, T. Prout and C.E. Taylor one of us (JSJ)
has been attempting to assess whether alleles at this locus also show
selective neutrality in their natural environment, or whether the apparent absence of selection reported by Yamazaki was an artefact of
laboratory conditions.

    We hoped to be able to find populations of *D. pseudoobscura* which
were well isolated and which could be used as "natural population cages"
in which the frequency of initially rare alleles could be greatly increased by the release of homozygous laboratory bred flies. Monitoring
of the perturbed population over a period of time might give some indication of the role of selection in determining the natural frequency of
the allele under investigation. In the low deserts of southern California (such as Death Valley) small oases support *Drosophila* populations which appear to be well isolated, as no flies can be collected
in the arid flats which surround them for many miles in every direction. These sparse and apparently isolated populations seemed suitable
candidates for use as natural population cages. Two oases, Salt Creek
and Saratoga Spring, which are about 15 kilometres apart at the south
end of Death Valley were chosen for more detailed study (see Figure 1).

    Gene frequencies at the Esterase-5 locus were perturbed using the
0.85 allele, which has a frequency of 1-3% in southern California. It
would of course be preferable to use flies originating from the oasis
populations themselves as the origin of the homozygous lines used in
perturbation to avoid artefacts due to selection at linked loci which
might differ from population to population. Unfortunately, not enough
flies could be collected in the oases to provide sufficient independent

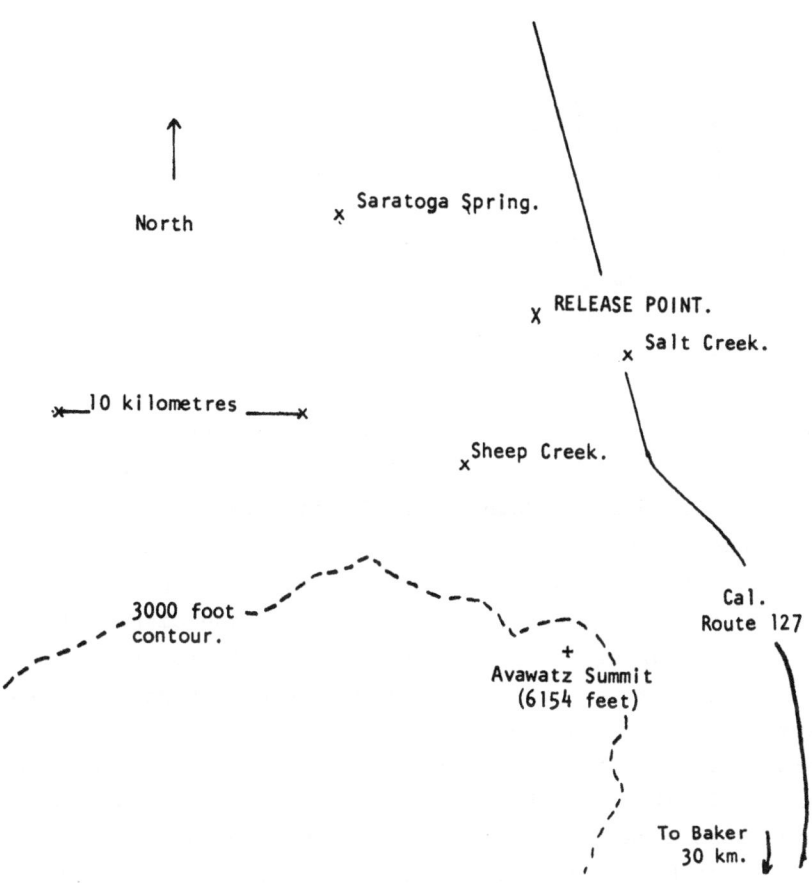

Figure 1. The *D. pseudoobscura* release site. Marked flies were recaptured in each of the three oases.

Est-5$^{0.85}$ alleles for use in founding the experimental populations. Flies from an abundant population at Desert Center, some 250 kilometres south of Death Valley were therefore used as a source of alleles. Approximately thirty independent replicates of the 0.85 allele were extracted; in this way we hoped to reduce the possibility of confound-

ing any effects of selection at linked loci from those operating at
the allozyme locus itself (see Jones and Yamazaki 1974).

Large numbers of flies homozygous for Est-5$^{0.85}$ were released at
each of the oases in March 1973. The laboratory flies survived and
bred with native flies, as was demonstrated by the recovery of many
individuals heterozygous for Est-5$^{0.85}$ and another esterase allele o-
ver the lollowing six weeks. The populations were resampled in March
1974. No flies could be found at Saratoga Spring, and, somewhat to our
dismay, the frequency of the 0.85 allele at Salt Creek had returned to
the 1.5% which is characteristic of this region.

It seems unlikely that natural selection could act to restore the
frequency of this allele to its natural level within such a short period. It is more probable that the effect is due to large scale immigration from other *D. pseudoobscura* populations. As the nearest vegetation capable of supporting large numbers of *Drosophila* is the forest
and scrub near the summits of the Avawatz mountains (which are some 10
km from Salt Creek) this would necessitate much more rapid fly movement
than that found by Dobzhansky and Wright (1943) in their classic release experiments. We have carried out some studies (which, together
with the perturbation experiments already described, will be discussed
at more length in a future publication) on the dispersal of *Drosophila*
across unfavourable habitats such as those found in the California
deserts. Initial experiments involving the release of wild-caught
flies marked with fluorescent dust at the centre of a cross of traps
similar to that used by Dobzhansky and Wright and by Powell and Dobzhansky (1976) showed that *D. pseudoobscura* moved much more rapidly
in the desert than they did in the forest habitats studied by these
authors and by Crumpacker and Williams (1973). The speed of movement was
such that flies reached a trap 500 metres from the release point in
large numbers within twelve hours of release. Rapid long distance movement of this kind was further investigated by using the oases themselves as "natural *Drosophila* traps". In one such experiment, approximately 5000 flies were released in the desert one evening, and re-
captures made over the following two days at three oases with different directions and distances from the release point (Figure 1). Marked
flies in significant numbers were recaptured at each of the oases, including the most distant (which was 10 km from the point of release).
This rate of dispersal is considerably greater than that previously
found for *D. pseudoobscura* and exceeds even that of the desert-adapted
species *D. nigrospiracula* (Johnston and Heed 1976). Dispersal was un-

likely to have been due simply to movement of the released flies by
winds, as the weather was calm and the flies seemed to move more or
less equally in each direction. The distinction between active dispersal over short distances and passive transport causing gene flow
on the larger scale made by Dobzhansky (1973) may therefore be un-
necessary, in desert habitats at least. Rapid movement over unfavourable habitats of the type described here may also mean that much of the
D. pseudoobscura population of North America is a single interbreeding
unit, and might be interpreted as favouring the view that random processes and gene flow are largely responsible for the observed pattern
of allozyme distribution in this species (Kimura and Ohta 1971).

DISCUSSION

The main conclusion to be drawn from the experiments described here
may be that there are great practical difficulties involved in measuring the strength of selection in natural populations of animals by
means of the experimental manipulation of populations. However, this
technique still represents perhaps the least equivocal method of establishing the action of selection on a polymorphic locus. One approach
which might have practical advantages would be to utilise a species
which is released in great numbers as a biological control agent. Relatively little extra effort would be involved in carrying out genetic
manipulations of the released population and later in monitoring gene
frequency changes in nature.

One group is at present attempting to manipulate allozyme frequencies in an island population of D. pseudoobscura using simultaneous
perturbation at two loci in the hope that comparison of the rate of
change of gene frequency (if any) at each locus will separate the effects of selection from those of migration (which would be expected
to affect each locus to the same degree). We hope that future publications will not consist solely of an explanation of the problems encountered.

Summary.

   The significance of naturally occurring variation has been investigated in two
instances by means of perturbation experiments. In the snail *Cepaea vindobonensis*
a correlation between shell colour and habitat is found in parts of northern Yugoslavia. Faint coloured snails are found on hillsides and dark coloured snails in basins.
The significance of this was investigated by transplanting populations between and
within these habitats. No consistent correlation between the viability of the morphs
and habitat could be found in the transplanted populations. — In *Drosophila pseudoobscura* a similar investigation was carried out in some "island" populations in the
Mohave desert. Here a rare variant at the Est-5 locus was artificially brought to a
high frequency in the populations. Subsequent sampling demonstrated that the island
populations are not persistent but founded every winter. The suitability of this explanation was investigated by mark-release-recapture experiments.

Acknowledgements.

   The *C. vindobonensis* research was supported by a grant from the Royal Society.
We thank the Yugoslav Academy of Science and the Croatian Institute for Nature Protection for permission to work in Yugoslavia, numerous colleagues for their help in
the field, and Drs. B.C. Clarke and G.E. Thomas for useful discussions. Work on *D.
pseudoobscura* was carried out with the support of grants from the National Science
Foundation and the Ford Foundation.

REFERENCES

Benson, W.W. 1972. Natural selection for Mullerian mimicry in *Heliconius erato* in
      Costa Rica.
      Science 176: 936-939.

Brower, L.P., Cook, L.M., and Croze, H.J. 1967. Predator responses to artificial
      Batesian mimics released in a neotropical environment.
      Evolution 21: 11-23.

Bundgaard, J., and Christiansen, F.B. 1972. Dynamics of polymorphisms. I. Selection
      components in an experimental population of *Drosophila melanogaster*.
      Genetics 71: 439-460.

Cain, A.J., and Currey, J.D. 1963. Area effects in *Cepaea*.
      Phil. Trans. Roy. Soc. Lond. B 246: 1-81.

Clarke, B.C. 1966. The evolution of morph ratio clines.
      Amer. Natur. 100: 389-402.

Crumpacker, D.W., and Williams, J.S. 1973. Density, dispersion and population
      structure in *Drosophila pseudoobscura*.
      Ecol. Monogr. 43: 449-538.

Dobzhansky, Th. 1973. Active dispersal and passive transport in *Drosophila*.
      Evolution 27: 565-575.

Dobzhansky, Th., and Wright, S. 1943. Genetics of natural populations. X. Disper-
      sion rates in *Drosophila pseudoobscura*.
      Genetics 28: 304-340.

Endler, J.A. 1973. Gene flow and population differentiation.
Science 179: 243-250.

Fienberg, S.E. 1970. The analysis of multi-dimensional contingency tables.
Ecology 51: 419-433.

Goodhart, C.B. 1963. "Area effects" and non-adaptive differentiation between populations of *Cepaea* (Mollusca).
Heredity 18: 459-465.

Johnson, M.S. 1976. Allozymes and area effects in *Cepaea nemoralis* on the western
Berkshire Downs.
Heredity 36: 105-122.

Johnston, J.S., and Heed, W.B. 1976. Dispersal of desert-adapted *Drosophila*: the
Saguaro-breeding *D. nigrospiracula*.
Amer. Natur. 110: 629-651.

Jones, J.S. 1973. Ecological genetics and natural selection in molluscs.
Science 182: 546-552.

Jones, J.S. 1974. Area effects in the snail *Cepaea vindobonensis* in the Lika region of Yugoslavia.
Heredity 32: 165-170.

Jones. J.S., and Selander, R.K. 1976. Population differentiation at the molecular
level in *Cepaea*.
In preparation.

Jones, J.S., and Yamazaki, T. 1974. Genetic background and the fitness of allozymes.
Genetics 78: 1185-1189.

Kettlewell, H.B.D. 1961. The phenomenon of industrial melanism in the Lepidoptera.
Ann. Rev. Entomol. 6: 245-262.

Kimura, M., and Ohta, T. 1971. Theoretical Aspects of Population Genetics. Princeton
University Press.

Krebs, C., and Myers, J. 1974. Population cycles in small mammals.
Adv. Ecol. Res. 8: 267-399.

Lomnicki, A. 1969. Individual differences among adult members of a snail population.
Nature 223: 1073-1074.

Nelder, J.A. 1974. General Linear Interactive Modelling. Available from the Numerical Algorithms Group, 13 Banbury Road, Oxford, England.

Powell, J.R. 1973. Apparent selection of enzyme alleles in natural populations of
Drosophila.
Genetics 75: 557-570.

Powell, J.R., and Dobzhansky, Th. 1976. How far do flies fly?
Amer. Sci. 64: 179-185.

Prout, T. 1971. The relation between fitness components and population prediction
in *Drosophila*. I. The estimation of fitness components.
Genetics 68: 127-149.

Richardson, A.M.M. 1974. Differential climatic selection in natural populations of
the land snail *Cepaea nemoralis*.
Nature 247: 572-573.

Slatkin, M. 1973. Gene flow and selection in a cline.
Genetics 75: 733-756.

Wolda, H., and Kreulen, D.A. 1973. Ecology of some experimental populations of the
land snail *Cepaea nemoralis* (L.). II. Production and survival of eggs and juveniles.
Neth. J. Zool. 23: 168-188.

Yamazaki, T. 1971. Measurement of fitness at the esterase-5 locus of *Drosophila*
     pseudoobscura.
     Genetics 67: 579-603.

1. STUDY OF SELECTION

GENETIC VARIANCE FOR VIABILITY AND LINKAGE DISEQUILIBRIUM IN NATURAL POPULATIONS OF *DROSOPHILA MELANOGASTER*

Terumi Mukai

Genetic load and genetic variance are the most common measures of gene-tic variability in random mating populations. Although a large number of investigations have been carried out for the former, only few papers have been published for the latter. Recently, some studies have been done to detect linkage disequilibrium in order to test whether or not epistasis is common between polymorphic isozyme genes (Mukai, Mettler, and Chigusa 1971; Langley, Tobari, and Kojima 1974, and others). In the present paper, an attempt was made to find the mechanisms whereby the genetic variability has been maintained in natural populations of *Droso-phila melanogaster*, using the estimates of the above three genetic pa-rameters and the cumulative heterozygous effects on viability and fecun-dity of polymorphic isozyme gene loci.

THEORETICAL CONSIDERATIONS

The following Wrightian fitness model for a locus with two alleles are used:

Genotype	AA	Aa	aa
Frequency	p^2	2pq	q^2
Relative viability	1	1 - hs	1 - s

where p and q are the gene frequency of *A* and *a*, respectively; s stands for selection coefficient and h for the degree of dominance.

With this model, the total genetic variance for a single locus, σ_t^2, the additive (genetic) component, σ_a^2, and the dominance component, σ_d^2, are given by:

$$\sigma_t^2 = pqs^2 \; [2(1 - 2pq)h^2 - 4q^2h + q(1 + q)] \tag{1}$$

$$\sigma_a^2 = 2pqs^2 \; [(p - q)h + q]^2 \tag{2}$$

$$\sigma_d^2 = p^2q^2s^2 \; (1 - 2h)^2 \tag{3}$$

Under the condition of h >> $\sqrt{\mu/s}$ (μ is the mutation rate from *A* to *a*), the equilibrium gene frequency ($\hat{q}$) becomes: $\hat{q} \approx \mu/hs$. That is to say, $\hat{q}$ is very small quantity as compared with 1. Then, almost the whole genetic variance becomes additive, $\sigma_a^2 \approx 2pqs^2h^2$, and the dominance variance becomes nearly zero.

In the case of overdominance (h < 0), the equilibrium gene frequency is $\hat{q} \approx h/(2h - 1)$. Then, the whole genetic variance becomes dominance variance. Accordingly additive variance is zero.

If we accumulate spontaneous mutations in an initially homozygous population without natural selection, the gene frequency becomes $\hat{q} = n\mu$ where μ is the mutation rate from *A* to *a* and n represents the number of generations during which mutations were accumulated. In this situation, the gene frequency is assumed to be much less than 1, if n is not extremely large. Then, almost the whole variance becomes additive variance, regardless of the sign of h, and the dominance variance is nearly zero.

Thus, if we estimate the components of genetic variance for viability in an equilibrium population and in a population in which only

newly arisen mutations have been accumulated at a minimum prssure of
natural selection, we may obtain a considerable amount of information
for the mechanism of the maintenance of genetic variability in the
equilibrium population.

MATERIALS AND METHODS

We have extracted 702 and 475 second chromosomes from the Raleigh, N.C.
population (Mukai and Yamaguchi 1974) and the Orlando, Fla. population
(Mukai, unpublished), respectively, using the marked inversion technique
(Wallace 1956). These chromosomes were maintained in the laboratory as
individual stocks with the $SM1(Cy)$ chromosomes which help to maintain
chromosomes carrying lethal and/or semi-lethal genes, and used in the
following experiments.

The relative viabilities of homozygotes and heterozygotes of these
chromosome lines were estimated by using the Cy-method (Wallace 1956).
The relative viability was estimated by (the number of wild-type flies)/
(the number of the Cy flies + 1) (Haldane 1956). The homozygous load
was estimated and it was partitioned into the detrimental load, D, and
lethal load, L, by the technique developed by Greenberg and Crow (1960).
This method is as follows: Let A, B, and C be the average viability of
random heterozygotes of all homozygotes, and of non-lethal homozygotes,
respectively, then T, D, and L values can be estimated as follows under
the assumption of multiplicative gene action:

$$T \simeq \ln A - \ln B \qquad (4)$$
$$D \simeq \ln A - \ln C \qquad (5)$$
$$L \simeq \ln C - \ln B \qquad (6)$$

It should be noted here that the standard of the comparison for viability is not the optimum genotype in the population (Crow 1958) but the
average viability of heterozygotes in the population.

In order to estimate the components of genetic variance for viability, a partial diallel cross method [Design II of Comstock and Robinson (1952)] with 7 rows and 7 columns was employed (14 chromosome lines
were used for a partial diallel cross). Four simultaneous replications
were made within cells. Additive, σ_A^2, and dominance, σ_D^2, variances were

estimated by the following formulae considering that *SM1(Cy)* chromosomes are not completely dominant over the deleterious genes located
in the second chromosomes extracted from the natural populations (Mukai, Cardellino, Watanabe, and Crow 1974):

$$\hat{\sigma}^2_A = 4(\sigma^2_R + \sigma^2_C - \sigma^2_{RXC}/2) \tag{7}$$

$$\hat{\sigma}^2_D = \sigma^2_{RXC} \tag{8}$$

where σ^2_R, σ^2_C, and σ^2_{RXC} are the variance components of Rows, Columns,
and Row × Column interactions, respectively.

The same type of analysis was conducted using 144 second chromosomes
on which newly arisen mutations were accumulated at a minimum pressure
of selection for about 160 generations (Mukai, Chigusa, Mettler, and
Crow 1972). In this accumulation of mutations, the *Cy-Pm* technique was
used. Nine diallel crosses with 8 rows and 8 columns were conducted with
four simultaneous replications. The genetic background was arranged so
as to be heterozygous and homogeneous. The components of genetic variance were estimated by using the above method.

Comparing the viabilities of the chromosomes that were extracted
from male flies with those from female flies that were captured in the
same equilibrium population, it is possible to test the existence of
epistasis and linkage disequilibrium. The chromosomes from males have
experienced natural selection but crossing over did not occur after
this selection, while those from females are produced after recombination following the natural selection. Thus, if there is a difference
between them, it may indicate epistasis and linkage disequilibrium.
We have estimated homozygous and heterozygous viabilities of these two
types of chromosomes.

Linkage disequilibrium between isozyme genes, *inter se*, [αglycerol-
3-phosphate dehydrogenase (α*Gpdh-1*, 2-17.8), malate dehydrogenase-1
(*Mdh-1*, 2-35.3), alcohol dehydrogenase (*Adh*,2-50.1), hexokinase-C
(*Hex-C*, 2-74.5, detected by F. M. Johnson and located by Mukai and
Voelker 1977) and two types of polymorphic inversions [*In(2L)t*
In(2R)NS, see Lindsley and Grell (1967)] were tested for the purpose of
detecting epistasis. Six hundred and seventeen second chromosomes were
extracted from the Raleigh, N. C. population in 1974 and used for the
test of linkage disequilibrium. Furthermore, linkage disequilibrium between the octanol dehydrogenase locus (3-49.2) and the esterase-C locus

(3-49±) in the third chromosome was tested using 526 third chromosomes
extracted from the same population as the above at the same time (Mukai
and Voelker 1977). Indeed, the actual distance between these two loci
estimated by Mukai and Voelker (1977) is 0.0058 ± 0.0022.

Relative viabilities and fecundities of homozygotes and heterozygotes with respect to the four isozyme genes [αGpdh-1, Mdh-1, Adh, and
α amylase (Amy, 2-77.3)] were estimated using the data for estimating
homozygous load and linkage disequilibrium. In this case, genetic backgrounds were heterozygous since only random heterozygotes for the second chromosomes were employed for this analysis.

RESULTS AND ANALYSES

Homozygous load.

The homozygous loads of the Raleigh, N. C. population and the Orlando, Fla. population were estimated using formulae (4), (5), and (6).
The results are shown in Table 1 (Mukai and Yamaguchi 1974; Mukai unpublished). For the sake of comparison, the D and L values for newly
arisen mutations accumulated for 40 generations (Mukai et al. 1972) are
also shown in Table 1. In this case, the A value (standard) is the homozygote viability of the original chromosome on which mutations were accumulated after replication. The D/L ratios were calculated. For the
equilibrium populations, the standard for the homozygous load is the
average viability of random heterozygotes. For the sake of comparison
between the homozygous loads of the equilibrium populations and newly
arisen mutations, the standard for the equilibrium populations was
changed to the hypothetical optimum genotype assuming that (1) the mutant genes are deleterious both in homozygotes and heterozygotes (Mukai
and Yamaguchi 1974) and (2) a mutant gene for viability is deleterious,
to the same degree, to fertility. The D/L ratios under these assumptions
were calculated and were shown in the parentheses in Table 1. It should
be pointed out that the D/L ratios for the equilibrium populations and
the newly arisen mutations are approximately the same. This means that
the deleterious effects of lethal genes and mildly deleterious genes in

Table 1.   Homozygous loads and the components of genetic variance for
            viability in various populations of *Drosophila melanogaster*
            (From Mukai *et al.* 1972; Mukai and Yamaguchi 1974; Mukai *et al.* 1974;
            Mukai, to be published)

		Raleigh, N. C. population	Orlando, Fla. population	Newly arisen mutations
Homozygous load:				
D		0.3337	0.4029	0.2247[a]
L		0.5006	0.4499	0.2355[a]
D/L		0.667 (0.966)	0.896 (1.229)	(0.954)[a]
Additive genetic variance × 10	A	96 ± 25	202 ± 38	---
	B	85 ± 27	275 ± 33	209 ± 37[b]
Dominance genetic variance × 10^5	A	115 ± 46	63 ± 57	---
	B	98 ± 54	62 ± 50	24 ± 50[b]

a) The results at generation 40.
b) The results at generation 160.
A: Including inversions.
B: Excluding inversions.

heterozygous condition are very similar to each other. In fact, average
degrees of dominance were estimated to be 0.01 - 0.02 and 0.43, for the
effects of lethal genes and mildly deleterious genes (= viability polygenes with $\bar{s} \simeq$ 0.02 - 0.03), respectively (Mukai and Yamaguchi 1974).
Thus, $\overline{hs}$ for lethals is 0.01 - 0.02 and that for mildly deleterious
genes is about 0.01; they are approximately the same. Lethal genes are
known to be heterozygously deleterious in viability as well as in fertility. Thus, viability polygenes should be deleterious to fertility

in heterozygous condition. Another point to be stressed here is that
the detrimental load in the Florida population is larger than that in
the N. C. population in contrast to the lethal load. This will be discussed in detail in the DISCUSSION.

Variance component analyses.

    The results of 46 of the 7 × 7 partial diallel crosses with four
simultaneous replications and 9 of the 8 × 8 partial diallel crosses
with four simultaneous replications are summarized in Table 1. From this
table, the following findings can be ontained: (1) For both the newly
arisen mutations and the equilibrium populations, the additive variance
is much larger than the dominance variance, which is quite negligible.
(2) A slight, and not significant, amount of increase in dominance variance can be seen in the equilibrium populations in comparison with
the newly arisen mutations. These two findings imply that *absolute*
overdominance in viability is rare in natural populations if viability
is proportional to the total fitness of an individual. This assumption
will be discussed in the DISCUSSION.

Effects of recombination on viability.

    The experimental results that were used for the analysis of homozy-
gous load of the Raleigh, N. C. population were employed (Mukai and
Yamaguchi 1974):
    1) Homozygotes: There were 367 and 324 second chromosomes extracted
from females and males, respectively. The difference in the distribution
of viabilities was examined using the χ^2 contingency table method. The
result is $\chi^2_{11} = 20.38$ ($P < 0.05$), which indicates a significant difference between them. The overall mean values are 0.4271 for the chromosomes from female parents and 0.4422 for the chromosomes from male pa-
parents. Thus, it may be concluded that there is a small but significant epistasis among viability genes in homozygous condition.

2) Heterozygotes: Random heterozygotes can be classified into three
categories. (1) Both homologous chromosomes originated from females,
(2) one homologous chromosome originated from a female and the other
from a male, and (3) both homologous chromosomes originated from males.
In the second category, one unusual cross that showed an extremely low
viability was excluded from the analysis. The distribution of the pooled
result of categories 1 and 2 was compared with that of category 3 but
no significant difference was detected. Analyses of variance were performed in order to examine the differences in mean viabilities and to
estimate the genotypic variances in the above respective categories.
The results are as follows:

Origin	N	Mean viability	Genotypic variance
Female and female	277	0.9954 ± 0.0059	0.005859 ± 0.000852
Female and male	176	1.0151 ± 0.0074	0.005689 ± 0.001062
Male and male	234	0.9969 ± 0.0058	0.003579 ± 0.000788

An adjustment for genetic variance which becomes necessary due to the
incomplete dominance of the Cy chromosomes (in the Cy method) has not
been done for the above results. The results of the analyses show (1)
that there are significant differences among mean viabilities of individuals belonging to categories 1, 2, and 3, $i.e.$ the mean viability
of category 2 is higher than those of categories 1 and 2, and (2) that
the genotypic variance in category 1 is significantly larger than in
category 3 ($P < 0.05$); in category 2 it is larger than in category 3,
but not significantly so. The genotypic variance pooled over categories
1 and 2 is significantly larger than that of category 3 ($P < 0.05$).
Probably part of this inflation of the variance is due to slight but
significant epistasis among viability genes. This means, insofar as
the population is in genetic equilibrium, that there is a small amount
of linkage disequilibrium among viability genes.

Linkage disequilibria including isozyme genes and polymorphic inversions.

The results of the surveys of linkage disquilibria in the Raleigh,
N. C. population from 1968 to 1970 have been reported by Mukai, Watanabe, and Yamaguchi (1974). A further test was made for the 1974 population by Mukai and Voelker (1977). The results are presented in Table 2.

Table 2. The result of the test for linkage disequilibrium
            (1974 Raleigh, N. C. population)
            (From Mukai and Voelker 1976)

Component	χ_1^2	Component	χ_1^2	Component	χ_1^2
AB	1.69	ABC	0.09	ABCH	2.92
AC	0.84	ABH	5.52*	IABC	0.05
AH	0.00	ACH	0.00	IABH	0.40
BC	0.22	BCH	2.26	IACH	0.00
BH	2.21	IAB	0.02	IBCH	0.14
CH	0.39	IAC	1.57	ABCJ	1.71
IA	2.67	IAH	0.01	ABHJ	2.54
IB	0.92	IBC	0.21	ACHJ	0.05
IC	17.44***	IBH	0.12	BCHJ	0.59
IH	0.22	ICH	0.21	IABJ	0.19
AJ	1.62	ABJ	3.14	IACJ	0.12
BJ	0.43	ACJ	0.32	IAHJ	0.04
CJ	0.02	AHJ	0.10	IBCJ	0.03
HJ	7.39**	BCJ	0.51	IBHJ	0.04
IJ	0.04	BHJ	0.21	ICHJ	0.06
		CHJ	0.05		
		IAJ	0.00		
		IBJ	0.04		
		ICJ	0.01		
		IHJ	0.06		

I = *In(2L)t*,   A = *αGpdh*,   B = *Mdh*,   C = *Adh*,   H = *Hex-C*,   J = *In(2R)NS*

*Significant at the 5% level
**Significant at the 1% level
***Significant at the 0.1% level

From this table, it may be seen that linkage disequilibrium between isozyme genes, *inter se*, is rare, although significant linkage disequilibria could be seen between isozyme genes and polymorphic inversions. In particular, higher order interactions, which are expected in the case of overdominance of many genes with multiplicative gene action (Franklin and Lewontin 1970), is negligible.

Linkage disequilibrium for the two extremely close loci (*Est-C* and *Odh*) in the third chromosome was tested (Mukai and Voelker 1977). The results are as follows:

Genotype	FF	FS	SF	SS	Total
Frequency	416	49	57	4	526
Expected frequency	418.0	47.0	54.9	6.1	526

The χ_1^2 value becomes 0.9 (0.3 < P < 0.4). Thus, no evidence for linkage disequilibrium was detected between two loci whose distance is about 0.005. Thus, it may be concluded that linkage disequilibrium between isozyme genes is a very rare phenomenon in a large random mating population. In fact, the effective size of the Raleigh population was expected to be 20,000 - 100,000 (Mukai and Yamaguchi 1974).

Estimation of relative viability and fecundity.

Cumulative heterozygous effects of isozyme genes located in the second chromosome on viability and fecundity were tested (Tables 3 and 4), but no significant effect was detected in both fitness components. Thus, it is likely that isozyme genes are very nearly neutral at least in the laboratory condition.

Table 3. Cumulative effect of heterozygous loci on viability at the
αGpdh-1, Mdh-1, Adh and Amy loci

(From Mukai, Watanabe, and Yamaguchi 1974)

Number of heterozygous loci	Number of crosses	Average viability (or fecundity)
0	264	1.0000
1	350	0.9991 ± 0.0081
2	185	1.0013 ± 0.0096
2 or more	215	0.9998 ± 0.0092
(3 or more)	30	0.9837 ± 0.0187

Table 4. Cumulative effect of heterozygous loci on fertility (or fecundity) of females at the αGpdh-1, Mdh-1, Adh and Amy loci

(From Mukai, Watanabe, and Yamaguchi 1974)

Number of heterozygous loci	Number of crosses	Average fertility (or fecundity)
0	174	1.0000
1	471	0.9894 ± 0.0111
2	161	1.0094 ± 0.0134
2 or more	186	1.0088 ± 0.0128
(3 or more)	25	1.0060 ± 0.0209

DISCUSSION

Assumptions and conditions for the analysis of populations.

For the analysis of natural populations, it is usually assumed that
the population in question is in genetic equilibrium. This was carefully
examined for the Raleigh, N. C. population, and it was concluded that
the above assumption holds true approximately (Mukai and Yamaguchi 1974).
The effective size of the population becomes very important for the
studies of linkage disequilibrium. That of the Raleigh population was,
as described before, estimated to be 20,000 - 100,000 (Mukai and Yamaguchi 1974) and that of the Florida population was found to be effectively infinite (Mukai, unpublished).

The second question is whether or not viability is proportional to
total fitness. In the present investigation, the natural populations
were analyzed mainly for viability. If this component of fitness is not
proportional to the total fitness, many of the results obtained in the
present work lose their population-genetical significance. Fortunately,
we have good reasons to assume that viability is approximately proportional to total fitness, since, as described above, the D/L ratio of
newly arisen mutations is approximately the same as that of equilibrium
populations and since lethal genes are deleterious both in viability
and fertility. If all or a significant fraction of the viability polygenes were heterotic with respect to fertility, the D/L ratio of equilibrium populations would have to be very large as compared with that
of newly arisen mutations.

Lack of evidence for predominance of overdominance.

The D/L ratios of newly arisen mutations and equilibrium population
are approximately the same, especially in the Raleigh, N. C. population.
This indicates that the average degree of dominance of mildly deleterious genes (viability polygenes) in equilibrium populations, $\bar{h}_E$, is large
(Greenberg and Crow 1960), since lethal genes are heterozygously deleterious both for viability and fertility, and the product of selection
coefficient, s, and the degree of dominance must be approximately the
same for lethal genes and viability polygenes. In fact, the $\bar{h}_E$ value
was estimated to be 0.2 - 0.3 for both the N. C. and the Florida populations.

As described earlier, overdominant loci contribute to dominance variance, but not to additive variance in equilibrium populations. It is
necessary to accumulate spontaneous mutations for about 80 generations
to reach the same amount of genetic variability as in equilibrium populations [compare the observed lethal load in Table 1 with the lethal
mutation rate (0.006/second chromosome/generation, Mukai *et al.* 1972)].
The expected amount of dominance variance at generation 80 is approximately 1/4 times that at generation 160, *i.e.* ca. 0.00006. This amount
is smaller than the observed dominance variance, which is also a very
small quantity. Thus, it is expected that at only a small number of loci
is overdominance manifested, or, at least, that gene frequencies are
intermediate for mildly deleterious genes.

The effective number of overdominant loci, n_e, can be estimated as
follows (Mukai *et al.* 1974):

$$n_e = L'^2/[(k - 1) \, \sigma_D'^2] \tag{9}$$

where L' is the homozygous load at overdominant loci (the standard is
the average heterozygote viability), $\sigma_D'^2$ is the dominance variance due
to overdominant loci, and k is the number of alleles per locus.

Applying this formula to the Raleigh, N. C. population, the effective number of overdominant loci was estimated to be less than 10 in
the second chromosome. This finding agrees with the above speculation.

Selection on polymorphic isozyme gene loci.

Although it is concluded that mildly deleterious genes or viability
polygenes rarely manifest overdominance, this finding cannot always be
extended to isozyme genes which are supposed to be selectively nearly
neutral or neutral. No significant differences in relative viability
and fecundity were detected between homozygotes and heterozygotes by
our direct measure, involving approximately 1.5 million flies.

Franklin and Lewontin (1970) states that a large amount of linkage
disequilibrium is maintained at equilibrium if a number of overdominant
loci with multiplicative gene action are tightly packed in a chromosome,
even if the selection coefficient at each locus is extremely small. In
the present experiment, no linkage disequilibrium was detected between
any pair of isozyme genes even in the case of an extremely short dis-

tance (0.0058 ± 0.0022). Furthermore, no higher order interactions were
detected. Thus, these findings are unfavorable to overdominance of isozyme genes. Incidentally, linkage disequilibria between isozyme genes
and polymorphic inversions may be due to founder effect. These polymorphic inversions occurred uniquely and increased in frequency without
recombination.

Too large additive variance.

In both the North Carolina population and the Florida population,
the estimated additive variances are too large as compared with the
predicted value, 0.003, on the basis of mutation rate, μ, selection coefficient, s, and the average degree of dominance, $\bar{h}$ [$\Sigma\mu$ = 0.17/second
chromosome/generation, $\bar{s}$ = 0.023, and $\bar{h}$ = 0.4 (Mukai *et al.*1972)]. Furthermore, in the Florida population, the D value is significantly larger
than that in the N. C. population although the L value in the Florida
population is smaller than that in the N. C. population. These phenomena might be expected if there is diversifying selection. A typical case
of diversifying selection is as follows: there are several niches and
the order of the genotypes for fitness is different in the different
niches, (Levene 1953), although random mating occurs over all niches.
In such cases, the gene frequency becomes intermediate and, in a constant laboratory environment, additive variance will be inflated, as
in the present experiment.

Since linkage disequilibrium was detected on a whole chromosome
basis, there is some possibility that the increase in estimated additive variance is due to epistasis. However, this possibility may be rejected since an estimated dominance variance that includes some fraction of epistatic variance is very small. Incidentally, this linkage
disequilibrium may be due to mildly deleterious genes or viability polygenes because no significant difference in viability and fecundity
was detected between homozygotes and heterozygotes for isozyme genes
and since no significant linkage disequilibrium was found between isozyme genes, *inter se*.

In the Florida population, the additive variance was estimated to
be twice as large as that in the N. C. population although the D value
in the former population is only 1.2 times larger than that in the
latter population. Furthermore, the average gene frequency of isozyme
genes at 7 loci of the Florida population was very close to that of the

N. C. population (Johnson and Schaffer 1973). These phenomena suggest
that the inflation of the additive variance in the Florida population
is due to the diversifying selection mainly for mildly deleterious genes
with detectable effects at a small number of loci. This may also suggest that mildly deleterious genes are important for adaptation of the
organisms.

Summary.

Using the second chromosomes, homozygous genetic loads and genetic variance components were estimated for the Raleigh, N. C. population and the Orlando, Fla. population. The Florida population showed a larger additive variance and homozygous detrimental load than the N. C. population, although the lethal load in the former population was slightly less than that in the latter.

Linkage disequilibrium was detected for the chromosome as a whole but not so between the isozyme genes, *inter se*, in the N. C. population.

No significant difference in relative viability and fertility was detected between
homozygotes and heterozygotes for isozyme genes.

From these results it was speculated that diversifying selection was operating on
mildly deleterious genes especially in the Florida population. However, no evidence
for selection was detected for isozyme gene loci.

This contribution is paper No. 1 from the Laboratory of Population Genetics, Department of Biology, Kyushu University.

REFERENCES

Comstock, R.E. and H.F. Robinson. 1952. Estimation of average dominance of genes.
In *Heterosis* (J.W. Gowen, ed.). Iowa State College Press, Ames. Pp. 494-516.

Crow, J.F. 1958. Some possibilities for measuring selection intensities in man.
Hum. Biol. 30: 1-13.

Franklin, I. and R.C. Lewontin. 1970. Is the gene the unit of selection?
Genetics 65: 707-734.

Greenberg, R. and J.F. Crow. 1960. A comparison of the effect of lethal and detrimental chromosomes from Drosophila populations.
Genetics 45: 1154-1168.

Haldane, J.B.S. 1956. Estimation of viabilities.
J. Genetics 54: 294-296.

Johnson, F.M. and H.E. Schaffer. 1973. Isozyme variability in species of the genus
Drosophila. VII. Genotype-environment relationships in populations of *D. melano-
gaster* from the Eastern United States.
Biochemical Genetics 10: 149-163.

Levene, H.   1953.   Genetic equilibrium when more than one ecological niche is available.
Am. Natur. 87: 311.

Lindsley, D.L. and E.H. Grell.   1967.   Genetic variations of *Drosophila melanogaster*.
Carnegie Inst. Washington Publ. 627.

Langley, C.H., Y.N. Tobari and K. Kojima.   1974.   Linkage disequilibrium in natural
populations of *Drosophila melanogaster*.
Genetics 78: 921-936.

Mukai, T. and R.A. Voelker.   1977.   The genetic structure of natural populations of
Drosophila melanogaster. XIII. Further studies on linkage disequilibrium.
Genetics (in press).

Mukai, T., R.A. Cardellino, T.K. Watanabe and J.F. Crow.   1974.   The genetic variance for viability and its components in a local population of *Drosophila melanogaster*.
Genetics 78: 1195-1208.

Mukai, T., S.I. Chigusa, L.E. Mettler and J.F. Crow.   1972.   Mutation rate and dominance of genes affecting viability in *Drosophila melanogaster*.
Genetics 72: 335-355.

Mukai, T., L.E. Mettler and S.I. Chigusa.   1971.   Linkage disequilibrium in a local
population of *Drosophila melanogaster*.
Proc. Natl. Acad. Sci. U. S. 68: 1065-1069.

Mukai, T., T.K. Watanabe and O. Yamaguchi.   1974.   The genetic structure of natural
populations of *Drosophila melanogaster*. XII. Linkage disequilibrium in a large
local population.
Genetics 77: 771-793.

Mukai, T. and O. Yamaguchi.   1974.   The genetic structure of natural populations of
Drosophila melanogaster. XI. Genetic variability in a local population.
Genetics 76: 339-366.

Wallace, B.   1956.   Studies on irradiated populations of *Drosophila melanogaster*.
J. Genetics 54: 280-293.

1. STUDY OF SELECTION

SEXUAL SELECTION AND THE EVOLUTION OF TERRITORIALITY IN BIRDS

P. O'Donald

Many birds defend territories. Territorial birds are often insectivo-
rous or raptorial: their territories are assumed to be the places where
they collect their food. Territoriality will thus evolve if natural selection gives an advantage to individuals who defend their feeding
areas. Horn (1968) suggested that a more or less uniform distribution
of food may favour territoriality: if food is distributed patchily, it
may be better to forage in groups. Lack (1966, 1968) pointed out, however, that there is no direct evidence for a selective advantage of
feeding territories, although comparisons between species with different diets strongly support the theory. For example, cardueline finches
nest in groups of small territories. They feed their young mainly on
seeds, foraging in groups. The fringilline chaffinch defends a large
territory, feeding its young on insects caught in the territory.

Many birds defend large territores although they do not feed within
them. Crook (1962, 1964, 1965), in his study of pair formation and social organisation of weaver birds, suggested that a large territory
may help to conceal the nest. This would explain the different sizes
of territories of the different species of *Euplectes*: species which
nest in taller and denser vegetation have smaller territories (Emlen
1957).

Territory may also have selective advantage as an area for sexual
display and breeding. This is likely if territories are used exclusive-

ly for display and breeding and give no protection from predators. Wilson (1975) has listed examples from a large number of studies in which the functions of territoriality have been deduced from activities that take place within the territories: territories may be feeding areas, areas for sexual display and mating, and areas for shelter and concealment. It is argued that territoriality gains a selective advantage from one or more of these functions, but no direct evidence of the selective advantage has been obtained. The argument is always inductive, based on comparisons between species. Such arguments, which were the only arguments Darwin himself could use, are very convincing when a particular character like territoriality differs between species, genera, or subfamilies in relation to differences in ecology.

In this paper, I wish to suggest that a larger territory will often be of direct selective advantage to a defending male: it will increase his chances of mating. Territoriality will thus evolve by sexual selection. Davis and O'Donald (1976b) obtained data on territoriality in the Arctic Skua, *Stercorarius parasiticus*. From these data, the possible selective advantage of territory size can be calculated.

TERRITORY SIZE AND BREEDING TIME IN THE ARCTIC SKUA

The Arctic Skua is a polymorphic seabird with three phenotypes in plumage - pale, intermediate and dark. Sexual selection favours the intermediate and dark males. Computer models of mate selection give estimates of the proportions of females preferring to mate with intermediate and dark males (O'Donald, Wedd and Davis 1974; O'Donald 1976; Davis and O'Donald 1976a). O'Donald and Davis (1975, 1976) estimated the overall selection acting on the polymorphism from demographic data of survival and fledging success. These data were obtained from a study of the Arctic Skua colony on Fair Isle in the Shetland Islands. Since 1973, this study has included the accurate surveying of the positions of the nests of all pairs of birds in the colony. Davis and O'Donald (1976b) mapped the nests on paper. Then, by drawing the lines bisecting and perpendicular to the lines separating one nest from its neighbours, polygons were fitted round the nests. Parts of the Arctic Skua colony are occupied by Great Skuas. The Great Skuas' nests were also mapped, since Great and Arctic Skuas show interspecific territoriality. Davis and O'Donald used the area of each polygon to measure the size of each pair's territory. When territories are sharply bounded in a dense colony, they naturally tend to become polygonal (Wilson 1975).

The pairs of breeding birds were divided into three classes: those in which the male was breeding for the first time; those in which the male had changed his mate from the previous year; and those in which both birds of the pair were unchanged from the previous year. Behavioural observations suggested that birds meet and form pairs in different ways, depending on whether the males had previous breeding experience or not. Experienced males who have lost their mates remain on their territories of the previous year and take their new mates there. The new mate is often a female who bred nearby in the previous year. Inexperienced birds collect in groups, or "clubs", however, and seem to form pairs in these clubs: thus they may establish their territories after they have paired. Davis and O'Donald found a significant correlation between territory size and breeding time among those males with previous breeding experience who were taking a new mate. Males with larger territories breed earlier in the season. This effect is reduced, and is not quite significant statistically, if the males are breeding for the first time. There is no correlation at all in pairs which had bred together in previous years. Figure 1 shows the data of territory size and breeding time when experienced males mate with a new female.

In the Arctic Skua, as in many birds, pairs breeding earlier in the season have a higher reproductive success: they lay more eggs on average and are more successful in fledging their chicks. A male who mates early in the breeding season thus gains a selective advantage. Larger territories are directly favoured by sexual selection. O'Donald (1972) and O'Donald, Wedd and Davis (1974) gave data of the relation between breeding time and reproductive success in the Arctic Skua. Table 1 shows the average number of chicks fledged during successive weeks of the breeding season.

On average, the intermediate and dark males find their mates before the pale males when mating for the first time or when mating with a new female. This effect, like the correlation of territory size and breeding time, disappears if the pair remain together to breed in subsequent years. O'Donald, Wedd and Davis (1974) and O'Donald (1976) used the observed distributions of breeding times of the three phenotypes to estimate the female mating preferences according to number of different models. The males' chances of mating may be determined by factors such as gonadotrophin and testosterone levels. Watson and Moss (1971) found that

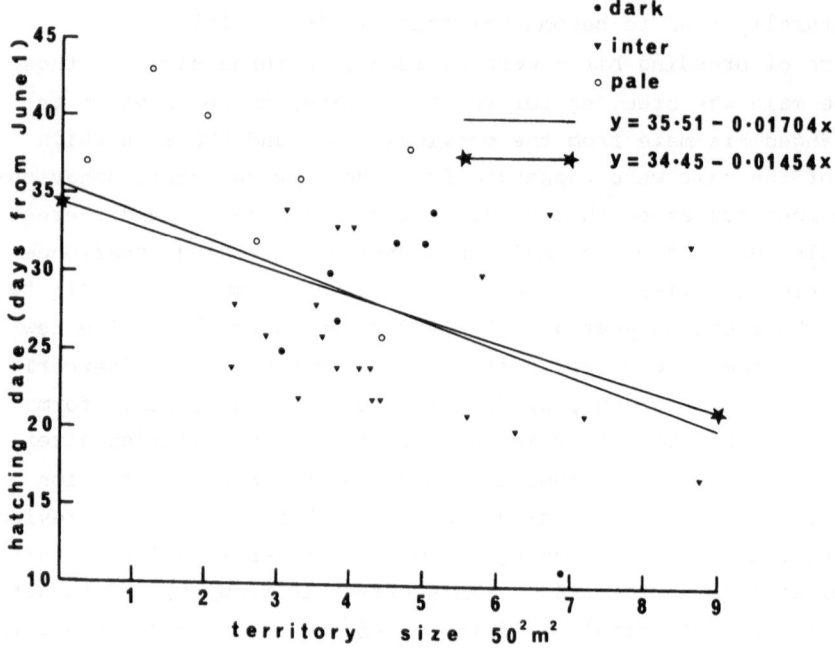

Figure 1. Territory sizes and breeding times of male Arctic Skuas who have bred in previous years and who are taking new mates.

male red grouse implanted with androgen defended much larger territories, devoted more time to courtship and greatly increased their mating success. Territory size and mating success may thus be determined by a common cause such as hormone levels: larger territories would then evolve if mating success were determined by genes that pleiotropically increased territory size. Intermediate and dark males do in fact have larger territories than pale males; but the difference is not statistically significant given such a small sample of pale birds.

As an alternative hypothesis, Davis and O'Donald (1976b) suggested that size of territory is the prime cause of mating success. Suppose the females arrive and land at random on the territories of the males. They will be more likely to land on the larger territories. Males with larger territories will find a mate sooner than those with smaller territories: they will breed earlier on average in the breeding season and

thus gain a selective advantage from the greater reproductive success
of earlier pairs. This selection would not by itself explain the earlier
breeding of the intermediate and dark males. The difference in the
breeding times is largest when the males are mating for the first time:
the correlation of territory size and breeding time is largest and statistically significant only when males who are taking a new mate have
previous breeding experience. More data are required to establish the
facts reliably. Particularly, we need to know if the phenotypes differ
significantly in size of territory and if experienced and inexperienced
males differ significantly in the correlations of territory size and
breeding time. It seems more likely on the present evidence that territory size and mating success are both determined by a common cause,
possibly by physiological differences in hormone levels. Territory size
may still determine mating success to some extent, however: there must
be some random element in the arrival of the females on the males' territories. A male with a larger territory will thus have a slightly
better chance of mating earlier in the season and some additional advantage must therefore accrue to him. And whatever the cause of the
correlation between territory size and breeding time, males with larger
territories should gain a selective advantage.

SEXUAL SELECTION AND TERRITORIALITY

A larger territory must always increase a male's chances of mating.
This must therefore be a component of selection in favour of larger
territories. If the females arrive in succession on the breeding
grounds and land at random in the males' territories, the males with
the larger territories have the greater chance of mating and therefore
on average find their mates earlier in the breeding season, thereby
gaining a selective advantage. This is the model I wish to analyze.
Of course, other factors may outweigh the selective advantage gained
by the males, especially if the random element in the females' arrivals
is only small. Presumably a balance will be reached between sexual selection for a larger territory and natural selection against expending
more energy defending it.

The following symbols will be used.

p_{ij} is the probability that the jth male is unmated when the ith female arrives ($p_{1j} = 1$).

$p_{i+1,j}$ is the probability that the jth male is unmated when the (i+1)th female arrives.

P_{ij} is the probability that the jth male mates with the ith female.

x_j is the size of the jth male's territory.

w_j is the average number of chicks fledged by the jth male.

W_i is the average number of chicks fledged by the ith female.

b_i is the breeding date of the ith female.

The probability that the ith female mates with the jth male is then assumed to be proportional to the size of the territory and the probability that the male is still unmated. Thus

$$P_{ij} = p_{ij}\, x_j / \sum_j p_{ij}\, x_j.$$

$$\text{If } P_{ij} \leq p_{ij} \text{ then } p_{i+1,j} = p_{ij} - P_{ij}.$$

$$\text{If } P_{ij} > p_{ij} \text{ then } P_{ij} = p_{ij} \text{ and } p_{i+1,j} = 0.$$

$$w_j = \sum_i P_{ij}\, W_i.$$

The values of W_i are determined by the times in the breeding season when the females arrive on the breeding grounds. Table 1 gives actual mean numbers of chicks fledged by pairs hatching their eggs at different times in the breeding season. These values can be used for the values of W_i. Thus any female who hatches her first chick from 10 to 16 June produces an average of 1.625 fledged chicks, and so on. More generally, W_i may be given by a quadratic fitness function of the form $W_i = 1 - (\theta - b_i)^2/\phi$. This has been found to give values of relative fitness that fit the Arctic Skua data. Using the method described by O'Donald (1972), the values of the parameters were found to be $\theta = 4.6768$ and $\phi = 1624.7$ for data in which the values of b_i are days from June 1. The quadratic fitness function can then be used to investigate the effect of different values of θ on the selection for territory. For example, when $\theta = \bar{b}$, fledging success decreases symmetrically on both sides of the mean breeding date: then it is as disadvantageous to breed

Table 1.   Breeding time and fledging success of the Arctic Skua
            The data are condensed from Table 1 in O'Donald, Wedd
            and Davis (1974).

Date of hatching first chick	Number of chicks fledged
10-16 June	1.6250
17-23 June	1.5826
24-30 June	1.5128
1-7 July	1.1212
8-15 July	0.7419

early as to breed late.

This model of mating success can be used to construct the hypothetical bivariate distribution of territory size and breeding time. A simple example can be used to illustrate how the model works. Suppose there are males with territories of 5, 25, 40, 60, and 85 m^2; females arrive to breed on days 17, 22, 28, 34, and 39. Table 2 shows the probabilities of mating and male fitnesses according to the model.

For the Arctic Skua the data for Figure 1 can be used to calculate the probability P_{ij} that the jth male mates with the ith female. There are 35 males and females whose territory sizes and breeding times are shown in the figure. According to the model, the correlation coefficient of territory size and breeding time should be -0.3941: the actual correlation coefficient is - 0.4622, which is not significantly different from the hypothetical value. The regression of breeding date on territory size is Y = 35.51 - 0.01704x and the regression according to the model should be Y = 34.45 - 0.01454x. These actual and hypothetical regressions are plotted on the figure to show how closely they agree. But we know that females do not in fact arrive to land at random on the territories as the model assumes: the females have often come from nearby territories where they mated in previous years. No doubt there must

Table 2. Simple example of probabilities of mating and resultant fitnesses when probability of mating is proportional to territory size

The correlation between territory size and breeding time is
- 0.6320 showing that the males with the larger territories
are much more likely to mate with the early females. The male
with the smallest territory is most likely to mate with the
last female to arrive.

Breeding date of females	Territory size of males					Chicks fledged by females
	5	25	40	60	85	
17	0.0233	0.1163	0.1860	0.2791	0.3953	1.6250
22	0.0317	0.1433	0.2112	0.2805	0.3333	1.5826
28	0.0488	0.1912	0.2490	0.2729	0.2382	1.5128
34	0.0991	0.3036	0.3129	0.1675	0.0332	1.1212
39	0.7972	0.2457	0.0409	0.0	0.0	0.7419
Chicks fledged by males	0.8643	1.2276	1.3943	1.498	1.5674	

be some random element in the females' arrivals and hence some advantage gained by males with larger territories as a result of the increased chance of mating that their larger territories bring. But in
the Arctic Skua this may be only a small component of the males' overall selective advantage. Their mating success may be determined by some
other cause, such as hormone level, which pleiotropically increases
territory size. In this paper, however, I am analyzing the effect of
territory size only as far as it directly determines mating success.

SAMPLING DISTRIBUTION OF CORRELATION OF
TERRITORY SIZE AND BREEDING TIME

The model of territory as a factor which determines mating success was used in the previous section to calculate the probabilities P_{ij} that the jth male mates with the ith female. These probabilities can be considered as the bivariate frequency density of the variates of territory size and breeding time. When this frequency density is computed using the territory sizes and breeding times of the Arctic Skua, the two variates have an hypothetical, or expected, correlation coefficient of - 0.3941, compared to the actual correlation coefficient of - 0.4622. This correlation coefficient was calculated from a sample of 35 observations. The expected correlation coefficient will have some unknown sampling distribution, of which, by hypothesis, the actual coefficient will be a particular value. The sampling distribution can be obtained by computer simulation.

From the bivariate frequency density, the cumulative frequency of matings was calculated for each female. Thus

$$C_{ij} = \sum_{k=1}^{j} P_{ik}$$

Then for each male a random number, r, was generated in the range $0 < r < 1$ and compared with C_{ij}. If r lies in the range $C_{i,j-1} < r \le C_{ij}$ the ith female mates with the jth male. This procedure was carried out for each of the 35 females, thus generating samples taken from the hypothetical bivariate frequency density. The cumulative frequency of the distribution of sample correlation coefficients is shown in Figure 2 drawn on a normal probability scale. For comparison, the corresponding normal distribution is shown as a straight line. The values of probability are greater than the normal values at the lower values and less at the higher values showing negative skewness in the distribution of sample correlations. However, the distribution is approximately normal around the central values.

From the sampling distribution, it can be seen that the probability of observing a value equal to or greater than the observed value is 0.323. The variance of the correlation coefficient is 0.01673. We have $r = - 0.4622$, $\varepsilon(r) = - 0.3941$, $var(r) = 0.01673$, and hence

$$\chi^2 = [r - \varepsilon(r)]^2/var(r) = 0.2774.$$

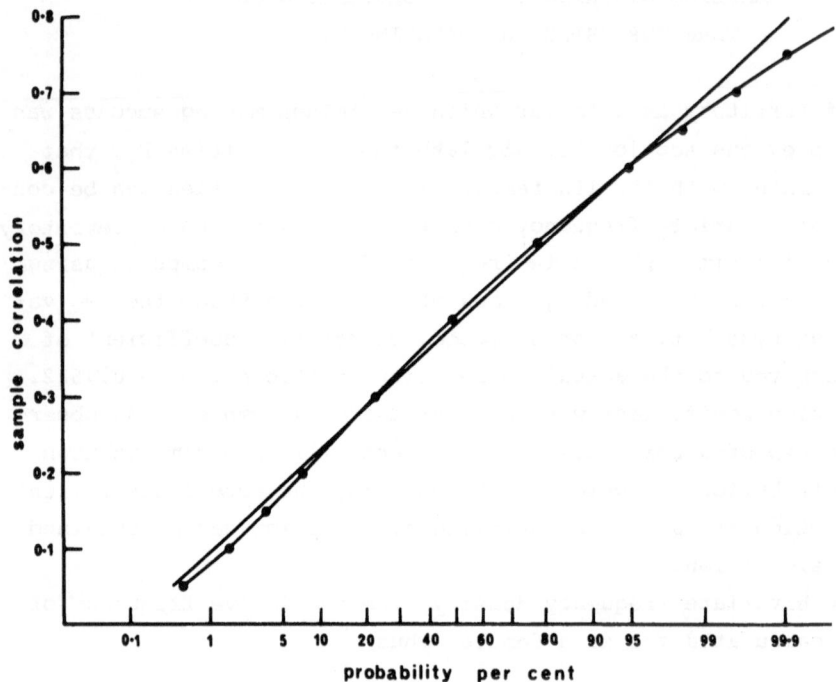

Figure 2. Distribution of the correlation coefficient in random samples of 35 males taken from the hypothetical bivariate distribution of territory size and breeding time.

The fit is obviously very good. The hypothesis would certainly explain the observed correlation.

COMPUTER SIMULATION OF THE EVOLUTION OF TERRITORY

As Table 2 shows, increased territory size gives increased chances of mating early in the breeding season and hence increased reproductive success. Territory should increase in size by sexual selection. Computer simulations confirm this qualitative argument.

The simulations were carried out using populations of 200 males and 200 females. In the computer model, variation in territory size is determined by genetic and environmental factors. The genetic variance is determined by pairs of alleles at n loci acting additively on territory size. The environmental variance is fixed by the initial heritability assumed to be 0.5 in most of the simulations. Each simulation was started with the observed distribution of territory sizes of the Arctic Skua. The mean and standard deviation of the observed sample are $\bar{x}$ = 4.28 and s = 1.84 where territory size is given in units of 50^2 m². From this distribution, a normally distributed sample of 200 values was obtained using a random number generator. The mean and standard deviation of the breeding times were used to obtain a sample of 200 values of the females' breeding times. Using the model, the bivariate distribution of territory size and breeding time was then computed and hence the reproductive success of the males. This was determined either by the actual values for the Arctic Skua as given in Table 1, or by the fitness function $W_i = 2 [1 - (\theta - b_i)^2/\phi]$. Using the fitness function, the value of θ can be varied to investigate the effect of using different relationships between breeding time and fitness.

If breeding time is measured by hatching date in days from June 1, then values θ = 4.677 and ϕ = 1625 fit the data given by O'Donald, Wedd and Davis (1974). Early pairs have a much higher reproductive success. If $\theta = \bar{b}$, the reproductive success decreases symmetrically round the mean breeding date. Figure 3 shows the fitness functions of territory size when θ = 4.677 and when $\theta = \bar{b}$. The normal distribution of territory size is also shown. The relative increase in reproductive success with territory size is much greater when θ = 4.677 than when $\theta = \bar{b}$. However, larger territories still have an overall selective advantage when $\theta = \bar{b}$, even though it is then just as disadvantageous to mate early as to mate late. The males with the largest territories suffer only a slight disadvantage, but the males with the smallest territories are always at a great disadvantage. An analysis of the probabilities of mating shows that the males with the larger territories are only slightly more likely to mate early and usually mate around the mean breeding date, which is the best time to mate if $\theta = \bar{b}$. At the start of the breeding season, all the males are unmated; the largest territory is only a small proportion of the total area; and the male with the largest territory has only a small chance of mating at the beginning of the breeding season. But the males with the smallest territories have a negligible chance of mating until most of the other males have mated: they are almost always left to mate with the last females to arrive. The example in Table 2 illustrates this effect.

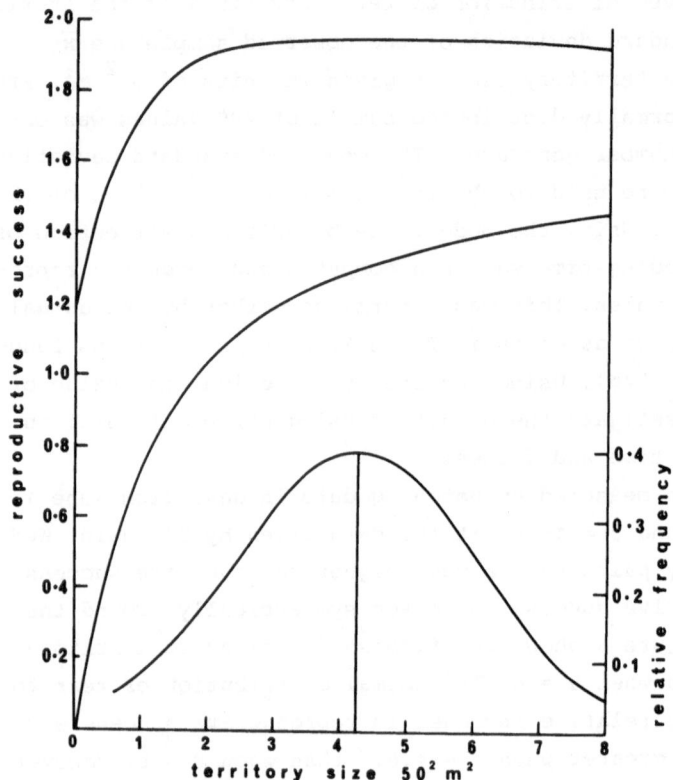

Figure 3. Fitness functions of territory size. The lower curve shows the normal distribution of territory size. The middle curve shows the fitness function when the
optimum breeding date is at the start of the breeding season: fitness increases over
the whole range of territory sizes. The upper curve shows the fitness function when
the optimum breeding date is the mean: fitness only increases at the smaller territory sizes. The difference in the average fitness in the two cases is arbitrary and
caused by the difference in the proportion of pairs breeding at about the optimum
date.

Thus even when $\theta = \bar{b}$, sexual selection still favours increased territory size. But the selection is then very slow: only a small proportion of males are at a selective disadvantage. In Figure 3, only about
10% of the males suffer an appreciable selective disadvantage. This is
shown by the variance in fitness. The intensity of selection can be

measured by calculating $V_w/(\bar{w})^2$ where V_w is the variance in fitness and $\bar{w}$ the mean fitness. When $\theta = 4.677$, we find that the average values of $V_w/(\bar{w})^2$ for 20 samples of 200 individuals is 0.01186. But when $\theta = \bar{b}$, the corresponding average value is 0.0001994. Selection should be about 60 times faster when the optimum breeding date is the start of breeding season rather than the mean breeding date. Some selection for increased territory size will still occur, however, even when early and late breeding are both equally disadvantageous.

Figure 4 shows the progress of selection over 50 generations in a population of 200 males and females. In each simulation it was assumed

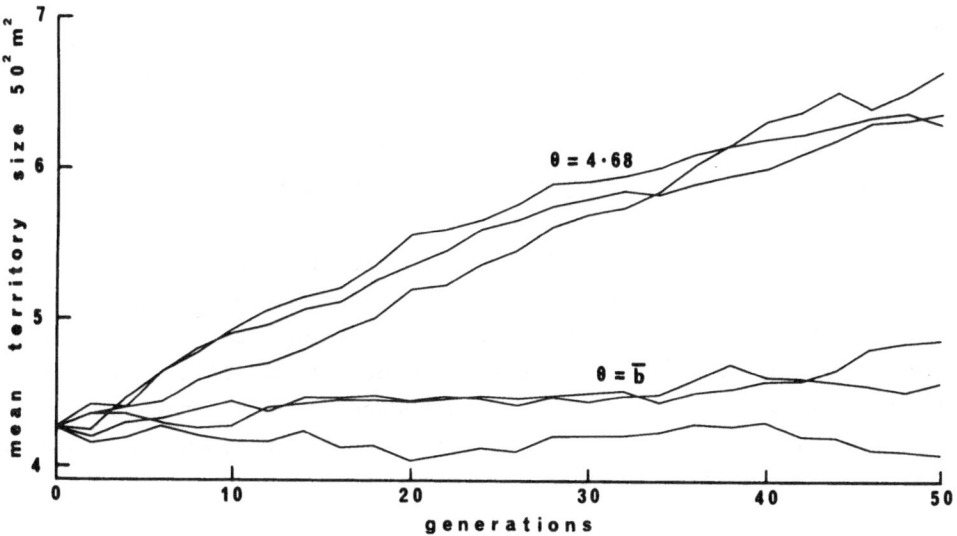

Figure 4. Computer simulation of the evolution of territory in a population of 200 males and 200 females. Selection started with the mean and variance observed in the Arctic Skua and a heritability of 0.5. Genetic variance was determined by 10 loci with equal effects and gene frequencies.

that 10 loci determined the genetic variance in territory size with a heritability of 0.5. The mean territory size rapidly increases when

θ = 4.677, but as expected, little progress is made when θ = $\bar{b}$. The selection is so slight that it is negligible compared with the genetic
drift.

The fitness function of territory size is only slightly dependent
on the relative frequency of individuals with different territories. Figure 5 shows the fitness functions determined by the actual values of
the reproductive success of the Arctic Skua as given in Table 1. Starting
with a uniform distribution of territory sizes, the larger territories
increase in frequency with selection. As they increase in frequency,
their relative advantage declines slightly. The male with the largest
territory starts with a reproductive success of just over 1.5 chicks
fledged in a season, compared with 0.8 chicks fledged by the male with
the smallest territory. However, if 50 per cent of males have the largest
territories, they fledge 1.4 chicks compared to 0.8 fledged by the very
few males with the smallest territory. The relative disadvantage of the
males with the smallest territory declines from 47 per cent to 43 per
cent, showing that there is a slight negative frequency-dependence in
the selection for increased territory size. When the males with the
largest territories are rare, they are more likely to mate with the
earlier, fitter females: when they have become common, they mate with
females arriving over a long period of the breeding season, and lose
some of their advantage.

The number of loci which give rise to the genetic variance in territory size has an effect on the rate of selection because genetic variance and heritability decline more rapidly with fewer loci. After 50
generations, the heritability has been reduced from 0.5 to 0.45 with 10
loci. With 5 loci, the heritability is reduced to 0.43 after 50 generations, and with 2 loci to 0.21. This produces a corresponding reduction
in the rate of increase of territory size.

CONCLUSION

If females arrive in succession on their breeding ground, landing at
random within a particular area, then males holding larger territories
in the area will have a greater chance of mating earlier in the breeding season, thus gaining a selective advantage. In many birds, early
breeding increases the average number of chicks fledged. The larger

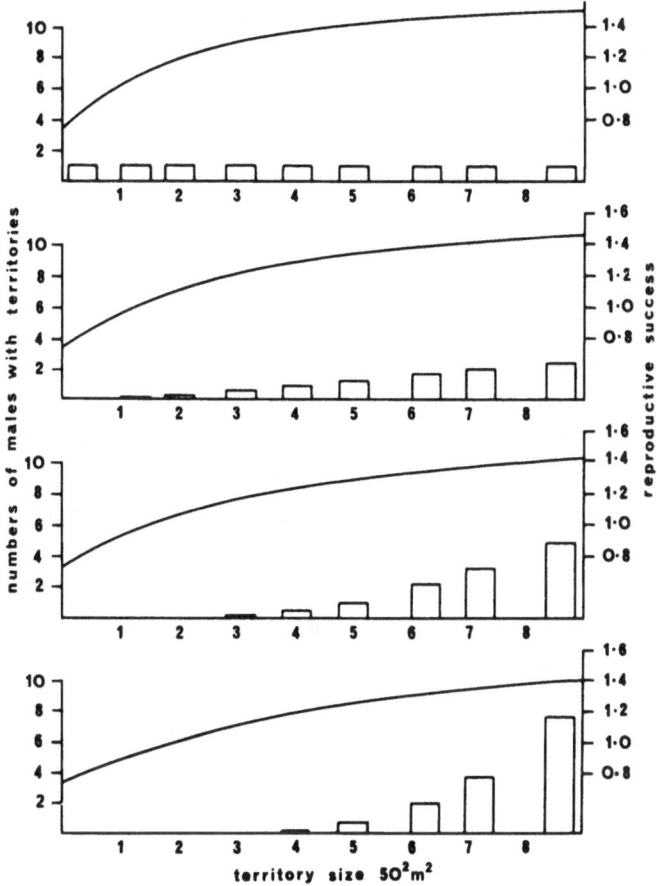

Figure 5. Fitness functions of territory size for different frequency distributions of territory size.

the territory, the earlier a male will find a mate and the greater the advantage he will gain. Some advantage will still be gained even if early breeding is just as disadvantageous as late breeding. For example, if the optimum date for breeding is the mean date and fledging success decreases with both earlier and later breeding, sexual selection still favours the evolution of larger territories since the males with the smallest territories are always the last to find mates.

Territoriality may originally have evolved for a number of reasons, including the sexual selection that results from early breeding. If

there is some random element in the females' arrivals on the territories,
then males with larger territories will always gain some advantage. If
the random element is small, the advantage must also be small. Eventually the advantage of the larger territory for finding a mate must be
counterbalanced by the disadvantage of having to expend more energy to
defend it: the advantage from sexual selection is then opposed by the
disadvantage from natural selection. If territory had evolved primarily
as an area for feeding, sexual selection might increase territory size
beyond the optimum at which the energy saved for feeding was counterbalanced by the energy spent defending the territory: solely as a place
to feed, the territory would be larger than it needed be.

In a seabird like the Arctic Skua, territory cannot be a place for
feeding. It is unlikely to be an area for concealment since predation
is negligible until the chicks start flying: by the time the chicks are
flying, the territories are no longer strongly defended and the chicks
blunder from one territory to another. The main predator is then the
Great Skua. On the island of Foula in Shetland, Great Skuas are very
common and take a considerable proportion of the fledged chicks but
they do not appear to be an important predator elsewhere. The Great Skua
itself, which shows interspecific territoriality with the Arctic Skua,
appears to have no predators at all, except for man. Both these species
nest on large areas of open moorland. Their territoriality may therefore
have evolved primarily by sexual selection.

Summary.

In the Arctic Skua, males with larger territories, who are forming new pairs, mate
earlier in the breeding season. This gives them an advantage because early breeding
increases reproductive success. Territory size will thus evolve by sexual selection.

If the females land at random on the breeding grounds, the males' chances of mating
will be proportional to the sizes of their territories. The larger his territory, the
sooner a male will find a mate. The hypothetical correlation thus produced agrees closely with the observed correlation of territory size and breeding time in the Arctic Skua.

A model is analyzed in which additive effects of genes at a number of loci partly
determine variation in territory size. Computer simulation shows that sexual selection
increases the mean territory size for many generations. Larger territories are favoured
even when the pairs breeding in the middle of the season are the most successful and
early breeding is just as disadvantageous as late breeding: the males with the smallest
territories are about the last to find mates and are always at a disadvantage compared
to the others. Sexual selection must therefore be a factor in the evolution of territoriality in birds.

REFERENCES

Crook, J.H. 1962. The adaptive significance of pair formation types in weaver birds. *Symp. Zool. Soc.* Lond. 8: 57-70.

Crook, J.H. 1964. The evolution of social organisation and visual communication in the weaver birds (*Ploceinae*). *Behav. Suppl.* 10.

Crook, J.H. 1965. The adaptive significance of avian social organizations. *Symp. Zool. Soc.* Lond. 14: 181-218.

Davis, J.W.F. and O'Donald, P. 1976a. Estimation of assortative mating preferences in the Arctic Skua. *Heredity* 36: 235-244.

Davis, J.W.F. and O'Donald, P. 1976b. Territory size, breeding time and mating preference in the Arctic Skua. *Nature* 260: 774-775.

Emlen, J.T. 1957. Display and mate selection in the Whydahs and Bishop Birds. *Ostrich* 28: 202-213.

Horn, H.S. 1968. The adaptive significance of colonial nesting in the Brewer's blackbird (*Euphagus cyanocephalus*). *Ecology* 49: 682-694.

Lack, D. 1966. Population studies of birds. Clarendon Press, Oxford.

Lack, D. 1968. Ecological adaptations for breeding in birds. Chapman and Hall, London.

O'Donald, P. 1972. Natural selection of reproductive rates and breeding times and its effect on sexual selection. *Amer. Natur.* 106: 368-379.

O'Donald, P. 1976. Mating preferences and their genetic effects in models of sexual selection for colour phases of the Arctic Skua. In *Population genetics and ecology*: 411-430. (S. Karlin and E. Nevo, eds.). Academic Press, New York.

O'Donald, P., Wedd, N.S. and Davis, J.W.F. 1974. Mating preferences and sexual selection in the Arctic Skua. *Heredity* 33: 1-16.

O'Donald, P. and Davis, J.W.F. 1975. Demography and selection in a population of Arctic Skuas. *Heredity* 35: 75-83.

O'Donald, P. and Davis, J.W.F. 1976. A demographic analysis of the components of selection in a population of Arctic Skuas. *Heredity* 36: 343-350.

Watson, A. and Moss, R. 1971. Spacing as affected by territorial behaviour, habitat and nutrition in red grouse (*Lagopus l. scoticus*). In *Behaviour and environment: the use of space by animals and men*: 92-111. (A.H. Esser, ed.). Plenum Press, New York.

Wilson, E.O. 1975. Sociobiology. The Belknap Press of Harvard University Press, Cambridge, Massachusetts.

FUNCTIONAL ASPECTS OF GENETIC VARIATION

W. Scharloo, F. R. van Dijken, A. J. W. Hoorn, G. de Jong and
G. E. W. Thörig

Genetic variation is ubiquitous. When it is looked for it is found in
all characters in outbreeding species. This is so on three levels of
observation. These three levels can be thought to form the Magic Triangle of population genetics (Figure 1). At the apex there is fitness,
F, the ultimate character in relation to the evolutionary fate of the
genes involved. Differences in fitness must be based on variation on
the molecular level, M, either directly or via variation on the phenotypic level, P, variation in physiology, morphology, or behaviour.

We know much of genetic variation in the separate angles. For fitness there is the enormous amount of work done on lethals and viability genes in *Drosophila* (Wallace 1968; Dobzhansky 1970; Mukai and Yamaguchi 1974). Molecular variation has been revealed by the ever increasing evidence on protein polymorphism (reviews Lewontin 1974; Powell
1975). On the phenotypic level success in a large number of experiments
with artificial selection on physiological, morphological and behavioural characters attest the presence of large reservoirs of genetic variation.

But the sides of our Magic Triangle, the connections between the
different levels on which genetic variation is observed are largely unknown territory. The coherence of our triangle and therefore of evolutionary genetics is still rather weak.

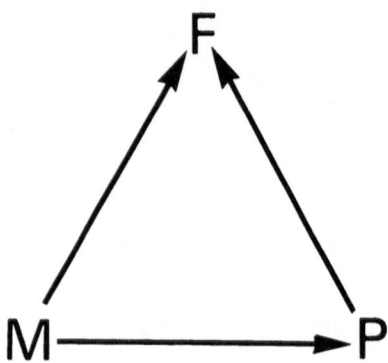

Figure 1.   The Magic Triangle of population genetics
             M: molecular variation; P: phenotypic variation; F: variation in fitness.

PHENOTYPE VARIATION AND FITNESS

The first side to be explored was the relation between phenotypic variation and fitness. Some twenty years ago it was recognised that this relation is expressed in the genetic architecture of the character involved; in the composition of its phenotypic variance (Robertson 1955). The
more remote the relation of its variation to fitness, the smaller is
the variance component representing genetic variation (Figure 2). The
length of the fourth vein of a ci^D-mutant has only a large additive genetic variance component and no variance due to genetic interaction.
Because the mutant was introduced into the background of a wild population just before the test the character had no history of natural selection (Scharloo *et al*. 1967). Abdominal bristles (Clayton and Robertson
1957) and thorax length (Robertson 1957a) are characters which are not
under simple directional selection. They show a large additive genetic
component and only a small interaction component. Ovariole number and
egg production, which are, of course, closely connected to fitness and
therefore subject to strong directional selection, have a large component due to genetic interaction (Robertson 1957a, 1957b).

A more direct approach in which the measurement of a character is
related to mortality is still older (Wheldon 1901; O'Donald 1971). In
several quantitative characters it was demonstrated that fitness de-

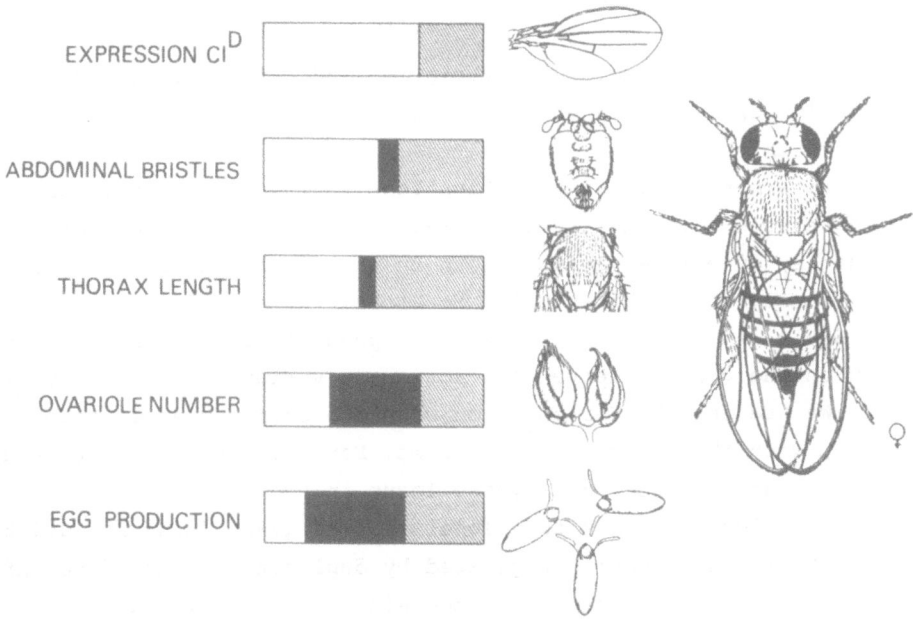

EXPRESSION CI^D

ABDOMINAL BRISTLES

THORAX LENGTH

OVARIOLE NUMBER

EGG PRODUCTION

Figure 2. The composition of the phenotypic variance in *Drosophila melanogaster* of characters which have a different connection to fitness.

creased when the deviation from the mean increases. But how this is related to effects on genes is not clear. It is characteristic of our ignorance that even for a character as sternopleural bristles in *Drosophila*, the evidence is still contradictory.

While Alan Robertson (1966) has given convincing evidence that variation in such bristle characters is selectively neutral, the Birmingham group has shown that genotypes which give extreme sternopleural bristle numbers are selected against (Kearsey and Barnes 1970). However, they showed that selection acts in the larval period, thus before the character is present.

This reveals the fundamental difficulty in this kind of research: the separation of the effects on the character itself and secondary effects via linked genes or pleiotropic effects. A large similar problem arises when we explore the relation between molecular variation and fitness.

MOLECULAR VARIATION AND FITNESS

The relation between molecular variation and fitness is a focus of discussion in population genetics and much effort is put into attempts to
elucidate this relation.

There are now several investigations in which a relation was established between some environmental factor and the frequency, or frequency changes, of enzyme variants both in laboratory and field populations
(*e.g.* Johnson and Schaffer 1973; Mc Kechnie *et al.* 1975).

However, in these cases it was not possible to distinguish between
action on the genes observed and action on variation of closely linked
genes. We are in a better position when the environmental factor has a
relation with the action of the enzyme. For this reason we started research on variation of the amylase locus in *Drosophila melanogaster*.
The amylase locus carries two closely linked genes (distance 0.008cM,
Bahn 1968) which probably originated by duplication. Therefore, after
electrophoresis homozygotes can show either one or two bands with amylase activity. In both wild and laboratory populations a large number
of electrophoretic variants are found (Doane 1967, 1969), which differ
in total amylase activity (Doane 1969).

Amylase is found predominantly in the gut and its main function
seems to be the digestion of starch in the food medium. Here, then, is
a possibility to change the environment in a way relevant to the function of the enzyme under consideration, *i.e.* by variation of the starch
content of the food medium.

First, we explored the possibilities in cage populations. We compared populations living on food media in which sucrose was the only carbohydrate, with populations living on food in which starch was an important component. These populations descended from wild populations
from widely different origin: Kaduna from Africa, Pacific from the U.S.A.
and Bogota from South America. The different amylase variants have different activities when total amylase activity is measured in vitro. Our
prediction is, then, that when starch is present the genotype with the
higher amylase activity will have an advantage, because they can use
the starch in the food medium more efficiently. From the variants present in our cage populations $Amy^{4,6}$ shows the highest amylase activity
and Amy^{1} the lowest activity (Doane 1967; de Jong and Scharloo 1976).
In agreement with our prediction in all four pairs of cages the active

$Amy^{4,6}$ variant has a higher frequency on the food medium with starch than on the food medium with sucrose as the only source of carbohydrate. This could be inferred from a survey of amylase phenotypes (De Jong, Hoorn, Thörig and Scharloo 1972) and was confirmed when gene frequencies were determined by crossing individual flies to a marked stock with known amylase type (Figure 3). We decided to concentrate on showing di-

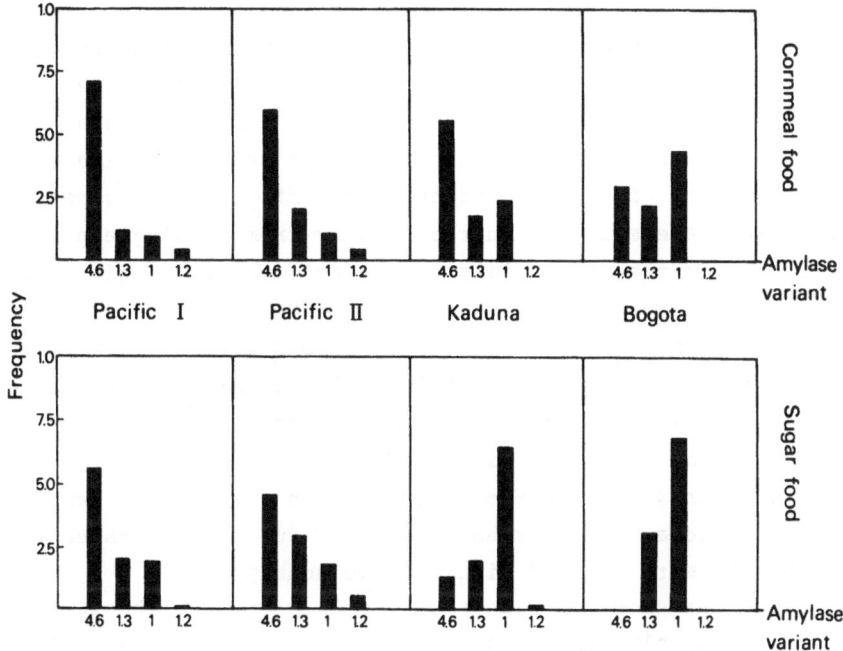

Figure 3. Frequencies of amylase variants in four pairs of populations on food medium in which starch is present and in which sugar is the only source of carbohydrate.

The Pacific I, Kaduna and Bogota populations were on different media for more than two years before the test.

The Pacific II populations were separated only 9 months before the test. In each population at least 200 chromosomes were tested.

rectly that natural selection can perceive the difference in activity between $Amy^{4,6}$ and Amy^{1} under appropriate circumstances. First we showed that the amylase substrate starch and its product maltose really contribute to survival and growth of *Drosophila melanogaster*. If this were not true, selection via a difference in digestive action of amylase variants could not be expected. This was of particular importance because

Sang's figures (Sang 1956) suggest that maltose is not used by *Drosophi-la* larvae and starch is not very effective either.

Our experiments with addition of starch and maltose show that both substances contribute to survival and growth. This was tested by addition to a sterile yeast medium (De Jong and Scharloo 1976) and by addition to chemical defined sterile media. But this effect is restricted to media with low concentrations of yeast and casein, respectively. Then, amounts of starch and maltose containing the same number of glucose residues were equivalent in their contribution to survival and growth.

Experiments were started in which a stock homozygous for $Amy^{4,6}$ was compared with a stock homozygous for Amy^{1}. These stocks were isolated from our Kaduna cage population. A large number of single pair cultures were made and after the production of offspring the genotype of the parents was determined. Then the progeny from 10 single pair cultures homozygous for the appropriate type were combined to give the stocks used in the experiments.

After some trial and error the stocks were compared under the following conditions. Culture vials (2.5 × 8 cm) which contained a bottom layer of 5 ml agar (2%) preventing desiccation of the food and a food layer of 2 ml agar (1%) with dead yeast and starch or maltose were stocked with 40 larvae from eggs laid in a four hour period and transferred immediately after hatching. All cultures compared in the same experiment were started within a two hour period.

Food was provided with four levels of yeast. The lowest yeast level was scarcely sufficient to sustain survival of any of the larvae, but the highest yeast level supported survival of the majority of the larvae. These four yeast levels were each combined with four levels of starch. Survival patterns for the lowest and the highest yeast levels show a striking difference; at the two highest levels, starch addition does not contribute to survival and at the highest level there is even a decline (Figure 4). But at the two lowest yeast levels starch addition contributes considerably to survival.

At the three lowest yeast levels there is no difference between the $Amy^{4,6}$ and the Amy^{1} stocks. At the highest level there is a difference but because it is present at all starch additions it cannot be attributed to a difference in starch digestion between the $Amy^{4,6}$ and Amy^{1} stocks. It must be a consequence of the genetic background and/or a difference in the non-digestive function of the *Amy* locus.

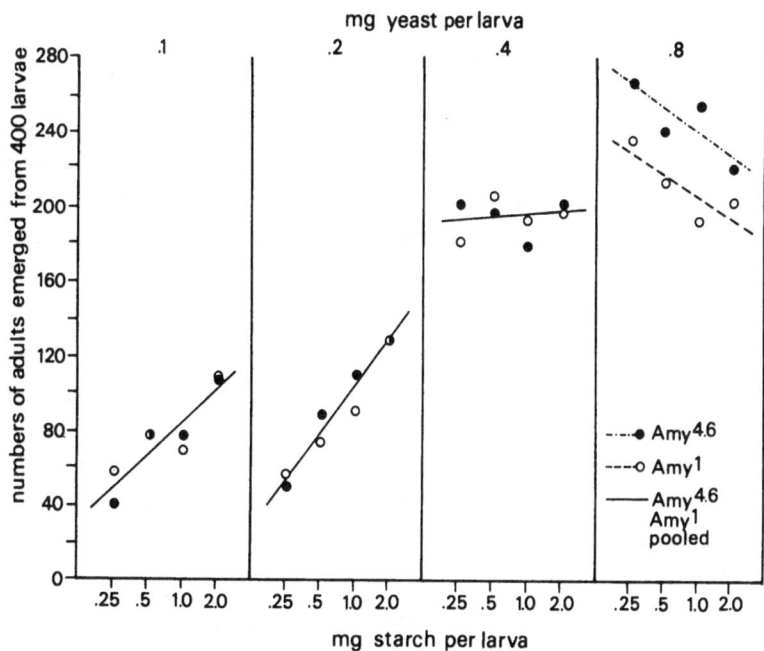

Figure 4. Numbers of adults hatching from pure cultures on different levels of yeast and starch, *Amy^1* o *Amy^{4,6}* ●. The regression of numbers of adults on starch level is significant (P < 0.01) at the two lowest yeast levels. Each point is based on 10 replicates each with 40 larvae.

In pure culture at the three lowest yeast levels there is no difference in survival and weight between the two stocks. This does not necessarily mean that there are no functional differences between *Amy^1* and the *Amy^{4,6}* allozymes, and that we must therefore admit that this is a case of selective neutrality for electrophoretic variants.

Performance in pure cultures does not predict performance in mixed cultures as was shown in the brillant experiments of Bakker (1961, 1969) on competition between Bar and wildtype in *Drosophila*. Therefore competition experiments were started with 20 *Amy^{4,6}* and 20 *Amy^1* larvae under the same conditions as in the pure culture experiments. Before dealing with the results we would like to emphasize the features of the system that permit the conclusion that selection is acting and moreover, that it acts on the gene locus under consideration. These features are 1.

The substrate of the enzyme is known: starch 2. the product of the enzyme is known: maltose 3. both substances contribute to survival and growth 4. the function - or at least an important part of the function - of the enzyme is digestion of starch 5. there is a prediction which variant will be selected for: under conditions where starch contributes to survival the most active Amylase variant will be favoured. Only at the two lowest yeast levels does addition of starch contribute to survival. Therefore, only under these conditions there is scope for viability selection between $Amy^{4,6}$ and Amy^1 with respect to a difference in digestive activity.

In Figure 5 it is shown that at the lowest yeast levels the frequency of the variant with the highest amylase activity $(Amy^{4,6})$ increases, when starch addition increases. Is it then justified to conclude that selection is favouring the $Amy^{4,6}$ allele because its relative survival is increased by addition of starch which enables the superior digestive activity of its product? The $Amy^{4,6}$ stock has a large disadvantage com-

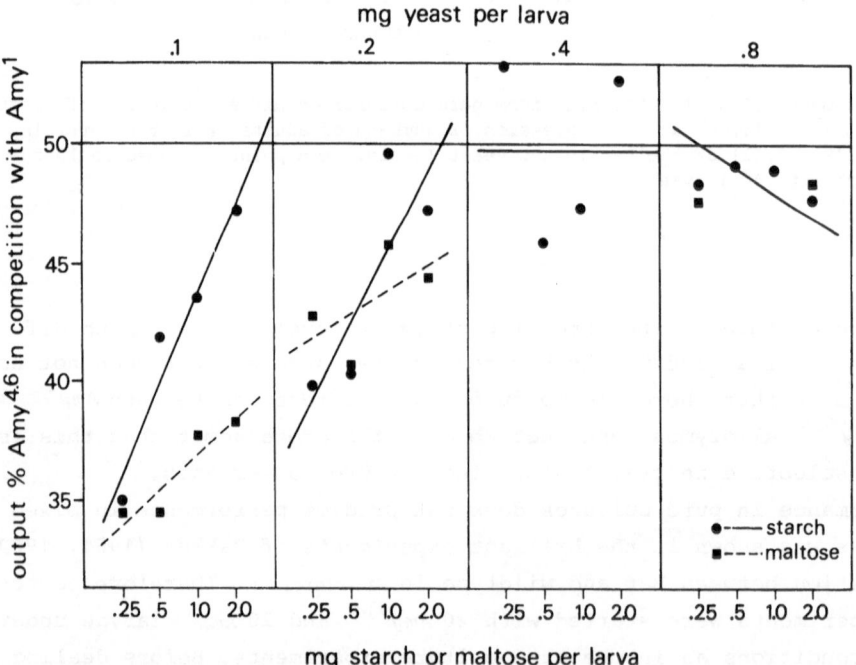

Figure 5. Percentage $Amy^{4,6}$ hatching from competition cultures with equal frequencies of $Amy^{4,6}$ and Amy^1. The difference in slope between the regression of percentage $Amy^{4,6}$ on starch level and on maltose level is significant (P < 0.05) at 0.1 mg yeast per larvae. At the two lower yeast levels each point is based on 30 replicates, at the two higher yeast levels 20 replicates for the starch media and 6 for the maltose media.

pared to the Amy^1 stock both at low starch and low maltose additions. This must be a consequence of differences in the genetic background of the two stocks and/or non-digestive functions of the *Amy* genes. It is often found that viability differences are expressed only under stress conditions. Therefore, the relative improvement of the $Amy^{4,6}$ stock could be a consequence of the general improvement of food conditions by adding starch. However, addition of maltose should give an equal general improvement as addition of starch and cause the same relative improvement of $Amy^{4,6}$ viability. Maltose addition seems to give some improvement but regression of survival (transformed to arcsine values) on starch addition is significantly higher than the regression of survival on maltose addition at the lowest yeast level.

This difference in regression represents the specific action of the difference between the two Amylase genes in the digestion of starch. It is independent of the genetic background and of the non-digestive function of the amylase genes. The comparison between regression of survival on addition of substrate and regression of survival on addition of equivalent amounts of product is essential for the conclusion that selection acts on the locus under consideration.

Our results show that the functional differences are only revealed under special conditions, *i.e.*, when addition of starch contributes to survival. But under these conditions the one-locus selection coefficient can be quite large; there is a difference in relative viability of $Amy^{4,6}$ on starch and on maltose of approximately 30%. At the higher yeast levels the difference on the *Amy* locus are selectively neutral with respect to viability selection. We have shown that selection acts on electrophoretic variants of the amylase locus. The only other case in which there is sufficient evidence that selection is acting on electrophoretic variants is the alcohol dehydrogenase locus in *Drosophila melanogaster*. There, high ethanol concentrations favour the fast allele which has a higher ADH-activity than the slow allele. Although it is not possible here to compare performance of Fast and Slow variants on food with different concentrations of both the substrate and the product of this enzyme, the consistent results of several authors with stocks of widely different origin make it very improbable that other variable closely linked loci are involved here (Gibson 1970; Van Delden *et al.* 1973; Bijlsma-Meelis and Van Delden 1974; Clarke 1975).

Neither in amylase, nor in ADH do we understand how the variants are
maintained in populations. Probably the best guess is at present that
they are maintained by selection in a heterogeneous and fluctuating
environment.

HIDDEN VARIABILITY

During this symposium several speakers mentioned the possible presence
of hidden alleles with identical electrophoretic mobility, but separable
by other properties, *e.g.* heat resistance. Differences in heat resistance were described by several authors (Bernstein *et al.* 1973; Sing *et al.*
1974). At best, it was shown that the difference is genetic because it
segregates as a Mendelian factor. This does not prove, however, that
the difference revealed by heat treatment is caused by a difference on
the enzyme locus. It could be caused by a genetic difference elsewhere
in the genome.

The only hidden allele which has been localised by appropriate genetic methods is the Adh^{71k} allele (Thörig *et al.* 1975). It has the
same electrophoretic pattern as the F-allele (Figure 6) but it is far
more resistant to heat inactivation (Figure 7). The Adh^{71k} enzyme is
not inhibited by high ethanol concentrations when the NAD concentration
is low. These properties are revealed both after electrophoresis and
with crude extracts in the spectrophotometer.

With the help of outside markers (*b* 48.5, *Adh* 50.1, *pr* 54.5) the
differences in biochemical properties could be localised on the *Adh* locus itself. One could perhaps expect that hidden alleles would represent selective neutral variation, while electrophoretic variation would
cause more easily functional changes. Yet the difference in properties
in vitro between the Adh^{71k} enzyme and the Adh^F enzyme are as great as
between the Adh^F and Adh^S enzymes. With the purpose to test whether selection can perceive these differences an experiment was started on food
medium with different concentrations of ethanol. The base population
was an F_2 from a cross between an Adh^{71k} stock and an Adh^F stock. At
each concentration there were ten bottle cultures. It is shown in Table
1 that the frequency of *FF* homozygotes did not change when no ethanol
is present, but that *F* decreases in frequency when ethanol is present.

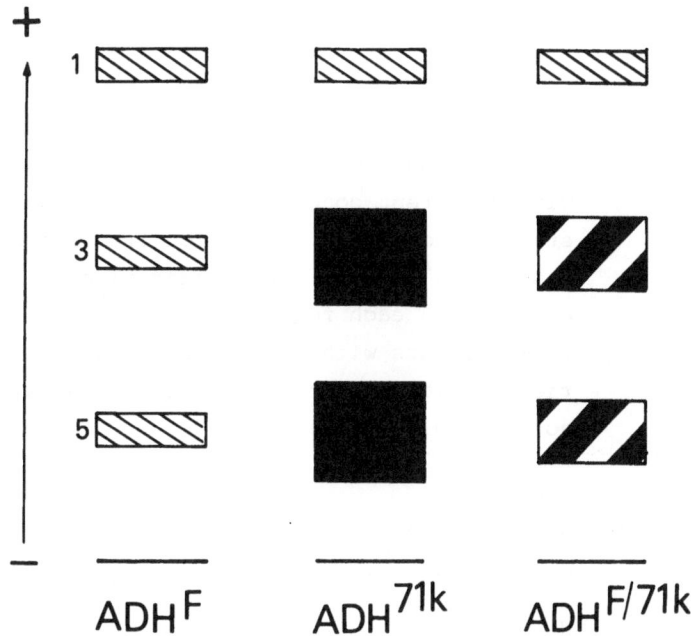

Figure 6. The electrophoretic pattern of the major bands of Adh^F, Adh^{71k} and their heterozygote at higher ethanol concentrations in the incubation medium and high temperature. In practice, there is overlap between homozygous and heterozygous Adh^{71k}.

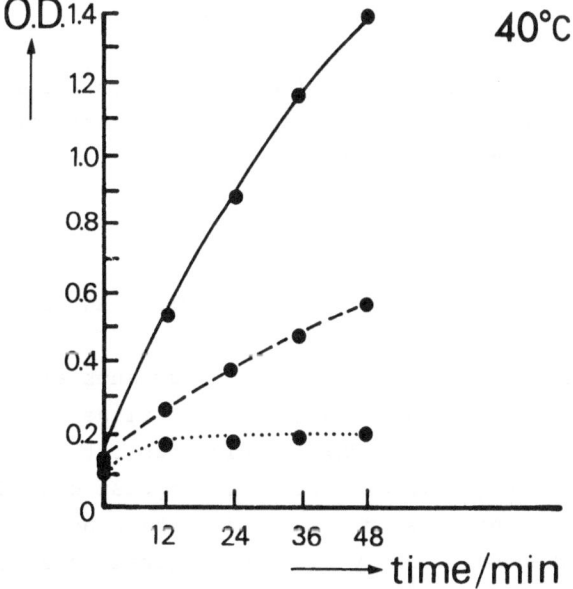

Figure 7. Activity of the different Adh allozymes at high temperature measured on a crude extract. Ordinate: activity in optical density units, abscissae time in minutes. Full line: Adh^{71k}, broken line Adh^S, dotted line Adh^F.

Table 1.　Selection on the hidden alleles Adh^{71k} and Adh^F by food media
with different concentrations of ethanol. The selection was
started from an F_2 generation from a cross between an Adh^{71k}
and an Adh^F stock. Every generation there were 10 bottle
cultures on each food medium. The flies were transferred to
fresh bottles with intervals of 14 days. In the test generations between 40 and 50 flies were assayed for Adh^F homozygotes from each culture.

Ethanol concentration	0%	4%	8%	6%
Test generation	28	35	32	20
N	445	482	473	405
FF homozygotes	114	81	74	48
% *FF* homozygotes	26	17	16	12

At the highest ethanol concentration the selection seems to be quite
strong.These hidden alleles can be affected by selection as well as electrophoretic alleles.

INTERACTION AND SELECTION

One would expect when enzyme variants are subject to selection by environmental conditions that their fitness would also be dependent on other
genetic factors. This problem has often been approached by investigation
of linkage disequilibrium in populations. These attempts have been generally unsuccessful. This is perhaps not surprising because the loci involved have mostly no physiological relation.
　The Adh^{71k} allele was found in a $Notch^8$ stock. $Notch^8$ is a large
deficiency in the X chromosome, which when homozygous is lethal. In heterozygous ♀♀ it has various morphological effects of which the most

prominent are the incisions of the wing border. On a biochemical level, Notch[8] affects the NADH-dehydrogenase activity. This mitochondrial enzyme is involved in the transfer of H from NADH to Coenzyme Q. Wildtype ♂♂ show a far lower activity than ♀♀ and Notch[8] heterozygotes have only marginally higher activities than the ♂♂ (Table 2). Introduction of a

Table 1.   Measurement of NADH-dehydrogenase activity in wildtype and Notch[8] ♀♀ and wildtype ♂♂ with two different genetic backgrounds. The rate of conversion of NADH to NAD is expressed as extinction values at 340 mM. The differences between experiments are at least partly a consequence of age effects.

Genetic background	+ ♂	N/+ ♀	+/+ ♀
HD-6	30	29.5	48
	28	35	74
	43	68	185
	60	74.5	120
LD-2	39	48	83
	67	65	156
	56	70.5	174

duplication of the Notch[8] region elsewhere in the genome compensates for the presence of the Notch[8] deficiency. Because ADH is NAD-dependent, interaction can be expected here. This could perhaps explain the fixation of Adh[71k] in the Notch[8] stock. The possibility of a physiological interaction between the Adh locus and the Notch locus resulting in a change of the selective values of Adh alleles is now under investigation. Alleles of the Adh[71k] type were found in low frequency in several wild populations.

With the relation between Notch, NADH-dehydrogenase and its phenotypic effects we are exploring the third side of our triangle, the relation between molecular differences and phenotypic effects.

MOLECULAR AND PHENOTYPIC VARIATION

This relation is touched upon again in our selection experiments on locomotory activity in *Drosophila*. Selection is performed in an apparatus
consisting of a row of twenty tubular compartments (30 mm × 70 mm) ending
in funnels with a passage of 3 mm. Samples of 100 flies are transferred
without narcosis to the first compartment. The flies pass almost exclusively in one direction. Activity of individual flies is determined after a suitable running time as the number of the compartment they have
reached.

Divergent directional selection from two different base populations
resulted in lines with widely different locomotor activity. They also
differ in NADH-dehydrogenase activity. The lines with high locomotor
activity show a higher NADH-dehydrogenase activity than lines with low
locomotor activity. This is so in three pairs of independently selected
lines. There are strong indications that these are primary genetic effects and not secondary effects of the difference in locomotor activity.
Localization of the difference in locomotor activity shows that in all
three pairs of high and low lines the X-chromosome has by far the largest effect. A more precise localization on the X-chromosome is now being performed and it remains to be seen whether the *Notch* locus is involved.

A difference in biochemical properties between the NADH-dehydrogenase of the high and low lines seems to map within the limits of the
$Notch^8$ deficiency, but this needs further conformation. It is interesting
in view of our ideas on interaction between $Notch^8$ and *Adh* alleles that
there are appreciable differences in the frequency of Adh^F and Adh^S between the lines with high and low locomotor activity, and that α-glyce-
rolphosphate dehydrogenase (α-GPD) is also affected. Both ADH and α-GPD
are NAD dependent and α-GPD has an important function in NADH-metabolism.
This suggests how frequencies of allozymes can be dependent on the situation at other gene loci.

CONCLUSION

Understanding of the forces which govern genetic variability, its
organisation, and its phenotypic expression, can only be based on an
approach which combines physiological, developmental and ecological

aspects. Then, perhaps the Magic Triangle will obtain some coherence, and can we go beyond the "aridities of Neo-Darwinism algebra" (Waddington 1969) and design a real biological evolutionary theory dealing with organisms which do not consist of genes alone, but have phenotypes with internal organisation and adaptations to the outer world.

REFERENCES

Bahn, E. 1968. Crossing over in the chromosomal region determining amylase isozymes in *Drosophila melanogaster*.
Hereditas 58: 1-12.

Bakker, K. 1961. An analysis of factors which determine success in competition for food among larvae of *Drosophila melanogaster*.
Arch. Neerl. Zool. 14: 201-281.

Bakker, K. 1969. Selection for the rate of growth and its influence on competitive ability of larvae of *Drosophila melanogaster*.
Neth. J. Zool. 19: 541-595.

Bernstein, S.C., L.H. Throckmorton and J.L. Hubby. 1973. Still more genetic variability in natural populations.
Proc. Nat. Acad. Sc. U.S.A. 70: 3928-3931.

Bijlsma-Meeles, E. and W. van Delden. 1974. Intra- and interpopulation selection concerning the alcohol dehydrogenase locus in *Drosophila melanogaster*.
Nature 247: 369-371.

Clayton, G.A. and A. Robertson. 1957. An experimental check on quantitative genetical theory. II. The long term effects of selection.
J. Genet. 55: 152-170.

Clarke, B. 1975. The contribution of ecological genetics to evolutionary theory: detecting the direct effects of natural selection on particular polymorphic loci.
Genetics 79: 101-113.

Delden, W. van, A. Kamping and H. van Dijk. 1973. Selection at the Alcoholdehydrogenase locus in *Drosophila melanogaster*.
Experientia 31: 418-419.

Doane, W.W. 1969. *Drosophila* amylases and problems in cellular differentiation (in R.W. Hanley, ed.).
RNA in development: 73-108.

Doane, W.W. 1969. Amylase variants in *Drosophila melanogaster*: linkage studies and characterization of enzyme extracts.
J. Exp. Zool. 171: 321-342.

Doane, W.W. 1967. Quantitation of amylases in *Drosophila melanogaster*: linkage studies and characterization of enzyme extracts.
J. Exp. Zool. 164: 363-378.

Dobzhansky, Th. 1970. Genetics of the evolutionary process.
Columbia University Press, New York and London.

Gibson, J.B. 1970. Enzyme flexibility in *Drosophila melanogaster*.
Nature 227: 959.

Johnson, F.M. and H.E. Schaffer. 1973. Isozyme variability in species of the genus *Drosophila* VII. Genotype environment relationships in populations of *Drosophila melanogaster* from the Eastern United States.
Biochem. Genet. 10: 149.

Jong, G. de, A.J.W. Hoorn, G.E.W. Thörig and W. Scharloo. 1972. Frequencies of amylase variants in *Drosophila melanogaster*.
Nature 238: 453-454.

Jong, G. de and W. Scharloo. 1976. Environmental determination of selective significance or neutrality of amylase variants in *Drosophila melanogaster*.
Genetics 84: 77-94.

Kearsey, M.J. and B.W. Barnes. 1970. Variation for metrical characters in *Drosophila* populations: II Natural selection.
Heredity 25: 11-21.

Mc Kechnie, S.W., P.R. Ehrlich and R.R. White. 1975. Population genetics of *Euphydras* butterflies I. Genetic variation and the neutrality hypothesis.
Genetics 81: 571-594.

Lewontin, R.C. 1974. The genetic basis of evolutionary change.
Columbia University Press, New York and London.

Mukai, T. and O. Yamaguchi. 1974. The genetic structure of natural populations of *Drosophila melanogaster*. XI. Genetic variability in a local population.
Genetics 76: 339-366.

O'Donald, P. 1971. Natural selection for quantitative characters.
Heredity 27: 137-153.

Powell, J.R. 1975. Protein variation in natural populations of animals.
Evolutionary Biology 8: 79-119.

Robertson, A. 1955. Selection in animals: synthesis.
Cold. Harb. Symp. quant. Biol. 20: 225-229.

Robertson, A. 1966. Artificial selection in plants and animals.
Proc. Roy. Soc. Series B. 164: 341-349.

Robertson, F.W. 1957. Studies in quantitative inheritance. XI. Genetic and environmental correlations between bodysize and eggproduction in *Drosophila melanogaster*.
J. Genet. 55: 428-432.

Robertson, F.W. 1957. Studies in quantitative inheritance. X. Genetic variation of ovary size in *Drosophila*.
J. Genet. 55: 410-427.

Sang, J.H. 1956. The quantitative nutritional requirements of *Drosophila melanogaster*.
J. Exp. Biol. 33: 45-72.

Scharloo, W., M.S. Hoogmoed and A. ter Kuile. 1967. Disruptive and stabilizing selection on a mutant character. I. The phenotypic variance and its components.
Genetics 56: 709-726.

Singh, R.S., J.L. Hubby and R.C. Lewontin. 1974. Molecular heterosis for heatsensitive enzyme alleles.
Proc. Nat. Acad. Sc. U.S.A. 71: 1808-1810.

Thörig, G.E.W., A.A. Schoone and W. Scharloo. 1975. Variation between electrophoretically identical alleles at the alcohol dehydrogenase locus in *Drosophila melanogaster*.
Biochem. Genet. 13: 721-731.

Waddington, C.H. 1972. Epilogue in: Towards a theoretical biology. 4. Essays. (C.H. Waddington, ed.). I.U.B.S. Symposium: 283-289.

Wallace, B.   1968.   Topics in population genetics.
    Northon, New York.

Wheldon, W.F.R.   1901.   A first study of natural selection in *Clausilia laminata*.
    Biometrika 1: 109.

Wallace, A. (Ed.) ... in population genetics ...
Lexington, New York.

Watson, W.T.R. (1973) ... final study of data ... Roc ... Mixed ... in computer.
Biometrika 39, 90.

1. STUDY OF SELECTION

ON CONDITIONAL INFERENCE FOR DEVIATION FROM HARDY-WEINBERG DISTRIBUTION

Ole Barndorff-Nielsen

In investigations of questions of possible deviations from Hardy-Weinberg distribution it is sometimes proposed to reason conditionally on the observed gene frequencies, thus disregarding the stochastic nature of these frequencies. (See Levene (1949)). The problem is thereby raised of whether the disregard implies any loss of information (or evidence) as to the particular question of deviation. The present essay treats this problem, in the simplest case, that of two alleles and a single sample. The interest in discussing this case is, of course, mainly theoretical, particularly in view of the wellknown insensitivity of the χ^2-test for Hardy-Weinberg distribution (Lewontin and Cockerham (1959), Ward and Sing (1970)) and difficulties in interpreting the test outcome genetically (Frydenberg (1956), Wallace (1958)).

For a random sample of n from a population in which the proportions of the genotypes AA, Aa and aa are p_1, p_2 and p_3, the probability of observing x_1, x_2 and x_3 individuals of genotypes AA, Aa and aa, respectively, is given by

$$\binom{n}{x_1\ x_2\ x_3} p_1^{x_1}p_2^{x_2}p_3^{x_3} , \tag{1}$$

where

$$\binom{n}{x_1 \, x_2 \, x_3} = \frac{n!}{x_1! \; x_2! \; x_3!} \quad .$$

The genotype probabilities p_1, p_2, p_3 will, throughout, be assumed to be positive.

It is convenient to introduce the quantity

$$\kappa = (1/2) \ln[\, p_2^2 \, / \, (4p_1 p_3)\,]$$

as a parameter expressing the degree of deviation from Hardy-Weinberg distribution. The hypothesis of Hardy-Weinberg distribution is equivalent to $\kappa = 0$, and κ is positive or negative according as there is an excess or deficiency of heterozygotes in the population. Furthermore, if the population sampled has arisen through selection from a population of zygotes in Hardy-Weinberg distribution with gene frequencies p for A and q for a, so that

$$p_1 = (w_1/\bar{w}) p^2, \quad p_2 = (w_2/\bar{w}) 2pq, \quad p_3 = (w_3/\bar{w}) q^2$$

where w_1, w_2, w_3 denote the selection coefficients and $\bar{w} = w_1 p^2 + w_2 2pq + w_3 q^2$, then

$$\kappa = (1/2) \ln[\, w_2^2 \, / \, (4 w_1 w_3)\,] \quad .$$

Finally, it should be noted that, letting z denote the observed number of A genes, i.e. $z = 2x_1 + x_2$, and writing ζ for the mean value of z, the pair (ζ, κ) is in one-to-one correspondence with (p_1, p_2, p_3), and the conditional probability of the sample given z depends on (ζ, κ) through κ only and has the form

$$\left[\binom{n}{x_1 \, x_2 \, x_3} 2^{x_2} e^{\kappa x_2}\right] \Big/ \left[\sum_{\substack{2x_1 + x_2 \\ = z}} \binom{n}{x_1 \, x_2 \, x_3} 2^{x_2} e^{\kappa x_2}\right] \qquad (2)$$

which, incidentally, for $\kappa = 0$ reduces to

$$\binom{n}{x_1 \, x_2 \, x_3} 2^{x_2} \Big/ \binom{2n}{2} \quad .$$

The expression (2) equals, of course, the conditional probability of x_2 given z.

The question of deviation from Hardy-Weinberg distribution may thus be formulated as that of drawing inference on the parameter κ. However, in case the issue is solely whether the hypothesis of Hardy-Weinberg distribution is fulfilled or not, the parameter of interest is not κ but

$$\psi = \begin{cases} 0 & \text{if } \kappa = 0 \\ \\ 1 & \text{if } \kappa \neq 0 . \end{cases}$$

In more precise terms, then, the problem to be discussed here is whether the conjunction of x_2 and its conditional distribution family given the observed value of z contains all the information about κ or ψ, as the case may be, which is accessible in the conjunction consisting of $x = (x_1, x_2, x_3)$ and its distribution family. In still other words, whether z is ancillary* with respect to κ or to ψ.

First suppose the parameter of interest is κ. Then z is ancillary if and only if z is nonformative with respect to κ, i.e. if the conjunction of z and its marginal distribution family contains no (accessible) information on κ. But, as pointed out in Barndorff-Nielsen (1973), observation of an odd value of z yields the information that κ cannot be negative and extremely large in numerical value because, with $p(z; \zeta, \kappa)$ denoting the probability of z and with z odd, $p(z; \zeta, \kappa) \rightarrow 0$ as $\kappa \rightarrow -\infty$, uniformly in ζ. Thus the random variate z is not ancillary with respect to κ. The possibility remains, however, that even values of z can be said to give no information on κ. More specifically, it seems likely that z can be shown to be pointwise M-nonformative - and hence M-ancillary - with respect to κ, at every even value of z. What is required for this is a verification that for every s in $\{0, 1, \ldots, n\}$ and every κ there exists a ζ such that $p(z; \zeta, \kappa)$ has 2s as a mode point.

---

* The concept of ancillarity originated with R.A. Fisher. A concomitant notion, referred to in the sequel, is that of nonformation. For discussions of the contents of these ideas and of vairous mathematical definitions of nonformation and ancillarity, in particular those of (pointwise) M-nonformation and M-ancillarity mentioned subsequently, see Barndorff-Nielsen (1976 b) and references given therein.

It is illuminating to have the following picture in mind. A plot of the possible values of (z, x_2) looks like Figure 1. If κ tends to $-\infty$, respectively $+\infty$, the total probability mass of 1, which is distributed over these values, "flows" downwards, respectively upwards, and is in the limit concentrated on the values on the base line of the triangle indicated in Figure 1, respectively the other two sides of that triangle.

From this picture it is obvious that odd and even values of z have a very different status in the present, inferential context.

Considering, next, ψ as the interest parameter, it may be noted that z together with its marginal distribution family does not, by itself, afford information on ψ, in the sense that z is M-nonformative with respect to ψ, as is simple to see. However, this does not necessarily imply that the conditional situation given the observed value of z contains all the evidence on ψ accessible in the original situation (consisting of the conjunction of the observed x and its distribution family). The conditional situation yields, in general, information not only on ψ but on κ, and it is conceivable that the extra information could be used, as it were, to release some latent information on ψ from the z-marginal situation. No established methods exist for deciding this and like questions. The kind of reasoning proposed by Kalbfleisch and Sprott (1973) could, however, be adapted to treat such questions from

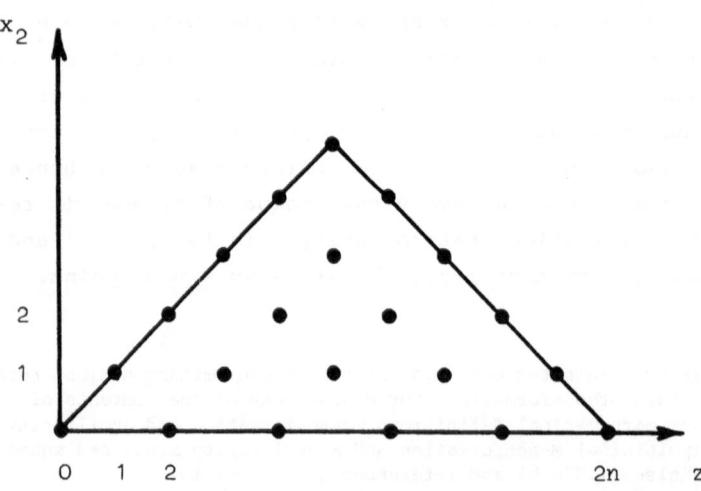

Figure 1.   The possible values of (z, x_2), marked by $\cdot$. (Here n = 5).

a likelihood viewpoint. In the following discourse the approach will
be via plausibility inference*.

Searching for an instance where not all of the information on ψ is
contained in the conditional situation, it is, in view of the above,
inviting to focus on cases where z is odd and the conditional situation
strongly indicates that κ is considerably smaller than 0. This, rather
uniquely, points to taking n odd, $z = n$ and $x_2 = 1$.

The plausibility functions for the marginal and conditional situations provide chartings of the evidence in these situations. By definition, the plausibility functions are, respectively,

$$\Pi(\zeta,\kappa;z) = p(z;\zeta,\kappa)/\sup_{z} p(z;\zeta,\kappa) \tag{3}$$

and

$$\Pi(\kappa;x_2|z) = p(x_2;\kappa|z)/\sup_{x_2|z} p(x_2;\kappa|z) \tag{4}$$

where $p(x_2;\kappa|z)$ denotes the conditional probability (2), of x_2 given z.

With $x_2 = 1$, the conditional plausibility function (4) is 1 for κ
less than or equal to a certain κ_n, to the right of which the function
decreases monotonically, towards 0. This κ_n is the largest value of κ
such that the conditional distribution of x_2 given z (= n) has mode at
$x_2 = 1$, and it is determined by the equation

$$\exp(2\kappa_n) = 6(n-1)^{-2}.$$

On the other hand, consider the partially maximised plausibility function

$$\tilde{\Pi}(\kappa;z) = \sup_{\zeta|\kappa} \Pi(\zeta,\kappa;z)$$

derived from (3). It seems certain, although a strict proof is not a-
vailable at present, that there exists a $\tilde{\kappa}_n < 0$ such that $\tilde{\Pi}(\kappa;n)$ in-

---

* Plausibility theory (Barndorff-Nielsen, 1976 a) is a scheme for inference, in many
respects similar to likelihood theory but also differing significantly from the latter. With data t, parameter ω and probability function $p(t;\omega)$, the role of the likelihood function $L(\omega) = L(\omega;t) = p(t;\omega)$ is in plausibility inference taken over by
the plausibility function $\Pi(\omega) = \Pi(\omega;t) = p(t;\omega)/\sup_t p(t;\omega)$. (M-nonformation and
-ancillarity have a natural place in plausibility theory.)

creases monotocally from 0 to 1 over the interval $(-\infty, \tilde{\kappa})$, while
$\tilde{\Pi}(\kappa;n) = 1$ for $\kappa \geq \tilde{\kappa}_n$. A sketch of $\Pi(\kappa;1|n)$ and of two hypothetical
versions of $\tilde{\Pi}(\kappa;n)$, differing only in location, is shown in Figure 2.
In case $\tilde{\Pi}(\kappa;n)$ is as suggested by the upper version, the inference on
whether κ is 0 or not, i.e. on ψ, which is possible from the conditional situation by itself, must be strongly affected by the known random
variation of z, whereas with the lower version the indication would be
that all the accessible evidence on ψ is provided by the conditional
situation. Thus it appears that, on the plausibility viewpoint, the
question of loss of information on ψ, by conditioning on z, is decided
by whether $\tilde{\Pi}(\kappa;n)$ is equal to 1 or less than 1.

To find a value of n such that $\tilde{\Pi}(\kappa_n;n) < 1$ it is sufficient (and
may well be necessary) to find an n and a ζ_n for which

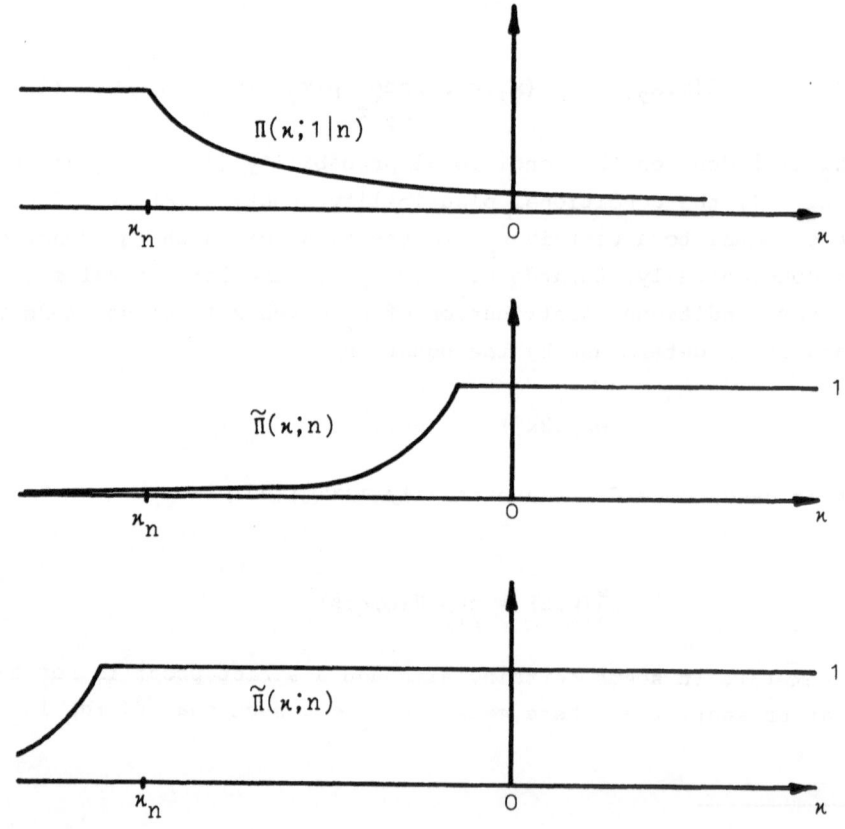

Figure 2. A sketch of the conditional plausibility function $\Pi(\kappa;1|n)$ and of two hy-
pothetical versions of the partially maximised marginal plausibility function $\tilde{\Pi}(\kappa;n)$.

$$p(n-1;\zeta_n,\kappa_n) > p(n;\zeta_n,\kappa_n) < p(n+1;\zeta_n,\kappa_n).$$

Such n and ζ_n do, in fact, exist, the smallest n possible being 43. A list of illustrative values is given in Table 1, where p_{n1} is the p_1 value corresponding to (ζ_n,κ_n). It is only of interest to consider moderate values of n since it is well-known from general results that the amount of information lost by the conditioning becomes negligible as $n \to \infty$. That the amount is never substantial is brought out by Figures 3 and 4. It follows, in particular, from Table 1 and Figure 4 that for n = 91 one has $\tilde{\Pi}(\kappa_n;n) \geq p(n;\zeta_n,\kappa_n)/p(n+1;\zeta_n,\kappa_n) \geq .989$.

Summary.

   A discussion is given of whether conditioning on observed gene frequencies, as a first step in making inference regarding possible deviations from Hardy-Weinberg distribution, implies a loss of statistical evidence. The question is treated from the viewpoint of plausibility inference.

Acknowledgements.

   I am grateful to Morten Frydenberg, Hanne Østergaard Kristensen and Arno Jensen for assistance with the numerial and graphical work.

Table 1.

n	P_{n1}	$10^2 \cdot p(z;\zeta_n,\kappa_n)$		
		z=n-1	z=n	z=n+1
43	.4725	6.1866	6.1848	6.1880
61	.481	5.1739	5.1559	5.1871
91	.488	4.2204	4.1969	4.2420
151	.49	3.2890	3.2405	3.2628
161	.49	3.1853	3.1345	3.1537

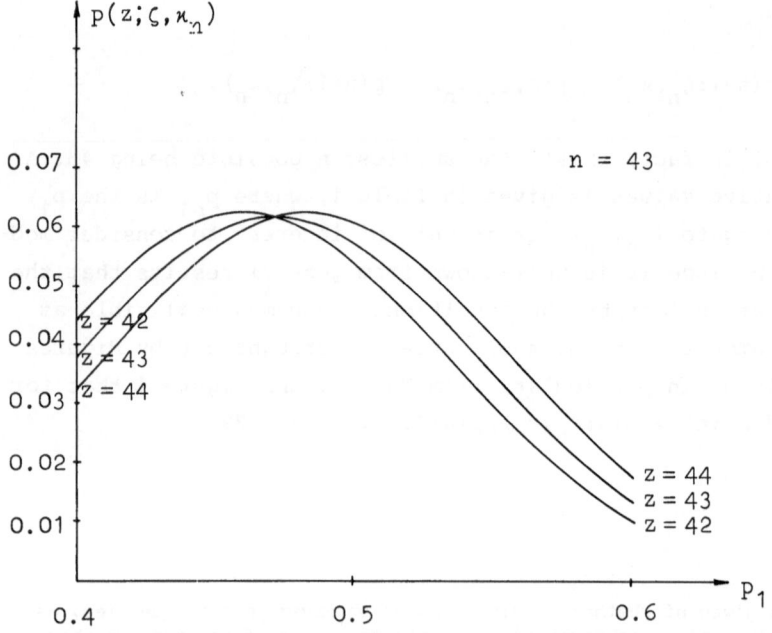

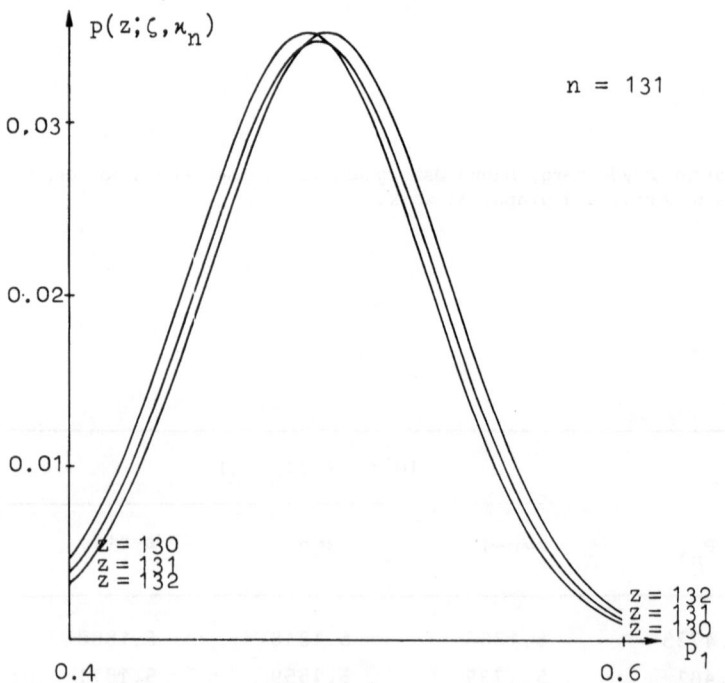

Figure 3. The graph of $p(z;\zeta,\kappa_n)$ considered as a function of p_1, for $n = 43$ (upper diagram) and $n = 131$ (lower diagram), and $z = n-1$, n and $n+1$.

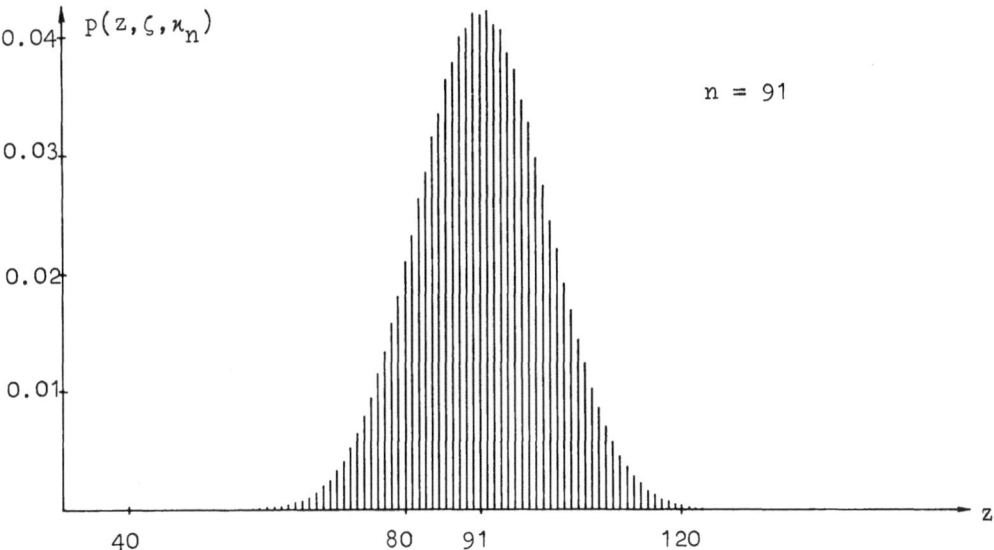

Figure 4.   The marginal distribution of z for n = 91, $\kappa = \kappa_n$ and $p_1 = p_{n1}$ = .488.

REFERENCES

Barndorff-Nielsen O. 1973. On M-ancillarity.
    Biometrika 60: 447-456.

Barndorff-Nielsen, O. 1976 a. Plausibility inference. (With discussion.).
    J. Roy. Statist. Soc. B 38: 103-131.

Barndorff-Nielsen, O. 1976 b. Nonformation.
    Biometrika 63. (To appear).

Frydenberg, O. 1956. The observation of heterosis in natural populations of *Droso-phila*.
    Semana de Genetica, Univ. São Paulo, 25.

Kalbfleisch, J.D. and Sprott, D.A. 1973. Marginal and conditional likelihoods.
    Sankhya A 35: 311-328.

Levene, H. 1949. On a matching problem arising in genetics.
    Ann. Math. Statist. 20: 91-94.

Lewontin, R.C. and Cockerham, C.C. 1959. The goodness-of-fit test for detecting
    natural selection in random mating populations.
    Evolution 13: 561-564.

Wallace, B. 1958. The comparison of observed and calculated zygotic distributions.
    Evolution 12: 113-115.

Ward, R.H. and Sing, C.F. 1970. A consideration of the power of the χ^2 test to de-
    tect inbreeding effects in natural populations.
    Amer. Natur. 104: 355-366.

Figure 4. The Aeynton distribution for $t=0$, $m=100$, $k=...$, $\rho=...$, $z_1=...$

REFERENCES

Chapter 2. Study of Polymorphism

SELECTION AND NEUTRALITY

Warren J. Ewens

Just over a year ago, Marc Feldman and I made an assessment of the then-current state of the art of using gene frequencies to test for selective neutrality. This assessment has recently appeared in a review paper (Ewens and Feldman. 1976. (EF in the following)). My purpose here is to summarize briefly the conclusions reached in that paper, and then to discuss the advances that have been made during the last 15 months, so as to give an up-to-date summary of present tests of neutrality, as well as of certain theoretical conclusions which impinge upon these tests. This paper concerns only theoretical problems and no attempt is made at data analysis. Nor is any view offered one way or the other on the neutrality question. The paper may be taken as concerning an extreme point on the spectrum defined by the question of measuring selection, namely, is there any selection at all at the locus we are looking at?

BACKGROUND

The purpose of this section is to summarize the conclusions reached by EF on the question of using gene frequency data to test for selective neutrality. These conclusions fall under several main headings.

Models.

There are now several classes of models of gene frequency change in current use in population genetics theory. These different models are designed to describe the behaviour of various forms of data. The four most commonly used models are the following. (i) Classical models. These have their origins in the classical population genetics theory of the first half of this century. It is assumed that at the locus of interest there exist two alleles (more generally, k alleles are sometimes allowed), sometimes with mutation between alleles. (ii) Charge-state models. These models are inspired by the technique of electrophoresis. It is assumed that we measure the electric charge on a protein and that the possible charge levels are discrete and equidistant (and thus may be taken for convenience as occupying one or other of the integer values $(..., -2, -1, 0, 1, 2, ...)$). Mutation increases the charge level by an amount i with probability v_i, where i can be positive, negative or zero. Since a number of different alleles can occupy the same charge level, and since the charge level of the protein is the only property measured, this model strictly concerns charge level frequencies of proteins rather than allele frequencies. (iii). Infinite alleles models. These models suppose that a mechanism exists which enables us to tell unambiguously whether or not two given genes are of the same allelic type. Such mechanisms do not currently exist but may possibly be approached through a combination of electrophoresis, heat denaturation, *etc*. To this extent the model is the same as the classical model. However, it departs from that model in an essential way in that it assumes that all mutations lead to entirely novel allelic types. Thus there is no upper bound to the total number of alleles possible at the locus over the course of time. (iv). Infinite sites models. Here we assume a mechanism giving the complete nucleotide sequence of any gene. While we cannot expect much data of this nature for some time, this model is of some interest since it represents in a sense an ultimate state of knowledge of the gene.

Unfortunately, theoretical results for one or other of these models have often been used to analyze gene (or charge level) frequencies which really relate to some other model. While there is some disagreement upon which model properly describes certain sets of data, some effort at least should be made to reconcile the data with the form of model used.

Stationarity.

Practically all of present theory assumes stationary behaviour in
the model considered. But it is well known that stationarity is achieved
very slowly in all the models listed above (typically, the leading non-
unit eigenvalue is of the form $1-cN^{-1}$, where c is a constant of order
unity and N is the population size). Considerable caution must be exer-
cised in using stationary theory for a population which has not been
stable for some time.

Parameters.

The theoretical results for all models involve unknown parameters,
usually N (the population size) and u (a mutation rate). Often these
parameters occur only through their product Nu, which is of course
highly variable from one case to the next. Problems of estimating Nu
and in incorporating these estimates into the testing procedure are not
easily solved.

Model-free tests.

Within any class of test (*e.g.* the infinite alleles class), a number
of specific models is available (for example generalized Wright models
and generalized Moran models). We can have little if any knowledge of
which of these models nature actually "follows". Thus any testing procedure which is model-specific must be used with some caution. It is
sometimes possible to devise testing procedures that are model-free,
at least within a certain broad range of models. These are clearly to
be preferred to model-specific tests.

Method of testing.

The neutral theory is one that fundamentally involves stochastic
variation, so that any test of that theory which is based on gene frequencies must be statistical. On the other hand, the tests used can
range from the formal to the rather informal and non-quantitative. Formal tests may initially seem preferable, but have several drawbacks.
Thus a precise alternative hypothesis is often difficult to write down
and this leads to problems concerning the choice of test statistic and

an assessment of the power of the tests. Moreover, a full mathematical
treatment may be possible only in simple cases, whereas it is likely
that more information can be gained from complex situations (involving
perhaps several loci and geographical patterns), which must be treated
in a more informal manner. Another problem concerns the nature of the
tests and the form of the conclusion reached. Some tests attempt to
assess neutrality on a locus by locus basis, while others are "overall"
tests and make a single assessment about neutrality on the basis of the
gene frequencies at a number of loci. Problems arise with both
approaches.

The above is a very brief outline of some of the general comments
made in EF: more detailed comments, including the details of the mathematical procedures, may be found by consulting that reference. The second part of EF concerns properties of specific tests and we now turn
to a brief summary of that discussion.

The first test considered in EF was that of Ewens (1972). This applies to data obeying the infinite alleles model. The data is of the
form $(2n; k; n_1, \ldots, n_k)$, where $2n$ is the number of genes sampled, k
is the number of different alleles observed in the sample and $n_1, \ldots,$
n_k are the respective numbers of genes of each observed type. Results
for this test are well known and are not reproduced here, except that
we note for future reference that the neutral-theory probability distribution of $n_1, \ldots, n_k$, given k, is of the form

$$f(n_1 \ldots n_k | k) = \text{const } (n_1 n_2 \ldots n_k)^{-1}, \tag{1}$$

where the constant does not involve any unknown parameters. The testing
procedure is carried out by comparing the observed heterozygosity in the
data with the amount expected under the neutral theory for the observed
values of k, this latter quantity being computed from (1).

A procedure similar to the above, in that it relates observed to expected heterozygosity for a given value of k, is that of Johnson and
Feldman (1973). This test does not proceed on a locus by locus basis
(as does that of Ewens) but makes rather an overall assessment of neutrality based on a number of loci. A further test (Yamazaki and Maruyama (1972)) also considers a number of loci simultaneously. The test is
based on a classical model with two alleles, with no mutation. Although
ostensibly the test is carried out using heterozygosity measures, in
fact its basis is to compare the number of alleles observed in various
frequency ranges with the number predicted by the neutral theory. We

discuss such tests in more detail later.

A second class of tests discussed in EF concerns geographical sub-division. These tests stem from the observation of Cavalli-Sforza (1966) that standarized variances in allele frequency might be used as a basis for tests of neutrality. Two such tests were discussed in EF, both put forward by Lewontin and Krakauer (1973). The first test proceeds by com-puting the standarized variances at a number of loci and testing whether the values obtained differ in distribution significantly from the chi-square form. The second test in effect involves the estimation of the kurtosis of the gene frequency distribution and attempts to test whether this falls within the range of values associated with selective neutra-lity.

All of the above procedures involve difficulties, both of a general kind (*e.g.* the problem of stationarity) and specific to individual tests. These difficulties are discussed at some length in EF and apart from those difficulties which have emerged during the last 15 months, will not be reconsidered here.

Except for a number of informal procedures discussed in EF and again not reconsidered here, one other form of testing was available in 1975 but was not discussed in EF. This is the method of Langley and Fitch (1973, 1974), who constructed a tree of evolution from the protein se-quences in certain contemporary species. Langley and Fitch then tested whether the form of this tree supported the prediction (made by the neutral theory) of equal rates of evolution between species. We do not review here the mathematical details of this test nor the conclusions reached on the neutrality question, and remark only that although this test differs from others we have considered, it properly falls within the ambit of this review since its starting point is in a sense the ob-served gene frequencies in contemporary populations.

RECENT ADVANCES

The aim of this section is to review the advances made in the last 15 months concerning the above-mentioned tests and theoretical material.

Stationarity.

It was remarked above that non-stationarity could be a serious problem in the application of tests of selective neutrality. Several recent papers have considered this question. Kirby (1975) has used the
spectral representation of the infinite alleles process to devise a method for testing neutrality which in part overcomes the non-stationarity
problem. Specifically, her approach is to use as test statistic a function which returns to its stationary distribution at a rate faster than
that prescribed by the leading eigenvalue of the process. Computer simulations suggest that some improvement can be made in this way although,
since the second-largest eigenvalue is still close to unity, it is not
clear that really substantial improvements along these lines are possible.

It was noted above that several tests for neutrality in the
infinite alleles model in effect reduce to a comparison of the observed
amount of heterozygosity with that expected under neutrality. The theory
assumes stationarity in a population of constant size. Recently, Nei
et al. (1975) have studied the effect of a bottleneck in population size
on the amount of heterozygosity and the number of alleles present in the
infinite alleles model. They find that the size of the bottleneck and
the subsequent rate of increase in population size after the bottleneck
is passed through determined rather substantially the behaviour of these
two quantities. Specifically, even for comparatively small bottleneck
sizes, the heterozygosity is not strongly affected by the bottleneck
provided that the subsequent increase in population size is rapid. On
the other hand, the number of alleles is severely affected by the bottleneck size but not so strongly affected by the subsequent growth rate.
Once the original population size is restored the mean number of alleles
increases more rapidly than does the heterozygosity. It is clear that
tests of neutrality which compare the observed heterozygosity with the
amount expected under neutrality for the observed number of alleles can
be seriously affected by such a bottleneck.

A parallel paper by Chakraborty and Nei (1976) considers the same
problem for the charge-state model and reaches conclusions very similar
to those just described for the infinite alleles model.

Hitch-hiking.

Almost all theoretical studies concerning the neutral alleles hypothesis have been based on single locus theory. But the general theory
of popualtion genetics reveals that linkage between loci is an important
determinant of the behaviour of gene frequencies and it is desirable to
consider the effect of a selected locus on a linked neutral locus. Work
on this question has recently been carried out by Kimura and Ohta (1975),
Maynard Smith and Haigh (1974) and Thomson (1976). The general conclusion reached by these authors is that the amount of heterozygosity (and
no doubt also the number of alleles) at a neutral locus is affected by
a linked selective locus only if the recombination fraction between the
two loci is of the same order of magnitude as the selective differences
at the selected locus. This conclusion is for the moment qualitative
rather than quantitative so far as tests of neutrality are concerned in
that little is known about the extent to which single locus theory is
affected by this phenomenon.

Infinite alleles test.

Tests of selective neutrality based on data from an infinite alleles
model have been referred to, and the neutral theory conditional distribution upon which this test is based is given by (1) above. The test
reduces to assessing whether the observed frequency vector $(n_1, \ldots, n_k)$
conforms reasonably closely to that prescribed by (1). To do this we
construct a test statistic which for convenience we may take as the heterozygosity measure $H = 1 - \Sigma (n_i/2n)^2$. In principle the distribution of
H can be found from (1) and hence significance levels calculated. Values
of H that are significantly small suggest selection for one specific
allele, while values of H which are significantly large suggests heterotic selection. I have recently become somewhat uncertain of the usefulness of this test so far as small values of H are concerned. The reason
for this is that the most likely neutral configurations happen to lead
to very small H values. Take, for example, the case $2n = 500$, $k = 7$.
Then the most likely neutral theory configuration is (1, 1, 1, 2, 2, 3,
490) and this is very difficult to distinguish from the case where one
allele is selected for against the remaining alleles. It is therefore
quite possible that this test can be used successfully only to test for
heterotic selection and furthermore that this deficiency is shared by

many tests of selective neutrality.

In connection with the test for heterotic selection, I understand
that G.A. Watterson has recently developed the distribution of H under
the assumption that all heterozygotes (which are assumed equally fit)
enjoy a selective advantage over all homozygotes (which again are all
assumed equally fit). If this is so, some indication of the power of
this test might soon become available. In this connection it has long
been suspected that the power of the test increases considerably with
k (the observed number of alleles) but very slowly with n (the sample
size). Recent advances in experimental techniques which reveal values
of k considerably in excess of those previously found will therefore
lead to much more powerful tests of neutrality.

Lewontin-Krakauer tests.

In EF, some doubt was cast on the validity of the Lewontin-Krakauer
(1973) tests for selective neutrality. Similar doubts have recently been
expressed by Nei and Maruyama (1975) and by Robertson (1975a, 1975b).
More recently, Nei *et al.* (1976) and Nei and Chakravarti (1976) have considered certain properties of the standardized variance in completely
and in incompletely isolated populations. In both cases it is shown that
the doubts cast on the validity of the Lewontin-Krakauer procedure are
verified.

It may be asked whether there is any way at all in which standardized
variances can be used to form valid tests of the neutral theory. The
view of this author is that it is most unlikely that a rigorously valid
test of neutrality can ever be found in this way. The values of the mutation and migration parameters and the sizes of the subpopulations, about
which we are probably rather uncertain, will almost certainly influence
the distribution of the standardized variance or indeed of any other test
statistic. Even if populations are completely isolated, have known sizes
and there is no mutation, the unknown evolutionary history of the process
leading to the current configurations will affect this distribution. It
is quite possible that more is lost (through the addition of unknown parameters, in particular migration parameters) than is gained by basing
tests of neutrality on gene frequencies at a number of incompletely isolated subpopulation.

NEW TESTS

In this section we review tests which have appeared during the last
15 months and thus were not considered in EF. These will be treated in
somewhat greater detail than the tests outlined above.

Tests based on the frequency spectrum.

We consider first data arising from the infinite alleles model. Specifically, we consider a diploid population of fixed size N, and suppose
that any given gene in the population mutates with probability u, all
new mutants being entirely novel allelic types. Put $\theta = 4Nu$ and assume
that all alleles are selectively equivalent. Then it is a classical result of population genetics theory that at equilibrium, the mean number
of alleles to occur at the locus in question whose frequency is between
the values x_1 and x_2 is

$$\int_{x_1}^{x_2} f(x)\,dx \tag{2}$$

where

$$f(x) = \theta x^{-1}(1-x)^{\theta-1}. \tag{3}$$

Suppose now L loci are considered, all with the same value of u and all
selectively neutral. Taking all L loci into account, the mean number of
alleles to occur with frequency between x_1 and x_2 is then clearly

$$L\,\theta\int_{x_1}^{x_2} x^{-1}(1-x)^{\theta-1}\,dx. \tag{4}$$

One way of assessing whether we can assume selective neutrality at all
L loci is to compare the observed number of alleles whose frequency is
between x_1 and x_2 with the expected number computed from (4). Note that
strictly speaking this requires a common value of θ (equivalently, a
common value of u) for all L loci, and further will require estimation
of θ before the procedure can be carried out.

Soon after the data of Lewontin and Hubby (1966) appeared, I attempted (Ewens (1969)) to do this. Lewontin and Hubby estimated the
proportion of individuals who were homozygotes to be .885. Equating

this with the standard theoretical value $(1+\theta)^{-1}$ for this model, I estimated $\theta = .130$. Since 85 loci were examined by Lewontin and Hubby, the
estimate for the total number of alleles in the entire sample to have
frequency in any range (x_1, x_2) is

$$85(.130) \int_{x_1}^{x_2} x^{-1}(1-x)^{-.870} dx \qquad (5)$$

The values of x_1 and x_2 may be chosen more or less arbitrarily. Choosing
(x_1, x_2) to be $(.05, .1)$, $(.1, .9)$ and $(.9, 1.0)$, we obtain from (5) expected numbers 8.2, 43.5 and 63.7, respectively. The observed numbers
were 13, 42 and 67. It seems intuitively clear that there is no significant difference between observed and expected numbers, so that the hypothesis of neutrality cannot be rejected. Note, however, that the standard
chi-square test is probably not appropriate here and that, in particular,
the sum of the observed numbers is not necessarily equal to the sum of
the expected numbers. This point is referred to again in a moment.

    There are several problems associated with the above procedure. Per-
haps the main one is that a joint decision is made for or against neutrality on the basis of a number of loci. Further problems concern the
assumption of a common value of u at all loci, the lack of a formal testing procedure, and the method of estimating θ. The method used above is
to estimate θ by equating observed and expected homozygosity, yet optimal methods (Ewens 1972) estimate θ by equating observed and expected
numbers of alleles. Note that under the latter method, a chi-square test
for significance might prove to be correct.

    The procedure of Yamazaki and Maruyama (1972), based on a comparison
of observed and expected frequency spectra, has already been referred
to. This test was criticized in EF (see also Ewens and Feldman 1974).
More recently, Maruyama and Yamazaki (1974) have put forward a revised
procedure which is essentially equivalent to that described above. The
only difference between the two procedures is that Maruyama and Yamazaki multiply observed and expected numbers in the frequency range $(x - d,$
$x + d)$, by $2x(1-x)$, and thus in this sense compare observed and expected
heterozygosity measures.

    Recently, Ohta (1975) has suggested a parallel procedure for data
conforming to the charge-state model. This approach depends on an approximate formula developed by Kimura and Ohta (1975) for the frequency

spectrum, namely

$$f(x) = const\ x^{\beta-1}\ (1-x)^{\theta-1} \tag{6}$$

with

$$\beta = [1+\theta-(1+2\theta)^{\frac{1}{2}}]/[(1+2\theta)^{\frac{1}{2}}-1].$$

Here $\theta = 4Nu$, where the probability of a mutation which increases the charge level by 1 is $u/2$, as is the probability of a mutation which decreases the charge level by 1. The procedure now is as outlined above: a number of loci are considered simutaneously, and a common value of θ assumed for all loci and estimated by equating observed and expected homozygosities. This form of testing is subject to the same qualifications as those outlined above for the corresponding infinite allele test.

Tests based on autocorrelations.

Two rather similar testing procedures, put forward independently by Weir *et al.* (1976) and Wehrhahn (1975), without doubt provide the best approach for testing for selective neutrality in data conforming to the charge-state model. Before considering these tests it is useful to consider first some theoretical conclusions relating to this model, arrived at by Moran (1975), Weir *et al.* and Wehrhahn.

The essential features of the model have been discussed above. The permissible charge state levels are assumed to be the integers $(\ldots, -2, -1, 0, 1, 2, \ldots)$, and a gene mutates and changes the charge level by i with probability v_i. Let x_j be the fraction of genes at charge level j in any generation and put

$$C_k = \Sigma x_j x_{j+k}. \tag{7}$$

Then C_k is a random variable: in particular, C_0 can be viewed as the (random) current amount of "observed homozygosity". It is convenient to consider first the case $v_{-2} = v_{+2} = 0$ (*i.e.* one-step transitions only). If we define $\theta = 4N\ (v_{+1} + v_{-1})$, then at equilibrium (Ohta and Kimura 1973), and assuming selective neutrality,

$$E\{C_k\} = z^k/\{1+2\theta\}^{\frac{1}{2}},\tag{8}$$

where

$$z = \theta^{-1}\{1+\theta -(1-2\theta)^{\frac{1}{2}}\}.\tag{9}$$

Note in particular that (8) and (9) show that

$$E\{C_0\} = (1+2\theta)^{-\frac{1}{2}}.\tag{10}$$

Suppose now that two-step mutations are also possible. We put $\Psi = 4N(v_2+v_{-2})$ and consider the polynomial equation

$$\Psi z^4 + \theta z^3 - (2+2\theta+2\Psi) z^2 + \theta z + \Psi = 0.\tag{11}$$

Clearly, if z is a solution on (11), then z^{-1} is also a solution: let z_1 and z_2 be the two solutions of (11) less than unity in absolute value. Put $w_1 = z_1 + z_1^{-1}$ and $w_2 = z_2 + z_2^{-1}$. Then, in this more general case, at equilibrium,

$$E\{C_k\} = z_1^k \ [\tfrac{1}{2}\theta+\Psi w_1 (w^2-4)^{\frac{1}{2}}]^{-1} - z_2^k \ [\tfrac{1}{2}\theta+\Psi w_2 (w_2^2-4)^{\frac{1}{2}}]^{-1};\tag{12}$$

[an equivalent formula for $E(C_k)$, in terms of an integral, is given by Moran (1975)].

We now consider variance properties. Suppose that at time t, $n_i(t)$ genes exist at charge level i, and put

$$M_1(t) = \Sigma i \ n_i(t),\tag{13}$$

$$M_2(t) = \Sigma i^2 n_i(t).$$

The random variable $M_2(t)$ does not converge to any finite value or distribution as $t\to\infty$. However, the expected value of the variance-like quantity $M_2(t)-M_1^2(t)$ does converge as $t\to\infty$, due to the tendency for the genes to cluster at closely located charge levels. We have (Moran 1976)

$$E [M_2(t) - \{M_1(t)\}^2] = (1-(2N)^{-1}E [M_2(t-1)-\{M_1(t-1)\}^2]\tag{14}$$

$$+ 2u_1 + 8u_2.$$

Taking the limit as t→∞ gives the equilibrium value

$$\sigma^2 = E\ [M_2-M_1^2] = \theta+4\Psi.\qquad\qquad(15)$$

Note that the value of σ^2 does not determine the expected values of the C_k explicitly, since these are not functions of σ^2 alone. If on the other hand we are willing to assume, on *a priori* grounds, some plausible value for the ratio θ/Ψ, further progress can be made, since then both σ^2 and the $E(C_k)$ will be functions of the single parameter θ. One test of the neutrality theory is to estimate σ^2 from the data and then, with the assumed value for θ/Ψ, to estimate θ from (15). The test of this hypothesis is carried out by assessing whether the observed values of the C_k conform reasonably to what is expected from (12), using the value of θ so estimated.

   Another approach is possible. Consider a random variable taking the value i(i=0, ±1, ±2, ...) with probability p_i. Assume $p_i = p_{-i}$. Then if γ_k is defined by

$$\gamma_k = \sum_i\ p_i p_{i+k},\qquad\qquad(16)$$

the values of the γ_k uniquely determine the values of the p_i. Again, we do not know in reality the values of the γ_k, but if these are estimated by using the observed values C_k, estimates of the p_i can be obtained and compared to observed values. (Note that certain theoretical problems arise with this procedure. We do not pursue these problems here).

   We now consider specifically how testing for selective neutrality is carried out, and consider first the procedure of Weir *et al.* Note first that C_k (defined by (7)) is identically equal to C_{-k}. It is thus convenient to define new quantities

$$D_k = 2C_k,\ (k{\neq}0),\ D_k = C_k,\ (k=0),$$

and to carry out the test in terms of the D_k's. Weir *et al.* carry out their test by assessing whether the observed values of D_k conform sufficiently closely to the neutral-theory mean values defined explicitly by (12). To do this they form the chi-square-like statistic

$$\chi^2 = n\sum_k\ [D_k-E(D_k)]^2/E(D_k)\qquad\qquad(17)$$

whose distribution under neutrality depends on the unknown parameters θ and Ψ. In the summation in (17), some amalgamation of classes might be necessary for large values of k: details of this are given by Weir *et al.* It is possible, by numerical methods, to minimize χ^2 with respect to θ and Ψ: the value of χ^2 calculated by using these minimizing values in (17) is denoted χ^2_{min} and is a measure of the extent to which the functions C_k depart, as a group, from what is expected under the closest-fitting neutral model.

A standard statistical question now arises: how large a departure from the neutral theory expectation is tolerable? The answer can be found only by determining the distribution of χ^2_{min} when the hypothesis of selective neutrality is true. We may anticipate that this distribution is approximately chi-square but no immediate assumption can be made on this point because of the rather complex definition of the D_k. To determine the distribution of χ^2_{min}, Weir *et al.* performed an extensive Monte Carlo simulation and found that the distribution of χ^2_{min} under the null hypothesis can be taken, with a bias in the conservative direction, to be approximately chi-square with m-3 degrees of freedom, where m is the number of terms in the sum (17). It is now possible to compute χ^2_{min} using real data and to accept or reject the neutral theory in the usual sense of statistical hypothesis testing. This is a most valuable and objective procedure for testing the neutral theory using data conforming to the charge-state model.

We consider next the procedure of Wehrhahn (1975). Wehrhahn first assumes a particular value (*viz.* .1) for the ratio Ψ/θ, and then estimates θ from (15), using the data to estimate σ^2. Using this estimated value for θ, the expectations (12) are calculated for various values of k. One possible way of proceeding is to compare these expectations with the sample values C_k; this is in effect the approach of Weir *et al.* However, Wehrhahn proceeds along a slightly different line by assuming that the distribution $\{p_i\}$ of class frequencies is symmetric (this amounts to assuming $u_{-1} = u_{+1}$, $u_{-2} = u_{+2}$), and then (cf. the argument following equation (16)) by estimating this distribution from the sample values C_k. The test proceeds by a visual comparison of the values so calculated with the observed frequencies. In doing this, a location adjustment will normally be necessary because the sample mean position of electrophetic frequencies will not usually be zero. Wehrhahn considers this procedure to be more interesting visually, but perhaps less efficient, than that which compares observed and expected values of C_k, in particular because it requires a further assumption of symmetry of mutation rates not required in the latter test. Note also that the assumption that a constant value for Ψ/θ can be assumed is not supported by the calculations of Weir *et al.*, and an extension of Wehrhahn's method which avoids such an assumption is probably desirable.

Again, this provides a most useful method of testing for selective neutrality for charge-level data.

GENERAL CONSIDERATIONS

At the moment, most "gene frequency" data are really charge level frequencies. For such data, the tests of Weir *et al.* and Wehrhahn are undoubtedly the best currently available. We may soon expect, on the other hand, that further techniques will lead to a breakdown of each charge-level frequency into separate allele frequencies, so that the data will have the characteristics corresponding to both charge-level and infinite alleles models.

How should such data be tested for neutrality? One approach would be to ignore the charge-level of each allele entirely and to carry out an infinite-alleles test.In the extreme limit $v_o = 1$ (*i.e.* no charge-level change through mutation) this would be appropriate, and so presumably this would be a useful approximation when v_o is large. Another circumstance under which we might ignore charge levels is when the distribution $\{v_i\}$ is quite unknown and when in particular no assumptions (*e.g.* symmetry) can be made about this distribution. In this case the information in the charge levels is probably of little value. On the other hand, if the v_i ($i \neq 0$) are large and assumptions such as symmetry can be made, the charge level information is presumably valuable. Apart from these rather general comments, little is known about how infinite alleles and charge-state model testing procedure should be combined for such data.

These comments illustrate the point that experimental techniques are still in a state of development, that theoretical conclusions are incomplete, and that several years will pass until both have reached a form where rather reliable tests of the neutral theory, based on gene frequencies and population genetics theory, will have been arrived at.

REFERENCES

Cavalli-Sforza, L.L. 1966. Population structure and human evolution.
Proc. Roy. Soc. Lond. **164**: 362.

Chakraborty, R. and M. Nei. 1976. Bottleneck effects on average heterozygosity and
genetic distance with the stepwise mutation model.
To appear in *Evolution*.

Ewens, W.J. 1969. The transient behavior of stochastic processes, with applications
in the natural science.
Bull. I.S.I. XLII: 603.

Ewens, W.J. 1972. The sampling theory of selectively neutral alleles.
Theoret. Pop. Biol. **3**: 87.

Ewens, W.J. and M. Feldman. 1974. Analysis of neutrality in protein polymorphism.
Science **183**: 446.

Ewens, W.J. and M. Feldman. 1976. The theoretical assessment of selective neutrality.
In *Population genetics and ecology* (S. Karlin and E. Nevo, eds.). Academic Press,
New York. Pp. 303-337.

Johnson, G. and M.W. Feldman. 1973. On the hypothesis that polymorphic enzyme alleles are selective neutral. 1. The evenness of allele frequency distribution.
Theoret. Pop. Biol. **4**: 209.

Kimura, M. and T. Ohta. 1975. Distribution of allelic frequencies in a finite population under stepwise production of neutral alleles.
Proc. Nat. Acad. Sci. **72**: 2761.

Kirby, K. 1975. Statistical results for the theory of selectively neutral alleles.
Ph.D. Thesis, Princeton University.

Langley, C.H. and W.M. Fitch. 1973. The constancy of evolution: A statistical analysis of the a and b hemoglobins, cytochrome c, and fibrinopeptide A.
In *Genetic structure of populations*: 246-262. (N.E. Morton, ed.). University Press
of Hawaii, Honolulu.

Langley, C.H. and W.M. Fitch. 1974. An examination of the constancy of the rate of
molecular evolution.
J. Mol. Evol. **3**: 161.

Lewontin, R.C. and J.L. Hubby. 1966. A molecular approach to the study of genic
heterozygosity in natural populations. II. Amount of variation and degree of heterozygosity in natural populations of *drosophila pseudoobscura*.
Genetics **54**: 595.

Lewontin, R.C. and J. Krakauer. 1973. Distribution of gene frequency as a test of
the theory of the selective neutrality of polymorphisms.
Genetics **74**: 175.

Maruyama, T. and T. Yamazaki. 1975. Analysis of heterozygosity in regard to the
neutrality theory of protein polymorphisms.
J. Molec. Evol. **4**: 195.

Maynard Smith, J. and J. Haigh. 1974. The hitch-hiking effect of a favorable gene.
Genet. Res. **23**: 23.

Moran, P.A.P. 1975. Wandering distribution and the electrophoretic profile.
Theoret. Pop. Biol. **8**: 318.

Nei, M., T. Maruyama and R. Chakraborty. 1975. The bottleneck effect and genetic variability in populations.
Evolution 29: 1.

Nei, M. and T. Maruyama. 1975. Lewontin-Krakauer test for neutral genes.
Genetics 80: 395.

Nei, M. and A Chakravarti. 1976. Drift variances of F_{st} and G_{st} statistics obtained from a finite number of isolated populations.
(to appear).

Nei, M., A. Chakravarti and Y. Tateno. Mean and variance of F_{st} in a finite number of incompletely isolated populations.
(to appear).

Ohta, T. 1975. Statistical analyses of *Drosophila* and human protein polymorphisms.
(to appear).

Ohta, T. and M. Kimura. 1975. The effect of selected linked locus on heterozygosity of neutral alleles (the hitch-hiking effect).
Genet. Res. Camb. 25: 313.

Robertson, A. 1975a. Remarks on the Lewontin-Krakauer test.
Genetics 80: 396.

Robertson, A. 1975b. Gene frequency distributions as a test of selective neutrality.
Genetics 81: 775.

Thomson, G. 1976. The effect of a selected locus on linked neutral loci.
(to appear).

Wehrhahn, C.F. 1975. The evolution of selectively similar electrophoretically detectable alleles in finite natural populations.
Genetics 80: 375.

Weir, B.S., A.H.D. Brown, and D.R. Marshall. Testing for selective neutrality of electrophoretically detectable protein polymorphisms.
(to appear).

Yamazaki, T. and T. Maruyama. 1972. Evidence for the neutral hypothesis of protein polymorphisms.
Science 178: 56.

2. STUDY OF POLYMORPHISM

PROTEIN EVOLUTION: NONRANDOM PATTERNS IN RELATED SPECIES

Francisco J. Ayala

SCIENTIFIC HYPOTHESES AND THEIR FALSIFICATION

Empirical science employs a hypothetico-deductive method consisting of two interdependent exercises or episodes, one imaginative, the other critical. To have an idea, advance a hypothesis, or suggest what might be true is an imaginative or creative exercise. But the idea or hypothesis must be subject to critical examination and test. The "criterion of demarcation" which separates the empirical sciences from other realms of discourse, such as logic or metaphysics, is that empirical hypotheses or theories are subject to the possibility of empirical falsification (Popper, 1959). Empirical tests of a hypothesis consist of finding out whether predictions about the world of experience derived as logical consequences from the hypothesis agree or not with the states of affairs observed in the natural world.

Hypotheses and other imaginative exploits are the basis of scientific inquiry. It is the imaginative preconception of what might be true that provides the incentive to seek the truth and a clue as to where we might find it (Medawar, 1967). Hypotheses guide observation and experiment by reducing the domain of relevant observations to something less than the whole universe of observables. Whether empirical observations lead to corroboration or falsification of a hypothesis, the latter will have contributed to the progress of science if it has incited relevant empirical investigations.

The role that critical tests - *i.e.*, attempts to falsify a hypothesis - play in empirical science can be best appreciated if it is
realized that the informative (or empirical) content of a scientific
hypotheses is measured by its a priori probability of being falsified
by relevant observations (Popper, 1959; Ayala, 1977). A scientific
hypothesis divides all singular statements of fact into two non-empty
subclasses: first, the class of all statements with which it is inconsistent - this is the class of potential falsifiers of the hypothesis;
and second, the class of all empirical statements which it does not
contradict, *i.e.* with which it is consistent (the "permitted" statements).A hypothesis is scientific only if the class of its potential
falsifiers is not empty, because the hypothesis makes assertions only
about its potential falsifiers; namely, it asserts that they are false.
About the permitted statements it says nothing, since these permitted
statements may also be consistent with other hypotheses. Therefore, the
truth of empirical statements with which it is consistent does not *verify* a hypothesis, although it may contribute to provisional corroboration of the hypothesis. This point is well recognized in the statistical methodology of testing hypotheses. The hypothesis subject to
test, the *null* hypothesis, may be rejected if the observations are incompatible with it. If the observations are consistent with the predictions derived from the hypothesis, the proper conclusion is that the
test fails to falsify the null hypothesis, not that its truth has been
established.

One important point is that the empirical content of a hypothesis
is measured by the class of its potential falsifiers; the larger this
class the greater the information content of the hypothesis. A hypothesis or theory compatible with all possible states of affairs in the
world of experience is uninformative. Another important point, however,
is that the only way to test a hypothesis is to make observations or
experiments that are likely to be incompatible with the hypothesis if
this is false. Observations that a priori are (or are likely to be)
consistent with a given hypothesis do not contribute (or contribute
very little) to the corroboration of the hypothesis. The more severe
a test is - *i.e.* the more likely it is a priori that the outcome will
be incompatible with the hypothesis - the more it contributes to the
corroboration of a hypothesis if this passes the test.

The requirement that scientific hypotheses be falsifiable also has
a parallel in the rules of statistical inference, namely in the re-

quirement that the power of the test be greater than zero. Two kinds
of errors are recognized in statistical testing: type I error, the probability of rejecting the null hypothesis if it is true, usually represented as α; and type II error, the probability of not rejecting the
hypothesis if it is wrong, symbolized as β. Scientists pay considerable
attention to type I errors, and thus choose values sufficiently low.
It is unfortunate that scientists often pay little attention to type
II errors. Yet the power ot the test depends on the probability, $1-\beta$,
of rejecting the null hypothesis if this is wrong. Thus, small β's as
well as small α's are required to have a meaningful test. It is the
case, of course, that for any given test the magnitudes of α and β are
inversely related. Given a chosen value of α, the value of β may nevertheless be reduced by increasing the sample size or the number of observations.

Scientific hypotheses are accepted only in a provisional manner;
their truth is never definitively established. This is because there
is asymmetry between verification and falsification. It is possible to
show the falsity of a universal statement by showing that it logically
contradicts a true singular statement of fact. But no matter how many
true singular statements of fact are consistent with a universal statement, the truth of this never becomes conclusively established. However,
as stated above, a hypothesis becomes more and more corroborated (and
thus it is held with greater and greater confidence) as it passes more
tests or more severe ones. Science advances when an important hypothesis
passes a critical test. But science makes a quantum advance indeed when
a hypothesis well formulated, relevant, and accepted by experts becomes
falsified by empirical tests. When a relevant hypothesis is falsified
we learn definitively something about the subject matter covered by the
hypothesis. And the ground becomes open for the formulation of new
hypotheses or the correction of the old one.

The history of science gives testimony ot impressive advances that
occurred when a hypothesis was falsified and became replaced by a different, more nearly correct hypothesis. The substitution of general re-
lativity for Newtonian mechanics is a classical example. Another example
is relevant to the discussion to follow. As the study of protein polymorphisms provided evidence apparently inconsistent with the neutrality
hypothesis advanced by Kimura (1968), King and Jukes (1969), and others
Ohta(1973, 1974) was able to advance a new hypothesis which is more consistent with the patterns of polymorphism observed in natural populations. The new theory recognizes an important role for stochastic pro-

cesses (random genetic drift) and mutation rates in evolution like the
original neutrality theory, but incorporates small selective differences
counteracting mutation pressure and limiting the scope of chance effects. The new theory has provided new insights to understanding protein polymorphisms (and presumably the overall course of evolution at
the molecular level) and, perhaps even more important, provides the
stimulus for new experiments and observations. It may also be pointed
out that Ohta's hypothesis contains the previous neutrality hypothesis
as a limiting case.

PROTEIN POLYMORPHISMS AND THEIR ADAPTIVE SIGNIFICANCE

Lewontin (1974) has insightfully reviewed the efforts of population
genetics to ascertain the degree of genetic variation in natural populations. By the 1950's and early 1960's it had become apparent to most
evolutionary geneticists that populations of sexually reproducing organisms are genetically very polymorphic. The evidence supporting this
conviction derived from various sources, such as the success of plant
and animal breeders with artificial selection, and the pervasiveness of
"hidden" variation affecting morphological and physiological traits
observed by inbreeding (see, *e.g.*, Dobzhansky *et al.*, 1959). These
studies had, nevertheless, failed to provide quantitative estimates of
how many, or what proportion of the gene loci are polymorphic in natural populations.

The conceptual breakthrough that made possible to provide numerical estimates of genetic variation in natural populations came in 1966
through the brilliant insights of Harris (1966), Lewontin and Hubby
(1966), and others. The degree of polymorphism at a gene locus may be
determined by electrophoretic separation and selective staining of the
protein (polypeptide) products coded by the locus. The study of a moderate number of randomly selected loci provides an estimate of the a-
mount of genetic variation over the whole genome of a species. Estimates
based on electrophoretic studies are subject to various biases; not all
amino acid substitutions in a polypeptide are detectable by electrophoresis. Yet approximate estimates og genetic variation are possible with
a technique that is neither unduly laborious nor prohibitively expensive.

A great variety of organisms have been studied by electrophoresis
in the last decade. A staggering amount of genetic variation has been

discovered (Table 1). In most populations of sexually reproducing organisms, the proportion of heterozygous loci per individual ranges between 5 and 15 percent; and the proportion of polymorphic loci per population ranges from about 20 to more than 50 percent. Since these values
are underestimates, and perhaps grossly so, the true degree of genetic
variation in natural populations appears to be enormous. Human beings
are heterozygous at about 6.7 percent of their loci for protein variants
detected by electrophoresis (Harris and Hopkinson, 1972).

Various patterns have emerged from the electrophoretic surveys.
Whether these patterns will persist remains to be ascertained. For e-
example, Selander and Kaufman (1973) pointed out that invertebrates
are on the whole genetically more polymorphic than vertebrates. This
difference is still apparent in Table 1 - the average heterozygosity is
13.4 percent among invertebrates, but only 6.0 percent among vertebrate
species. Yet there are indications that marine invertebrates living at
higher latitudes may have low levels of polymorphism. The average heterozygosity detected by electrophoresis is 3.9 ± 1.9 percent in *Liothyrella notorcadensis*, a brachiopod from Antarctica (Ayala *et al.*, 1975),
while it is 16.9 ± 4.6 percent in *Freileia halli*, a brachiopod from the
deep-sea (Valentine and Yala, 1975). *Euphausia superba*, a krill living
in circumantarctic waters has an average heterozygosity of 5.7 ± 1.9
percent, while two other krill species of the same genus but from trop-
ical waters, *E. mucronata* and *E. distinguenda*, have heterozygosities
of 14.1 ± 2.5 and 21.3 ± 3.4 percent, respectively (Valentine and Ayala,
1976).

What is the adaptive significance of all this variation? It has
been suggested that most protein variation, and the underlying genetic
variation, observed in natural populations may be adaptively neutral.
King and Jukes (1969), Kimura and Ohta (1971), and others have proposed
that most protein variants found in natural populations of the same
species or in different species are functionally equivalent. If so,
protein polymorphisms are evolutionary noise; genetic variants encoding
protein variation would change in frequency not by the adaptive process
of natural selection but rather as a consequence of random sampling
through the generations. This hypothesis has come to be known as the
"neutralist hypothesis" or the neutrality theory of protein evolution.

The "neutralist" hypothesis recognizes that the morphological, physiological, and behavioral features of organisms evolve by natural selection. Neutralists argue, nevertheless, that evolution at the molecular level (i.e. in informational macromolecules) largely occurs through

Table 1. Genic variation in natural populations of some major groups
of animals and of plants. (After Selander, 1976)

Organisms	Number of species	Mean number of loci per species	Proportion of loci polymorphic per population[a]	Proportion of loci heterozygous per individual
Invertebrates:				
Drosophila	28	24	0.529	0.150
Haplodiploid wasps	6	15	0.243	0.062
Other insects	4	18	0.531	0.151
Marine	14	23	0.439	0.124
Land snails	5	18	0.437	0.150
Vertebrates:				
Fish	14	21	0.306	0.078
Amphibians	11	22	0.336	0.082
Reptiles	9	21	0.231	0.047
Birds	4	19	0.145	0.042
Mammals	30	28	0.206	0.051
Plants, outcrossing	8	8	0.464	0.170
Mean values				
Invertebrates	57	21.8	0.469	0.134
Vertebrates	68	24.1	0.247	0.060
All animals	125	23.0	0.348	0.094

a: The criterion to decide whether or not a locus is polymorphic is not the same for
all species.

stochastic events in the form of genetic drift. This statement needs
to be qualified. The neutrality hypothesis acknowledges that a large
fraction of all newly arising mutations are unconditionally deleterious.
These harmful variants are eliminated or kept at very low frequencies by
natural selection. The neutralist hypothesis proposes that at most gene
loci there are a number of mutants that are effectively equivalent with
respect to adaptation. These are functional variants, anyone of which is
favorably selected relative to the deleterious ones. However, carriers of
alternative genotypes for the adaptively equivalent variants do not differ in their adaptedness to the environment. The frequencies in populations of adaptively equivalent, or "neutral", variants are, therefore,
not affected by natural selection. Because natural populations consist of
finite numbers of individuals, the frequencies of neutral mutants would
change from generation to generation due to the accidents of sampling.
Differences between species in informational macromolecules (proteins and
the DNA sequences coding for them) are consequently due to random processes of chance, not to natural selection. And the pervasive protein
polymorphisms observed in natural populations represent transient conditions in populations going from random fixation for one gene allele towards fixation for another, adaptively equivalent allele.

The epistemological characteristics of the neutralist theory are
quite interesting. According to the theory, protein variants and the
genic variants coding for them, evolve for the most part by random genetic drift. Since natural selection is not involved, environmental variations as well as other parameters that might affect selective values
can be ignored. It is therefore possible to advance evolutionary models
that include a very few parameters, notably mutation frequencies (which
determine the rates at which allelic variants arise in a population),
population size (which determines the magnitude of the sampling errors
from generation to generation), and time (for situations not in equilibrium). The presence of very few parameters makes it possible to derive
precise predictions about evolutionary patterns. The neutralist theory
can thus be readily subject to empirical tests by ascertaining whether
its empirical predictions agree with the states of affairs observed by
experiment. In epistemological terms, the neutralist hypothesis is a
theory with large empirical (informative) content. This is a most wel-
come situation in the field of evolution, where theories and models are
notoriously difficult to test in their general form. The neutralist hypothesis is a most valuable evolutionary theory precisely because it is
readily amenable to empirical testing. My colleagues and I have dedicated

considerable effort in recent years to test the neutrality theory (see,
e.g., Ayala, 1974, and Ayala *et al.*, 1974a).

One powerful method to test the neutrality theory is to compare the
configurations of allelic frequencies between different species. Species are reproductively isolated groups of populations, *i.e.*, independently evolving entities. According to the neutrality theory, genetic
frequencies change at random, namely as a consequence of the stochastic
process of sampling the gene pool from generation to generation. Long
independent sequences of random events are unlikely to be identical.
The configurations of allelic frequencies can be compared among different species to assess whether they might have come about by independent
sequences of random events.

PROTEIN VARIATION IN DROSOPHILA

The *willistoni* group of *Drosophila* consists of about 15 species endemic
to the New World. I shall here be concerned primarily with five species
extensively studied in my laboratory. Four species, *D. willistoni*, *D.
tropicalis*, *D. equinoxialis*, and *D. paulistorum*, are siblings, *i.e.* morphologically nearly indistinguishable; the fifth species, *D. nebulosa*,
is closely related to the other four but easily distinguishable from
them. All five species live in the forests of the New World tropics,
and their geographic distributions overlap over an enourmous territory
which includes Central America and the northern half of continental
South America (Spassky *et al.*, 1971; Ayala *et al.*, 1974b).

Protein variation has been studied in the *Drosophila willistoni*
group using the techniques of starch gel electrophoresis and selective
assay of enzymes (Ayala *et al.*, 1972a). Thousands of individuals of each
species collected in scores of localities have been sampled. A summary
of the amount of genetic variation observed is given in Table 2. The
species are somewhat more polymorphic than the mean value observed in
invertebrate organisms. On the average, an individual of the *D. willistoni* group is heterozygous at 17.7 ± 0.8 percent of its gene loci, while
50 percent or more loci are polymorphic in a given population. There is,
however, considerable heterogeneity in the degree of polymorphism per
locus (Ayala *et al.*, 1974a). At a large fraction of loci (*i.e.* about
25 percent of the total) there is virtually no genetic variation, while
the rest of the loci have proportions of heterozygotes ranging from two

Table 2. Summary of the amount of variation at 36 gene loci in natural
populations of five closely related species of *Drosophila*.
A locus is considered polymorphic: (1) when the frequency of
the second most common allele is at least 0.01; (2) when the
frequency of the most common allele is no greater than 0.95

Species	Genomes sampled per locus	Polymorphic loci per population (1)	(2)	Heterozygous loci per individual
D. *willistoni*	4,983±636	0.738±0.064	0.491±0.072	0.179±0.037
D. *tropicalis*	1,731±229	0.647±0.067	0.474±0.076	0.152±0.031
D. *equinoxialis*	2,356±238	0.792±0.061	0.540±0.075	0.165±0.030
D. *paulistorum*	1,277±258	0.742±0.062	0.511±0.071	0.194±0.027
D. *nebulosa*	412±43	0.692±0.071	0.530±0.069	0.195±0.035
All species	10,759±341	0.722±0.025	0.509±0.012	0.177±0.008

to more than fifty percent.

However, for the present purposes we are more interested in the *pattern*, than in the amount of genetic variation. Particularly we are interested in the patterns of genetic variation observed in different species. In order to compare species, we must first establish how much differentiation exists among populations of the same species. The allelic frequencies at many loci in scores of local populations have been given elsewhere (*e.g.* Ayala, 1972; Ayala *et al.*, 1972a, b, 1974a). The situation can be simply stated: different local populations of the same species are genetically very similar. When two populations of a given species are compared, the same alleles and in virtually identical frequencues are found at all, or nearly all loci. Ocassionally, however, a pair of populations may have somewhat different, or even quite different, configurations of allelic frequencies at a locus.

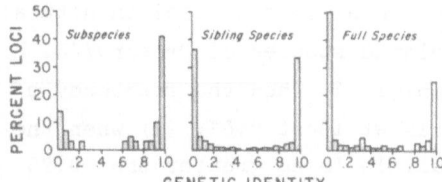

Figure 1. Distribution of *polymorphic* loci relative to genetic identity between subspecies (left), sibling species (middle) and nonsibling species (right) of the *Drosophila willistoni* group. A locus is considered polymorphic if the frequency of the most
common allele is no greater than 0.95. Only those loci that are polymorphic in both
populations being compared are included in these distributions. The number of pair-
wise comparisons are 29, 160, and 101, for subspecies, sibling species, and full species respectively. Genetic identity is calculated using Nei's (1972) *I* statistic.

GENETIC DIFFERENTIATION BETWEEN SPECIES

The situation is quite different when different subspecies or different
species are compared - the distribution of genetic identities is ʋ-
shaped (Ayala *et al.*, 1974a, b). That is, any two species (or subspecies) are genetically virtually identical at a large proportion of the
loci, while they are completely different at most other loci. It is worth
pointing out that the distributions of genetic identities are ʋ-shaped
whether all loci or only polymorphic loci are considered. The distributions of genetic identities among polymorphic loci for comparisons between populations of different subspecies, sibling species, and morphologically distinct species are shown in Figure 1. A locus is considered
polymorphic when the frequency of the most common allele is no greater
than 0.95 in both taxa being compared; if the frequency of the most common allele is greater than 0.95 in at least one of the two taxa the locus is considered monomorphic for the present purposes.

The distributions of genetic distance shown in Figure 1 contains
only a limited amount of the information available in the genetic configuration of the species. Table 3 gives allelic frequencies at 30 gene
loci in five species of the *D. willistoni* group. The average number of
alleles per locus discovered in the species of this group is 13, but the
alleles with low frequencies in these five species have been omitted to
avoid making the table unnecessarily large. Also for reasons of simpli-
city, the table gives data for only five species, although 14 species
and incipient species have been studied in the group. The allelic fre-

Table 3. Allelic frequencies at 30 gene loci coding
for enzymes in five species of *Drosophila*[a]

Locus	Allele	*Willistoni*	*Tropicalis*	*Equinoxialis*	*Paulistorum*	*Nebulosa*
Acph-1	94	0.011	0.959	0.013	0.000	0.000
	100	0.969	0.023	0.172	0.000	0.000
	104	0.016	0.004	0.811	0.035	0.008
	106	0.001	0.000	0.000	0.050	0.096
	108	0.000	0.000	0.004	0.671	0.619
	112	0.001	0.000	0.000	0.224	0.272
Adh	98	0.019	0.021	0.034	0.000	0.000
	100	0.967	0.961	0.961	0.994	0.008
	106	0.002	0.002	0.001	0.000	0.000
	110	0.000	0.000	0.000	0.000	0.968
Adk-1	100	0.387	0.410	0.386	0.573	0.000
	106	0.537	0.526	0.547	0.402	0.189
	112	0.065	0.062	0.059	0.024	0.601
	118	0.002	0.000	0.000	0.000	0.029
	124	0.004	0.000	0.001	0.000	0.164
Adk-2	96	0.011	0.033	0.003	0.000	0.000
	100	0.944	0.948	0.038	0.985	0.003
	104	0.019	0.019	0.941	0.015	0.971
	108	0.005	0.000	0.007	0.000	0.023
Ald-1	96	0.003	0.021	0.000	0.000	0.000
	98	0.011	0.009	0.061	0.000	0.011
	100	0.741	0.700	0.833	0.685	0.885
	102	0.238	0.259	0.091	0.241	0.104
	104	0.002	0.005	0.000	0.074	0.000
Ao-1	96	0.005	0.028	0.000	0.003	0.000
	98	0.091	0.038	0.000	0.041	0.032
	100	0.733	0.849	0.002	0.000	0.774
	101	0.000	0.000	0.000	0.043	0.003

Table 3 (continued)

Locus	Allele	*Willistoni*	*Tropicalis*	*Equinoxialis*	*Paulistorum*	*Nebulosa*
	102	0.108	0.057	0.050	0.131	0.120
	103	0.005	0.000	0.029	0.022	0.042
	104	0.010	0.000	0.796	0.655	0.004
	105	0.016	0.028	0.044	0.000	0.000
	106	0.001	0.000	0.072	0.200	0.000
Aph-1	98	0.029	0.004	0.002	0.000	0.000
	100	0.871	0.050	0.020	0.053	0.038
	102	0.060	0.902	0.919	0.921	0.885
	104	0.036	0.043	0.057	0.026	0.070
Est-2	94	0.000	0.001	0.000	0.000	0.087
	96	0.000	0.357	0.000	0.000	0.062
	98	0.009	0.034	0.014	0.001	0.814
	100	0.062	0.574	0.032	0.106	0.016
	102	0.915	0.026	0.939	0.724	0.000
	104	0.012	0.000	0.013	0.165	0.000
Est-3	95	0.000	0.049	0.000	0.000	0.045
	97	0.000	0.000	0.000	0.000	0.893
	98	0.022	0.002	0.093	0.014	0.006
	100	0.959	0.879	0.780	0.026	0.045
	102	0.015	0.011	0.088	0.951	0.010
	105	0.000	0.055	0.030	0.000	0.000
Est-4	96	0.000	0.001	0.000	0.029	0.011
	98	0.007	0.006	0.149	0.000	0.979
	100	0.099	0.283	0.768	0.968	0.009
	102	0.877	0.704	0.081	0.001	0.000
Est-5	95	0.024	0.000	0.033	0.051	0.000
	100	0.962	0.000	0.940	0.909	0.000
	105	0.012	0.000	0.024	0.037	0.000
	Null	0.000	1.000	0.000	0.000	1.000

Table 3 (continued)

Locus	Allele	*Willistoni*	*Tropicalis*	*Equinoxialis*	*Paulistorum*	*Nebulosa*
Est-6	100	0.782	0.858	0.013	0.044	0.005
	102	0.010	0.000	0.007	0.781	0.020
	104	0.196	0.089	0.842	0.112	0.797
	108	0.006	0.000	0.111	0.000	0.104
Est-7	96	0.023	0.021	0.000	0.000	0.000
	98	0.158	0.111	0.000	0.000	0.000
	100	0.529	0.599	0.000	0.005	0.004
	101	0.000	0.000	0.000	0.105	0.000
	102	0.232	0.234	0.000	0.061	0.024
	105	0.049	0.030	0.000	0.649	0.096
	107	0.005	0.002	0.000	0.049	0.762
	Null	0.000	0.000	1.000	0.109	0.000
Got	94	0.006	0.992	0.965	0.987	0.984
	96	0.006	0.004	0.000	0.000	0.000
	98	0.000	0.000	0.021	0.013	0.000
	100	0.979	0.000	0.006	0.000	0.000
α-*Gpd*	94	0.003	0.003	0.011	0.002	0.002
	100	0.991	0.994	0.983	0.990	0.990
	106	0.004	0.002	0.003	0.000	0.006
G6pd	97	0.000	0.000	0.000	0.000	0.772
	98	0.040	0.000	0.009	0.009	0.227
	99	0.000	0.000	0.000	0.084	0.000
	100	0.894	0.931	0.954	0.811	0.000
	101	0.000	0.000	0.000	0.018	0.000
	102	0.062	0.069	0.037	0.070	0.000
Hbdh	92	0.002	0.003	0.990	0.000	0.000
	98	0.000	0.000	0.000	0.000	0.959
	100	0.997	0.990	0.003	0.979	0.000

Table 3 (continued)

Locus	Allele	*Willistoni*	*Tropicalis*	*Equinoxialis*	*Paulistorum*	*Nebulosa*
Hk-1	96	0.024	0.016	0.081	0.000	0.000
	100	0.933	0.964	0.913	0.028	0.100
	104	0.039	0.018	0.004	0.971	0.883
Hk-2	96	0.010	0.007	0.013	0.006	0.022
	100	0.916	0.896	0.919	0.592	0.906
	102	0.009	0.000	0.000	0.383	0.000
	104	0.034	0.052	0.004	0.012	0.011
	108	0.031	0.040	0.000	0.004	0.060
Hk-3	96	0.006	0.012	0.007	0.000	0.000
	100	0.974	0.974	0.954	0.000	0.976
	104	0.013	0.012	0.036	0.944	0.000
	108	0.001	0.000	0.003	0.044	0.000
Idh	94	0.000	0.000	0.000	0.079	0.000
	96	0.005	0.004	0.002	0.000	0.000
	100	0.956	0.974	0.960	0.881	0.948
	104	0.033	0.016	0.038	0.000	0.040
	106	0.000	0.005	0.000	0.039	0.000
Lap-5	98	0.076	0.016	0.000	0.000	0.009
	100	0.246	0.162	0.000	0.034	0.048
	101	0.001	0.000	0.000	0.000	0.027
	103	0.481	0.557	0.004	0.027	0.739
	105	0.184	0.250	0.205	0.068	0.092
	107	0.000	0.012	0.712	0.799	0.054
	109	0.000	0.000	0.075	0.071	0.011
Mdh-2	86	0.001	0.990	0.003	0.000	0.000
	94	0.008	0.004	0.994	0.994	0.000
	98	0.000	0.000	0.000	0.000	0.983
	100	0.983	0.006	0.004	0.005	0.009

Table 3 (continued)

Locus	Allele	*Willistoni*	*Tropicalis*	*Equinoxialis*	*Paulistorum*	*Nebulosa*
Me-1	90	0.000	0.026	0.000	0.000	0.000
	94	0.000	0.923	0.000	0.006	0.000
	96	0.008	0.000	0.000	0.000	0.002
	98	0.013	0.000	0.000	0.984	0.000
	100	0.968	0.045	0.006	0.000	0.017
	104	0.009	0.001	0.990	0.000	0.976
Me-2	96	0.054	0.052	0.036	0.124	0.000
	100	0.765	0.731	0.801	0.784	0.000
	104	0.166	0.156	0.161	0.085	0.012
	106	0.000	0.000	0.000	0.000	0.940
	108	0.012	0.057	0.000	0.000	0.024
Odh-1	96	0.016	0.017	0.104	0.014	0.004
	100	0.929	0.937	0.826	0.926	0.964
	104	0.040	0.037	0.057	0.044	0.08
Pgm-1	92	0.001	0.000	0.000	0.000	0.049
	96	0.014	0.002	0.011	0.042	0.852
	98	0.000	0.000	0.000	0.000	0.068
	100	0.911	0.042	0.353	0.900	0.029
	104	0.072	0.919	0.622	0.057	0.000
	108	0.000	0.012	0.011	0.000	0.000
To	98	0.002	0.009	0.034	0.000	0.000
	100	0.931	0.983	0.956	0.998	0.000
	102	0.004	0.002	0.006	0.000	0.000
	120	0.000	0.000	0.000	0.000	0.975
Tpi-2	100	0.987	0.982	0.020	0.984	0.009
	105	0.000	0.000	0.000	0.000	0.985
	106	0.010	0.006	0.978	0.000	0.000

Table 3 (continued)

Locus	Allele	*Willistoni*	*Tropicalis*	*Equinoxialis*	*Paulistorum*	*Nebulosa*
Xdh	91	0.000	0.031	0.000	0.000	0.000
	92	0.000	0.013	0.000	0.000	0.010
	93	0.000	0.387	0.000	0.000	0.000
	94	0.000	0.017	0.000	0.000	0.015
	95	0.000	0.524	0.000	0.000	0.053
	96	0.001	0.019	0.003	0.002	0.730
	97	0.006	0.005	0.007	0.001	0.158
	98	0.113	0.000	0.031	0.194	0.031
	99	0.068	0.000	0.048	0.031	0.000
	100	0.467	0.000	0.687	0.404	0.000
	101	0.322	0.000	0.218	0.312	0.001
	102	0.017	0.000	0.002	0.020	0.000

a: Alleles are represented by numbers that refer to the electrophoretic mobility of their enzyme products; null alleles do not code for active enzymes under the experimental conditions. The average number of alleles per locus found in this group of species is about 13; only a few of the known alleles are shown in the table. The enzymes coded by each locus are given in Ayala *et al.* (1974a). The allelic frequencies given for *D. willistoni* are those found in the subspecies *D. w. willistoni*; for *D. equinoxialis* those found in the subspecies *D. e. equinoxialis*; and for *D. paulistorum* those found in the Amazonian semispecies.

quencies shown in Table 3 exhibit a remarkable pattern, which may be summarized in an oversimplified manner as follows. At each locus there are groups of two, three, and even more, species with virtually identical genetic configurations (*i.e.* have the same alleles in very similar frequencies), which are however completely different from the configurations present in other groups of species; moreover, the species within each group, *i.e.* the species which share identical genetic configurations, are different from one locus to another.

For example, at the *Acph-1* locus *D. paulistorum* and *D. nebulosa* have very similar genetic configurations which are nevertheless very different from those observed in the other three species which themselves are very different one from the other. However, at the *Adh* locus the first four species are very similar to each other, but very different from *D. nebulosa*. At the *Ao-1* locus, *D. nebulosa* is very similar to *D. willistoni* and *D. tropicalis*, but these three are very different from the other two species. At the *Lap-5* locus, *D. willistoni* and *D. tropicalis* are very similar to each other and different from the others, while at the *Xdh* locus *D. willistoni* is very different from *D. tropicalis* but similar to *D. equinoxialis* and *D. paulistorum*. And so on.

In order to examine the pattern of genetic variation of different species at different loci, we may use the following geometric representation. Let the allelic frequencies of a species at a given locus be represented by the position of the tip of a vector of unit length, in a set of coordinates. Let each coordinate represent an allele which may of course range in frequency from zero to one; the position of each species-vector is determined by the allelic frequencies. For example, it is assumed in the left of Figure 2 that there are only two alleles, A_1 and A_2, at a certain locus; the vectors S_1, S_2, and S_3 correspond to genetic constitutions for which the frequency of A_1 is 0.3, 0.5 and 0.8, respectively. The number of coordinates increases with the number of alleles, which results in a greater "area" for the hypersurface on which the tips of the species-vectors may lie.

When this kind of geometric representation is adopted for the species of the *D. willistoni* group, the following picture emerges. A total of 14 taxa (species, semispecies and subspecies) have been studied. The average number of alleles observed per locus is 13; *i.e.* the hypersurface on which the tips of the species vectors may lie has 12 dimensions, on the average. It is, however, the case that for any one locus the vectors occur in about three or four clusters, each consisting of three to five vectors in most cases. The vectors within a cluster are close together (they share in the same alleles with very similar frequencies), while different clusters are nearly orthogonal to each other, *i.e.* share virtually no alleles in common. This situation is represented for a hypothetical locus with only three alleles in the center of Figure 2 - the vectors for S_1, S_2, and S_3 are nearly parallel but lie at right angles with respect to two other clusters, one with S_4 to S_7 and the other with S_8 and S_9 (only nine rather than 14 species-vectors are represented).

A situation such as represented in the clusters of Figure 2 might be explained by the neutrality theory as the result of the history of the taxa, by assuming that S_1, S_2, and S_3 split from each other in recent evolutionary history, while they separated from the other taxa in the remote past. Thus, there would not have been sufficient time for the divergence of S_1, S_2, and S_3 from each other by random drift, but enough

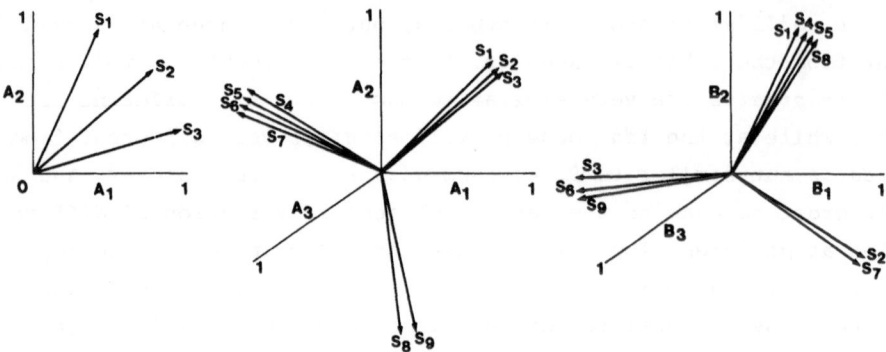

Figure 2. Geometric representation of the genetic constitution of species at a giv-
en locus. The genetic composition of each species is represented by a vector of unit
length in a set of Cartesian coordinates each representing a given allele. The position of the vector is determined by the allelic frequencies. *Left*: it is assumed that
there are only two alleles A_1 and A_2, at a given locus. The vectors S_1, S_2 and S_3 represent three different species in which the frequency of the A_1 allele is 0.3, 0.5
and 0.8, respectively. *Center*: vectors representing nine different species at a locus
with three alleles, A_1, A_2 and A_3. It is often found in the *Drosophila willistoni*
group that vectors occur in clusters. Vectors within a cluster are virtually parallel and represent species with very similar genetic constitutions. Vectors in different clusters are often at right angles, *i.e.* represent species with totally different genetic constitutions. *Right*: vectors representing the same nine species as in
the center of the Figure, but at a different locus, B, which also has three alleles.
As in the center of the Figure, the vectors occur in clusters; however, the sets of
species within each cluster are different from those at locus A. It is the case in
the *Drosophila willistoni* group that the species associated in a cluster are different from one gene locus to another.

time would have elapsed for the complete divergence of that set of species-vectors from the others. However this explanation would break down
if at a second locus, the position of the vectors might be as represented
in the right of Figure 2. In this case, S_1, S_4, S_5 and S_8 form a cluster,
S_2 and S_7 a second cluster, and S_3, S_6, and S_9 a third cluster. The evolutionary history that would satisfy the requirements of the neutrality
theory at the second locus (right of Figure 2) is incompatible with that
required for the first locus (center of Figure 2).

The clustering of 14 taxa of the *D. willistoni* group at 21 gene loci is shown in Table 4 using a principal components analysis (Gilpin and
Ayala, 1975). This analysis has been carried out by rotating the Cartesian coordinates so that they pass through the position of the various
clusters of species-vectors. In the representation of Figure 2, the a-
xes are single alleles; the rotated axes used for principal component
analysis are polymorphic combinations of the original axes. Orthogona-
lity is maintained, *i.e.* the rotated axes are at right angles. The table
shows for each taxon at each locus the single principal component which
explains most of the experimental variance. As would be expected from

Table 4. Principal components and proportion of variance explained for each of 21 gene loci in 14 taxa of the *Drosophila willistoni* group. The mean number of alleles per locus is 12.8±0.8

Loci	Taxa*														Percent explained of total variance
	t	*ww*	*wq*	*ee*	*ec*	*AM*	*IN*	*OR*	*AN*	*CA*	*T*	*pv*	*i*	*n*	
Acph-1	4	3	3	2	4	1	1	2	1	1	1	2	1	1	89.4
Adh	3	1	1	1	1	1	1	1	1	1	1	1	1	2	99.9
Adk-1	1	1	1	1	1	1	1	1	1	1	1	1	2	3	99.3
Adk-2	1	1	1	2	2	1	1	1	1	1	1	1	2	2	99.5
Ald-1	1	1	1	1	1	1	1	1	1	1	1	1	1	1	96.0
Est-2	1	2	1	2	1	1	1	1	1	1	1	3	4	3	97.0
Est-4	2	2	1	1	3	1	1	1	1	1	4	1	5	3	99.9
Est-5	2	1	1	1	3	1	1	1	1	1	1	1	2	2	99.9
Est-7	2	2	5	3	3	1	1	1	1	1	1	3	3	4	85.3
α-*Gpdh*	1	1	1	1	1	1	1	1	1	1	1	1	1	1	98.0
Hk-1	2	2	2	2	3	1	1	1	1	1	1	1	3	1	90.1
Hk-2	1	1	1	1	1	1	1	1	1	1	1	1	1	1	98.0
Hk-3	1	1	1	1	1	2	2	2	1	1	1	1	3	1	99.8
Idh	1	1	1	1	1	1	1	1	1	1	1	1	1	1	99.6
Lap-5	2	2	2	1	1	1	1	3	1	1	1	3	4	2	98.3
Mdh-2	3	2	2	1	4	1	1	1	1	1	1	1	3	4	86.9
Me-2	1	1	1	1	1	1	2	2	1	1	1	2	1	3	89.5
Pgm-1	2	1	1	2	2	1	1	1	1	1	1	1	2	3	99.9
To	1	1	1	1	1	1	1	1	1	1	1	1	2	2	92.0
Tpi-2	1	1	1	2	2	1	1	1	1	1	1	1	1	3	99.9
Xdh	3	1	3	1	1	1	1	1	1	1	1	1	2	2	77.3

Mean number of principal components per locus 2.86±0.25

Mean percent explained of total variance 95.0±1.4

*Symbols for the taxa are: *t* = *Drosophila tropicalis*, *ww* = *D. willistoni willistoni*, *wq* = *D. w. quechua*, *ee* = *D. equinoxialis equinoxialis*, *ec* = *D. e. caribbensis*; *Drosophila paulistorum* semispecies: *AM* = Amazonian, *IN* = Interior, *OR* = Orinocan, *AN* = Andean, *CA* = Centroamerican, *T* = Transitional; *pv* = *D. pavlovskiana*, *i* = *D. insularis*, *n* = *D. nebulosa*.

Table 3 and from what has been said above about the genetic configurations of the other species, several taxa lie on a given principal component. At the *Acph-1* locus, seven taxa share principal component 1,
three taxa share component 2, two taxa share component 3, and two other
taxa share component 4; these four components account for 89 percent
of the variance observed at this locus in all 14 taxa.

The important points emerging from this principal component analysis are as follows. (1) Groups of taxa share in a given principal
component. The mean number of alleles per locus is 12.8, but most of
the experimental variance (95 percent) can be explained by an average
of only 2.9 principal components per locus. (2) The sets of taxa that
share in a given principal component vary from locus to locus. For e-
xample at the *Est-5* locus *D. tropicalis* (*t*) and *D. insularis* (*i*) share
in component 2, while *D. willistoni* (*ww*) and *D. pavlovskiana* (*pv*) share
in component 1; however, at the *Est-7* locus *D. tropicalis* and *D. willistoni* are similar to each other (they share in component 2, but are different from *D. pavlovskiana* and *D. insularis* which share in component
3). The heterogeneity of the clustering of taxa from locus to locus excludes the possibility that the clustering simply results from the phylogenetic relationships among the taxa.

The occurrence of different clusters of species at different loci
might seem compatible with the neutrality theory at least in the case
of monomorphic loci. Species that have evolved independently for long
periods of time and that at some time might have been genetically different at a polymorphic locus, might become genetically similar when
they become fixed (perhaps by passing through a bottleneck in population
size) for the same allele. We might have thus a situation with any two
species fixed for exactly the same allele at some loci but for different alleles at other loci. However, this explanation is unsatisfactory
even for monomorphic loci. As time proceeds any two independently evolving populations become genetically different. If they become fixed for
a given allele as a consequence of random drift, the a priori probability that they would become fixed for the same allele is $1/k$, and the
probability that they would become fixed for different alleles is
$(k-1)/k$, where k is the number of alleles (see Ayala and Gilpin, 1974).
The average number of alleles observed per locus is 13 in the *D. willistoni* group. Therefore, any two species are expected to become fixed
for the same allele (and thus be genetically similar) at only 1/13 of
all fixed loci. However, pairs of species are similar at a proportion
of loci much greater than 1/13.

Table 5.   Genetic distances among five *Drosophila* species based
            on the allelic frequencies at the *Ao-1* locus (above the
            diagonal) or at the *Xdh* locus (below the diagonal)

	willistoni	tropicalis	equinoxialis	paulistorum	nebulosa
D. *willistoni*	–	0.014	3.612	3.016	0.007
D. *tropicalis*	8.948	–	4.343	4.135	0.014
D. *equinoxialis*	0.058	8.517	–	0.015	3.963
D. *paulistorum*	0.020	9.210	0.117	–	3.244
D. *nebulosa*	4.343	2.430	4.828	3.963	–

RANDOM DRIFT AND THE PHYLOGENY OF SPECIES

There is another way of looking at the genetic configurations of the
species of the D. *willistoni* group which makes further apparent the
inconsistency between such configurations and the expectations derived from the neutrality theory. The argument now to be presented is particularly compelling in the case of polymorphic loci. Consider the allelic frequencies at the *Ao-1* locus as shown for five species in Table
3. The genetic distance, D, between these species calculated as $D = -\log_e I$, where I measures genetic similarity (Nei, 1972), is shown above
the diagonal in Table 5. (Other measures of genetic differentiation,
for example 1-I, would have given similar results for the present purposes; we are interested only in the *relative* amount of differentiation between pairs of species, not in the absolute values of D.).

   If we assume, following the neutrality theory, that allelic frequencies change by random sampling from generation to generation, then
the phylogeny of the five species based on the *Ao-1* locus should be
as shown in the left side of Figure 3. Note that the average degree of
genetic differentiation between the set consisting of D. *tropicalis*,
D. *willistoni*, and D. *nebulosa*, and the set consisting of D. *paulistorum* and D. *equinoxialis* is 3.72, *i.e.* several hundred times greater
than the genetic differentiation between species belonging to the same
set (0.007 to 0.015). The neutrality theory would account for the similar genetic configurations of D. *tropicalis*, D. *willistoni*, and D. *ne-*

bulosa at this highly polymorphic locus by assuming that their evolutionary divergence from each other occurred in the very recent past. If
these species had evolved independently for long periods of time under
random drift, they would have become genetically quite different. The
probability that the great similarity of the three species at this highly polymorphic locus would have been acquired by random convergence of
previously differentiated populations is effectively nil, given the large
number of alleles known at the locus, and the high degree of polymorphism of the species.

Similarly, according to the neutrality theory the great resemblance
between *D. paulistorum* and *D. equinoxialis* at the *Ao-1* locus would be
due to the recency of their evolutionary divergence. The alternative explanation, namely that these two species were genetically different in
the past but have become virtually identical by independent random
changes in allelic frequencies, is incredible. On the other hand, the
great genetic differentiation between one set of species (*D. tropicalis*,
D. willistoni, and *D. nebulosa*) and the other set (*D. paulistorum* and
D. equinoxialis) would be explained by the neutrality theory by postulating that the ancestral species of the two sets diverged from each other
in the remote evolutionary past.

Consider now the allelic frequencies at the *Xdh* locus (Table 3); the
genetic distances based on this locus are shown in Table 5, below the
diagonal. If it is assumed that the phylogeny based on the *Ao-1* locus is
nearly correct, the proponents of the neutrality theory must conclude
that *D. willistoni* has acquired the genetic configuration observed in
D. paulistorum and *D. equinoxialis* at this locus independently of those
species by random change over a long evolutionary span – a conclusion
that defies credibility. Alternatively one could claim that *D. paulistorum* and *D. equinoxialis* have converged by random change towards the
allelic frequencies observed in *D. willistoni* – an equally unlikely possibility. Moreover, the neutrality theory would force one to accept that
at the *Xdh* locus *D. willistoni* has become quite different from *D. nebulosa* and *D. tropicalis* (and the latter two from each other) by random
changes in spite of their recent evolutionary divergence – again, an
unacceptable possibility since these three species remain virtually i-
dentical at the *Ao-1* locus.

It might, of course, be proposed that the genetic configurations observed at the *Xdh* locus reflect the true evolutionary history of these
species, as shown on the right of Figure 3. But then, proponents of the
neutrality theory would have to accept that *D. willistoni*, *D. tropicalis*

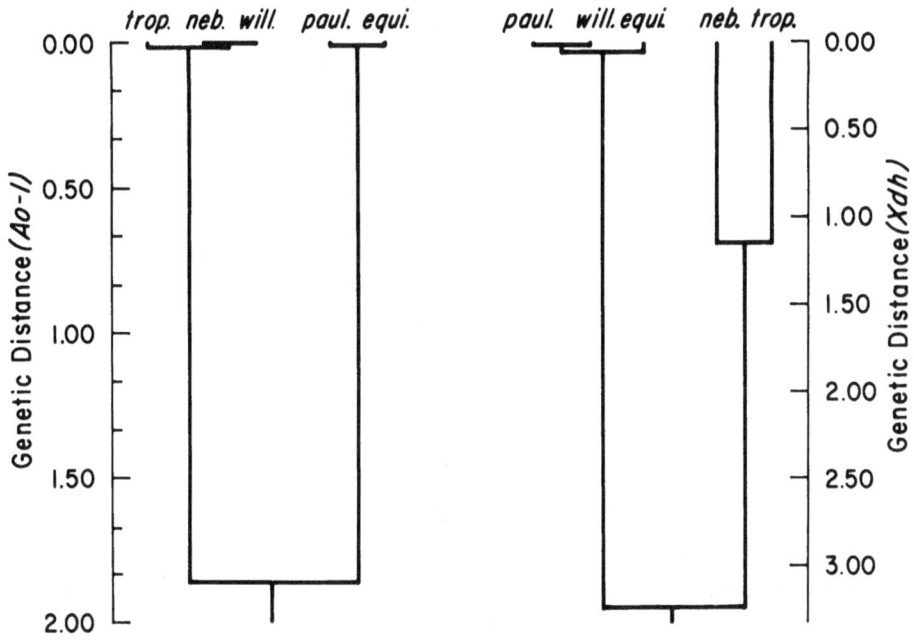

Figure 3. Phylogenies of five species of the *Drosophila willistoni* group at two gene loci, *Ao-1* and *Xdh*; *trop* = *D. tropicalis*, *neb* = *D. nebulosa*, *will* = *D. willistoni*, *paul* = *D. paulistorum*, *equi* = *D. equinoxialis*. The phylogenies are inferred from the genetic distances shown in Table 5 by assuming that the degree of genetic differentiation between two species at a given locus is proportional to the time since the divergence of the species. The neutrality theory may explain the great genetic similarity between *D. tropicalis*, *D. nebulosa* and *D. willistoni* at the *Ao-1* locus (See Tables 3 and 4) by assuming that they have diverged from each other only very recently. But then it becomes necessary to postulate that at the *Xdh* locus *D. willistoni* has acquired nearly identical genetic configuration as *D. paulistorum* and *D. equinoxialis* by an independent series of random changes over a long evolutionary period - a virtual impossibility.

and *D. nebulosa* have acquired their similar genetic configurations at the *Ao-1* locus through random convergence over long evolutionary periods - which is once again a totally unlikely possibility.

The argument presented in the previous paragraphs becomes considerably strengthened when the genetic configurations at other loci are taken into consideration (see Table 3 for five species, and publications referred to throughout this article for other taxa of the *D. willistoni* group). As stated earlier, the general situation is that any two species are found to have very similar genetic configurations at nearly half the loci and very different at nearly the other half. If the similarities at some loci are explained by recency of evolutionary divergence, it becomes necessary to account for the great differentiation at many other loci, which cannot be accomplished by random drift over short periods of time. Alternatively, if a fairly long period of evolutionary divergence is allowed for the species it becomes impossible to account for the similarities observed at many loci (particularly at the highly polymorphic ones) by random changes in allelic frequencies.

Any attempts to explain the genetic configurations observed in the *D. willistoni* group by the neutrality theory (or by any other hypothesis) must take into consideration simultaneously the patterns observed in all species of the group. It might be claimed that the species have diverged only recently, and that this accounts for the similarities observed at many loci between any two species. Differentiation at the other loci would then be explained by (diversifying) selection, but the loci with similar genetic configurations would need not be subject to selection. But when the five species shown in Table 3 are simultaneously considered, it is the case that only four or five loci (*Ald-1*, α*Gpd*, *Idh*, *Odh-1*, and perhaps *Hk-2*) are quite similar in *all* species. The number of loci not subject to natural selection would further be reduced when the other taxa in the group are considered.

Alternatively, one might suggest that the species diverged in a remote evolutionary past, and that only those loci in which any two species have virtually identical genetic configurations might be subject to (stabilizing) selection. But it is the case that there are *at least* two species with very similar genetic configurations at every locus studied (see Table 3).

It would seem most reasonable to assume that the evolutionary history of the *D. willistoni* group species is reflected on the *average* degree of differentiation observed at all loci studied (see Ayala *et al.* 1974b for the phylogeny inferred from genetic distances). Any sibling species is about equally different from all other siblings ($\bar{I}$ = 0.588 + 0.029 for the six pairwise comparisons between the four siblings), but *D. nebulosa* is more different from any of the siblings than these are from each other ($\bar{I}$ = 0.299 + 0.020 for the four comparisons between *D. nebulosa* and any of the siblings).

ADAPTIVE "FOCI" IN PROTEIN EVOLUTION

If we conclude from the various arguments presented in this paper that the genetic patterns observed in the *D. willistoni* group are incompatible (at least for most loci) with the neutrality theory of protein evolution, we must still propose an alternative explanation for such genetic patterns. The most obvious explanation (and the only alternative to neutrality available in current evolutionary theory) is natural selection. Similar genetic configurations in different populations may be explained by stabilizing natural selection; different configurations by diversifying selection. (Natural selection of course operates together with other processes including genetic drift, mutation, recombination, and migration). But there lies an epistemological difficulty. The fact that natural selection is in principle compatible with many different states of affairs in the living world, makes it all the more difficult to test the theory: appropriate testing of an empirical hypothesis requires to devise observations and experiments that might lead to the *falsification* of the hypothesis if this is not correct (see Ayala, 1974, 1975). There are methods to test empirically whether or not natural selection is occurring in a particular case. Ultimately, however, the evolutionary geneticist wants to ascertain the adaptive basis (at the physiological, morphological, or behavioral level) of any genetic polymorphisms maintained by natural selection. Unfortunately, this has proven to be a difficult task; we know the adaptive basis of very few polymorphisms (Lewontin, 1974).

It is worthwhile to advance a descriptive model of the genetic patterns observed in the *D. willistoni* group. This model underscores the occurrence of sets of similar genetic configurations at some, but not other loci in different species. The points to be made here are facilitated by reference to the geometric representation outlined in Figure 2 or to the results of principal component analysis given in Table 4. It is observed at any one locus that there are sets or clusters of species, so that the species in a cluster have virtually identical genetic configurations, but species in different clusters have quite different genetic constitutions. In the representation of Figure 2, this implies that the vectors representing some species (*e.g.* S_1, S_2, and S_3 in the center of the figure) are virtually parallel to each other, but are at right angles relative to the vectors representing other species (*e.g.* S_4, S_5, S_6, S_7, S_8, or S_9). That is, the tips of vectors

lie in clusters each with a fairly small area. These small areas where
clusters of vectors end may be referred to as "adaptive foci".

The picture that emerges from Tables 3 and 4 (and from additional
data published elsewhere) is as follows. In spite of the large "area"
of the hypersurface on which the species-vectors can end (remember
that about 13 alleles = 12 dimensions are known, on the average, per
locus), the tips of the vectors are found concentrated in a few highly
dense regions. It is these regions (and the corresponding configurations
of allelic frequencies) that are called "adaptive *foci*". It is the case
that these adaptive foci are reached by different species independently
from one another. This must be so, since different *sets* of species share
adaptive foci at different loci, and moreover several adaptive foci
exist at most loci. Thus at most loci, some species at least have moved
from whatever may have been the ancestral genetic configuration since
several such configurations (adaptive foci) exist now. Species that are
found in a given new adaptive focus cannot have reached there together
(*i.e.*, as descendants of a common ancestor that moved there before the
species diverged) because at many loci species sharing a given focus
diverged from each other earlier than some of them diverged from species in different foci.

The genetic patterns apparent in Figure 1 (and Tables 3 and 4)
imply that at a given locus species evolve from one adaptive focus to
another at very fast rates. This follows from the observation that individual species that do not share an adaptive focus with any other species are rare events. At most loci, all species are clustered in groups
of two or more species in each adaptive focus (or principal component -
see Table 4). If species would evolve slowly from one to another adaptive focus, we would often find individual species not associated with
any other species, because they were in the process of moving from one
focus to another.

Three final comments. First, it is not presently known why a few
particular genetic configurations are adaptive, while other intermediate configurations are not. Second, the arguments advanced in this
paper are not affected even if it is shown that each electrophoretically detectable allele consists in fact of several alleles that might
be distinguishable by other methods; different electrophoretic alleles
would in any case be different. What has been said herein about "alleles" could be reworded in terms of "sets of alleles". Third, the expression "adaptive foci" is inspired by Sewall Wright's "adaptive
peaks". However, Wright was speaking about particular genetic combina-

tions between alleles at *different loci*. The "adaptive foci" are equilibrium configurations of allelic frequencies at a *single gene locus* in a given population.

Summary.

Techniques such as gel electrophoresis have shown that a great deal of genetic variation exists in natural populations of most organisms. From 20 to more than 50 percent of all gene loci are polymorphic in a given population; diploid individuals are heterozygous at between 5 and 20 percent of their gene loci.

It has been suggested that most of the protein variation observed in natural populations is adaptively neutral. This proposition, known as the neutrality theory of protein evolution, implies that protein evolution occurs by the random process of sampling from generation to generation rather than by natural selection.

Evidence obtained from the study of a group of species of *Drosophila* appears to be incompatible with the neutrality theory. Thirty-six gene loci coding for enzymes have been studied in thousands of individuals belonging to 14 closely related species and incipient species. At a given locus there are several clusters each with several species; species within a cluster have virtually identical genetic configurations (*i.e.* the same alleles in similar frequencies), while species in different clusters have completely different genetic constitutions. This pattern cannot be attributed to the evolutionary history of the species because the clusters contain different species at different loci.

The patterns of genetic variation in the species of the *Drosophila willistoni* group cannot be explained by a random process such as genetic drift. Protein evolution, therefore, appears to be subject to natural selection. It seems that at any given gene locus there are a few particular configurations of allelic frequencies that are adaptive for a population, while intermediate or alternate combinations of allelic frequencies are inadaptive. The adaptive genetic combinations may be called "adaptive foci". Species evolve by moving from one adaptive focus to another at very fast rates.

Acknowledgements.

This research was supported by Grant DEB74-0892612 from the National Science Foundation and by Contract Pa200-14 Mod4 with the Energy Research and Development Administration.

REFERENCES

Ayala, F.J. 1972. Darwinian versus non-Darwinian evolution in natural populations of *Drosophila*.
Proc. 6th Berkeley Symp. Math. Stat. Prob. V: 211-236.

Ayala, F.J. 1974. Biological evolution: natural selection or random walk?
Amer. Sci. 62(6): 692-701.

Ayala, F.J. 1975. Scientific hypotheses, natural selection and the neutrality theory of protein evolution. In *The Role of Natural Selection in Human Evolution*, Francisco M. Salzano (ed.), pp. 19-42. American Elsevier Publ. Co., Inc. New York.

Ayala, F.J. 1977. Philosophical Issues. In *Evolution*, T. Dobzhansky, F.J. Ayala,
G.L. Stebbins, J.W. Valentine (eds.), ch. 16. Freeman, San Francisco.

Ayala, F.J., and Gilpin, M.E. 1974. Gene frequency comparisons between taxa: support for the natural selection of protein polymorphisms.
Proc. Nat. Acad. Sci. USA 71: 4847-4849.

Ayala, F.J., Powell, J.R., Tracey, M.L., Mourão, C.A., and Pérez-Salas, S. 1972a.
Enzyme variability in the *Drosophila willistoni* group. IV. Genic variation in
natural populations of *Drosophila willistoni* group.
Proc. Nat. Acad. Sci. USA 71: 999-1003.

Ayala, F.J., Powell, J.R., and Tracey, M.L. 1972b. Enzyme variability in the *Drosophila willistoni* group. V. Genic variation in natural populations of *Drosophila
equinoxialis*.
Genet. Res. 20: 19-42.

Ayala, F.J., Tracey, M.L., Barr, L.G., McDonald, J.F., and Pérez-Salas, S. 1974a.
Genetic variation in natural populations of five Drosophila species and the hy-
pothesis of the selective neutrality of protein polymorphisms.
Genetics 77: 343-384.

Ayala, F.J., Tracey, M.L., Hedgecock, D., and Richmond, R.C. 1974b. Genetic differentiation during the speciation process in *Drosophila*.
Evolution 28: 576-592.

Ayala, F.J., Valentine, J.W., DeLaca, T.E., and Zumwalt. G.S. 1975. Genetic variability of the antarctic brachiopod *Liothyrella notorcadensis* and its bearing on
mass extinction hypotheses.
Soc. Econ. Paleont. Mineral. (1975): 1-9.

Dobzhanksy, Th., Levene, H., Spassky, B., Spassky, N. 1959. Release of genetic variability through recombination. III. *Drosophila prosaltans*.
Genetics 44: 75-92.

Gilpin, M.E., and Ayala, F.J. 1975. Adaptive foci in protein evolution.
Nature 253: 725-726.

Harris, H. 1966. Enzyme polymorphisms in man.
Proc. Roy. Soc. Ser. B 164: 298-310.

Harris, H., and Hopkinson, D.A. 1972. Average heterozygosity in man.
J. Human Genet. 36: 9-20.

Kimura, M. 1968. Evolutionary rate at the molecular level.
Nature 217: 624-626.

Kimura, M., and Ohta, T. 1971. Protein polymorphism as a phase of molecular evolution.
Nature 229: 467-469.

King, J.L., and Jukes, T.H. 1969. Non-Darwinian evolution.
Science 164: 788-798.

Lewontin, R.C. 1974. *The Genetic Basis of Evolutionary Change*. Columbia Univ. Press,
New York.

Lewontin, R.C., and Hubby, J.L. 1966. A molecular approach to the study of genic
heterozygosity in natural populations. II. Amount of variation and degree of heterozygosity in natural populations of *Drosophila pseudoobscura*.
Genetics 54: 595-609.

Medawar, P. 1967. *The Art of the Soluble*. Methuen, London.

Nei, M. 1972. Genetic distance between populations.
    Amer. Nat. 106: 283-292.

Ohta, T. 1973. Slightly deleterious mutant substitutions in evolution.
    Nature 246: 96-97.

Ohta, T. 1974. Mutational pressure as the main cause of molecular evolution and
    polymorphism.
    Nature 252: 351-354.

Popper, K. 1959. *The Logic of Scientific Discovery.* Hutchinson, London.

Selander, R.K. 1976. Genic variation in natural populations. In *Molecular Evolu-
    tion,* F.J. Ayala (ed.), pp. 21-45. Sinauer Associates, Mass.

Selander, R.K., and Kaufman, D.W. 1973. Genic variability and strategies of adap-
    tation in animals.
    Proc. Nat. Acad. Sci. USA 70: 1875-1877.

Spassky, B., Richmond, R.C., Pérez-Salas, S., Pavlovsky, O., Mourão, C.A., Hunter,
    A.S., Hoenigsberg, H., Dobzhansky, T., and Ayala, F.J. 1971. Geography of the
    sibling species related to *Drosophila willistoni,* and of the semispecies of the
    Drosophila paulistorum complex.
    Evolution 25: 129-143.

Valentine, J.W., and Ayala, F.J. 1975. Genetic variation in *Frieleia halli,* a
    deep-sea brachiopod.
    Deep-Sea Research 22: 37-44.

Valentine, J.W., and Ayala, F.J. 1976. Genetic variability in krill.
    Proc. Nat. Acad. Sci. USA 73: 658-660.

2. STUDY OF POLYMORPHISM

POLYMORPHISM, SELECTION, AND MULTI-LOCUS HETEROZYGOSITY
IN THE PLAICE, *PLEURONECTES PLATESSA* L.

J.A. Beardmore and R.D. Ward

The central problem of evolutionary genetics concerning the nature of
the forces primarily responsible for the generally abundant polymorphism demonstrated by electrophoretic methods in both plant and animal species forms the basis of a vigorous and continuing debate. Despite the considerable body of data now available on gene frequencies
at many enzyme loci in many populations, critically unequivocal evidence that natural selection discriminates between genotypes at such
loci is still sparse. There are a number of studies, both of natural
populations and of laboratory populations, which indicate, frequently
indirectly, that this is probable (for example, Koehn 1969; Bryant 1974;
McDonald and Ayala 1974; Nevo 1976), and in the laboratory several
workers have shown that strong selection is operative either on the
loci themselves or on tightly linked complexes which, if they exist,
must themselves display considerable polymorphism (Gibson 1970; Huang,
Singh and Kojima 1971; Birley and Beardmore 1972a; van Delden, Kamping and van Dijk 1975). The best evidence of this sort is drawn from
Drosophila, but it must be stated that some carefully controlled experiments failed to produce any evidence of selection operating upon
a polymorphic esterase locus in *D. pseudoobscura* (Yamasaki 1971).

While the potential importance of linkage in considering the effects of selection upon multi-locus variation is generally recognised,
and some examples of linkage disequilibrium between pairs of enzyme

loci have been recorded (Charlesworth and Charlesworth 1973; Zouros
and Krimbas 1973), the extent to which natural populations actually
display systematic linkage disequilibrium as a result of fitness interactions between polymorphic loci is largely unknown. Lewontin (1974)
points out that different species may adopt radically different strategies in this respect. Although several models have been put forward
which relate fitness to multi-locus heterozygosity (King 1967; Milkman 1967; Sved, Reed and Bodmer 1967), relatively little experimental
effort has so far been expended on the collection of relevant data. To
carry out a suitable analysis requires ideally that individual genotypes are determined at various stages of the life cycle so that any
differences in viability, mating success, gametic transmission and
zygote production can be determined much as in the work by Frydenberg
and co-workers on an esterase locus in *Zoarces viviparus* (Christiansen and Frydenberg 1973; Christiansen, Frydenberg and Simonsen 1973;
Christiansen, Frydenberg, Gyldenholm and Simonsen 1974).

The systematic study of the population genetics of marine organisms
has, until recently, been a neglected area despite the pioneering work
of, for example, Battaglia (1958). It is fitting, however, to note
that Ove Frydenberg was one of those who, before the mass adoption of
electrophoretic methods of analysis, recognised that such organisms
are interesting as components of important ecosystems about which we
are still highly ignorant and instructive in that they offer particular features advantageous for studies in population genetics (Frydenberg *et al.* 1965)

The work described in this paper concerns a marine teleost, the
plaice *Pleuronectes platessa*. This is a flat-fish, the general biology and life-cycle of which is reasonably well known (Wimpenny 1953).
Our aim in this work is to investigate in general terms the adaptive
significance of enzyme polymorphism and in particular to see to what
extent the multiple genotype determined by examining the products of
a number of loci in the same individual influences Darwinian fitness.

The plaice is a widespread and commercially important species,
spawning in well defined areas and usually reaching sexual maturity
at two to three years of age. At fertilisation a planktonic zygote is
formed which gives rise to a bilaterally symmetrical larva which meta-
morphoses at between 10 and 14 mm into a flattened fish. The small fish
enter shallow water on sandy beaches for a period of about six to nine
months and then gradually move off into progressively deeper water.
The age of adult individuals can be simply and accurately determined

by counting the number of rings on the otoliths (ear bones). We have
studied the Bristol Channel population of plaice, which is thought by
fisheries biologists to be a largely discrete entity although some
migration in and out of neighbouring populations has been recorded
(Macer 1972). The spawning ground at which mating takes place in late
January or February is off Trevose Head, Cornwall, and the young fish
are carried by ocean currents to the nursery grounds along the South
Wales coast.

RESULTS AND DISCUSSION

We have described in broad terms the extent of enzyme polymorphism in
this species (Birley and Beardmore 1972b; Ward and Beardmore, in preparation). Of some 45 loci examined, 23 are polymorphic by the criterion of the most common allele having a frequency less than 0.99.
Eight polymorphic loci have been typed in large numbers of fish, but
three of these control liver-specific enzymes and cannot be reliably
typed in very young individuals. In this paper we are concerned with
the five most polymorphic loci expressed in muscle tissue (PGM-1,
phosphoglucomutase; ADA, adenosine deaminase; αGPDH-1, αglycerophos-
phate dehydrogenase; MDH-2, malate dehydrogenase, and PG1-2, phospho-
glucose isomerase or phosphohexose isomerase). These loci give highly
stable products which maintain activity for several years in material
stored at -20°C. With the exception of about 20 fish, which for a variety of reasons could not be typed at one or more loci, all of the
fish collected (ca. 2,300) were assayed at each locus. Allele frequencies at each of these five loci are given in Table 1.

Figure 1 shows the distribution of mean heterozygosity over all
the five loci in fish of different age classes. The data for 0-group
fish (i.e. those less than one year old, points 1-7 in Figure 1) are
taken from the extensive samplings of 1975, the data for the older fish
are from the 1973, 1974 and 1975 samples combined. Sample sizes and
approximate ages of the fish represented in Figure 1 are given in
Table 2. The age estimates assume a nominal fertilization date of
February 14th in each year. Heterozygosity is calculated by pooling
together all observed heterozygotes, and thus takes no account of any
differences in frequencies of different heterozygotes at any locus
nor of differences between loci. Two distinct trends are evident in

Table 1. Allele frequencies at five polymorphic loci in plaice (allele 1 codes for the allozyme with least mobility in each case)

Locus	No. of fish sampled	ALLELE							
		1	2	3	4	5	6	7	8
ADA	2272	0.007	0.128	0.013	0.740	0.110	0.002	-	-
αGPDH-1	2275	0.003	0.001	0.001	0.066	0.872	0.002	0.055	-
MDH-2	2275	0.118	0.876	0.005	-	-	-	-	-
PG1-2	2275	0.001	0.013	0.943	0.038	0.005	-	-	-
PGM-1	2273	0.002	0.005	0.611	0.360	0.002	0.002	0.006	0.012

Table 2. Estimated ages and numbers of fish included in the points plotted in Figure 1

Age Group	0							1	2	3+
Sample Point Number=	1	2	3	4	5	6	7	8	9	10
Estimated Age (days)	98-130	131-160	161-190	191-220	221-250	251-280	281-310	ca550	ca900	>ca1250
Sample Size	217	114	198	168	175	144	100	57	314	124

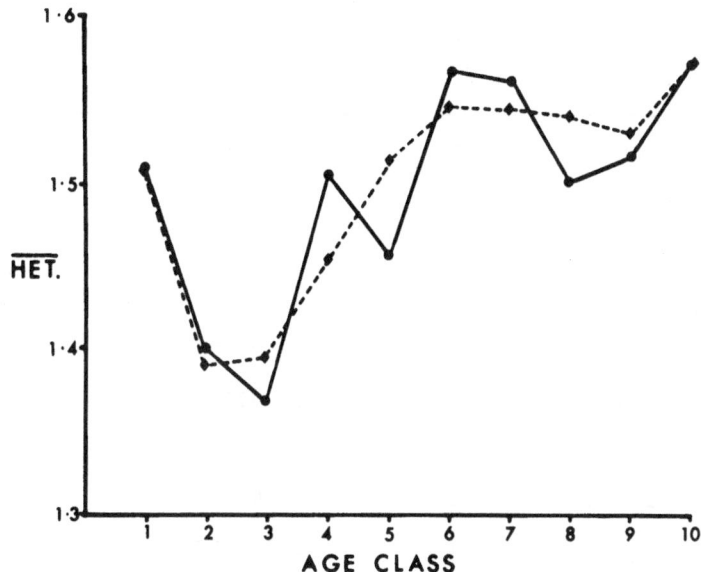

Figure 1. Changes in mean heterozygosity over five loci in different age classes solid line). Table 1. gives approximate ages of fish and total sample sizes. The 0 group fish (pts 1–7) are from 1975 only, the older fish are from 1973, 1974 and 1975. The expected line plotted is the quartic orthogonal polynomial fit (dashed line): $Y = 1.503 + 0.007255\xi_1 + 0.001174\xi_2 - 0.000798\xi_3 + 0.002184\xi_4$.

Figure 1. The earliest samples show a progressive fall in heterozygosity to about 190 days of age which is then followed by a steady increase in the older 0 group fish. Heterozygosity in adult fish is similar to that of the oldest 0 group. Sample sizes are substantial but statistical analysis of these data presents problems. A polynomial regression analysis is probably appropriate and we are indebted to Dr. R.I. Gilbert for such an analysis (Table 3). The linear component borders on statistical significance and the quartic component is significant at P = 0.034.

   The decrease in heterozygosity in the very young 0 group fish is examined more closely in Figure 2. Here the heterozygosity determinations included in points 1–3 of Figure 1 are replotted as individual

Table 3.   Variance analysis for regression of heterozygosity
             on age of plaice (Figure 1)

Item	d.f.	mean square	F	P
Linear	1	0.017367	5.08	0.054
Residual	8	0.003418		
Quadratic	1	0.000182	< 1	N.S.
Residual	7	0.003881		
Cubic	1	0.005458	1.51	N.S.
Residual	6	0.003618		
Quartic	1	0.013636	8.45	0.034
Residual	5	0.001614		
Quintic	1	0.000028	< 1	N.S.
Residual	4	0.002011		

samples from three separate South Wales nursery beaches. These data
cover the period from the time of arrival of the first new 1975 0-
group fish to (probably) the end of arrival of such fish. A linear regression analysis (Table 4) shows that the joint regression coefficient is significantly negative ($P < 0.02$). The initial decline in
heterozygosity is a real effect.

In 1973 and 1974 no very young fish were collected, but for both
years data for the older 0 group fish parallel the 1975 data in showing an increase in heterozygosity. (Figure 3). However, the lines are
displaced with respect to each other and to the 1975 line.

A linear regression using the mean heterozygosities of each individual sample from all three years of 0 group fish from age 133 days
onwards shows a highly significant positive regression ($P = 0.001$).

The changes in heterozygosity are not the result of statistically
detectable changes in allele frequencies at the loci concerned and at
each locus there is no significant deviation from Hardy-Weinberg expectation. Figure 4 shows that all loci contribute to the changes seen
in Figure 1 although the precise pattern of change is not the same

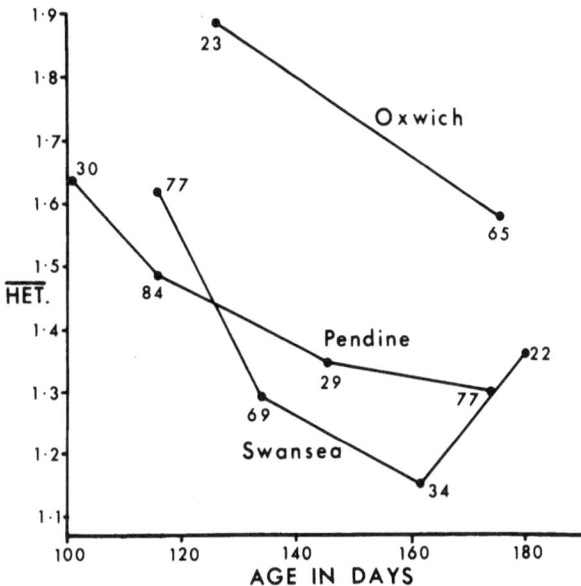

Figure 2. Mean heterozygosity over five loci in individual samples of very young
0-group fish (1975). Collections from three South Wales nursery beaches are plotted separately, and sample sizes given.

Table 4. Analysis of variance of data plotted in Figure 2

Item	d.f.	M.S.	t	P
Joint regression	1	0.1131	3.19[a]	< 0.02
Between regressions	2	0.0106	< 1	N.S.
Between means	2	0.1192	3.27[a]	< 0.02
Error	4	0.0113	–	–
Total	9	0.4181		

a: tested against pooled error with 6 d.f.

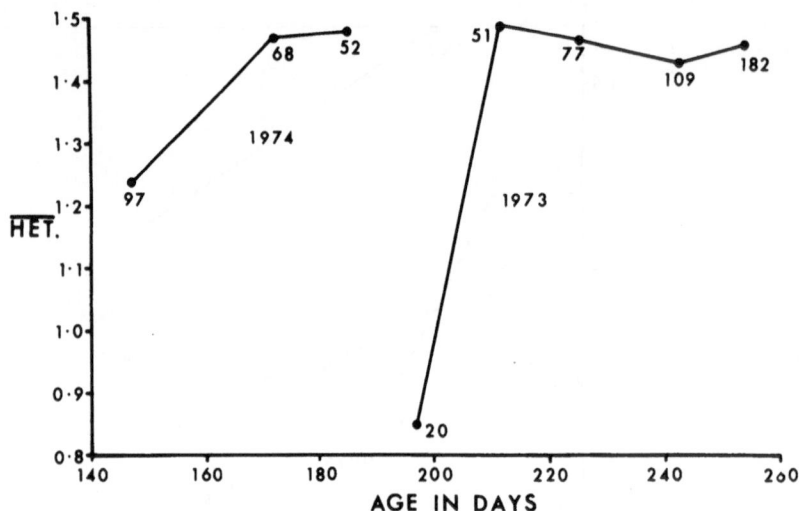

Figure 3. Mean heterozygosity over five loci of 0-group fish collected in 1973 and
1974. Individual samples are plotted, and sample sizes given.

from locus to locus. Exactly similar responses to selective processes
over the five functionally distinct loci would be extremely surprising.
    In Figure 5 the percentages of fish with 0 or 1 loci heterozygous
and those with 2 or more heterozygous are plotted against age class.
An initial fall in the frequency of the multiple heterozygote is followed by a rise which plateaus in the older classes. The simplest interpretation of these results is that heterozygosity influences survival in a positive way. How then do we account for the steep fall in
heterozygosity observed in the young fish on the nursery beaches? Our
provisional interpretation is that this fall is a reflection not of
the relative composition of the genotypes of the very young fish in
general, but rather of those which have actually arrived at the beaches. We have never been able to catch young fish on the beaches before
May, although attempts to do so have been made. The numbers of the earlier arrivals are small and the size of the population of young fish
builds up over a period of two to three months. The simplest explanation of our observations in that the more heterozygous fish, develop-

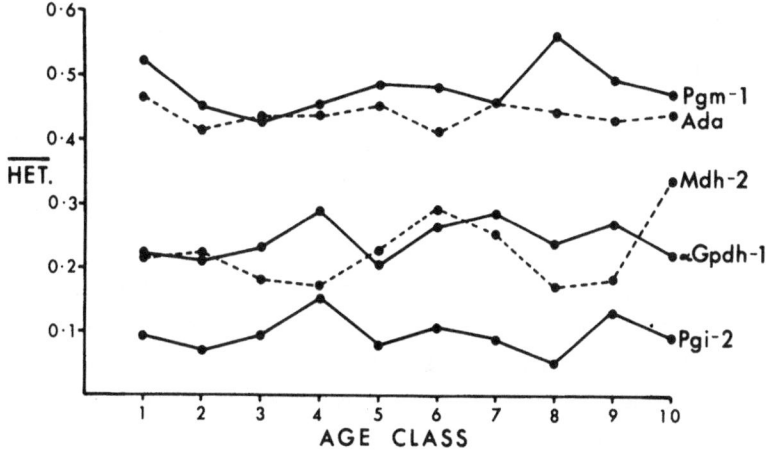

Figure 4.   Changes in mean heterozygosity at the five loci considered separately.
The samples are those depicted in Figure 1 and Table 1.

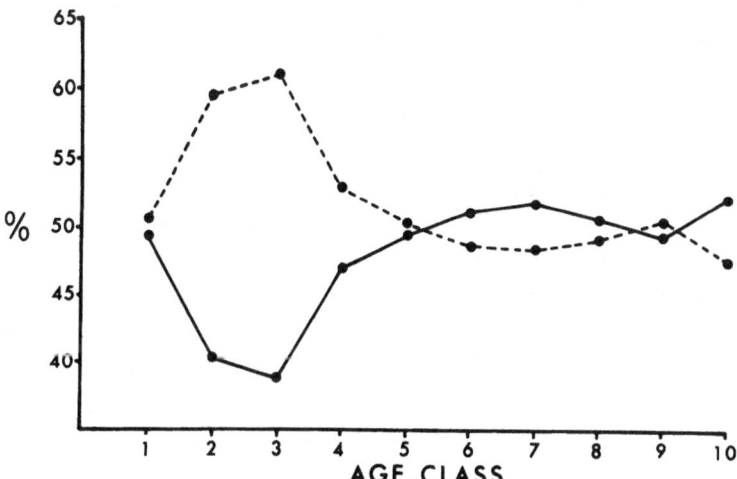

Figure 5.   Percentage of individuals with 0 or 1 loci heterozygous (dashed line) compared with the percentage heterozygous at 2, 3, 4 or 5 loci (solid line). The samples
are those depicted in Figure 1 and Table 1.

ing more rapidly, reach the beaches first, and are followed by the
less heterozygous individuals. As there is evidence of considerable
variation in rate of development to metamorphosis (Ryland 1966),
this seems a reasonable hypothesis.

It seems likely that the observed differences between the time of
the rise in heterozygosity in older 0 group fish in the years 1973,
1974 and 1975 owe something to differences in environmental conditions.
A very limited amount of data on water temperature at two sites in
the sampling area (Table 5) suggests that in the early part of the
year which is most important in influencing spawning and early development, 1974 was a warmer year than either 1973 or 1975. While many
factors other than temperature must be involved in growth and development, these findings correlate reasonably well with the data on heterozygosity. Temperature is known to be an important factor in controlling the rate of development of plaice (Ryland and Nichols 1975).

Table 6 displays data on the total incidence per locus, for 0
group and older (1+) group fish, of alleles with individual frequencies of less than 0.01. A proportion of such alleles might be the result of new or recent mutations, although the work of Harris (1974)
suggests that, at least in man, only very exceptionally are rare alleles expected to result from new mutations. Whether or not these uncommon genes are effective components of a balanced polymorphism is a
matter for speculation, but from Table 6 we can see that there is no
strong selection against this class of alleles. Indeed if any trend
is discernible, it is that the frequency of these alleles increases
slightly from 0 group to 1+ group fish.

The distribution of heterozygosity in relation to two fin ray
characters has also been calculated. These characters are of interest because they are used extensively in fishery population studies
(Garstang 1898; Schmidt 1930; Molander and Molander-Swedmark 1957)
and because in some cases they may be under appreciable stabilising
selection. Such evidence as we have in the plaice suggests that there
may be slight stabilising selection on caudal fin ray number (CFR)
but not on anal fin ray number (AFR). In laboratory populations of
the guppy, where very strong stabilising selection operates on CFR
(Beardmore and Shami 1976), central phenotypes are significantly more
heterozygous than more extreme phentoypes (Shami and Beardmore, in
preparation). In the plaice there is no significant association between heterozygosity and centrality of phenotype for either fin ray
count, but this is not inconsistent with the possibility that such

Table 5. The incidence of uncommon alleles (individual gene frequency
< 0.01) at 5 loci in 0 group and 1+ group plaice

Locus	Age group	Total genes examined	No. of different uncommon alleles	Total no. of uncommon genes	Total % uncommon alleles
PGM-1	0	3550	5	106	1.77
	1 +	996	5	26	1.51
ADA	0	3548	2	29	0.82
	1 +	996	2	13	1.31
αGPDH-1	0	3554	4	27	0.76
	1 +	996	2	10	1.00
MDH-2	0	3554	1	16	0.43
	1 +	996	1	8	0.80
PG1-2	0	3554	2	22	0.62
	1 +	996	2	6	0.60

Table 6.   The incidence of uncommon alleles (individual gene frequency
             < 0.01) at 5 loci in 0 group and 1+ group plaice

Locus	Age group	Total genes examined	No. of different uncommon alleles	Total no. of uncommon genes	Total % uncommon genes
ADA	0	3548	2	29	0.82
	1 +	996	2	13	1.31
αGPDH−1	0	3554	4	27	0.76
	1 +	996	2	10	1.00
MDH−2	0	3554	1	16	0.43
	1 +	996	1	8	0.80
PGI−2	0	3554	2	22	0.62
	1 +	996	2	6	0.60
PGM−1	0	3550	5	63	1.77
	1 +	996	5	15	1.51

associations may only be found when the meristic character is under
strong stabilising selection.

The role of migration in influencing the structure of the gene
pool will now be briefly considered. There is movement of adult fish
in and out of the Bristol Channel population (Macer 1972), but the
North Wales population with which some gene exchange may occur has
allele frequencies at the five enzyme loci not significantly different from those of the Bristol Channel population. Furthermore, practically all the heterozygosity changes occur in the 0 group fish, and
fish of this age do not migrate. They remain in the same nursery ground
until they move off into deeper water. It is also interesting to note
that it is in the 0 group fish that mortality is at its highest, and
therefore selection pressures must be severe.

We conclude that the age dependent changes in heterozygosity in

the plaice are most simply interpreted as the result of heterozygosity favouring survival. There are very few published data that can
be cited in support of the view that this crude measure of genotypic
variation is meaningful in fitness terms though Lerner (1954) and
others have argued strongly in support of it. Mitton and Koehn's
(1975) demonstration, based on twelve loci, of an increase in heterozygosity between age classes of the fish *Fundulus heteroclitus* is
the only relevant published evidence of which we are aware and broad-
ly agrees with our findings.

It is clear that in this study we have so far only examined in a
superficial way the relationship between enzyme variation and fitness.
It seems possible that natural selection is operating upon the enzyme
variants themselves rather than on polymorphic loci tightly linked
to the enzymic loci, since linkage disequilibrium, unless it is itself
maintained by selection, seems unlikely in the very large populations
of this species. However the loci may simply be acting as makers for
more complex arrangements of polymorphic genes subject to appreciable
selection and within which there is limited recombination. There are
also components of fitness that we have not yet assayed. For example,
the effects, if any, of variation in heterozygosity at these five
loci upon reproductive performance is as yet unknown, but information on this important component of Darwinian fitness is obviously
required.

Summary.

Polymorphism at five enzyme loci has been studied electrophoretically in a
natural population of the plaice *Pleuronectes platessa*. There are large and significant changes in average heterozygosity measured over all loci from age class to
age class. Most of the changes occur within the first year of life (0-group). In
very young fish greater heterozygosity is associated with more rapid achievement
of the stage at which young fish move onto nursery beaches. In somewhat older 0-
group fish the proportion of fish with lower heterozygosity scores decreases with
increasing age.

We conclude that variation in multi-locus heterozygosity significantly influences survival components of Darwinian fitness.

Acknowledgements.

The assistance of Geraldine Baugh and Jill Locke-Edmunds has been of much help
in this work. Financial support through Grant No. GR3/1558 from the Natural Environment Research Council to J.A. Beardmore is gratefully acknowledged.

REFERENCES

Battaglia, B. 1958. Balanced polymorphism in *Tisbe reticulata*, a marine Copepod. *Evolution* 12: 358-364.

Beardmore, J.A., and Shami, S.A. 1976. Parental age, genetic variation and selection. In: *Population Genetics* and *Ecology*, pp 3-22. ed. Karlin, S. and Nevo, E., Academic Press, Inc. New York, San Francisco, London.

Birley, A.J., and Beardmore, J.A. 1972a. Manifold large selective effects in an enzyme polymorphism. In: *Fifth Marine Biology Symposium*, pp 81-100. Piccin Edit., Padova.

Birley, A.J., and Beardmore, J.A. 1972b. Protein variability in the plaice *Pleuronectes platessa* L. *Anim. Blood Groups. biochm. Genet.* 3, supplement 1 (Abstr): 59-60.

Bryant, E.H. 1974. On the adaptive significance of enzyme polymorphisms in relation to environmental variability. *Amer. Natur.* 108: 1-19.

Charlesworth, B., and Charlesworth, D. 1973. A study of linkage disequilibrium in populations of *Drosophila melanogaster*. *Genetics* 73: 351-359.

Christiansen, F.B., and Frydenberg, O. 1973. Selection component analysis of natural polymorphisms using population samples including mother-offspring combinations. *Theoret. Populat. Biol.* 4: 425-445.

Christiansen, F.B., Frydenberg, O., and Simonsen, V. 1973. Genetics of *Zoarces* populations IV. Selection component analysis of an esterase polymorphism using samples including mother-offspring combinations. *Hereditas* 73: 291-304.

Christiansen, F.B., Frydenberg, O., Gyldenholm, A.O., and Simonsen, V. Genetics of *Zoarces* populations VI. Further evidence, based on age group samples, of a heterozygote deficit in the Est III polymorphism. *Hereditas* 77: 225-236.

van Delden, W., Kamping, A., and van Dijk, H, 1975. Selection at the alcohol dehydrogenase locus in *Drosophila melanogaster*. *Experientia* 31: 418-419.

Frydenberg, O., Moller, D., Naevdal, G., and Sick, K. 1965. Haemoglobin polymorphism in Norwegian cod populations. *Hereditas* 53: 257-271.

Garstang, W. 1898. On the variation, races and migrations of the mackerel. *J. mar. biol. Ass. U.K.* 5: 235-295.

Gibson, J.B. 1970. Enzyme flexibility in *Drosophila melanogaster*. *Nature* 227: 959-960.

Harris, H. 1974. Common and rare alleles. *Sci. Prog., Oxf.* 61: 495-514.

Huang, S.L., Singh, M., and Kojima, K. 1971. A study of frequency-dependent selection observed in the esterase-6 locus of *Drosophila melanogaster* using a conditioned medium method. *Genetics* 68: 97-104.

King, J.L. 1967. Continuously distributed factors affecting fitness. *Genetics* 55: 483-492.

Koehn, R.K. 1969. Esterase heterogeneity: Dynamics of a polymorphism.

    Science 163: 943-944.

Lerner, I.M. 1954. *Genetic homeostatis.* John Wiley, New York.

Lewontin, R.C. 1974. *The Genetic Basis of Evolutionary Change.* Columbia Univ. Press,
    New York.

Macer, C.T. 1972. The movements of tagged adult plaice in the Irish Sea.
    Fishery Invest., Lond., Ser 2, 27(6). 41 pp.

McDonald, J.F., and Ayala, F.J. 1974. Genetic response to environmental heteroge-
    neity.
    Nature 250: 572-574.

Milkman, R.D. 1967. Heterosis as a major cause of heterozygosity in nature.
    Genetics 55: 493-495.

Mitton, J.B., Koehn, R.K. 1975. Genetic organization and adaptive response of al-
    lozymes to ecological variables in *Fundulus heteroclitus*.
    Genetics 79: 97-111.

Molander, A.R., and Molander-Swedmark, M. 1957. Experimental investigations on va-
    riation in plaice (*Pleuronectes platessa* Linné).
    Rep.Inst.Mar.Res. Lysekil.Ser.Biol. 7, 45 pp.

Nevo, E. 1976. Adaptive strategies of genetic systems in constant and varying en-
    vironments. In: *Population Genetics and Ecology*, ed. Karlin, S., and Nevo, E.,
    pp 141-158. Academic Press, Inc. New York, San Francisco, London.

Ryland, J.S. 1966. Observations on the development of larvae of the plaice, *Pleuro-
    nectes platessa* L., in aquaria.
    J.Cons.perm.int.Explor.Mer. 30: 177-195.

Ryland, J.S., and Nichols, J.H. 1975. Effect of temperature on the embryonic deve-
    lopment of the plaice, *Pleuronectes platessa* L., (Teleostei).
    J.exp.mar.Biol.Ecol. 18: 121-137.

Schmidt, J. 1930. Racial investigations. X. The Atlantic cod (*Gadus callarias* L.)
    and local races of the same.
    C.r.Trav.Lab.Carlsberg 18 (6): 1-72.

Sved, J.S., Reed, T.E., and Bodmer, W.F. 1967. The number of balanced polymorphisms
    that can be maintained in a natural population.
    Genetics 55: 469-481.

Wimpenny, R.S. 1953. *The Plaice.* Edward Arnold and Co. London.

Yamasaki. 1971. Measurement of fitness at the esterase-5 locus in *Drosophila pseu-
    doobscura*.
    Genetics 67: 579-603.

Zouros, E., and Krimbas, C.B. 1973. Evidence for linkage diesequilibrium maintain-
    ed by selection in two natural populations of *Drosophila subobscura*.
    Genetics 73: 659-674.

2. STUDY OF POLYMORPHISM

HIDDEN HETEROGENEITY AMONG ELECTROPHORETIC ALLELES

George B. Johnson

Several lines of evidence have recently begun to suggest that allelic variants detected by electrophoresis may be heterogeneous, discrete electrophoretic variants actually representing a collection of alleles each migrating to the same position on a gel. In this paper I shall describe a series of investigations which I have carried out over the last six years employing gel sieving analysis to detect and characterize electrophoretically cryptic variation. These investigations have been reported in a series of publications (Johnson 1971, 1972, 1974, 1975a, 1975b, 1976a, 1977a, 1977b, and 1977c), which discuss in detail the theory and experimental methods of this approach.

VARIATION IN MOBILITY

Electrophoretic alleles of enzyme loci are typically detected and characterized by comparing their mobilities on starch or acrylamide gels. This has provided a particularly straight-forward and convenient approach to detect genetic polymorphism, as mobility classes appear clearly discontinuous in most systems. In discussing electrophoretically detected enzyme polymorphism of this sort, differences between alleles are commonly ascribed to amino acid substitutions involving charged residues.

The observed discontinuities in gel mobility would be an integral property of such "charge state" variation.

From a theoretical point of view, such a model is overly simple. In
ideal electrophoresis a protein migrates in a field at a rate (the free
electrophoretic mobility) determined by its net charge. Gel electrophoresis does not conform to ideal conditions, however, as the gel and protein interact during the course of the protein's migration. Whether one
visualizes the interactions as frictional (the protein "bumpts into"
fibers) or hydrodynamic (sheer forces are generated by the protein's
movement past fibers), the shape and size of proteins should also affect
their migration rates on gels.

Is allelic variation in the *shape* of proteins detectable by gel electrophoresis? Such variation might well be difficult to detect by electrophoresis if the changes in shape are subtle or have only minor effects
on mobility. Investigation of small differences in mobility requires
careful standardization in order to insure gel-to-gel reproducibility.
When small differences in mobility are seen, it is very important to
know how much of the difference is attributable to experimental error.
Rigorous standardization may be achieved by running two internal standard proteins along with the sample in each gel migration path. Standards
should be chosen so that one resembles the experimental protein of interest, while the other is quite different in both size and charge; the
two standards should thus respond quite differently to any alteration
of experimental conditions, and the ratio of their mobilities provides
a sensitive index of experimental error, (the approach is described in
detail in Johnson, 1975a).

This standardization technique indeed reveals the existence of reproducible small-scale differences in gel mobility. In studies of a natural population of *Colias eurytheme* (the common sulfur butterfly), the
gel-to-gel variation in enzyme mobility is much greater than what experimental error would have produced. A typical result is presented in
Figure 1: While the coefficients of variation ($V = [\sigma/\bar{x}] \cdot 100$) of the
internal standards were reproducible low (between 2 and 3), the corresponding values for enzyme variants were five times as high ($V = 10-15$).
This was not true for multiple runs of a single individual ($V = 2.0$),
nor for progeny of a single laboratory mating ($V = 2.8$). It was concluded
that this newly-detected variation is a property of the proteins themselves, and that it represents either post-translational modifications
or "an unresolved distribution of alleles with minor electrophoretic

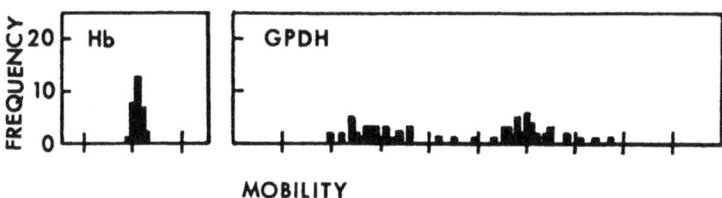

Figure 1. Distribution of electrophoretic mobilities of α-glycerophosphate dehydrogenase (αGPdH) in a California population of *Colias eurytheme* (from Johnson 1971).

mobility differences (Johnson 1971)."

To learn more of the nature of this variation in electrophoretic mobility, it was necessary to determine immediately the degree to which
it was heritable. The variation indicated the existence of either: 1.
extensive post-translational variation in nature, or 2. far more extensive allelic variation than had hitherto been reported. Variants of the
first sort should not segregate in a Mendelian fashion, while variants
of the second sort should. The variation proved heritable: when variants
representing the two extremes of a mobility distribution were crossed,
the progeny revealed a bi-model distribution of mobilities, as shown in
Figure 2.

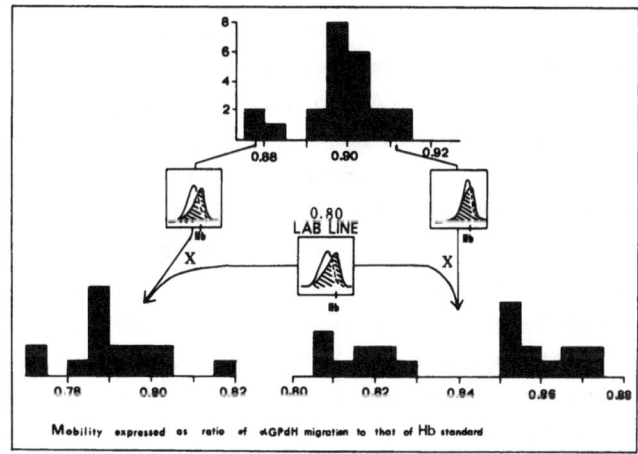

Figure 2. Inheritance of mobility variation in *Colias eurytheme* (after Johnson 1972).

GEL SIEVING ANALYSIS

In principle, the fine-scale variation in electrophoretic mobility detected in rigorously-standardized gels might reflect either partial differences in charge (perhaps the amino acid residue in question is only
partially exposed) or differences in conformation. To clearly differen-
tiate between these alternatives requires a more direct characterization
of the alleles. The mobility of individuals on 7% acrylamide gels is not
adequate for this purpose, because discrete discontinuous classes are
not obtained. What is required is a direct characterization of those
properties of shape and charge which are responsible for producing the
observed variation in electrophoretic mobility.

Such a characterization is by no means difficult to obtain, at least
in principle. The migration of proteins in polyacrylamide gels has been
studied by physical chemists in some detail (the matter is reviewed in
Chrambach and Rodbard 1971). The theory describing how a protein migrates
in polyacrylamide gel electrophoresis stems from the observation by Ferguson that a protein's mobility in gel electrophoresis is a *logarithmic*
function of gel pore size (Ferguson, 1964). This suggests a straightforward theoretical description:

$$R_f = (M_o/u_f)\exp(K_r \times T) \qquad\qquad (1)$$

where R_f is the protein mobility relative to front, u_f is a buffer constant, M_o is the free electrophoretic mobility, which is a function of
net charge, K_r is the retardation coefficient, which is a function of
molecular weight and protein conformation, and T is the per cent acrylamide, which determines pore size and is inversely proportional to it.
A considerable body of theory has been developed elaborating this simple
model (Fawcett and Morris 1966; Rodbard and Chramback 1970, 1971;
Chrambach and Rodbard 1971; Gonenne and Lebowitz 1975). In this model,
the rate of migration is seen to vary not only as a funtion of net
charge (expressed as free electrophoretic mobility, M_o, corrected by a
constant, u_f, for the buffer employed), but also, as suggested by Ferguson, as a logarithmic function of the gel pore size (expressed at per
cent acrylamide) and of the gel-protein interaction as the protein passes through these pores (expressed as the retardation coefficient, K_r).

The approach suggested by equation (1) seems ideally suited to the
experimental problem posed by the allelic variation in mobility (R_f)

discussed above, as it permits independent characterization of the contributions of charge and of interactive effects such as size or conformational differences to the electrophoretic behavior of a protein. These
parameters may be empirically estimated by the simple expendient of
taking the log of both sides of equation (1), yielding a function of
linear form:

$$\underbrace{\log R_f}_{y} = \underbrace{\log (M_o/u_f)}_{\text{intercept}} + \underbrace{K_r}_{\text{slope}} \times \underbrace{T}_{x} \qquad (2)$$

One runs replicate samples of an individual in parallel on several gels
of differing pore size, T, and determines for each gel the corresponding
mobility, R_f, thus characterizing directly the degree to which reducing
the pore size retards migration. Regressing $\log R_f$ on T, one obtains a
linear plot with a slope of K_r and an intercept whose anti-log is M_o
divided by a constant. A typical result is presented in Figure 3.

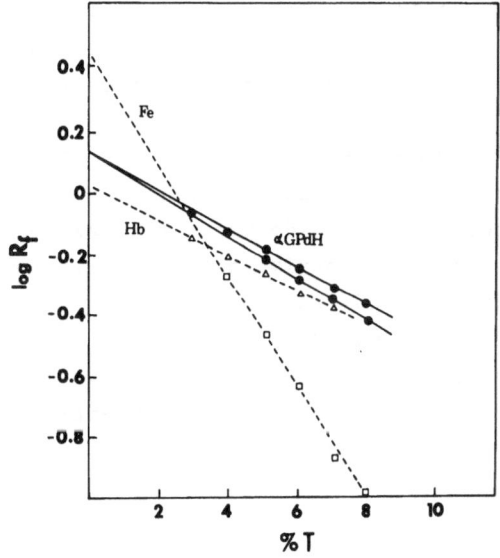

Figure 3. Gel sieving analysis of two alleles of α-glycerophosphate dehydrogenase
(α-GPdH) with identical charge. Hemoglobin (Hb) and ferritin (Fe) are run in the same
gels as internal standards (after Johnson 1977a).

One may standardize determinations of K_r and M_o for an enzyme of an individual by expressing the values relative to the corresponding value of the (similar) internal standard determined from the same gels. Because the standard is evaluated many hundreds of times in population survey work, the procedure permits standardization of any error affecting both proteins. Details of such standardization procedures are given by Johnson (1976a, 1977a).

In order to demonstrate that an enzyme characterized in one individual is electrophoretically different from that characterized in another, it is necessary to demonstrate that K_r, or M_o, or both, differ significantly between the two forms. In comparing estimates of K_r and M_o, it is important to note that K_r and M_o pairs are determined from a single linear regression as slope and intercept, and error in the estimation of an intercept is not independent of error in the estimation of the corresponding slope.

One simple approach to estimating the error associated with a mobility estimate independent of that associated with a corresponding K_r estimate is to express mobility in terms of the R_f observed at the mean value of T. A linear regression may be considered to rotate around such a midpoint as its slope varies, the midpoint remaining unchanged despite great changes in the Y intercept. The error in the estimate of this parameter is independent of the error in the estimate of the slope K_r.

When sampling from a natural population, there will be an error variance in K_r and in mid-Y associated with each protein type present in the sample. To document the existence of multiple classes requires an independent estimate of experimental error. This estimate may be readily obtained from the internal standards run in the same gels. In plotting K_r, mid-Y estimates from a natural population, points reflecting homologous proteins should have a distribution no greater than that seen for the standard. A significantly greater distribution is evidence of heterogeneity.

VARIATION AT THE α-GPdH LOCUS IN *COLIAS*

When gel sieving analysis such as described above was carried out on individuals sampled from natural populations of butterflies, it was immediately apparent that the variation in mobility observed previously reflected more than simple charge differences. If the only source of difference is net charge (presumably produced by amino acid substitutions involving charged residues), then one would expect variants to have similar retardation coefficients, K_r, and to differ primarily in free electrophoretic mobility, M_o. The range of variability in their K_r values would be expected to be limited to about that observed for the hemoglobin internal standard. In fact, the distribution of K_r values obtained is very much broader than the corresponding hemoglobin distribution, as seen in Figure 4. Non-charge differences clearly contribute to the differences seen in mobility on 7% acrylamide gels.

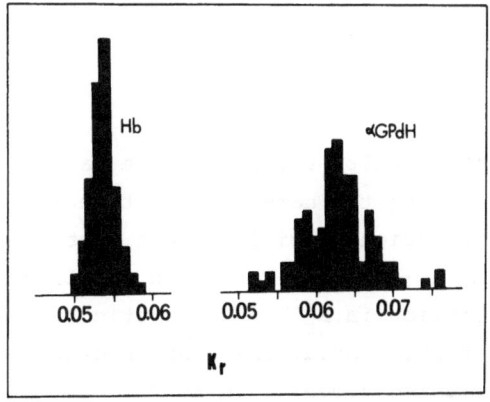

Figure 4. Retardation coefficients (K_r) of α-glycerophosphate dehydrogenase in *Colias* compared to hemoglobin standards run in the same gels (from Johnson 1976a).

This proved an important result, as it provides the basis for understanding the apparently continuous variation in mobility consistently seen in samples of natural populations. The difficulty in analysis on 7% acrylamide gels is due to the fact that charge and size/conformation *interact* in determining mobility, and these two protein properties prove to vary concordinately - as M_o values increase, so do the absolute values of K_r (*e.g.* bigger or more asymmetric proteins have greater net charge). The result is that the mobility functions described by equation (2) *intersect* at intermediate gel pore sizes. The nature of the mobility variation at the α-GPdH locus is now clear (Figure 5): A survey conducted at 5% acrylamide (equivalent to 10-11% starch) will not discriminate between variants, and will reveal a single uniform mobility type. Such a

survey would classify this locus as uniformly homozygous. A survey conducted at 7% acrylamide, as were my previous surveys, would report two
segregating alleles (see for instance Johnson 1972, or Johnson 1976b),
with considerable variation in the exact mobility observed. This is the
mobility variation we set out to analyze, and it reflects the fact that
there are a minimum of *five* alleles segregating at this locus.

Gel sieving analysis thus provides direct evidence of protein heterogeneity within electrophoretic classes. Note also that there are alleles which do not differ in net charge, differing only in K_r.

THE NATURE AND EXTENT OF HETEROGENEITY

The results reported above raise a variety of interesting questions concerning the nature and extent of electrophoretically-detectable protein
variation. I will address myself here to five: 1) To what extent is the
newly-detected variation genetic? Particularly, to what extent does variation in K_r have a genetic basis? 2) Does such electrophoretically-
cryptic variation occur in other organisms than *Colias*? Do intensively
studied groups such as *Drosophila* also exhibit this class of variation?
3) How much variation of this sort is there? Is it restricted to one or
a few loci, or is it typical of most loci? What is the allele frequency
distribution of these electrophoretically-cryptic alleles? 4) How are
the proteins different? What does it mean in terms of amino acid substitutions and protein tertiary structures when two proteins differ in
K_r? 5) Are the many newly-detected variant proteins significantly different in how they function? Particularly important, does a heterogeneous
electrophoretic class of uniform (5% T)R_f contain variant proteins with
differing K_m or V_{max} values?

The genetic basis of variation in K_r.

A variety of studies indicate that electrophoretically cryptic variants are in fact alleles in the accepted sense of the word, segregating in crosses just as electrophoretically distinct alleles do: 1. In
Colias butterflies, when variants of the five allelic classes shown in
Figure 5 are crossed and their progeny analyzed, the resulting ratios
are consistent with normal Mendelian segregation (Johnson 1976a). In pre-

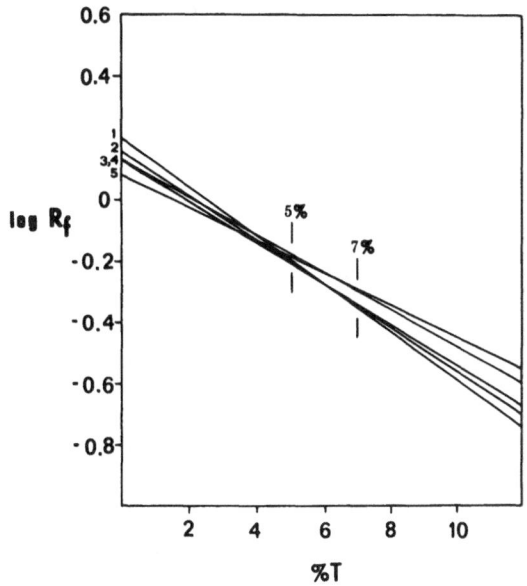

Figure 5. Gel sieving analysis of the five α-GPdH alleles in *Colias* (from Johnson 1976a).

sumptive backcrosses, involving a homozygote and a presumed heterozygote, 1:1 segregation was observed. In the single dihybrid cross, the results were consistent with a 1:2:1 pattern of segregation. 2. When the progeny of a gravid female homozygous for a "class 4" variant of α-GPdH are analyzed, only three classes of protein variant are seen among 20 progeny subjected to analysis (Johnson 1977a): female parental type, heterodimer, and presumptive male parental type (8 progeny were single-banded, "class 4", while 12 were triple-banded). These results are consistent with segregation from a a/a × a/b mating. It is important for the discussion below to note that no other variant types were observed in the analysis of these progeny. 3. More powerful genetic analysis is available in *Drosophila* than in difficult-to-raise *Colias*. As discussed in the following section, *Drosophila* species also exhibit the heterogeneity described in *Colias*. Genetic studies have been carried out on electrophoretically-cryptic variants within the standard 1.00 allele class at the esterase-5 locus of *D. pseudoobscura* (the strain was kindly provided by R. L. Lewontin). Two variants are revealed by gel sieving analysis which differ only in K_r, having identical M_o values. The results of crosses between these two variants support a gene-

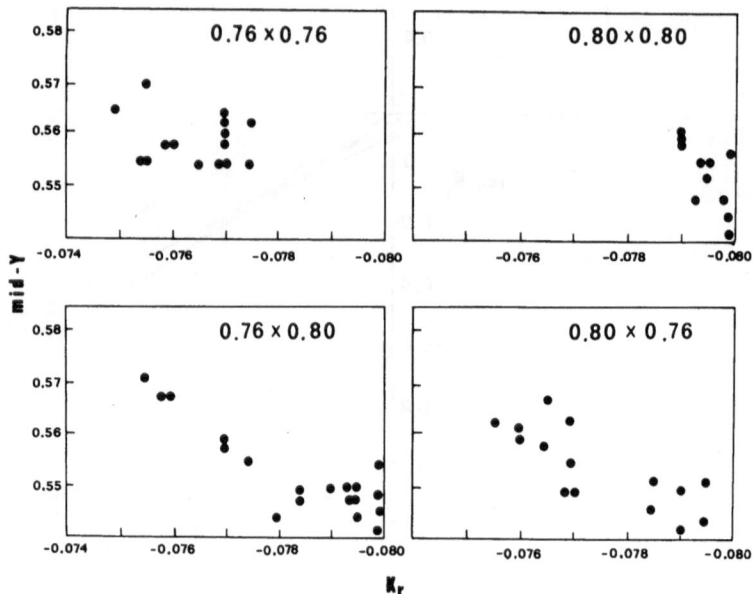

Figure 6.   Genetic crosses involving variants of the esterase 1.00 allele line of
D. pseudoobscura. Data represent F_2 progeny (from Johnson 1977b).

tic interpretation (Figure 6): When variant 0.76 is crossed with itself,
only this variant type is found among the F_2 progeny. Similarly, variant 0.80, when crossed with itself, yields only 0.80 progeny. However,
when the two variants are crossed with each other, *both* variant types
segregate among the F_2 progeny.

Thus preliminary genetic analyses support the hypothesis that at
least some of the electrophoretically cryptic variation represents allelic variation. This does not, of course, establish that all such variation need be allelic, although no evidence of post-translational or
other epigenetic effects have been obtained to date. The significant
deficiency in heterozygotes suggests that either multiple bands are in
some cases not resolved, or that epigenetic effects contribute to the
variation. The data do not permit the exclusion of the second possibility.

Experiments are currently underway to map K_r variants of alcohol
dehydrogenase in *Drosophila melanogaster* to the structural locus media-
ting synthesis of that enzyme. Close linkage to the AdH locus would

tend to rule out the possibility of generalized proteolytic or epigenetic modifications catalyzed by enzymes whose loci map elsewhere, and
which segregate in crosses.

Heterogeneity in *Drosophila*.

To assess the degree to which commonly-employed "homozygous" strains
of *Drosophila* may also carry hidden heterogeneity, gel sieving analysis
has been carried out on several species.

Drosophila pseudoobscura. Figure 7 presents a typical data set for
an allele of the esterase-5 locus of *D. pseudoobscura*. Gel sieving analysis indicates that at least two different variants are present. These
two variants are not routinely discriminated by electrophoresis, as
all analyzed individuals yield indistinguishable R_f values on 5% acrylamide gels. In all, seven different allelic lines were examined, and
all but one proved heterogeneous. A total of fourteen variants were detected by gel sieving analysis among the seven presumably homozygous
esterase-5 lines (Johnson 1977b).

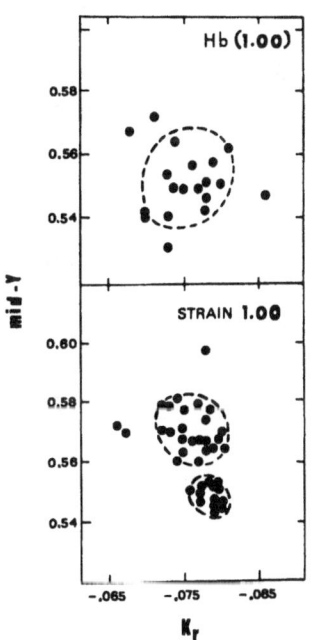

Figure 7. Electrophoretic variation at the esterase-5 locus in *D. pseudoobscura*: The
1.00 allele line. Values for the hemoglobin internal standards (Hb) were obtained from
the same gels (Johnson 1977b).

   Drosophila aldrichi.   Nineteen lines of *D. aldrichi*, representing
five alleles of the esterase-C locus, were provided by Dr. R. H. Richardson, and their gel sieving characteristics analyzed. Eleven distinct variants are observed, each presumptive allelic class proving
heterogeneous (Table 1).

Table 1.   Gel sieving analysis of five presumptive alleles at the este-
             rase-C locus of *Drosophila aldrichi* (from Johnson 1977c).
             K_r and M_o values are presented relative to the hemoglobin in-
             ternal standard

Presumptive allele	R_f 7%T	K_r Hb	M_o Hb	Variant
1	0.61	0.95	0.89	1-1
	0.61	1.04	0.95	1-2
2	0.64	1.10	0.94	2-1
	0.64	0.98	0.87	2-2
3	0.70	0.98	0.80	3-1
	0.70	0.96	0.79	3-1
	0.70	0.93	0.76	3-2
	0.71	0.95	0.77	3-2
	0.71	0.97	0.78	3-1
	0.70	0.93	0.76	3-2
4	0.75	1.00	0.76	4-1
	0.76	0.97	0.73	4-1
	0.75	0.95	0.73	4-1
	0.76	0.95	0.72	4-1
	0.76	0.89	0.69	4-2
	0.74	0.97	0.75	4-1
	0.75	0.70	0.57	4-3
5	0.75	0.82	0.66	5-1
	0.70	0.75	0.65	5-2

Total   11

Drosophila mulleri. Twelve lines of *D. mulleri* were examined (again provided by R. H. Richardson), representing two allelic classes of the esterase-C locus. Five variants were detected.

Gel sieving analysis thus appears to discriminate a significant amount of heterogeneity in each of the three *Drosophila* species examined to date. It seems quite likely that the finding of extensive electrophoretically-cryptic variation will prove a general result in surveys of natural populations. Indeed, preliminary examination of variation at the PEP and RuDP carboxylase loci of salt grass (*Distchlis spicata*) in my laboratory suggests electrophoretically-cryptic K_r variation (Enama 1977).

Evaluating patterns of hidden heterogeneity.

The discovery of extensive variation among α-GPdH and esterase alleles raises the question of whether these two enzyme loci are typical in this respect. Certain loci, particularly esterases, typically exhibit far more electrophoretically-detectable variation than others. Is the newly-detected variation concentrated among a particular sub-set of enzyme loci? To address this question, gel sieving analyses were carried out on 14 enzyme loci of the alpine butterfly *Colias meadii* (Johnson 1977a). For all loci but MdH, several common variants are detected which cannot be distinguished from one another on routine 7% acrylamide gels. In addition, other variants occur at low frequency at all the loci examined. These rare variants are clearly distinct from the common forms. A typical locus, G6PdH, is illustrated in Figure 8.

The common variants. Among the 14 loci examined, a total of 32 variants occur at frequencies of greater than 10%. Sixteen of the 32 variants would not have been distinguished on 7% acrylamide gels. In several cases (TPI, ME), variants occur which differ only in K_r, possessing the same free electrophoretic mobilities. Of the 32 common variants, 20% differ only in charge, 10% differ only in K_r, and 70% differ in both M_o and K_r. Thus *fully 80% of the common variants differ significantly in K_r*. If differences in K_r reflect conformational differences as seems likely (the matter is discussed below), then it seems quite unlikely that this widespread variation in shape does not effect the funtioning of the enzymes.

The rare variants. Among the 14 loci, a total of 97 "rare" variants occur! Ninety-one of them occur only once in a sample of twenty

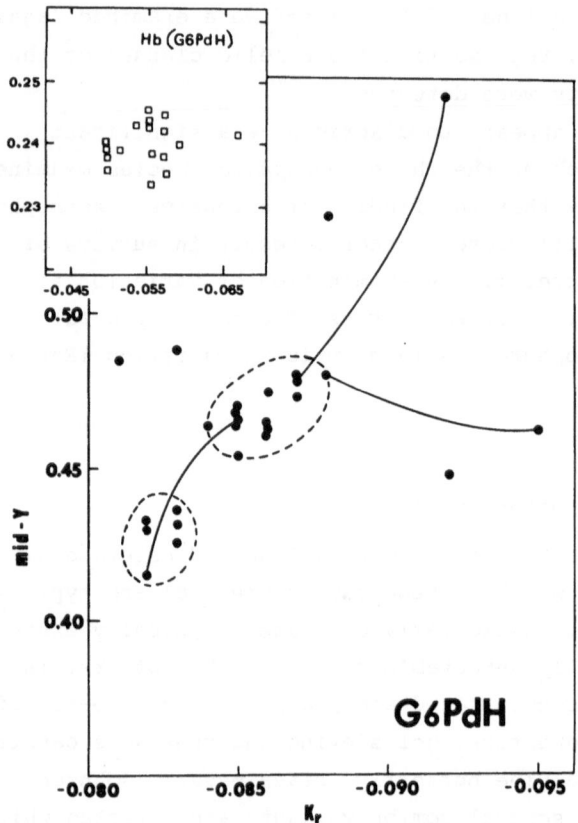

Figure 8. Electrophoretic gel sieving survey of glucose-6-phosphate dehydrogenase
(G6PdH) in a natural population of *Colias meadii*. Values for hemoglobin (Hb) were determined from the same gels (after Johnson 1977a).

individuals; six others occur twice. *Fully 30% of the genes analyzed in
this survey code for proteins which appear only once in the sample.* Perhaps these variants all occur typically at frequencies of 5%; alternatively, they may be unique alleles occurring only once. A larger sample
is required to resolve this issue. The variants appear allelic: note
that "class 4" of α-GPdH discussed earlier is heterogeneous and composed
of rare variants. Class 4 variants were observed to segregate in crosses
in a Mendelian fashion.

All 14 loci exhibit rare variants. Fully 70% of the rare variants
are not detected on 7% acrylamide gels. Of the 97 rare variants, 34%

differ solely in charge, 21% differ solely in K_r, and 45% involve differences in both M_o and K_r. Thus fully two-thirds of the rare variants
involve significant differences in K_r. Again, conformational variation
seems very prevalent.

   Allele frequency distributions. The frequency distribution of variants is
very skewed, with one or a few occurring commonly, and the others rarely
or uniquely (Fig. 9).This is similar to the distribution predicted under

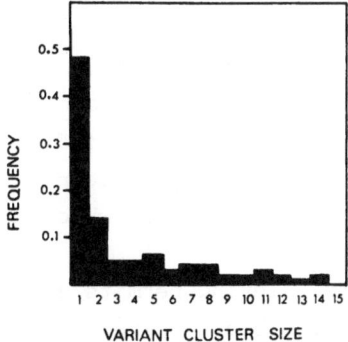

Figure 9.   Skew in allele frequency distribution.

the neutral hypothesis (Nei *et al.* 1977). Such a skewed distribution is also
consistent with a hypothesis of marginal overdominance of heterologous he-
terodimers:The rare variants may not function well themselves as homologous dimers, but in combination with another variant, the hybrid multi-
mer may be marginally advantageous. Perhaps hybrid dimers have slightly
differing kinetic properties from the common homologous type, thus extending the physiological range over which a suitable reaction takes
place (Johnson 1974). Alternatively, the skew in allele frequencies may
reflect uncharacterized temporal or spatial heterogeneity within the
population. In this regard it should be noted that, overall, one would
expcot oven more heterozygotes than are observed. This may reflect extensive inbreeding, as the sampled population was a very localized one.
Alternatively, some heterodimers may not be resolved. It is even possible that some heterodimers may have significantly tighter binding
constants than their constituent homodimers, so that these enzymes exist
predominantly as $\alpha\beta$ hybrids; one band would be seen rather than three

for such heterozygotes.

Electrophoretically-cryptic variation thus appears quite common at all the enzyme loci examined. Gel sieving analysis reveals three times the number of variants detected by routine electrophoretic survey.

The nature of cryptic variants.

There has been considerable recent speculations in the literature concerning cryptic variation at enzyme loci, and what sorts of molecules might be expected to exhibit similar mobilities: The "charge state" model of electrophoretic variation concerns families of alleles with identical net charge, and postulates that they thus migrate to the same band position despite individual differences between them. Polymorphism in heat stability is seen as reflecting internal differences between proteins, properties which have no influence on electrophoretic mobility.

Gel sieving permits a more pointed experimental analysis than has been possible to date, as it provides data on physical properties of proteins, ± explicit experimental error, rather than single-point rate measurements. The "charge state" model, for instance, may be evaluated directly by this approach, as the contribution of charge to mobility may be estimated independently of other gel-interactive effects. When this analysis is carried out for an electrophoretic mobility "ladder" of alleles at the esterase-5 locus of *D. pseudoobscura* (Johnson 1977b), it is seen that not only charge, but also conformation, plays an important role, and that even differences in R_f result from the interaction of these two effects (Figure 10). The gel sieving analysis thus clearly indicates that the charge state model is inappropriate for these data.

Other models may also be evaluated explicitly by gel sieving analysis. The prediction under the neutral hypothesis that larger genes should be more polymorphic, and thus that heterozygosity should be a function of subunit molecular weight (pointed out by Koehn at this symposium), may be conveniently evaluated, as MW may be determined from K_r values (Figure 11). The predicted relationship is not seen among *Colias* enzyme loci (Figure 12). The matter is discussed further in Johnson 1977d).

The gel-interactive component of electrophoretic mobility, K_r, is usually considered to reflect differences in either the size or shape of proteins. Thus the absolute value of K_r varies directly with molecu-

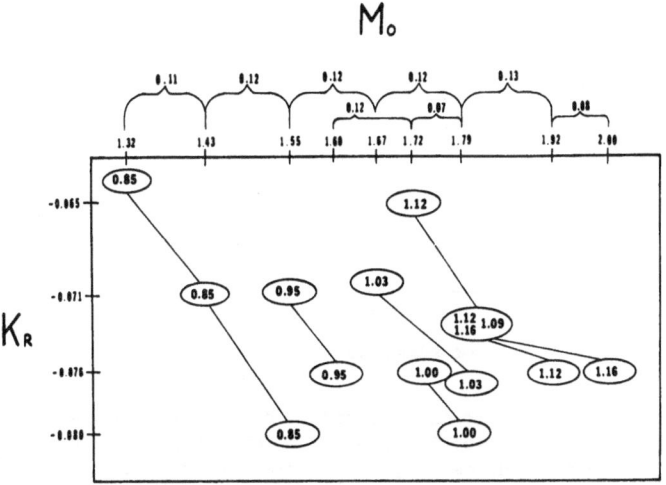

Figure 10. Conformational and charge properties of alleles at the esterase-5 locus of *D. pseudoobscura* (after Johnson 1977b). Lines connect forms with identical R_f values.

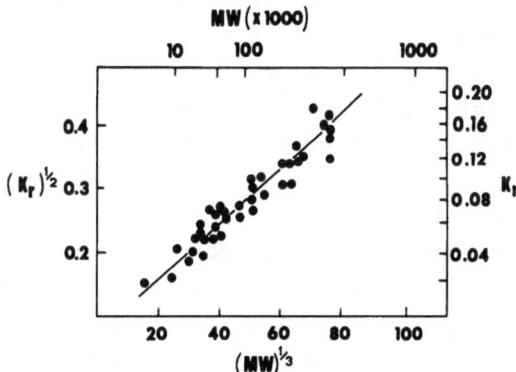

Figure 11. The retardation coefficient as a function of molecular weight (after Johnson 1976a).

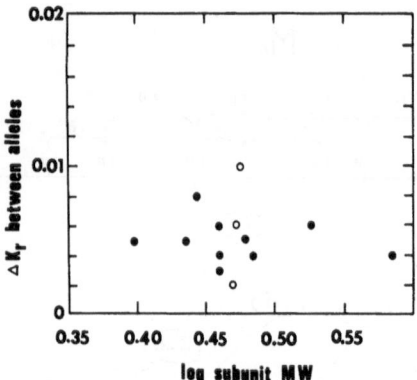

Figure 12. Allelic differences in K_r as a function of subunit molecular weight. Data
are those of Johnson 1977a, for 14 loci of *Colias meadii*. Open circles represent mo-
nomeric proteins, as judged by two-banded heterozygote banding patterns.

lar weight, although with significant scatter indicating the influence
of other factors (Figure 11). Note that the differences in K_r between
Colias alleles are far too small to be attributed to molecular weight
changes produced by the aggregation/dissociation of subunits.

The most straightforward interpretation of the K_r differences reported here is that they reflect conformational differences between the
proteins. That gel sieving analysis is capable of detecting such variants may be demonstrated directly by analyzing proteins such as bovine
fibrinogen, which are known to have very asymmetric shapes. If K_r is a
sensitive function of conformation, then it should not be a good pre-
dictor of molecular weight for very asymmetric molecules, (*e.g.* the K_r
values obtained by gel sieving should be significantly different than
predicted by the value corresponding to its molecular weight on Figure
11). For several molecules of known asymmetry, this proves to be the
case (myosin, fibrinogen, α-globulin - Johnson, unpublished). A particularly clear example is provided by α-chymotrypsin (Johnson 1976a).
This dimeric protein is known from single-crystal X-ray crystallography
to undergo an asymmetric conformational change (only one of the two subunits changes) at pH 8.0. The molecular weight of dimeric α-chymotrypsin (50,000) corresponds on Figure 11 to a K_r value of -0.058. When analyzed at pH 7.5, the value obtained is -0.061, in good agreement. When
the same sample is analyzed at pH 9.5, the value of K_r is -0.044. Thus
the asymmetric change in conformation produced a marked change in the

value of K_r.

Because a change in conformation *can* alter K_r does not, of course, imply that all differences in K_r need reflect conformational differences. Other hypotheses, while less likely, are being evaluated:

A. It is possible, at least in principle, that amino acid substitutions might significantly alter *the hydration shell* of the protein, producing a difference in K_r because of altered hydrodynamic interactions with the gel. In principle, one can test such a hypothesis by testing whether the value of K_r is affected by altering the degree of organization of the water solvent, as the hydration hypothesis would predict. When this is done, by carrying out electrophoresis in D_2O rather than H_2O, the value of K_r is not altered (Johnson 1977a).

B. Amino acid substitutions in the area of subunit interaction might affect the rate of dissociation/reassociation of dimers such as analyzed in these studies. Such rates are typically fast, so that dimers usually migrate in electrophoresis as single bands. *Small alterations in rapid dissociation rates* may have the effect of altering the *apparent* size of the molecules (*e.g.* the proportion of monomer to dimer will determine the realized K_r). Such a hypothesis is directly testable by the simple expedient of diluting the sample, as dissociation rates are concentration-dependent. This hypothesis would thus predict that dilution would alter K_r. It does not (Johnson 1977a).

Other classes of variation. It seems likely that cryptic variation reported in the thermal stability of proteins is quite different from the sorts of variation reported here. These variants may well represent polymorphism for conservative (polar$\rightleftharpoons$polar; non-polar$\rightleftharpoons$non-polar) amino acid substitutions within the interior of the protein which alter hydrogen bonding and thus affect stability. It is unlikely that such variants would be detected by gel sieving analysis.

The chances of detecting variation cryptic because of identical net charge (as described by the "charge state" model of electrophoretic mobility) are much better. A body of data indicate that the pK's of amino acids may be strongly influenced by an amino acid residue's position in a protein (the matter is discussed in detail in Johnson 1975b). Two proteins with different amino acid substitutions but with the same net charge at one pH would be very unlikely to exhibit identical *changes* in net charge at a different pH. Thus serial analysis at two different pH's (in my laboratory 7.0 and 8.9) should reveal any amino acid differences cryptic at one pH because of identical net charge. The finding substantial micro-heterogeneity in isoelectric point within M_o classes (Johnson 1976a) reflects this same underlying heterogeneity.

Are electrophoretic classes functionally heterogeneous?

Extensive K_r variation, if conformational, suggests that biochemical comparisons of polymorphic alleles may be difficult to interpret unless the alleles are demonstrably homogeneous. The extensive literature on functional variation within alleles of *D. melanogaster* AdH perhaps reflects to some degree undetected structural heterogeneity.

FUTURE DIRECTIONS

The major problems now being addressed by analysis of the cryptic variation reported here are: 1) What is the genetic basis of the highly skewed allele frequency distributions observed at most loci? 2) To what extent do differences in K_r reflect conformational differences? 3) Are these cryptic alleles functionally different in K_m, V_{max}, etc.? 4) Just how much variation is there? (Gel sieving analysis suggests three times that detectable on starch gels; to this must be added about 2 × detected by typical heat stability surveys, and perhaps another 3 × by varying pH, for a total of 8 times the currently reported levels!) Is heterozygosity practically universal?

Acknowledgment.

This work is the Publication No. 572 of Carnegie Institution of Washington - Department of Plant Biology.

REFERENCES

Chrambach, A., and D. Rodbard. 1971. Polyacrylamide gel electrophoresis. *Science* 172: 440-451.

Enama, M. 1977. Genetic variation in PEP and RuDP carboxylase in *Distchlis spicata*.
   Carnegie Institution Year Book 75, in press.

Fawcett, J., and C. Morris. 1966. Molecular-sieve chromatography of proteins on gra-
   nulated polyacrylamide gels.
   Separation Science 1: 9-26.

Ferguson, K. 1964. Starch gel electrophoresis: Application to the classification of
   pituitary proteins and polypeptides.
   Metabolism 13: 985-1002.

Gonnenne, A., and J. Lebowitz. 1975. Estimation of molecular weights of small proteins on polyacrylamide gel electrophoresis.
   Anal. Biochem. 64: 414-424.

Johnson, G. 1971. Analysis of enzyme variation in natural populations of the butterfly *Colias eurytheme*.
   Proc. Nat. Acad. Sci. USA 68: 997-1001.

Johnson, G. 1972. The selective significance of biochemical polymorphisms in *Colias*
   butterflies.
   Ph.D. thesis, Stanford University, Stanford, California, USA.

Johnson, G. 1974. Enzyme polymorphism and metabolism.
   Science 184: 28-37.

Johnson, G. 1975a. Use of internal standards in electrophoretic surveys of enzyme
   polymorphism.
   Biochemical Genetics 13: 833-847.

Johnson, G. 1975b. On the estimation of effective number of alleles from electrophoretic data.
   Genetics 78: 771-776.

Johnson, G. 1976a. Hidden alleles at the α-glycerophosphate dehydrogenase locus in
   Colias butterflies.
   Genetics 83: 149-167.

Johnson, G. 1976b. Polymorphism and predictability at the α-glycerophosphate dehydrogenase locus in *Colias* butterflies: Gradients in allele frequency within single
   populations.
   Biochemical Genetics 14: 403-426.

Johnson, G. 1977a. Characterization of electrophoretically cryptic variation in the
   alpine butterfly *Colias meadii*.
   Biochemical Genetics 15: 665-693.

Johnson, G. 1977b. Evaluation of the stepwise mutation model of electrophoretic mobility: Comparison of the gel sieving behavior of alleles at the esterase-5 locus
   of *Drosophila melanogaster*.
   Genetics, in press.

Johnson, G. 1977c. Hidden heterogeneity at the esterase C locus of cactus *Drosophila*.
   Submitted to *Nature*.

Johnson, G. 1977d. Isozymes, allozymes, and enzyme polymorphisms: Structural con-
   straints on polymorphic variation.
   In *Isozymes: Current Topics in Biological and Medical Research*. (Rattazzi, M.,
   J. Scandallos, and G. Whitt, eds.). Liss Publ., Inc., N.Y., in press.

Nei, M., R. Chakraborty, and P. Fuerst. 1977. The infinite allele model with varying
   mutation rate.
   Proc. Nat. Acad. Sci. USA, in press.

Rodbard, D., and A. Chrambach. 1970. Unified theory for gel electrophoresis and gel
    filtration.
    Proc. Nat. Acad. Sci. USA 65: 970-977.

Rodbard, D., and A. Chrambach. 1971. Estimation of molecular radius, free mobility,
    and valence using polyacrylamide gel electrophoresis.
    Anal. Biochem. 40: 95-134.

2. STUDY OF POLYMORPHISM

NATURAL SELECTION AND THE α-*GPDH* LOCUS IN *DROSOPHILIDAE*

Seppo Lakovaara, Anssi Saura and Pekka Lankinen

The approximate amount of variability at the gene level has now been known for about ten years. This knowledge is mainly based on the technique of gel electrophoresis of proteins and of enzymes in particular. It has been difficult, however, to demonstrate an unequivocal correlation between the physiological function of an enzyme and its electrophoretically detectable variants. Once a physiological difference has been established, it may be assumed to have an influence on the adaptive norm of an individual. There are, however, numerous difficulties with this approach. E.g. the substitution of a single amino acid for another may be thought to have an infinitesimally small effect on the total function of the enzyme molecule and the fitness of its bearer.

Linkage relationships between loci are a major obstacle in demonstrating the effects of selection at a single locus. The difficulty is easily demonstrated by numerous examples of shown overdominance involving the alleles of a certain locus. It may be namely always argued that there indeed is overdominance, but that it does not operate in the case of the locus under study - which, accordingly, may well be selectively as neutral as one wishes - but at some really important locus linked to this locus. Randomizing the genetic background of a locus is a prerequisite in selection studies involving the locus. This is accomplished by different means in different instances.

The enzyme α-glycerophosphate dehydrogenase (α-Gpdh) has been found
in electrophoresis studies to be invariable in very many species of the
genus *Drosophila*. The most important role of α-Gpdh in the metabolism
of an adult fly is its function in the flight muscle, where the soluble,
NAD-dependent enzyme together with a mitochondrial, particle-bound and
flavine-linked oxidase comprises the so-called α-glycerophosphate cycle
(e.g. Hansford and Sactor 1971). This cycle produces energy for flight
through electron transfer in the mitochondria. The two other functions
of α-Gpdh are maintenance of the balance between oxidized and reduced
NAD in the cytoplasm for glycolysis, and supplying glycerophosphate
for synthesis of lipids (e.g. O'Brien and MacIntyre 1972a,b).

We have studied the extent of variability or invariability of α-
Gpdh in the family *Drosophilidae*. The study has been based on an idea
that examining large number of closely or remotely related species
guarantees that the genetic background of the locus producing this enzyme is sufficiently randomized. This enzyme (α-Gpdh, L-glycero-3-phos-
phate NAD oxidoreductase E C 1.1.1.8.) is coded by the α-glycerophosphate dehydrogenase (α-*Gpdh*) locus.

MATERIALS AND METHODS

We have studied the variability of the enzyme coded by the α-*Gpdh* locus
by starch gel electrophoresis and by heat treatment. Large numbers of
flies belonging to different species have been collected from natural
populations for this study. We have, furthermore, searched the literature for any α-*Gpdh* allele frequency data. Whenever possible, we have
identified the α-*Gpdh* alleles published in the literature with the ones
we have observed ourselves. This has been done by electrophoresing enzymes of flies studied by other authors and of known genotype in a gel
together with all possible controls.

Flies collected for this study are mainly of northern European origin. In the following unpublished data are given without references. In
many cases, e.g. in many *Drosophila virilis* and *D. obscura* group species
the unpublished material is very large, originating form samples collected thousands of kilometers apart and comprising several thousand
flies. In most cases the conclusions are based on samples of more than
100 individuals from more than one population. Less than ten wild-
caught individuals were analyzed for *Drosophila lundstroemi*, *D. picta*,

D. pinicola, Chymomyza caudatula and *C. fuscimana.*

In this study we have established the electrophoretic identity of
the most common allozymes coded by the α-*Gpdh* locus in 175 species of
the family *Drosophilidae*. The species studied originate from five continents and from very different environmental conditions. They represent all life zones from tropics to the subarctic.

Soluble α-Gpdh activity for this study was assayed following starch
gel electrophoresis in a tris citric acid buffer, pH 7.1 (Shaw and Prasad 1970) by a staining solution containing 800 mg α-glycerophosphate,
25 mg NAD, 25 mg nitro blue tetrazolium and 3 mg phenazine methosul-
phate in 100 ml tris-HCl buffer, pH 8.5. Other wise the technique is
similar to that described by Ayala *et al.* (1972a). What we call α-glycerophosphate dehydrogenase in this paper is the intensively staining
adult isoenzyme, which in the described conditions migrates farthest
to the anode. The symbols for the electrophoretically detectable alleles at the α-*Gpdh* locus are devised in the same way as in our earlier papers (e.g. Lakovaara *et al.* 1972; Saura *et al.* 1973).

The regular electrophoresis method was supplemented by heat denaturation studies. Following electrophoresis the gels about 1 cm thick
were cut horizontally into seven slices of equal thickness. The upper-
most and lowermost of these were discarded, so that five identical
slices 1.5 mm thick were obtained. Four of these slices were immersed
into a 55°C warm 0.05 M tris-HCl (pH 8.5) buffer solution, one for two
minutes, another for four minutes, a third for six minutes and a fourth
for eight minutes. The heat treatment was discontinued by immersing the
gel into a similar buffer solution 20°C warm. The fifth slice served
as a control. It was incubated for ten minutes at 20°C in a similar
tris-HCl buffer solution. The gels were then stained overnight in the
α-Gpdh stainig solution given above. The denaturation of the enzyme
molecules was determined visually as less intense staining in the series of slices thus obtained (control, 2 min, 4 min, 6 min and 8 min)
by comparing samples containing electrophoretically identical forms
side by side in the same gel. The enzyme proteins with the same electrophoretic mobility but with differing thermostability properties have
been designated as products of isoalleles. The symbols of these isoalleles are devised so that the isoallele, which codes for the thermally
most stable enzyme molecule, is designated with the symbol number originally used for the electrophoretically identified allozyme. The next in
increasing thermolability scale is called *a* and the next one is *b*.

RESULTS

The proportion of flies observed by electrophoresis to be heterozygous
at the α-*Gpdh* locus in wild populations of any species in the family
Drosophilidae ranges in general from 0.01 to 0.02. Accordingly, there
is a clearly established predominant allele in each species and some
rare ones. The number of alleles electrophoretically observed at this
locus per species is somewhat variable. Three or four alleles have
been found in the extensively studied species (Ayala *et al.* 1972a;
Saura *et al.* 1973; Saura 1974).

Only two species are effectively polymorphic at this locus. These
species are *Drosophila melanogaster* (O'Brien and MacIntyre 1969; Johnson and Schaffer 1973) and *D. subarctica*. Each of these species has
been found in electrophoretic studies to have two common α-*Gpdh* alleles. The degrees of polymorphism at the locus are rather variable in
natural populations of *D. melanogaster* (Johnson and Schaffer, op. cit.)
with heterozygosities ranging from 0.05 to 0.35. Even higher proportions of heterozygotes detected by electrophoresis have been reported
from French populations (Anxolabehere, personal communication). *D. subarctica* populations are always highly polymorphic, with proportions of
heterozygous individuals between 0.40 and 0.50.

Table 1 gives for 168 species the most common allele at the α-*Gpdh*
locus as detected by electrophoresis. With the exception of *D. melanogaster* and *D. subarctica* the frequencies of the most common allele giv-
en in the Table seem to exceed 0.98 in these species. Several subspecies and semispecies, e.g. *D. equinoxialis caribbensis* (Ayala and
Tracey 1974), *D. willistoni quechua* (Ayala 1973) and five *D. paulistorum* semispecies (Richmond 1972), have also been studied. The subspecies and semispecies studied this far by electrophoresis have always been found to have the same allele in about the same frequency
as the most common one in the nominate species.

Extensive data on the α-*Gpdh* allele frequencies in Hawaiian *Drosophila* are presented by Rockwood *et al.* (1971). We have supplemented
these data with the new frequency data of Ayala (1975). Dr. Ayala has
also kindly supplied the information on certain other populations of
several species studied by Rockwood *et al.* (1971). The identity of the
Hawaiian alleles has been cross-checked with the exception of α-*Gpdh*[103].
Therefore it is not excluded that the allele designated α-*Gpdh*[103] might
not be identical with α-*Gpdh*[102].

In addition to the species listed in Table 1, we have electropho-

Table 1.   Prevalent alleles at the α-*Gpdh* locus in the species of the
family *Drosophilidae* as detected by electrophoresis (Modified from Lakovaara *et al.* 1976b)

Species	Alleles	References

Genus *Drosophila*

  Subgenus *Drosophila* s. str.

   repleta group

   mulleri subgroup

aldrichi	104	Zouros 1973
arizonensis	104	Hubby and Throckmorton 1968
meridiana	104	Ayala pers. comm.
mojavensis	104	Hubby and Throckmorton 1968
mulleri	104	Hubby and Throckmorton 1968

   mercatorum subgroup

carinata	104	Ayala pers. comm.

   melanopalpa subgroup

leonis	104	Ayala pers. comm.
melanopalpa	104	Ayala pers. comm.
nigrospiracula	104	Ayala pers. comm.
pachea	104	Ayala pers. comm.

   hydei subgroup

eohydei	104	Hubby and Throckmorton 1968
hydei	104	Hubby and Throckmorton 1968
neohydei	104	Hubby and Throckmorton 1968

  robusta group

robusta	104	Prakash 1973

  virilis group

americana	104
borealis	104
ezoana	104
flavomontana	104
lacicola	104
littoralis	104
lummei	104
montana	104
novamexicana	104
ovivororum	104
virilis	104

  nannoptera group

nannoptera	104	Ayala pers. comm.

  quinaria group

kunzei	99
limbata	104
phalerata	104
transversa	104

Table 1. (continued 1)

Species	Alleles	References
testacea group		
testacea	104	
funebris group		
funebris	100	
miscellaneous		
histrio	104	
picta	104	
unimaculata	101	
Subgenus *Sordophila*		
acanthoptera	104	Ayala pers. comm.
Subgenus *Lordiphosa*		
fenestrarum	100	
Subgenus *Hirtodrosophila*		
cameraria	97	
lundstroemi	97	
subarctoca	102,106	
pinicola group		
flavopinicola	104	
pinicola	104	
Hawaiian *Drosophila*		
ciliated tarsi group		
imparisetae	106	Ayala pers. comm.
latigena	106	Rockwood *et al.* 1971
bristle tarsi group		
melanopedis	106	Ayala pers. comm.
seclusa	106	Rockwood *et al.* 1971
spoon tarsi group		
dasycnemia	106	Ayala pers, comm.
disticha	106	Rockwood *et al.* 1971
neutralis	106	Ayala pers. comm.
percnosoma	106	Ayala pers. comm.
anomalipes group		
quasianomalipes	106	Rockwood *et al.* 1971
fungus feeders group		
cilifemorata	109	Rockwood *et al.* 1971
dolichotarsis	109	Rockwood *et al.* 1971
fungiperda	109	Ayala pers. comm.
inciliata	109	Ayala pers. comm.
nigella	112	Ayala 1975
nigra	109	Ayala 1975

Table 1. (continued 2)

Species	Alleles	References
modified mouthparts group		
asketostoma	*106*	Rockwood *et al.* 1971
comatifemora	*109*	Rockwood *et al.* 1971
eurypeza	*109*	Rockwood *et al.* 1971
kambysellisi	*106*	Rockwood *et al.* 1971
mediana	*109*	Rockwood *et al.* 1971
mimica	*106*	Rockwood *et al.* 1971
picture wings group		
grimshawi subgroup		
Standard 4 phylad		
balioptera	*109*	Rockwood *et al.* 1971
bostrycha	*109*	Rockwood *et al.* 1971
crucigera	*109*	Rockwood *et al.* 1971
disjuncta	*106*	Rockwood *et al.* 1971
engyochracea	*109*	Rockwood *et al.* 1971
gradata	*109*	Rockwood *et al.* 1971
grimshawi	*109*	Rockwood *et al.* 1971
hawaiiensis	*109*	Rockwood *et al.* 1971
limitata	*109*	Rockwood *et al.* 1971
murphyi	*106*	Rockwood *et al.* 1971
ochracea	*109*	Rockwood *et al.* 1971
orphnopeza	*106*	Rockwood *et al.* 1971
orthofascia	*109*	Rockwood *et al.* 1971
recticilia	*109*	Rockwood *et al.* 1971
silvarentis	*109*	Rockwood *et al.* 1971
sproati	*106*	Ayala 1975
villitibia	*109*	Rockwood *et al.* 1971
villosipedis	*109*	Rockwood *et al.* 1971
4b phylad		
discreta	*106*	Rockwood *et al.* 1971
distinguenda	*106*	Rockwood *et al.* 1971
fasciculisetae	*106*	Rockwood *et al.* 1971
inedita	*106*	Rockwood *et al.* 1971
lineosetae	*106*	Rockwood *et al.* 1971
odontophallus	*106*	Rockwood *et al.* 1971
pilimana	*109*	Rockwood *et al.* 1971
vesciseta	*109*	Rockwood *et al.* 1971
punalua subgroup		
basisetae	*106*	Rockwood *et al.* 1971
paucipuncta	*106*	Rockwood *et al.* 1971
prolaticilia	*106*	Rockwood *et al.* 1971
punalua	*103*	Rockwood *et al.* 1971
uniseriata	*103*	Rockwood *et al.* 1971

Table 1. (continued 3)

Species	Alleles	References
other subgroups		
adiastola	*106*	Rockwood *et al.* 1971
cyrtoloma	*106*	Rockwood *et al.* 1971
hanaulae	*106*	Rockwood *et al.* 1971
hemipeza	*106*	Rockwood *et al.* 1971
heteroneura	*106*	Rockwood *et al.* 1971
melanocephala	*106*	Rockwood *et al.* 1971
neopicta	*106*	Rockwood *et al.* 1971
nigribasis	*106*	Rockwood *et al.* 1971
oahuensis	*106*	Rockwood *et al.* 1971
ornata	*106*	Rockwood *et al.* 1971
picticornis	*106*	Rockwood *et al.* 1971
planitibia	*106*	Rockwood *et al.* 1971
primaeva	*106*	Rockwood *et al.* 1971
setosifrons	*106*	Ayala pers. comm.
setosimentum	*106*	Rockwood *et al.* 1971
silvestris	*106*	Rockwood *et al.* 1971
truncipenna	*106*	Ayala 1975
miscellaneous groups		
achyla	*106*	Rockwood *et al.* 1971
crassifemur	*106*	Rockwood *et al.* 1971
reducta	*106*	Ayala pers. comm.
Subgenus *Dorsilopha*		
busckii	*106*	
Subgenus *Sophophora*		
willistoni group		
capricorni	*104*	Ayala pers. comm.
equinoxialis	*104*	Ayala *et al.* 1972b
insularis	*104*	Ayala pers. comm.
nebulosa	*104*	Ayala and Tracey 1974
paulistorum	*104*	Richmond 1972
pavlovskiana	*104*	Ayala pers. comm.
tropicalis	*104*	Ayala and Powell 1972
willistoni	*104*	Ayala *et al.* 1972a
melanogaster group		
montium subgroup		
birchii	*104*	Ayala pers. comm.
serrata	*104*	Ayala pers. comm.
melanogaster subgroup		
erecta	*104*	McDonald pers. comm.
mauritiana	*108*	McDonald pers. comm.
melanogaster	*104,108*	O'Brien and MacIntyre 1969
simulans	*108*	O'Brien and MacIntyre 1969
teissieri	*108*	McDonald pers. comm.
yakuba	*108*	McDonald pers. comm.

Table 1.   (continued 4)

Species	Alleles	References
ananassae subgroup		
ananassae	*108*	Gillespie and Kojima 1968
bipectinata	*108*	Yang *et al.* 1972
malerkotliana	*108*	Yang *et al.* 1972
nigrens	*108*	Yang *et al.* 1972
pallens	*108*	Yang *et al.* 1972
parabipectinata	*108*	Yang *et al.* 1972
pseudoananassae	*108*	Yang *et al.* 1972
obscura group		
obscura subgroup		
alpina	*100*	Lakovaara *et al.* 1972
ambigua	*100*	Lakovaara *et al.* 1972
bifasciata	*104*	Saura 1974
eskoi	*104*	Lakovaara *et al.* 1976a
guanche	*100*	Lakovaara *et al.* 1976a
imaii	*104*	Lakovaara *et al.* 1972
lowei	*100*	Lakovaara *et al.* 1972
miranda	*101*	Lakovaara *et al.* 1972
obscura	*100*	Lakovaara and Saura 1971
persimilis	*100*	Lakovaara *et al.* 1972
pseudoobscura	*100*	Lewontin and Hubby 1966
subobscura	*100*	Saura *et al.* 1973
subsilvestris	*100*	Lakovaara *et al.* 1972
tristis	*100*	Lakovaara *et al.* 1972
affinis subgroup		
affinis	*104*	Kojima *et al.* 1970
algonquin	*104*	Lakovaara *et al.* 1972
athabasca	*104*	Kojima *et al.* 1970
azteca	*100*	Lakovaara *et al.* 1972
helvetica	*100*	Lakovaara *et al.* 1972
narragansett	*100*	Lakovaara *et al.* 1972
tolteca	*100*	Lakovaara *et al.* 1972
Genus *Antopocerus*		
aduncus	*100*	Rockwood *et al.* 1971
villosus	*106*	Rockwood *et al.* 1971
Genus *Scaptomyza*		
gilvivirilia	*109*	Ayala pers comm.
pallida	*104*	
Genus *Chymomyza*		
caudatula	*100*	
costata	*104*	
fuscimana	*104*	
Genus *Amiota*		
alboguttata	*102*	

retically checked the α-Gpdh allozymes of two unidentified European
species of the genus Scaptomyza. One of the species had the allele 100
and the other 104 as a prevalent one. Four unidentified Hawaiian species, all clearly belonging to the group of the fungus feeders, were
observed monomorphic for α-Gpdh^{109}. Eight flies of a Mexican undescrib-
ed species of the obscura subgroup of the subgenus Sophophora were monomorphic for α-Gpdh^{100} (Powell, personal communication).

Different authors cited in this study have used many different e-
lectrophoretic techniques (different buffers, polyacrylamide gels etc.)
in determining relative mobilities of allozymes and frequencies of corresponding alleles at the α-Gpdh locus. In general their results agree
with ours. Taken together, data on the alleles and their frequencies
at the α-Gpdh locus of 175 species of the family Drosophilidae are
available. With the exception of two species, they have very low amounts
of heterozygosity at this locus. In electrophoresis studies detected allele α-Gpdh^{104} occurs as the most common one in 55 species. This prevalent allele can be found in species belonging to branches of the family occurring on all continents. According to electrophoresis studies
another widespread allele is α-Gpdh^{106}. This allele is the most common
one in 46 species. A majority of these species inhabit the Hawaiian
Islands. The allele α-Gpdh^{109}, found as a most common one in 29 species,
is exclusively Hawaiian. These two alleles are geographically far less
widespread within the family than α-Gpdh^{100}. The latter has been found
by electrophoresis to be the prevalent allele in 19 species, most of
them belonging to the D. obscura group of the subgenus Sophophora. In
addition it is found in the subgenus Lordiphosa and in one species in
the subgenus Drosophila s. str. and the genera Scaptomyza and Chymomyza.
α-Gpdh^{108} characterizes the Drosophila melanogaster group of the subgenus Sophophora. It is the most common allele in 12 species, one of
which, D. melanogaster, however, is effectively polymorphic.

The other six common electrophoretically detected alleles are restricted to a certain species or species group. The allele 99 has been
found as a prevalent one in D. kunzei only, 102 seems to be restricted
to Amiota alboguttata and Drosophila subarctica (which, however, is a
polymorphic species). Similarly 112 has been found in D. nigella only.
A mobility position designated 101 has been found in D. unimaculata and
D. miranda, which belong to different subgenera. The two species having
the same allele, α-Gpdh^{97}, D. cameraria and D. lundstroemi, are definitely related, as well as are D.punalua and D. uniseriata, both of
which have α-Gpdh^{103}.

Table 2. The prevalent alleles at the α-*Gpdh* locus in 24 species of the family *Drosophilidae* as detected by electrophoresis and heat denaturation technique

Species	Alleles
Genus *Drosophila*	
Subgenus *Drosophila s. str.*	
virilis group	
ezoana	104
lacicola	104
littoralis	104
lummei	104
montana	104
novamexicana	104
ovivororum	104
virilis	104
quinaria group	
limbata	104a
phalerata	104a
transversa	104a
testacea group	
testacea	104a
funebris group	
funebris	100a
Subgenus *Hirtodrosophila*	
subarctica	102,102a,102b,106a
Subgenus *Dorsilopha*	
busckii	106
Subgenus *Sophophora*	
melanogaster group	
melanogaster	104,108
simulans	108
obscura group	
bifasciata	104a
eskoi	104a
guanche	100
obscura	100
pseudoobscura	100
subobscura	100b
Genus *Scaptomyza*	
pallida	104a

Electrophoresis has, as a technique, certain well-known limitations.
Therefore we have here used an independent measure of variability to
establish the true extent of polymorphism at the α-*Gpdh* locus. Heat
treatment of proteins is such an independent measure. Table 2 shows
the results of heat treatment and denaturation studies. The 24 species
studied may be assumed to represent a random sample of the more extensive electrophoretically studied material. This material contains
species closely related to each other as well as taxonomically more
distant forms. The material covered in Table 2 includes both of the
two polymorphic species, *D. melanogaster* and *D. subarctica*, too. The
Table likewise shows that the number of alleles at this locus indeed
is higher within the family than can be demonstrated by electrophoresis only, but the amount of polymorphism within a species increases
with the introduction of the new technique only in the case of *D. subarctica*. According to this technique this species has instead of two
at least four alleles at the α-*Gpdh* locus. Otherwise the amount of variation within a species remains unaffected, but what we have called
α-*Gpdh*^{100} is a cluster of three isoalleles. Likewise, α-*Gpdh*^{104} and
α-*Gpdh*^{106} consist of two different isoalleles. Therefore, the monomorphic species remain monomorphic and one of the two polymorphic species becomes even more polymorphic.

Figure 1 shows the distribution of different α-*Gpdh* alleles in the
phylogenetic tree of the family *Drosophilidae* and Table 3 the number
of species possessing each allele as well as their geographic distribution. When inspecting the Figure and the Table one should be kept in
mind that the total number of electrophoretically studied species is
175, whereas heat denaturation of α-Gpdh has been studied for 24 species only.

DISCUSSION AND CONCLUSIONS

Many enzyme loci of *Drosophila* are highly variable. The degrees of
heterozygosity at a locus often exceed 0.05 in natural populations of
flies with several alleles segregating in a single population (see e.g.
Ayala *et al.* 1972a,b). In some species a locus may be relatively monomorphic, whereas it may be highly polymorphic in closely related forms
or even other populations of the same species (see e.g. Lakovaara *et
al.* 1972). Again, different alleles are often established as predomi-

Table 3. α-*Gpdh* alleles found in the family *Drosophilidae*, the number of species possessing each allele as well as their geographic distribution

Alleles	Number of species	Geographic distribution
97	2	Northwestern Europe
99	1	Central Europe
100	17	North Africa, Asia, Europe, North and South America
100a	1	Cosmopolitan
100b	1	Europe, Asia Minor, North Africa
101	2	Europe, North America
102	2	Scandinavia
102a,102b	1	Scandinavia
103	2	Hawaiian Islands
104	48	Africa, Asia, Europe, North and South America, cosmopolitan
104a	7	Europe, Asia, North America
106	45	Hawaiian Islands, cosmopolitan
106a	1	Scandinavia
108	12	Europe, Africa, North and South America, Pacific Islands, cosmopolitan
109	29	Hawaiian Islands
112	1	Hawaiian Islands

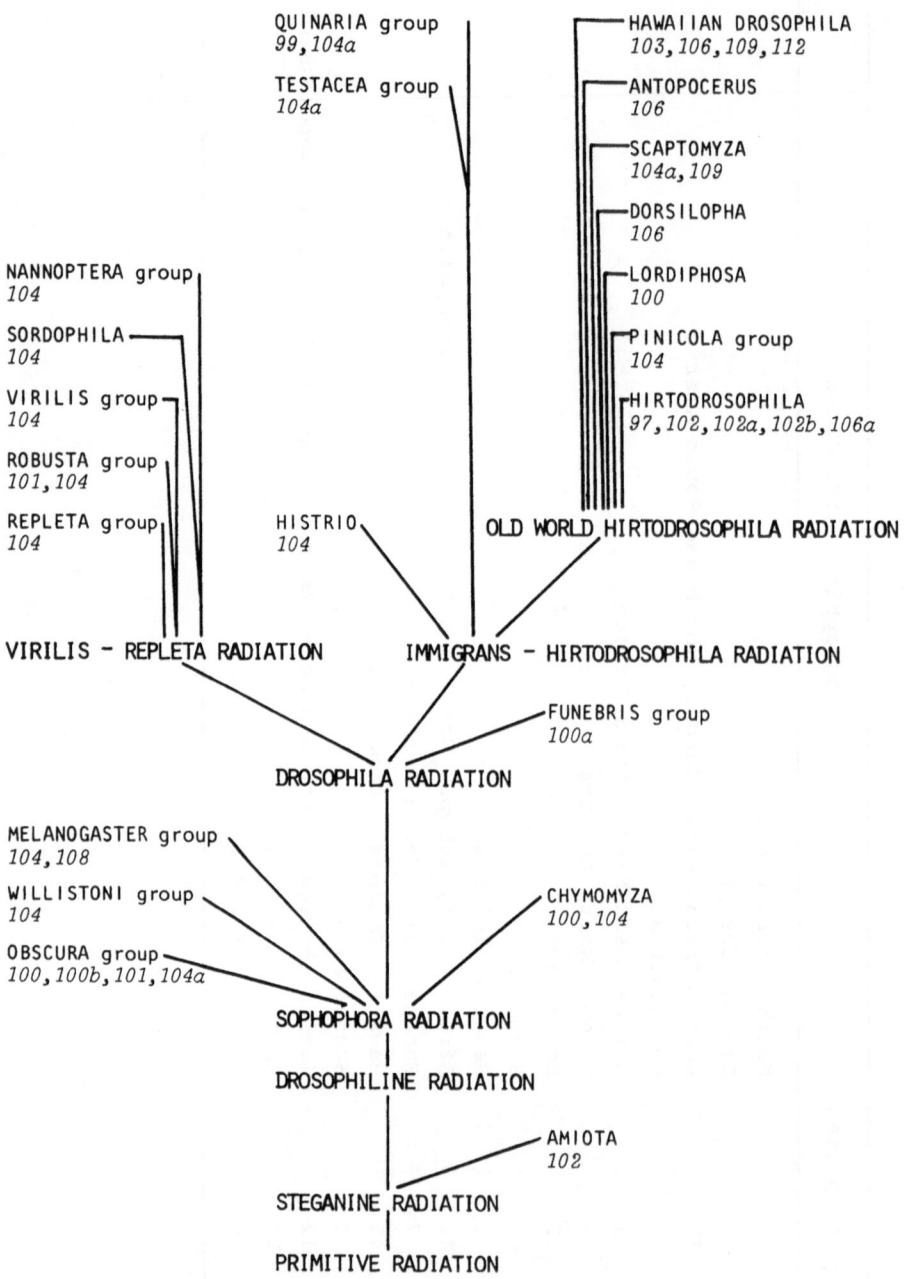

Figure 1. The distribution of different α-*Gpdh* alleles in the drosophilid phylogenetic tree.

nant ones in related species (see Ayala 1975 for a review). Compared
with *Drosophila* enzyme loci studied thus far, α-*Gpdh* is a remarkably
monomorphic locus. There are some other enzyme loci approaching com-
parably low values (e.g. a sophophoran cholinesterase) but the monomorphism seems to be peculiar to a certain species group only. Most
of the enzymes of the glycolytic pathway and the citric acid cycle
vary less than other enzymes. According to the argument of Kojima
et al. (1970) the amount of variation at these loci is related to the
important physiological function of the enzymes coded by them.

Electrophoretically detectable α-*Gpdh* allele frequency data are
available for 175 species of the family *Drosophilidae*. These species
represent several genera and subgenera. With the exception of a few
effectively polymorphic species the locus is nearly monomorphic in a
vast majority of species. Several electrophoretic techniques have also been used by different authors cited here with the same overall results. The main source of error is obviously the low power of resolu-
tion of electrophoresis in comparison with e.g. heat denaturation studies and amino acid sequencing. Applying the latter technique in this
study would, however, represent a formidable task for all the α-Gpdh's
of drosophilids. In practice one should be able to have samples of
many flies from different populations and analyze each fly separately
in order to establish whether there is polymorphism or not. The molecular weight of *Drosophila* α-Gpdh is estimated to be between 63,000
and 81,000 (Bewley *et al.* 1974).

The heat denaturation studies show that the monomorphism detected
at the α-*Gpdh* locus in the species of the family *Drosophilidae* is a
real phenomenon, i.e. there is really not much more variability hidden anywhere. In fact, certain groups, e.g. the *Drosophila virilis*
group are virtually devoid of both within and between species variability. Yet natural selection seems to enforce monomorphism at this
locus in almost any species. The heat denaturation technique shows, of
course, that the number of prevalent alleles is higher within the family than originally detected by electrophoresis.

Tentative explanation of the low amount of polymorphism and slow
rate of evolution at the α-*Gpdh* locus may be suggested. Whenever isoenzymes of an enzyme have different functions, the enzyme as a whole
should be subject to a multiple control. *Drosophila* α-Gpdh is made up
of three different isoenzymes, which differ with regard to their function and kinetic properties (Wright and Shaw 1969; Bewley *et al.* 1974).
All three isoenzymes of α-Gpdh are coded by the alleles of the α-*Gpdh*

structural locus, and accordingly the α-glycerophosphate polypeptide
is present in each isoenzyme (Grell 1967). Other factors are responsible for their differing properties. These factors, accordingly, exercise epistatic interactions with the α-*Gpdh* structural locus.

McDonal and Avise (1976) have noticed that there is remarkably
little variation in either pattern or level of α-Gpdh enzyme activity
between several species of *Drosophila*. We have also preliminarily
studied the temperature dependence of α-Gpdh activity in four species
of the *D. virilis* group and found it to be similar in each species
studied. It is very probable that activity levels and patterns in
the course of development are related to physiological function. Since
the physiological function of α-Gpdh is probably identical in all drosophilid species, the factors controlling this enzyme system may well
be similar throughout the family. The metabolic role of this enzyme
is crucial for insects. In other animals (including some insects) α-
Gpdh is not involved in the generation of energy, and in general the enzyme is more polymorphic in the animal kingdom as a whole (Johnson 1974;
Powell 1975).

Evolution of α-Gpdh of *Drosophilidae* is slow as compared with many
other enzymes in the family. This slow evolution can be interpreted in
terms of biochemical and physiological constraints operating upon the
enzyme. A low amount of heterozygosity implies that a majority of mutants at the α-*Gpdh* locus are harmful. Only when a selective constraint
is loosened, can polymorphism prevail for a while. Such an assumption
is compatible with both a selectionist and neutralist interpretation
(cf. Kimura and Ohta 1974). When we consider that only about one per
cent or two of the total number of species studied are polymorphic,
there is remarkably little evidence for this kind of polymorphism. Yet
polymorphism must exist, since several allele substitutions have taken
place in the evolution of *Drosophilidae* α-*Gpdh*.

Sing *et al.* (1973) have reported a heterozygote excess at the α-
Gpdh locus in laboratory populations of *Drosophila melanogaster*. Dr.
A. Grossman (personal communication) has noticed that α-*Gpdh* heterozygotes in the offspring of flies taken from natural populations of
D. melanogaster in Israel are more viable than homozygotes. Miller *et
al.* (1975) have shown that the two alleles of α-*Gpdh* of *D. melanogaster* show temperature-dependent differences in kinetic parameters. This
observation seems to fit nicely in with the latitudinal cline reported by Johnson and Schaffer (1973) in α-*Gpdh* heterozygosity of *D. melanogaster* in the temperate eastern United States. Accordingly, three

lines of evidence suggest that overdominance is the factor maintaining the heterozygosity at the α-*Gpdh* locus in *D. melanogaster* .

The relationships of α-*Gpdh* alleles within the Hawaiian species are most interesting, although the allele data are based on electrophoresis only. The allele α-*Gpdh*[106] appears to represent the ancestral condition in this group.The second allele, α-*Gpdh*[109], is found in three groups, namely in the modified mouthparts, picture wings and fungus feeders. It appears to be the prevalent allele in the fungus feeders group and in the species of the picture wings group having the "Standard" gene arrangement. Of the picture wings, with the exception of four species (*D. disjuncta*, *D. murphyi*, *D. orphnopeza* and *D. sproati*), the Standard inversion species (Clayton *et al.* 1972) have α-*Gpdh*[109]. Outside of this group the same allele has been found in two picture wing species having the 4b inversion (*D. pilimana* and *D. vesciseta*) only. Two related species having the 4b inversion have the allele α-*Gpdh*[103] as the most common one. The fungus feeders, as a group, are characterized by the allele α-*Gpdh*[109] with one species having the unique α-*Gpdh*[112]. We furthermore know that four unidentified fungus feeders species seem to be monomorphic for α-*Gpdh*[109]. The data are, unfortunately, too scant to allow for any conclusions to be made. It should be stressed that two genotypes with an identical electrophoretic mobility phenotype, occurring in groups taxonomically far apart do not necessarily indicate a common ancestry, but might have arisen from different ancestral forms. This can be tested by applying different analytical techniques. It is, however, obvious that a slowly evolving "conservative" enzyme like *Drosophilidae* α-Gpdh has much taxonomic value and is apparently useful as a key in classification.

In comparison with the *Drosophilidae* enzymes studied this far, α-Gpdh is a slowly evolving protein. It seems to have been canalized to exist in a few forms only. There is a "wild type" allele at the α-*Gpdh* locus in each species of the family *Drosophilidae*. Selection sweeps out recurrent, and in this case, mostly harmful mutants. When polymorphism exists, it is probably maintained by a balancing selection.

Summary.

   The variability at the α-*Gpdh* locus in 175 species of the family *Drosophilidae*
has been studied by using data obtained by electrophoresing flies collected from
natural populations as well as data presented in literature. With the exception of
two species the proportion of flies heterozygous at this locus in populations of
different species is in general less than 0.02. This observation is supported by
studying heat denaturation of the α-Gpdh enzyme. These studies, which are somewhat
preliminary, confirmed the contention that the locus is almost invariably monomorphic within a drosophilid species. The number of electrophoretically detectable
principal α-*Gpdh* alleles within the family is as high as eleven. Heat denaturation
studies have shown that the true number is higher, but that the amount of variability within a species is not appreciably higher than originally disclosed by electrophoresis. As the species studied and found monomorphic, in some cases even for the
same allele, originate from different environmental conditions from tropics to the
subarctic and they occupy widely different ecological niches, it has been concluded
that the environment of the flies can not alone select the products of this slowly
evolving locus. Stability of metabolism in the flies seems to be the main factor
limiting the polymorphism at this locus.

Acknowledgements.

   This study was supported by grants from the National Research Council of Sciences
of Finland and the Finnish Academy of Science.

REFERENCES

Ayala, F.J.   1973.   Two new subspecies of the *Drosophila willistoni* group (Diptera:
   Drosophilidae).
   The Pan-Pacific Entomol. 49: 273-279.

Ayala, F.J.   1975.   Genetic differentiation during speciation process.
   Evol. Biol. 8: 1-78.

Ayala, F.J., and Powell, J.R.   1972.   Enzyme variability in the *Drosophila willistoni* group. VI. Levels of polymorphism and the physiological function of enzymes.
   Biochem. Genet. 7: 331-345.

Ayala, F.J., Powell, J.R., Tracey, M.L., Mourao, C.A., and Péres-Salas, S.   1972a.
   Enzyme variability in the *Drosophila willistoni* group. III. Genic variation in
   natural populations of *Drosophila willistoni*.
   Genetics 70: 113-139.

Ayala, F.J., Powell, J.R., and Tracey, M.L.   1972b.   Enzyme variability in the *Drosophila willistoni* group. V. Genic variation in natural populations of *Drosophila
   equinoxialis*.
   Genet. Res., Camb. 20: 19-42.

Ayala, F.J., and Tracey, M.L.   1974.   Genetic differentiation within and between
   species of the *Drosophila willistoni* group.
   Proc. Nat. Acad. Sci. 71: 999-1003.

Bewley, G.C., Rawls, J.M., and Lucchesi, J.C. 1974. α-Glycerophosphate dehydrogenase in *Drosophila melanogaster*: Kinetic differences and developmental differentiation of the larval and adult isozymes.
J. Insect Physiol. 20: 153-165.

Clayton, F.E., Carson, H.L., and Sato, J.E. 1972. Polytene chromosome relationships in Hawaiian species of *Drosophila*. VI. Supplementary data on metaphases
and gene sequences. Studies in Genetics VII.
Univ. Texas Publ. 7213: 163-177.

Gillespie, J.H., and Kojima, K. 1968. The degrees of polymorphism in enzymes in-
volved in energy production compared to that in nonspecific enzymes in two
Drosophila ananassae populations.
Proc. Nat. Acad. Sci. 61: 582-585.

Grell, E.H. 1967. Electrophoretic variants of α-glycerophosphate dehydrogenase in
Drosophila melanogaster.
Science 158: 1319-1320.

Hansford, R.G., and Sacktor, B. 1971. Oxidative metabolism of insecta, p. 213-247.
In: M. Florkin and B.T. Scheer (eds.), *Chemical Zoology 6*. Academic Press, New
York.

Hubby, J.L., and Throckmorton, L.H. 1968. Protein differences in *Drosophila*. IV.
A study of sibling species.
Amer. Natur. 102: 193-205.

Johnson, G.B. 1974. Enzyme polymorphism and metabolism.
Science 184: 28-37.

Johnson, F.M., and Schaffer, H.E. 1973. Isozyme variability in species of the
genus *Drosophila*. VII. Genotype-environmental relationships in populations of
Drosophila melanogaster from the eastern United States.
Biochem. Genet. 10: 149-163.

Kimura, M., and Ohta, T. 1974. On some principles governing molecular evolution.
Proc. Nat. Acad. Sci. 71: 2848-2852.

Kojima, K., Gillespie, J., and Tobari, Y. 1970. A profile of *Drosophila* species'
enzymes assayed by electrophoresis. I. Number of alleles, heterozygosities, and
linkage disequilibrium in glucose-metabolizing systems and some other enzymes.
Biochem. Genet. 4: 627-637.

Lakovaara, S., and Saura, A. 1971. Genetic variation in natural populations of
Drosophila obscura.
Genetics 69: 377-384.

Lakovaara, S., Saura, A., Lankinen, P., and Lokki, J. 1972. Evolution of enzymes
and genetic distance in *Drosophila obscura* and *affinis* subgroups. 17 Congr.
Int. Zool. (5), p. 1-18.

Lakovaara, S., Saura, A., Lankinen, P., Pohjola, L., and Lokki, J. 1976a. The use
of isoenzymes in tracing evolution and in classifying *Drosophilidae*.
Zoologica Scripta 5: 173-179.

Lakovaara, S., Saura, A., and Lankinen, P. 1976b. Evolution at the α-*Gpdh* locus
in *Drosophilidae*.
Evolution (in press).

Lewontin, R.C., and Hubby, J.L. 1966. A molecular approach to the study of genic
heterozygosity in natural populations. II. Amounts of variation and degree of
heterozygosity in natural populations of *Drosophila pseudoobscura*.
Genetics 54: 595-609.

McDonald, J., and Avise, J. 1976. Adaptive significance of enzyme activity levels;
Interspecific variation in α-GPDH and ADH in *Drosophila*.
Biochem. Genet. (in press).

Miller, S., Pearcy, R.W., and Berger, E. 1975. Polymorphism at the α-Glycerophos-
phate dehydrogenase locus in Drosophila melanogaster. I. Properties of adult allozymes.
Biochem. Genet. 13: 175-188.

O'Brien, S.J., and MacIntyre, R.J. 1969. An analysis of gene-enzyme variability
in natural populations of Drosophila melanogaster and D. simulans.
Amer. Natur. 103: 97-113.

O'Brien, S.J., and MacIntyre, R.J. 1972a. The α-glycerophosphate dehydrogenase in
Drosophila melanogaster. II. Genetic aspects.
Genetics 71: 127-138.

O'Brien, S.J., and MacIntyre, R.J. 1972b. The α-glycerophosphate dehydrogenase in
Drosophila melanogaster. I. Biochemical and developmental aspects.
Biochem. Genet. 7: 141-161.

Prakash, S. 1973. Patterns of gene variation in central and marginal populations of
Drosophila robusta.
Genetics 75: 347-369.

Powell, J.R. 1975. Protein variation in natural populations of animals.
Evol. Biol. 8: 79-119.

Richmond, R.C. 1972. Enzyme variability in the Drosophila willistoni group. III.
Amounts of variability in the superspecies, Drosophila paulistorum.
Genetics 70: 87-112.

Rockwood, E.S., Kanapi, C.G., Wheeler, M.R., and Stone, W.S. 1971. Allozyme changes
during the evolution of Hawaiian Drosophila. Studies in Genetics VI.
Univ. Texas Publ. 7103: 193-212.

Saura, A. 1974. Genic variation in Scandinavian populations of Drosophila bifasciata.
Hereditas 76: 161-172.

Saura, A., Lakovaara, S., Lokki, J., and Lankinen, P. 1973. Genic variation in
central and marginal populations of Drosophila subobscura.
Hereditas 75: 33-46.

Shaw, C.R., and Prasad, R. 1970. Starch gel electrophoresis of enzymes - a compilation of recipes.
Biochem. Genet. 4: 297-320.

Sing, C.F., Brewer, G.J., and Thirtle, B. 1973. Inherited biochemical variation
in Drosophila melanogaster: Noise or signal? I. Single-locus analyses.
Genetics 75: 381-404.

Wright, D.A., and Shaw, C.R. 1969. Genetics and ontogeny of α-glycerophosphate
dehydrogenase isozymes in Drosophila melanogaster.
Biochem. Genet. 3: 343-353.

Yang, S.Y., Wheeler, L.L., and Bock, I.R. 1972. Isozyme variations and phylogenetic
relationships in the Drosophila bipectinata Species complex. Studies in Genetics
VII.
Univ. Texas Publ. 7213: 213-227

Zouros, E. 1973. Genic differentiation associated with the early stages of speciation in the mulleri subgroup of Drosophila.
Evolution 27: 601-621.

2. STUDY OF POLYMORPHISM

NONRANDOM ASSOCIATIONS BETWEEN ALLOZYMES IN
NATURAL POPULATIONS OF *DROSOPHILA MELANOGASTER*

Charles H. Langley

During the last several years I have been involved in determining the
amount of linkage disequilibrium to be found in natural populations of
Drosophila melanogaster. Several of the participants of this symposium
have also put much effort in this direction (Charlesworth and Charlesworth 1973; Kojima, Gillespie, and Tobari 1970; and Mukai *et. al.* 1971,
1974, and 1976). In this paper I address myself to the question of linkage disequilibrium between allozymes. The amount of linkage disequilibrium between these loci has been considered relevant, if not critical, to the detection of natural selection of enzyme polymorphisms.

Previous reports have not found enormous amounts of linkage disequilibrium between allozyme loci. Although several isolated instances
were detected and some general phenomena were suggested (Charlesworth
and Charlesworth 1973; Langley, Tobari, and Kojima 1974) the overall
picture is *little detectable linkage disequilibrium between allozymes
and significant linkage disequilibrium between allozymes and closely
linked polymorphic inversions*. The first part of this paper deals with
the review of the available data on gametic linkage disequilibrium (*i.e.*
disequilibrium estimated from genetically isolated gametes). The second
portion of the paper reviews linkage disequilibrium data estimated from
wild-caught zygotes. And the last section is a discussion of the value
of these data with special attention given to electrophoretically silent variation and to polymorphic inversions. All the data are or will

be published soon so I will make little effort to be thorough in this
area and will put most effort into condensing as clear a picture of the
overall situation as possible.

REVIEW OF ALLOZYME-ALLOZYME DISEQUILIBRIUM FROM
ISOLATED GAMETES OF *D. MELANOGASTER*

My criteria for inclusion in this review are (1) a survey study, (2)
of *D. melanogaster*, and (3) by extraction of gametes from wild-caught
flies. The following reports are sources: Mukai, Mettler, and Chigusa
(1971); Charlesworth and Charlesworth (1973); Mukai, Watanabe, and Yamaguchi (1974); Langley, Tobari, and Kojima (1974); Mukai and Voelker
(1976); Voelker, Mukai, and Johnson (1976); and Langley, Ito, and Voelker (1976). Since virtually all the data involve either two alleles per
locus or nearly so, rarer alleles are pooled. The data are 133 sets of
four gametes. The sum of the individual χ^2's for these 133 is 176.9
(p < 0.004). By this criterion one might conclude that there is indeed
some linkage disequilibrium. By dividing the distribution of these χ^2's
into deciles and testing for goodness of fit, we find the resulting χ^2
is 9.93 with 9 degrees of freedom. Furthermore, if we discard the largest single χ^2 (*Odh* - *Ao*; χ^2 = 21.2; from the 1970 collection from Katsunuma, Japan; Langley, Tobari, and Kojima 1974) the probability value
drops (p < 0.08). As a group these data suggest no linkage disequilibrium other than a rare large exception.

Almost all population genetics theories indicate increasing linkage disequilibrium with increasing linkage. To pursue this possibility a measure of effective recombination is needed. To estimate effective recombination, I assumed that there is no recombination in chromosome arms heterozygous for inversions and no recombination in males.
Using the population estimates of inversion heterozygosities and the
known map positions each gametic set is given an effective recombination value, c'. Figure 1 is a plot of χ^2 against c'. No relationship
is readily apparent.

In order to discern any possible relationship that might be "hidden" in the data the χ^2's are summed in groups of increasing recombination. The sums are of groups of 22. The first circle (0) is the a-
verage χ^2 of the most closely linked 22 gametic sets. The second circle
is the average χ^2 of the next 22 gametic sets, etc. The sum of the

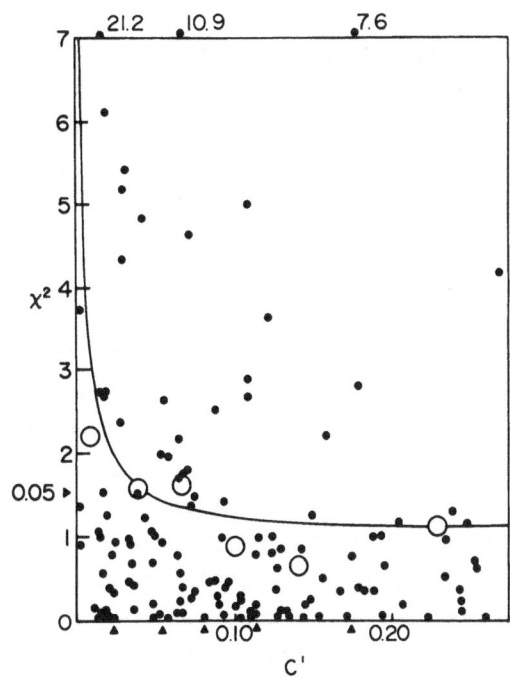

Figure 1. Combined linkage disequilibria from gametic sampling. Individual observations of allozyme x allozyme linkage disequilibrium. χ^2's plotted against effective recombination, C' ($\bullet$). $\blacktriangle$ along the C' axis indicates the divisions of the six groups of 22 observations in ($\circ$). $\blacktriangleright$ along the χ^2 axis indicates the $P < 0.05$ values for $\chi^2/$ 22 degrees of freedom. The curve is explained in the text.

first group is 48.8 with 22 degrees of freedom ($P < 0.001$). The next two groups have sums such that $P < 0.05$. I conclude that there is a modest negative relationship between χ^2 and recombination.

Pursuing this analysis one precarious step further I can ask what effective population size would be consistent with these results. There is no proper statistical means of estimating N_e since the formula for the relationship between the squared correlation coefficient and N_e is only approximate (Hill and Robertson 1968; and Ohta and Kimura 1969). To a similarly crude order of approximation I can write

$$E(\chi^2) = 1 + n_s/(4N_e c') \qquad (1)$$

where n_s is the sample size. Substituting the average χ^2 for $E(\chi^2)$ and
the averages of c' and n_s in this relationship, the following values of
N_e are obtained: 2964, 1962, and 1733 for the first three groups of 22
respectively. The mean of these is 2220 and is graphed in Figure 1.

This might be a maximum; in the sense that only the statistically
significant groups are averaged. If the following relationship is utilized

$$\chi^2 c'/n_s = c'/n_s + 1/(4N_e) \qquad (2)$$

all the groups can be averaged. This yields a value for N_e equal to 7042.
These numbers are not unrealistic for many of the populations sampled.
Yoshikawa and Mukai (1970) estimated the effective population size of
the Katsunuma population to be 1880 based on lethal allelism rates. The
same calculation estimated the Raleigh, North Carolina effective population size at 22,000 (Yoshikawa and Mukai 1970). This analysis indicates
that these data are at least consistent with reported effective population sizes and a pure random genetic drift model. It is likely that
these data are also consistent with alternative selective models and
this I will discuss below.

ESTIMATES OF LINKAGE DISEQUILIBRIUM BETWEEN ALLOZYMES FROM ZYGOTIC DATA

D. Smith, F.M. Johnson, and I are completing a detailed analysis of
a large collection of zygotic allozyme data at this time. In this section, I summarize the allozyme-allozyme linkage disequilibrium results.
These data were collected by F.M. Johnson in 1970, 1971, 1972, and 1973
in North Carolina. They represent some 13,000 individual wild-caught
D. melanogaster scored for 7 or 8 polymorphic allozyme loci on chromo-
somed II and III (*Gpdh*, *Mdh*, *Adh*, *Est-6*, *Pgm*, *Est-C*, *Odh*, and *Acph*).

P. Burrows suggested the following measure of nonrandom association
between loci in zygotes (per com. P. Burrows):

$$\Delta = (D + T) = 1/2\{f(\tfrac{\cdot\cdot}{1\,1}) + f(\tfrac{1\,1}{\cdot\cdot}) + f(\tfrac{1\,\cdot}{\cdot\,1}) + f(\tfrac{\cdot\,1}{1\,\cdot})\} - 2pq$$

where p and q are the observed gene frequencies at the two loci (also
see Bodmer, Bodmer, Ihde, and Adler 1969), $f(\tfrac{\cdot\cdot}{1\,1})$ is the frequency of
paternal gametes having the number 1 alleles at both loci. $f(\tfrac{\cdot\,1}{1\,\cdot})$ is the

frequency of zygotes whose maternal gamete carries the number 1 allele
at the second locus and whose paternal gamete carries the number 1 allele at the first locus. These individual frequencies can not be estimated from simple zygotic frequencies but their sum can. Δ, the covariance of A and B loci in zygotes is the sum of linkage disequilibrium,
D (covariance in gametes) and what might be termed "transdisequilibrium", T (covariance in uniting gametes). Corresponding to this is
a 2×2 χ^2 with total number of observations equal to four times the sample
size. The entries in the cells of the 2×2 χ^2 are the number of times
A_i occurs with B_j. "Transdisequilibrium" is expected to be small since
it goes to zero under random mating in one generation. This χ^2 can be
compared to the gametic 2×2 χ^2:

$$\chi^2 = \frac{D^2 N}{p(1-p)q(1-q)} \quad \text{to} \quad \chi^2 = \frac{4(D+T)^2 N}{4p(1-p)q(1-q)} = \frac{D^2 N}{p(1-p)q(1-q)}$$

These are then equivalent if $\Delta = D$ and $\Delta^2 = D^2$. These seem like reasonable assumptions. One can, however, consider the discussion in this section more properly as that of the zygotic, interlocus allelic covariance.

There are 305 such χ^2 for three gametic types formed between the
three loci in chromosome II. The reason the 305 is not a multiple of 3
is that those samples with size less than 50 were excluded and not all
loci were scored in each sample. Secondly, those χ^2 in which one cell
had expected value less than one and one or more was observed were also
excluded. This is a conservative policy with respect to the 2×2 χ^2. The
summed χ^2 for these second chromosome gametes is 375 (P < 0.002). The
summed χ^2 for the 846 2×2 χ^2's involving two of the 5 loci on the third
chromosome is 987 (P < 0.0003). For the linked comparisons as a whole,
the summed χ^2 is 1363 (P < 0.0001). For the unlinked comparisons (1388)
the summed χ^2 is 1613 (P < 0.0001). These large χ^2's indicate nonrandom association among allozymes in zygotes. The large significant values reflect the large number of samples rather than large interactions.

Again it is interesting to look for a relationship to linkage. In
this case I use representative inversion heterozygosities (2L, 0.05;
2R, 0.09: 3L, 0.07; 3R, 0.21) to estimate the effective recombination.
Figure 2 is a plot of the average χ^2 for each pair of linked loci a-
gainst its effective recombination, c'. Those that are statistically
significant (squares) are among the more closely linked comparisons.

By substituting the average χ^2 for $E(\chi^2)$ and the average sample size

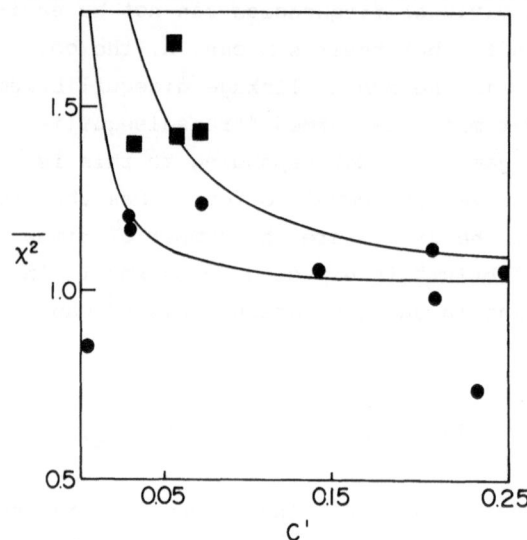

Figure 2.   Linkage disequilibria estimates from zygotic samples. Average χ^2's for the linked comparisons plotted against effective recombination, C'. ■ indicates comparisons with statistically significant ($P < 0.05$) summed χ^2's. The curves are explained in the text.

for n_s in (1) we obtain the following value for N_e associated with the four statistically significant comparisons: 818, 1440, 665, and 1071. The mean is 999. The upper line in Figure 2 is the graph of $\chi^2 = 1 + \dfrac{n_s}{4N_e c'}$ with $N_e = 999$ and $n_s = 92$ (the mean).

The second formula, (2), yields the value, 5263 for N_e when the average over comparisons of $\overline{\chi^2} c'/n_s$ and c'/n are substituted. This relationship is also plotted on Figure 2 (the lower line). These data also suggest a modest negative relation between nonrandom associations and recombination. These tenuous values for effective population size are also within the range of other estimates.

DISCUSSION

In the zygotic analysis the amount of nonrandom association among unlinked loci is similar to that for linked loci. One reason for this
might well be interactions among inversions. These allozyme loci are
often in linkage disequilibrium with inversions in the same chromosome arm. This could lead to associations among loci that are unlinked
if inversions interact among chromosomes. Stalker (1976) has reported
significant non-random associations between these same inversions. One
possible mechanism is nondisjunction in individuals with high inversion
heterozygosity. These nonrandom associations could possibly be due to
linkage disequilibrium between inversions generated in meiosis or other
life cycle states and noncommitant linkage disequilibrium between allozymes and polymorphic inversions. The gametic data show no such association among unlinked elements. Furthermore, the linked gametic data
indicate that inversions may not play a role in allozyme-allozyme linkage disequilibrium. In fact, it is rare that a nonrandom association
between linked allozymes can be attributed to associations with inversions.

The gametic data indicate little interchromosomal or inter-arm interaction between allozymes or inversions, while the zygotic data clearly indicate nonrandom association among unlinked allozyme loci. Perhaps the gametic data are too limited (four collections where both
chromosomes were extracted together).

At face value these data are consistent with random genetic drift
of selectively neutral allozymes. There are no strong, consistent non-
random associations among allozymes analogous to the situation envisaged
by Franklin and Lewontin (1970) or that in the "balancing theory" (Lewontin, 1974). This could be due to two causes. The first is that there
is little nonrandom association along the chromosome. And the second is
that allozymes are not the simple genetic entities of population genetics theory. It is clear from molecular genetics that allozymes are
likely to be phenotypic classes. The actual amount of electrophoretically unseen variation may well be much greater than that predicted
based on the assumption of 1/3 of the mutations being detectable. One
has an intuitive feeling that this ambiguity could obscure real associations with fitness or other loci. Consider that as the number of undetected alleles increases, the disequilibrium is the sum of these small
individual disequilibria, $d_{ij} = 0_{ij} - p_i q_j$

$$D = 0_{IJ} - P_I 0_J = \Sigma\Sigma[0_{ij} - P_i Q_j] = \Sigma\Sigma d_{ij}.$$

Where 0 is the observed gamete frequency; P, p, P, q, are observed allele frequencies; I and J represent allozymic classes and i and j are
the unseen alleles in I and J. The sum $\Sigma\Sigma d_{ij}$ will tend toward 0, since
some will be positive and some negative. The resulting 2×2 χ^2 is
$[\Sigma\Sigma d_{ij}]^2 N/[P_I Q_J(1-P_I)(1-Q_J)]$ which also tend to 0 (where N is the
sample size). This treatment is clearly too cursory. But I think it
leaves the correct impression that undetected variation within electrophoretic classes could well lead us to conclude that there is little
linkage disequilibrium among allozymes. My view is that the understanding of the natural selection of molecular genetic variation must follow
a more basic understanding of molecular genetic variation within populations.

Summary.

     Available data on allozyme-allozyme linkage disequilibria in natural populations
of *Drosophila melanogaster* are reviewed. Little linkage disequilibrium is apparent.
A slight negative association between observed linkage disequilibrium and recombination is suggested. No definite statement about the interactions of natural selection
and these loci can be drawn.

REFERENCES

Bodmer, W.R., Bodmer, J., Idhe, O., and Adler, S. 1969. Genetic and seralogical
     associations analysis of the HL-A leucocyte system. *In Computer Applications in
     Genetics*. Ed. N.E. Morton, University of Hawaii. Press, Hawaii.

Charlesworth, B., and Charlesworth, D. 1973. A study of linkage disequilibrium in
     populations of *Drosophila melanogaster*.
     Genetics 73: 351-359.

Franklin, I., and Lewontin, R.C. 1970. Is the gene the unit of selection?
     Genetics 65: 707-734.

Hill, W.G., and Robertson, A. 1968. Linkage disequilibrium in finite populations.
     Theoret. and Applied Genet. 38: 226-231.

Kojima, K.J., Gillespie, J.H., and Tobari, Y.N. 1970. A profile of Drosophila spe-
     cies' enzymes assayed by electrophoresis 1. Number of alleles, heterozygosities,
     and linkage disequilibrium in gluclose-metabolizing systems and some other en-
     zymes.
     Biochem. Genet. 4: 627-637.

Langley, C.H., Tobari, Y.N., and Kojima, K.   1974.   Linkage disequilibrium in natur-
    al populations of *Drosophila melanogaster*.
    Genetics 78: 921-936.

Langley, C.H., Ito, K., and Voelker, R.A.   Linkage disequilibrium in natural popula-
    tion of *Drosophila melanogaster*. Seasonal variation.
    To appear in Genetics.

Lewontin, R.C.   1974.   The genetic bases of evolutionary change.
    Columbia University Press, New York.

Mukai, T., Mettler, L.E., and Chigusa, S.I.   1971.   Linkage disequilibrium in a local
    population of *Drosophila melanogaster*.
    Proc. Natl. Acad. Sci. U.S.A. 68: 1065-1069.

Mukai, T., Watanabe, T.K., and Yamaguchi, O.   1974.   The genetic structure of natur-
    al populations of *Drosophila melanogaster*. XII. Linkage disequilibrium in a large
    local population.
    Genetics 77: 771-793.

Mukai, T., and Voelker, R.A.   The genetic structure of natural populations of *Droso-
    phila melanogaster* XIII. Further studies on linkage disequilibrium. Submitted.

Ohta, T., and Kimura, M.   1969.   Linkage disequilibrium due to random genetic drift.
    Genet. Res. 13: 47-55.

Stalker, H.D.   1976.   Chromosome studies in wild populations of *Drosophila melano-
    gaster*.
    Genetics 82: 323-347.

Voelker, R.A., Mukai, T., and Johnson, F.M.   Genetic variations of population of
    Drosophila melanogaster from the western United States.
    To appear in Genetica.

Yoshikawa, I., and Mukai, T.   1970.   Heterozygous effects on viability of spontaneous
    lethal genes in *Drosophila melanogaster*.
    Japan J. Genet. 45: 443-455.

2. STUDY OF POLYMORPHISM

POLYMORPHISM FOR THE NUMBER OF GENES CODING FOR SALIVARY AMYLASE IN THE BANK VOLE *CLETHRIONOMYS GLAREOLA*

J. Tønnes Nielsen

Mainly as a result of improved electrophoretic techniques and the development of new specific enzyme stains, the amount of detected genetic variation in natural populations of both plants and animals has increased dramatically during the last decade. The extension of electrophoretic variation together with the considerable number of alleles that have been found in many of the polymorphisms has often caused this group of alleles to be considered the most significant numerically, and sometimes also as far as selection is concerned.

More recently a number of observations, obtained by means of other techniques, have indicated that electrophoretic variation may only be a (minor?) part of the total genetic variation in populations. By means of gel sieving Johnson (1977) has shown molecular heterogeneity within several electrophoretic alleles of different enzymes from a number of species. Also surveys of heat stability (Singh *et al.* 1976) have disclosed variation within electrophoretically uniform alleles, and amino acid sequence analysis of human hemoglobins has shown that many of the rare variants, having the same electrophoretic pattern, are due to different mutations.

The quantitative variation, as may be seen in the relative enzyme activity of electrophoretic alleles from different heterozygotes, adds one more dimension to the variation spectrum. In man, for example, many

of the heterozygotes for hemoglobin α-chain variants have about 80 per
cent of the normal form of the hemoglobin and only about 20 per cent of
the mutant form. These observations caused Lehmann and Carrell (1968) to
suggest the existence of two structural loci for the hemoglobin α-chain.
According to such a model, a mutation in one of the four α-genes would
result in the production of only 1/4 of the abnormal hemoglobin. The model was later confirmed for Caucasian populations by Hollan *et al.* (1972).
In other ethnic groups, populations monomorphic for only one Hb-α locus
are known (Abrahamson *et al.*1970; Beaven *et al.* 1972), as are populations polymorphic for the number of Hb-α loci (Rucknagel and Winter
1974; Rucknagel and Rising 1975). In mammalian hemoglobins polymorphism
for the number of structural genes is quite common (Kitchen and Boyer
1974) but other published examples of such genetic variation are rare.

The absence of examples (outside the hemoglobins) of polymorphisms
for gene number may be due to difficulties in analyzing such quantitative variations. The overall content of a certain protein is frequently
determined not only by the genotype of the specimens, but also by other
factors such as age, sex and feeding conditions. Since it is often
impossible to get an estimate of the relative importance of the individual componets, it is difficult to determine the genetics of a given
quantitative variation by comparing different specimens. Lehmann and
Carrell (1968) based their model on observations of the relative amounts
of gene products in single heterozygous individuals. I have used the
same approach in an investigation of a polymorphism in the salivary
amylase from the bank vole, *Clethrionomys glareola*. The polymorphism
has both qualitative (electrophoretic) and quantitative elements. Previous studies of a bank vole population from Jutland (Nielsen 1977)
have demonstrated the existence of a polymorphism for three electrophoretic alleles and a polymorphism for the number of amylase loci on the
chromosomes. Three different chromosome types with one, two or three
closely linked salivary amylase loci were found.

In this paper, I present an extended study of both polymorphisms.
The electrophoretic alleles have a heterogeneous distribution within
Denmark and large monomorphic areas are found. The polymorphism for the
number of salivary amylase genes per chromosome, on the other hand,
was recorded in all the investigated populations.

MATERIALS AND METHODS

With the purpose of surveying bank vole populations for electrophoretic
variations in the salivary amylase, a total of 1,520 animals were sampled
from 33 different Danish localities.

Saliva was taken from the etherized animals by flushing their mouth
with a drop of destilled water. The samples were centrifuged for a few
minutes at 15,000 g, and electrophoresis was carried out on microscope
slides covered with 2 ml. of a 0.9 per cent agar gel. The same phosphate
buffer (pH 7.3, i.s. = 0.02) was used both in the gel and in the buffer
tanks and a satisfactory separation of the isoenzymes was obtained
after one hour, using a current intensity of 5 mA per slide. Technique
and apparatus used have been described in detail by Sick and Nielsen
(1964), Sick (1965) and Nielsen (1977).

From the visual inspection of the zymograms one can determine the
electrophoretic allele composition of the single animals but it is not
possible to obtain information on the actual number of amylase genes
on the chromosomes. To identify the virtual genotype (chromosome type)
of the wild animals, 22 voles from Lading, 29 from Hov, 23 from Lammehave, 35 from Langå plus 15 voles from other localities were mated to
laboratory animals. A total of 1,534 F_1's were reared from these 124
pairs. Knowing the electrophoretic phenotype of the wild animals (typically the male), the mate could be chosen so that the two parents
had no electrophoretic alleles in common. In most of the crosses we
have used homozygous laboratory animals from a stock with two closely
linked amylase loci, both carrying the electrophoretic B allele, $i.e.$
a stock with the single-banded electrophoretic pattern, type B, and
the genotype $Amy-1^B$, $Amy-2^B/Amy-1^B$, $Amy-2^B$. From crosses where the
wild parent had the electrophoretic A type, the offspring were all
double-banded with the A and the B band. Substantial variation in the
relative enzyme activity of the two bands was seen among the zymograms
from the single F_1 individuals. Since these F_1's had the test chromosome, $Amy-1^B$, $Amy-2^B$, in common this variation is likely to be caused
by the allele contents of the chromosome from the wild parent. To measure the amounts of enzyme in the amylase bands from heterozygotes,
the animals were sacrificed at approximately 6 weeks of age. The salivary glands were removed and treated as described by Nielsen (1977).
After electrophoresis and fixing, the slides were stained with Amido
black and the density of the amylase bands were measured with an inte-

grating scanner (Joyce Loebl). The relative amounts of stained enzyme
in the individual bands were used as a measure of the production of
the amylase alleles present in the animals. Since our test chromosome
contained two amylase loci, both with the allele of the B form, the
ratio of the amount of type A enzyme divided by half the amount of type
B enzyme was used to identify the number of loci on the chromosome originating from the wild parent. By means of amylase activity measurements
we have previously determined the amounts of enzyme in the single electrophoretic bands and compared the results with those obtained by spectrophotometric protein scanning (Nielsen 1977). The relative proportions
found by the two methods are virtually identical, so protein scanning,
the most simple of the two methods, is used in this study.

AMYLASE PHENOTYPES IN NATURAL POPULATIONS

Clethrionomys glareola is not found in western Jutland or north of the
Limfjord, so the sampling locations given in Fig. 1 more or less represent the distribution of the animal within Denmark. Results from studies
of two of the populations, *i.e.* Ishøj (No. 30, Fig. 1) and Langå (No. 3)
have previously been published (Nielsen 1969 and 1977). In the sample
from Zealand (No. 30) two electrophoretic alleles, designated *A* and *B*,
were found. By breeding experiments it was shown that the chromosomes
carried two closely linked amylase loci, occupied either by two *A* alleles, by one *A* and one *B* allele or by two *B* alleles. The population
from Jutland (No. 3) was polymorphic for three alleles: *S*, *A* and *H*
and also here cosegregation of two alleles was demonstrated in several
cases. These four electrophoretic alleles were the only ones seen among
the remaining 1,312 individuals tested.

A number of homozygous stocks have been established from wild animals. Their electrophoretic patterns and their genotypes are indicated
in Fig. 2. Two of the stocks have both the S and the A band, but the
relative intensity of the bands is different. Their phenotypes are designated "SA" and "sA" respectively, using upper and lower case letters
to indicate the *relative* band strength in the individuals. It is not
possible, in a simple way, to measure the absolute quantities of amylase in an animal, so the use of upper case letters in the phenotypes
of all the single-banded stocks is an arbitrary decision.

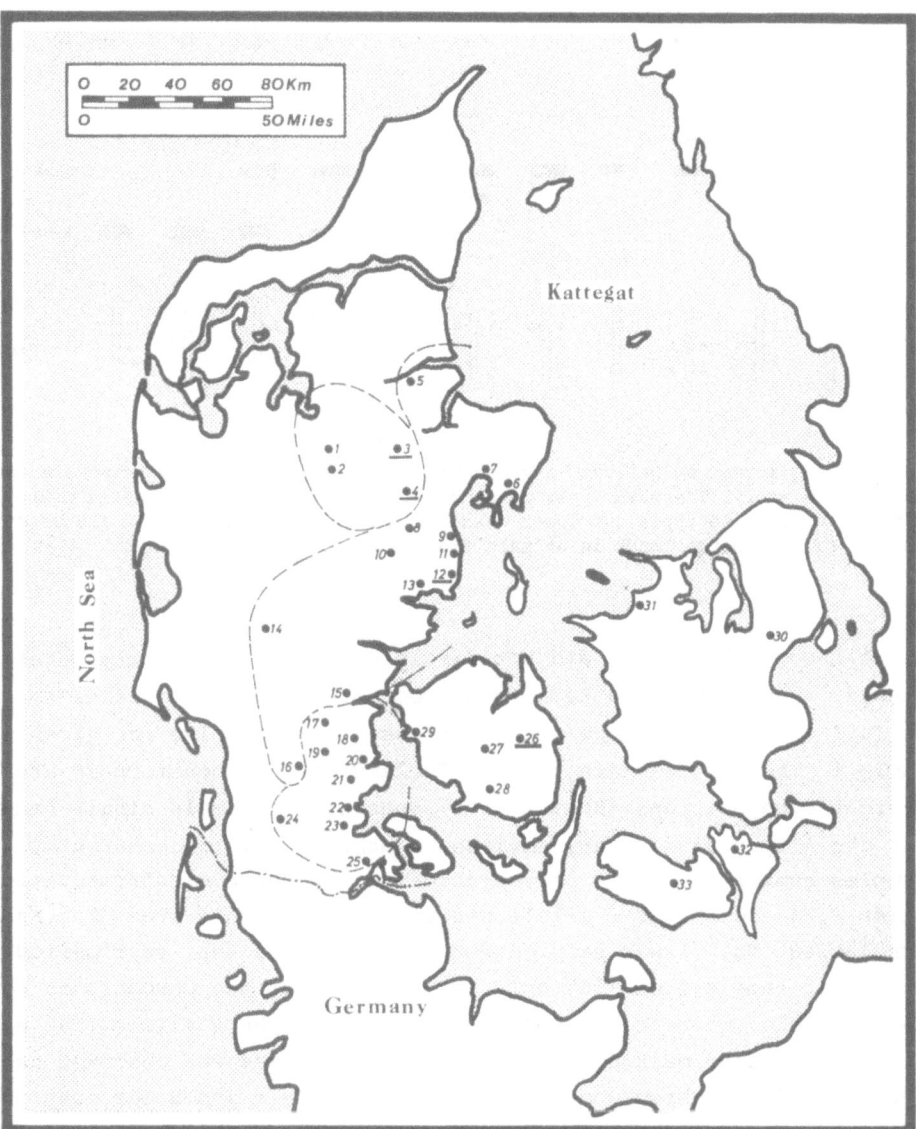

Figure 1.   Sampling localities in Denmark. The dotted lines in Jutland delineate areas with different phenotype compositions. The underlined numbers indicate populations which are described in detail in the text.

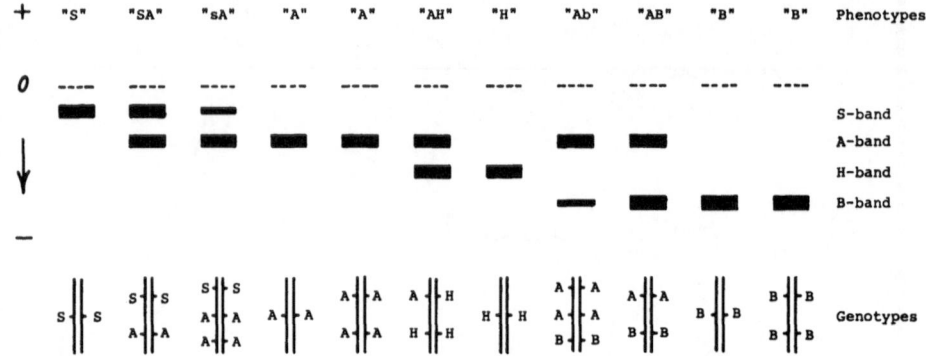

Figure 2. Diagrams of salivary amylase patterns from pure breeding stocks established in the laboratory. The phenotypes of single-banded stocks are designated with upper case letters, while upper and lower case letters are used to indicate the *relative* intensity of the enzyme bands in animals from the double-banded stocks.

Although the survey did not disclose additional electrophoretic variants, a geographic pattern in the occurrence of the alleles emerged in Jutland. The phenotype distributions in all samples are given in Table 1. The northeastern part of Jutland forms a monomorphic area, where the populations (Nos. 5 - 16) consist of animals single-banded for the A type of salivary amylase. Another area, represented by the samples number 1 - 4, is polymorphic for the three electrophoretic alleles S, A and H. The H allele was seen only in this area. The Langå population (No. 3) was sampled four times over a four year period. In the first sample from 1969 only 1.4 per cent of the chromosomes carried the S allele, whereas 17.4 per cent contained the H allele (Table 2). In 1971 a rather marked change in the frequencies was observed as 11 per cent of the chromosomes carried the S allele and 3 per cent the H allele. Another change was seen between 1971 and 1972 but since nothing is known about migration to or from the population, or about fluctuations in the number of animals, the reasons for the variation in the frequencies remain unknown.

A third area could be delineated in southern Jutland (Nos. 17 - 25), consisting of populations polymorphic for the A and B alleles. The same pattern characterized three of the four samples from Funen (Nos. 26 - 28) and both samples from Zealand (Nos. 30 and 31). One population from Funen (No. 29) was polymorphic for the three alleles S, A and B. On ba-

Table 1. Salivary amylase phenotypes in natural populations of bank voles

Locality	"SA"	"sA"	"sAh"	"A"	"Ah"	"AH"	"Ab"	"AB"	"aB"	"B"	"SAB"	"sAB"	"sAb"	"sB"	"saB"	Total
1. Hald	1			76	6	1										84
2. Aunsbjerg	1		1	46	20	4										72
3. Langå (1969)	1	16	3	65	3											88
3. - (1971)		4		56	3	1										64
3. - (1972)		5		38	8	1										52
3. - (1973)		3		26	4											33
4. Lading				5	1											6
5. Mariager				15												15
6. Mols				54												54
7. Kalø				69												69
8. Pinds Mølle				47												47
9. Hørret				15												15
10. Boes				55												55
11. Fløjstrup				57												57
12. Hov				55												55
13. Søvind				10												10
14. Grindsted				42												42
15. Kolding				15												15
16. Hjortbro				11												11
17. Vamdrup				16				6								22
18. Christiansfeld				9				1								10
19. Sommersted				12				4								16
20. Ladegård				39				10	1							50
21. Pamhule				69				56	12							137
22. Lerskov				16				18	2							36
23. Søst				2				1								3
24. Draved				8				6	1	1						16
25. Rinkenæs								1		1						2
26. Lammehave				30				15								45
27. Torpeløkke				20				10	2							32
28. Brahetrolleborg				15				9	4							28
29. Wedellsborg	1	8		17				27	11	6	1	1	2	5	1	80
30. Ishøj				43				51	38	3	1					136
31. Havnsø				8				10	5							23
32. Alslev										8				1		9
33. Maribo										31						31
																1520

sis of these alleles and the relative intensity of the amylase bands, 11 different phenotypes could be sorted out (Table 1). The *S* allele also occurred in one out of 9 animals from Falster (location No. 32), where the rest of the individuals were single-banded type B. The last sample (No. 33) was monomorphic for the B-band.

Table 2.   Estimates of frequencies of chromosomes with an *S* or *H* allele
             in the Langå population over a four year period. The estimates
             are based on the electrophoretic phenotypes of wild animals.

Sampling year	Frequencies of chromosomes containing: the *S* allele	the *H* allele	Number of animals
1969	2/144 = .014	25/144 = .174	72
1971	20/176 = .114	6/176 = .034	88
1972	4/128 = .031	4/128 = .031	64
1973	5/104 = .048	9/104 = .087	52

GENE NUMBERS IN NATURAL POPULATIONS

In addition to the polymorphism for electrophoretic alleles, the 1969
Langå sample (No. 3) was also polymorphic for the number of amylase
loci per chromosome (Nielsen 1977). We have now analyzed the chromosome
content in another population from the same geographic area. Seventeen
bank voles from Lading (No. 4), exhibiting the single-banded phenotype
"A", were mated to "B" animals with the genotype $Amy-1^B$, $Amy-2^B/Amy-1^B$,
$Amy-2^B$. The protein stained electrophoretic slides from two F_1 individuals are shown in Fig. 3 together with the curves drawn by the scanner.
The quantities of amylase are given as the integrals of the single
peaks. In one animal the amounts of enzyme in the two bands are almost
identical and the estimated number of *A* alleles on the "wild" chromosome is 195/(182/2) = 2.14. In the other animal the gene number estimate is 97/(203/2) = 0.96. In Fig. 4 each rectangle gives the measurement of a single F_1 individual and the progeny from each wild parent
are shown alternatively as open and filled symbols. Six of the matings
gave offspring which segregated in a group with an estimated number of
one *A* allele and another with two *A* alleles, thereby indicating that
the wild parents were heterozygotes for two types of chromosomes, $Amy-1^A/$
$Amy-1^A$, $Amy-2^A$. Three pairs segregated offspring with two or three *A*
alleles on their "wild" chromosomes. The progeny from 9 matings all

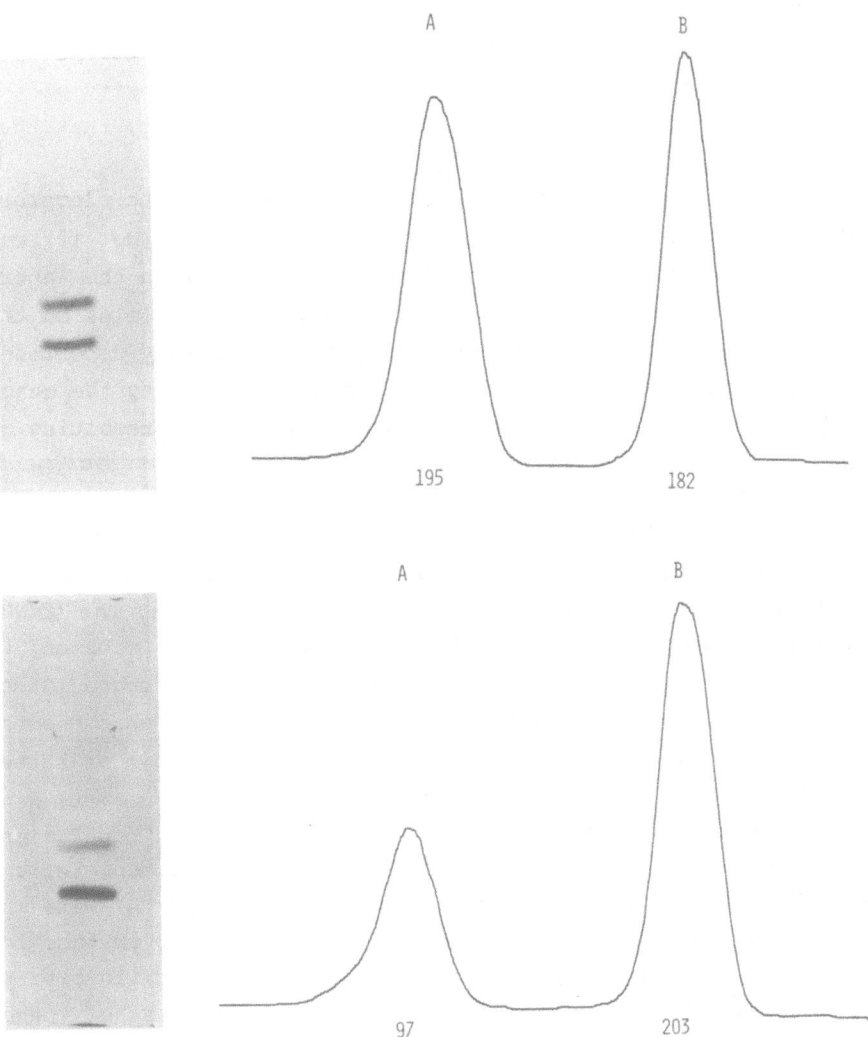

Figure 3. Protein stained electrophoretic slides from two F_1 individuals and the curves drawn by the integrating scanner.

clustered around an estimated gene number of two, *i.e.* the 9 wild parents were homozygotes for the chromosome type *Amy-1*^{A}, *Amy-2*^{A}. The remaining 5 specimens tested from Lading had the phenotype "Ah" and all
of them gave two types of offspring, one double-banded with the A and
the B band and one triple-banded with the phenotype "ahB". The double-
banded progeny were measured as before and the results are given in

Fig. 4. By protein scanning the H and B peaks could not be delineated
satisfactorily in all triple-banded animals, so only their phenotypes
and numbers are given in the Figure. The relative intensity of the
amylase bands in the zymograms makes it likely, however, that they all
had the genotype $Amy-1^A$, $Amy-2^H/Amy-1^B$, $Amy-2^B$.

To represent the geographic area monomorphic for the electrophoretic
A form of the salivary amylase, a population from Hov (No. 12) was
studied. Of the 29 animals analyzed, 21 were mated with the laboratory
stock, type "B", as used before and the number of A alleles on the
chromosomes was estimated from the offspring as already described. The
remaining 8 individuals were crossed with animals having the genotype
$Amy-1^S/Amy-1^S$. Since the reference chromosome in the resulting offspring had only one amylase locus, the amount of amylase enzyme in the
A peak divided by the amount of enzyme in the S peak directly estimated
the number of A alleles on the chromosomes derived from the wild parent.
The results are given in Fig. 5 (Hov), where the triangles are the average values of the progeny groups from these 8 matings. The squares represent the values of the offspring groups from the 21 crosses, wild
"A" × lab "B". It is clear that the averages cluster around integer
values of estimated gene numbers, indicating that the chromosomes in
this "monomorphic" population also carry either one, two or three amylase loci.

The same three chromosome types were recorded in a population from
Lammehave on Funen (No. 26), polymorphic for the electrophoretic A and
B alleles. Fourteen wild specimens with the phenotype "A" were mated
to laboratory animals of either the "B" type, with two amylase loci,
or the "S" type, with only one locus, while 9 wild individuals showing
the phenotype "Ab" were crossed to "S" animals. From one of these
crosses, wild "Ab" × "S", a progeny group with the phenotype "SB" was
found. The equality of the amylase bands was confirmed by protein
scanning, a result which strongly indicates that the chromosome from
the wild parent carried only one B allele, $Amy-1^B$. Other matings of the
same kind produced offspring groups with three bands; S, A and B. The
chromosome constitution of the triple-banded F_1's was determined from
visual inspection of the zymograms. The types and numbers of these chromosomes carrying both A and B alleles are given in Fig. 5 where the average values of the estimated gene numbers from all measured progeny
groups are also diagrammed. The data include those from Langå (No. 3)
previously published (Nielsen 1977). As before, squares are the values

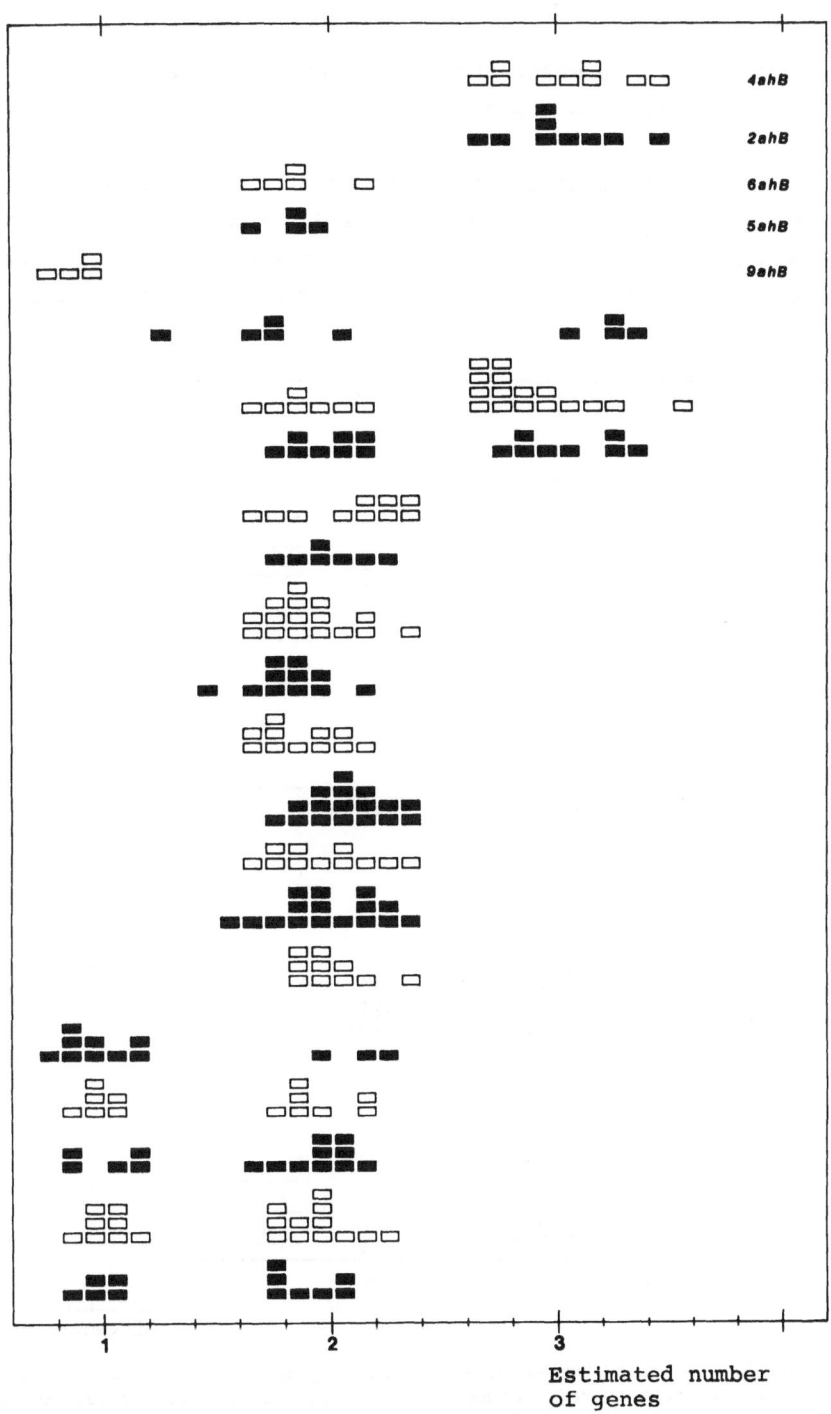

Figure 4. Estimated number of *A* alleles in the offspring from 22 crosses. Each rectangle represents the measurment of one F₁ individual. Progeny from the single wild
parents are shown by alternating open and filled symbols. In the crosses where triple-
banded offspring occurred, their phenotypes and numbers are given to the right.

of progeny groups where the test chromosome had two loci, $Amy-1^B$, $Amy-2^B$, while the triangles represent those having a chromosome with one locus, $Amy-1^S$.

The estimates of the frequencies of chromosomes with the different gene numbers are presented in Table 3. It is characteristic that the type with two amylase loci is the most frequent in all four populations, making up 55 to 75 per cent of the chromosomes. The Table also gives

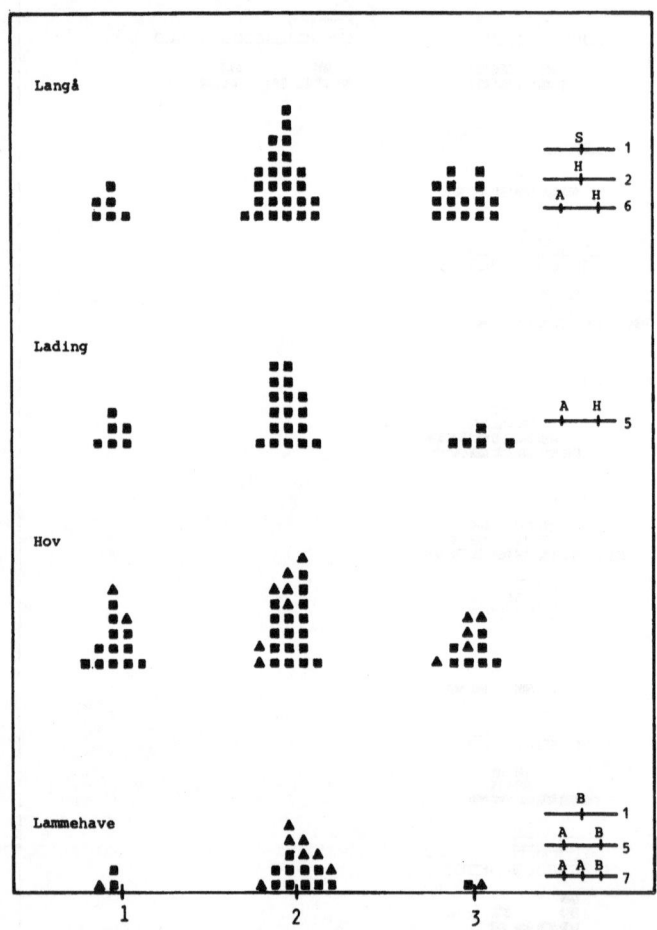

Estimated number
of genes

Figure 5. Estimated number of A alleles in progeny groups from four populations. The squares represent average values from offspring groups with a test chromosome containing two amylase loci while the triangles indicate those having a test chromosome with only one locus. The types and numbers of chromosomes carrying other alleles than A are given for each population.

the observed number of animals with the different chromosomes. Within
the four populations, there is good agreement between the observed
numbers and their Hardy-Weinberg expectations, with the reservation
that the number of tested animals from single populations is small.

Finally, a few animals of phenotype "B" from the monomorphic population recorded in Maribo (No. 33) were tested. Also here the three
chromosomes were found, so the polymorphism for the number of amylase
loci per chromosome exists in all Danish bank vole populations studied.

Table 3. Frequencies of chromosomes with different number of amylase
loci from four populations and the distribution of "genotypes"
among wild animals according to the number of *Amy* loci on the
chromosome.

Locality	Chromosome frequencies			Number of *Amy* loci on chromosomes						
	1(*Amy*)	2(*Amy*)	3(*Amy*)	1/1	1/2	1/3	2/2	2/3	3/3	Σ
Langå	.13	.64	.23	1	5	2	13	0	0	35
Lading	.14	.75	.11	0	6	0	11	0	0	22
Hov	.24	.55	.21	0	11	3	7	1	1	29
Lammehave	.07	.74	.20	0	3	0	14	2	2	23

DISCUSSION

The overall content of amylase in the salivary glands from a number of
rodents undergoes significant fluctuations. During resting periods the
animals build up stores of the enzyme which is then partly released in
connection with food uptake. Other non-genetic factors have an influence
on the system as well, so it is comprehensible that the amounts of amylase found in the glands from different bank voles vary considerably.
To overcome this type of variation I have based the results presented
here on measurements of the relative amounts of gene products in single
individuals. We have, however, carried out some investigations of the
amylase content in salivary glands from laboratory stocks homozygous
for one, two or three loci. The relative enzyme levels in these stocks

are near 1:2:3 (Hjorth, unpublished results), and in animals with three
loci the amylase made up about half of the protein in the gland.

It seems likely that amylase is an enzyme of major importance in a
rodent such as the bank vole. When one compares the two polymorphisms
it also seems likely, to the author, that the variation in enzyme levels
ranging from one to three is more important to the animals than the variation which causes changes in the electrophoretic mobility of the
alleles (the specific activities of these alleles are the same [Hjorth
unpublished results]). The quantitative polymorphism was recorded in all
populations investigated, independent of their electrophoretic allele
composition. The chromosome type containing two amylase loci was the
most frequent, while the frequencies of the other types were more or
less symmetrically distributed in the single populations. Such a distribution could indicate that two amylase loci per chromosome is the most
favourable condition and that the other types are generated by, for instance, unequal crossing over. If this is a valid explanation for the
existence of the chromosome polymorphism, it accounts for the observations that no particular electrophoretic allele was associated with one
chromosome type and that populations can be monomorphic for different
electrophoretic alleles and still contain three chromosome types. The
explanation is, however, based only on speculation, but by describing
the salivary amylase variation in the bank vole, I hope to have demonstrated that quantitative enzyme variation may have a rather simple
genetic basis on which selective forces can act.

Summary.

In a survey of Danish bank vole populations, four different electrophoretic alleles of salivary amylase were observed and three geographic areas could be delineated
in Jutland. One area, monomorphic for a single allele, was surrounded on the south by
populations polymorphic for two alleles and on the northwest by populations harbouring
three electrophoretic alleles. In addition to the qualitative variation, many of the
animals with more than one amylase band showed a substantial variation in the relative enzyme activity of these bands. The quantitative variation is due to a polymorphism for the number of amylase loci per chromosome. Several crosses have been made
among laboratory animals homozygous for chromosomes with, e.g. two loci, both carrying
the electrophoretic allele B, and wild animals with the electrophoretic phenotype A.
In the double-banded offspring the relative amounts of amylase protein were determined
by spectrophotometric measurements. Since the known chromosome, in this example,
carried two amylase loci, the amount of A-amylase divided by half the amount of B-amy-
lase was used as an estimate of the number of A alleles on the chromosome from the
wild parent. Animals from four populations, representing different parts of Denmark,
were studied. In all four cases, the gene number estimates clustered around integer
values of one, two or three, indicating that these populations were polymorphic for
three chromosome types with either one, two or three closely linked amylase loci.

Acknowledgment.

   This study was supported by a grant from the Danish Natural Science Research Council.

REFERENCES

Abrahamson, R.K., D.L. Rucknagel, D.C. Shreffler and J.J. Saave. 1970. Homozygous Hb J-Tongariki: Evidence for only one alpha chain structural locus in Melanesians. *Science* 169: 194-196.

Beaven, G.H., R.W. Hornabrook, R.H. Fox and E.R. Huehns. 1972. Occurrence of hetero-zygotes and homozygotes for the α-chain hemoglobin variant Hb-J (Tongariki). *Nature* 235: 46-47.

Hollan, S.R., J.G. Szelenyi, B. Brimhall, M. Duerst, R.T. Jones, R.D. Koler and Z. Stocklen. 1972. Multiple alpha chain loci for human hemoglobin: Hb J-Buda and Hb G-Pest. *Nature* 235: 47-50.

Johnson, G.B. 1977. Hidden heterogeneity among electrophoretic alleles. Presented in this volume.

Kitchen, H. and S. Boyer (Eds.). 1974. Hemoglobins: comparative molecular biology models for the study of diseases. *Ann. N.Y. Acad. Sci.* 241.

Lehmann, H. and R.W. Carrell. 1968. Differences between α- and β-chain mutants of human haemoglobin and between α- and β-thalassaemia. Possible duplication of the α-chain gene. *Brit. Med. J.* 4: 748-750.

Nielsen, J. Tønnes. 1969. Genetic studies of the amylase isoenzymes of the bank vole *Clethrionomys glareola*. *Hereditas* 61: 400-412.

Nielsen, J. Tønnes. 1977. Variation in the number of genes coding for salivary amy-lase in the Bank vole, *Clethrionomys glareola*. *Genetics* 85: 155-169.

Rucknagel, D.L. and W.P. Winter. 1974. Duplication of structural genes for hemoglo-bin α and β chains in man. *Ann. N.Y. Acad. Sci.* 241: 80-92.

Rucknagel, D.L. and J.A. Rising. 1975. A heterozygote for Hb_{β}^{S}, Hb_{β}^{C} and Hb_{α}^{G} Philadelph. in a family presenting evidence for heterogeneity of hemoglobin alpha chain loci. *Amer. J. Med.* 59: 53-60.

Sick, K. and J. Tønnes Nielsen. 1964. Genetics of amylase isozymes in the mouse. *Hereditas* 51: 291-296.

Sick, K. 1965. Hemoglobin polymorphism of cod in the Baltic and the Belt Sea. *Hereditas* 54: 19-48.

Singh, R.S., R.C. Lewontin and A.A. Felton. 1976. Genetic heterogeneity within elec-trophoretic "alleles" of xanthine dehydrogenase in *Drosophila pseudoobscura*. *Genetics* 84: 609-626.

2. STUDY OF POLYMORPHISM

A GEOMETRIC FORMULATION OF THE STABILITY CONDITION AT A TRI-ALLELIC LOCUS

A.W.F. Edwards

Mandel (1959) gave the necessary and sufficient conditions for an internal stable gene-frequency equilibrium at a tri-allelic autosomal locus in the following form. Let W be the 3×3 matrix of genotypic viabilities, $W = (w_{ij})$, with determinant $\Delta = |W|$, and let D be the determinant of the matrix W extended by an additional right column of 1's and an additional lower row of 1's, except for the additional diagonal element of 0. Let σ_{ij}, $i \neq j$, be the determinant D without the row and column *not* corresponding to i or j. Thus

$$\sigma_{13} = \begin{vmatrix} w_{11} & w_{13} & 1 \\ w_{13} & w_{33} & 1 \\ 1 & 1 & 0 \end{vmatrix} = 2w_{13} - w_{11} - w_{33}$$

Then the necessary and sufficient conditions for an equilibrium to be stable are

$$\sigma_{23}, \ \sigma_{13}, \ \sigma_{12} > 0, \ D < 0 \tag{1}$$

For an equilibrium to be internal it is necessary and sufficient that M_1, M_2 and M_3 must all have the same sign, where M_i is the determinant

of the matrix W with its ith row replaced with 1's. These two sets of
conditions are jointly necessary and sufficient for an internal stable
equilibrium according to Mandel.

In this paper we derive a set of conditions, namely

$$M_1, M_2, M_3, \sigma_{23}, \sigma_{13}, \sigma_{12} > 0, \tag{2}$$

by an essentially geometric route deriving from the work of Feller
(1969), and show them to be equivalent to Mandel's conditions. We
take the opportunity, in passing, to derive some of Feller's other results by using homogeneous coordinates.

THE SHAPE OF THE MEAN VIABILITY SURFACE

For alleles with frequencies p_1, p_2, p_3 the mean viability is,

$$w = w_{11}p_1^2 + w_{22}p_2^2 + w_{33}p_3^2 + 2w_{23}p_2p_3 + 2w_{13}p_1p_3 + 2w_{12}p_1p_2, \tag{3}$$

which, since $(p_1 + p_2 + p_3)^2 = 1$, may be thrown into homogeneous form
as

$$(w_{11} - w)p_1^2 + (w_{22} - w)p_2^2 + (w_{33} - w)p_3^2 +$$

$$2(w_{23} - w)p_2p_3 + 2(w_{13} - w)p_1p_3 + 2(w_{12} - w)p_1p_2 = 0. \tag{4}$$

For any particular value of w from among the set of possible values
(that is, those values generated by all p_1, p_2, p_3 such that $p_1 + p_2 + p_3 = 1$, negative values allowed) this equation represents a conic. (It
will already be obvious to readers familiar with homogeneous coordina-
tes that, for varying w, (4) represents a family of concentric similar
conics, but we shall show this in stages). In what follows w is to be
interpreted as any one of the possible values other than the maximum
or minimum of (3) (if such exist, and unless otherwise indicated).

By a standard result, the centre (x,y,z) is the pole of the line
at infinity $p_1 + p_2 + p_3 = 0$, which is

$$(w_{11} - w)x + (w_{12} - w)y + (w_{13} - w)z$$

$$= (w_{12} - w)x + (w_{22} - w)y + (w_{23} - w)z$$

$$= (w_{13} - w)x + (w_{23} - w)y + (w_{33} - w)z$$

or, adding $wx + wy + wz$ (the constant w) to all sides,

$$w_{11}x + w_{12}y + w_{13}z = w_{12}x + w_{22}y + w_{23}z = w_{13}x + w_{23}y + w_{33}z. \quad (5)$$

Since this does not involve w, (4) represents a concentric family of conics.

If in (3) we put the σ_{ij} as being measures of pair-wise dominance in an additive sense,

$$w = \sigma_{23}p_2p_3 + \sigma_{13}p_1p_3 + \sigma_{12}p_1p_2 + w_{11}p_1 + w_{22}p_2 + w_{33}p_3. \quad (6)$$

If all the σ_{ij} are zero (6) is a straight line and, representing w by the dimension perpendicular to the plane of the reference triangle, the w-surface is a plane, inclined in general. We defer the cases of one or two σ_{ij} zero until later.

Next, in order to show that each conic (6) is of the same shape and orientation, or *homothetic*, we use the proposition that if any two conics can be expressed in inhomogeneous form with identical quadratic terms, they are homothetic. This is easily seen to be the case, for if in each we put $p_3 = 1 - p_1 - p_2$ the two conics still have identical quadratic terms, and on a particular choice of reference triangle p_1 and p_2 are simply Cartesian coordinates, in terms of which the proposition is known to be true. Other reference triangles induce linear transformations which will affect the two conics' shape and orientation equally.

It follows that the conic

$$\sigma_{23}p_2p_3 + \sigma_{13}p_1p_3 + \sigma_{12}p_1p_2 = 0 \quad (7)$$

has the same shape and orientation as (6), but it is more convenient to work with, since it clearly passes through the three vertices of the reference triangle. Since all the conics (4) formed by varying w thus have the same shape, orientation, and (as we have seen) centre, the contours of the mean viability surface are a concentric family of homothetic conics.

Theorem 1.   Contours of the mean viability surface.

The contours of the mean viability surface are a concentric family of homothetic conics, being ellipses, parabolae or hyperbolae according as to whether

$$\sigma_{23}^2 + \sigma_{13}^2 + \sigma_{12}^2 - 2\sigma_{13}\sigma_{12} - 2\sigma_{23}\sigma_{12} - 2\sigma_{23}\sigma_{13}$$

is less than, equal to, or greater than, zero.

Proof: That they are concentric conics homothetic to (7) has already been shown. To find when (7) is an ellipse, etc., we consider its intercept with the line at infinity $p_1 + p_2 + p_3 = 0$. It is a standard result that the line at infinity intersects the conic $ap_1^2 + bp_2^2 + cp_3^2 + 2fp_2p_3 + 2gp_1p_3 + 2hp_1p_2 = 0$ at no, one or two points according as to whether the determinant

$$\begin{vmatrix} a & h & g & 1 \\ h & b & f & 1 \\ g & f & c & 1 \\ 1 & 1 & 1 & 0 \end{vmatrix}$$

is less than, equal to, or greater than, zero; conditions which indicate an ellipse, a parabola, and a hyperbola, respectively. Applying this to (7) we find (as $a = b = c = 0$, $h = \sigma_{12}$, $g = \sigma_{13}$, and $f = \sigma_{23}$) the determinant as $\sigma_{23}^2 + \sigma_{13}^2 + \sigma_{12}^2 - 2\sigma_{13}\sigma_{12} - 2\sigma_{23}\sigma_{12} - 2\sigma_{23}\sigma_{13}$, and the theorem is proved. □

An alternative proof is to apply the above standard result directly to (3), obtaining D as the determinant. Substituting for the w_{ij} ($i \neq j$) in terms of the σ_{ij} and w_{ii}, the w_{ii} and w_{ij} are all eliminated, leading to $4D = \sigma_{23}^2 + \sigma_{13}^2 + \sigma_{12}^2 - 2\sigma_{13}\sigma_{12} - 2\sigma_{23}\sigma_{12} - 2\sigma_{23}\sigma_{13}$.

It is natural that the condition should be expressible in terms of the σ_{ij} alone, since adding a constant to each viability changes only w in (3), and does not change the σ_{ij}.

Another form for D is in terms of the cofactors of $|W|$:

$$M_{11} = w_{22}w_{33} - w_{23}^2, \quad M_{12} = w_{13}w_{23} - w_{12}w_{33}$$

$$M_{13} = w_{12} w_{23} - w_{22} w_{13}, \quad M_{22} = w_{11} w_{33} - w_{13}^2$$

$$M_{23} = w_{12} w_{13} - w_{11} w_{23}, \quad M_{33} = w_{11} w_{22} - w_{12}^2. \tag{8}$$

Straightforward algebra leads to a result of Mandel's:

$$-D = M_{11} + M_{22} + M_{33} + 2M_{23} + 2M_{13} + 2M_{12} = M_1 + M_2 + M_3, \tag{9}$$

where $M_i = M_{i1} + M_{i2} + M_{i3}$.

If any two σ_{ij} are zero, say σ_{23} and σ_{31}, (7) becomes $\sigma_{12}p_1p_2 = 0$ which, by Theorem 1., is a hyperbola. Indeed it is the degenerate hyperbola consisting of the lines $p_1 = 0$ and $p_2 = 0$; thus for varying w (6) must then be a family of hyperbolae with asymptotes parallel to $p_1 = 0$ and $p_2 = 0$. If only one σ_{ij} is zero, say σ_{12}, (7) becomes $\sigma_{23}p_2p_3 + \sigma_{13}p_1p_3 = 0$ which, again by Theorem 1 , is a hyperbola, this time the degenerate hyperbola $p_3 = 0$ and $\sigma_{23}p_2 + \sigma_{13}p_1 = 0$. Hence for varying w (6) must then be a family of hyperbolae with asymptotes parallel to these two lines.

The case of all σ_{ij} zero has already been mentioned, but it may also be noted that then $|W| = 0$, and W is singular.

Theorem 2. Feller's Theorem, Part I.

The contours are ellipses if and only if the σ_{ij} are of the same sign and there exists a triangle with sides L_i such that

$$L_1^2 : L_2^2 : L_3^2 = \sigma_{23} : \sigma_{13} : \sigma_{12}.$$

Proof: By Theorem 1 the contours are ellipses if and only if

$$\sigma_{23}^2 + \sigma_{13}^2 + \sigma_{12}^2 - 2\sigma_{13}\sigma_{12} - 2\sigma_{23}\sigma_{12} - 2\sigma_{23}\sigma_{13} < 0. \tag{10}$$

If one or two σ_{ij} are zero (10) cannot hold, as we have seen, and neither can there be a non-degenerate triangle. If one is negative, say σ_{12}, and the others positive, (10) is

$$\sigma_{23}^2 + \sigma_{13}^2 + \rho_{12}^2 + 2\sigma_{13}\rho_{12} + 2\sigma_{23}\rho_{12} - 2\sigma_{23}\sigma_{13} < 0,$$

where $\rho_{12} = -\sigma_{12}$ and all the values are now positive. If we define

$\tau_{ij} = \sqrt{\sigma_{ij}}$, it follows that $(\sigma_{23} + \sigma_{13} + \sigma_{12})^2 < 4\sigma_{23}\sigma_{13}$, whence $\sigma_{23} + \sigma_{13} + \rho_{12} < 2\tau_{23}\tau_{13}$, the positive root being taken, and $(\tau_{23} - \tau_{13})^2 < -\rho_{12}$, an impossibility. If just two σ_{ij} are negative the argument can be repeated, leading to a similar impossibility. Thus a necessary condition for an ellipse is that all the σ_{ij} be the same sign. Without loss of generality, we assume them all positive.

Now the left side of (10) may be factored into

$$(\tau_{23} - \tau_{13} - \tau_{12})(\tau_{13} - \tau_{23} - \tau_{12})(\tau_{12} - \tau_{23} - \tau_{13})(\tau_{23} + \tau_{13} + \tau_{12}),$$

$$(11)$$

positive roots being taken. For this to be negative, either all the first three factors must be negative, or just one of them.

Suppose only the first negative. Then $\tau_{13} > \tau_{23} + \tau_{12}$ and $\tau_{12} > \tau_{23} + \tau_{13}$, a joint impossibility. Therefore all three would have to be negative: $\tau_{23} < \tau_{13} + \tau_{12}$, $\tau_{13} < \tau_{23} + \tau_{12}$, $\tau_{12} < \tau_{23} + \tau_{13}$. But these are simply the triangle inequalities for sides τ_{23}, τ_{13}, τ_{12}, and Feller's Theorem, Part I, is proved. □

Theorem 2 is a slightly more graphic, but also more complex, statement of Theorem 1.

By virtue of the factorization (11), and assuming that not all the σ_{ij} are zero (a case already dealt with), the contours will be parabolae when, for example,

$$\sqrt{\sigma_{23}} = \sqrt{\sigma_{13}} + \sqrt{\sigma_{12}} .$$

If the contours are ellipses we can ask with respect to what reference triangle they would become circles. In contemplating a reference triangle which is not equilateral we must decide between the use of trilinear coordinates (the lengths of the perpendiculars from the given point to the sides of the reference triangle) and the use of areal coordinates (the areas of the triangles formed by the lines from the given point to the vertices of the reference triangle). We choose the latter because areal coordinates are unaffected by orthogonal projection, which we shall use to project the ellipse into a circle and establish

Theorem 3. Feller's Theorem, Part II.

Elliptical contours may be rendered circular if, using areal co-ordinates, the reference triangle is chosen with sides L_1, L_2, and L_3. *Proof:* By Theorem 2. we note that such a reference triangle exists. A known result in areal coordinates is that if the reference triangle has sides a, b and c the circumscribing circle is

$$a^2 p_2 p_3 + b^2 p_1 p_3 + c^2 p_1 p_2 = 0 \qquad (12)$$

Identifying (7) and (12) we see that $a^2 = \sigma_{23}$, $b^2 = \sigma_{13}$, $c^2 = \sigma_{12}$, ensures the required result. For the ellipse (7) circumscribing the reference triangle with equal sides is the same shape and orientation as the elliptical contours, and the orthogonal projection implied by the new reference triangle takes this ellipse into a circle and the equilateral triangle into an inscribed triangle with sides L_1, L_2 and L_3. □

If the contours are parabolae then the reference triangle is degenerate; as is well known, no orthogonal projection will take parabolae into circles. If the contours are hyperbolae, then by Theorem 2, a reference triangle with the necessary sides does not exist. We can, however, always render a hyperbola rectangular by orthogonal projection, but we shall not pursue this, first because an element of arbitrariness enters into the choice of the projection, and hence the reference triangle, and secondly because our primary interest is in cases of stable equilibrium, which have elliptical contours. There is the further factor that, notwithstanding the availability of these interesting projections, we prefer to continue to use an equilateral triangle of reference for graphical presentation. Feller (1969) examines the hyperbolic cases in more detail.

Additionally we may note that a transformation to new viabilities w'_{ij}, where $w_{ij} = \lambda w'_{ij} + \mu$, $w'_{ij} \geq 0$, leads to the same geometrical figure with the contour lines relabelled, since substitution in (3) gives $w = \lambda w' + \mu$.

Theorem 4. Sections of the mean viability surface.

Along any line the graph of the mean viability w (3) is a parabola.

Proof: Solving for p_2 and p_3 in terms of p_1 between the equation for the line and $p_1 + p_2 + p_3 = 1$, and substituting the results in (3), (3) becomes a quadratic in p_1, which represents a parabola. □

Theorem 5. Curvature of the mean viability surface in the elliptical case.

With elliptical contours, if the σ_{ij} are positive the mean viability surface is concave, and if the σ_{ij} are negative it is convex.

Proof: By Theorem 4 every section of the surface is a parabola, and with elliptical contours if any section is concave so is the whole surface. Consider $\sigma_{12} > 0$, which implies $2w_{12} > w_{11} + w_{22}$ and hence $(1/4)w_{11} + (1/2)w_{12} + (1/4)w_{22} > (w_{11} + w_{22})/2$. The left side is the mean viability at (1/2, 1/2, 0), which thereby exceeds the mean of the mean viabilities at (1, 0, 0) and (0, 1, 0). Thus along $p_3 = 0$ the surface is concave, and hence everywhere. The reverse holds if $\sigma_{12} < 0$. □

We may note in passing that the stationary point on $p_1 = 0$ is at $p_2 = (w_{23} - w_{33})/\sigma_{23}$, and similarly for the other boundaries $p_2 = 0$ and $p_3 = 0$, whence, using Feller's Theorem, Part II, the circular contours in the transformed space may be located geometrically, the centre of the family of circles being at the join of the perpendiculars to the three sides of the reference triangle at these stationary points.

CONDITIONS FOR STABLE EQUILIBRIUM

The Equivalence Theorem states that at an internal equilibrium point the gene frequencies identify with the values at the turning point of w, and that the equilibrium is stable if and only if the turning point is a maximum.

It immediately follows that for an internal stable equilibrium to exist it is necessary and sufficient that

(1)   the centre of the family of conics (5) be within the reference triangle;

(2)   the family of conics be ellipses; and

(3)   w be a maximum and not a minimum at the centre.

Taking these points in order, we first require that x, y and z in (5) be all positive. Now from the definition of the average effects: $\alpha_i = w_{1i}P_1 + w_{2i}P_2 + w_{3i}P_3 - w$, we see that the centre is where the average effects are all equal. But each is then zero, whence at the centre (x,y,z), where

$$w_{11}x + w_{12}y + w_{13}z = \hat{w},$$

$$w_{12}x + \bar{w}_{22}y + w_{23}z = \hat{w}$$

$$w_{13}x + w_{23}y + w_{33}z = \hat{w}$$

the condition that x, y and z be all positive is that M_1, M_2, and M_3 must all have the same sign.

The second point, that the conics be ellipses, is met if and only if (10) is fulfilled, or

$$4D = -4(M_1 + M_2 + M_3) < 0 \qquad\qquad (13)$$

by Theorem 1 and the subsequent arguments, and the third point, that the mean viability be a maximum rather than a minimum at the centre, is met if and only if the σ_{ij} are all positive (Theorem 5).

Thus we have the following theorem:

Theorem 6. Conditions for stable equilibrium.

The necessary and sufficient conditions for an internal stable equilibrium are that each of M_1, M_2, M_3, σ_{23}, σ_{13}, and σ_{12} be positive. □

In view of equation (13) these six quantities represent only five independent parameters, in spite of being functions of the six independent viabilities w_{ij}. This is because both the σ_{ij} and the M_i are insensitive to the addition of a constant to all the w_{ij}, so that the smallest of the w_{ij} may be assigned the value zero without loss of generality.

That these conditions are equivalent to those of Mandel given in the introduction is immediate from $-D = M_1 + M_2 + M_3$; for if, as Mandel asserts, M_1, M_2 and M_3 are all to be of the same sign and D < 0, then they must all be positive, and the above conditions are obtained.

If, on the other hand, M_1, M_2 and M_3 are all positive, they are all of the same sign and $D < 0$, and Mandel's conditions are obtained.

Summary.

Mandel (1959) gave the necessary and sufficient conditions for an internal stable gene-frequency equilibrium at a tri-allelic autosomal locus. In this paper we derive a set of conditions by an essentially geometric route deriving from the work of Feller (1969) and show them to be equivalent to Mandel's conditions.

Acknowledgement.

The substance of this paper forms part of Chapter 5 of the author's book *Foundations of Mathematical Genetics* to be published by Cambridge University Press.

REFERENCES

Feller, W.   1969.   A geometrical analysis of fitness in triply allelic systems.
    Math. Biosci. 5: 19–38.

Mandel, S.P.H.   1959.   The stability of a multiple allelic system.
    Heredity 13: 289–302.

2. STUDY OF POLYMORPHISMS

A GENERAL MODEL TO ACCOUNT FOR ENZYME VARIATION IN NATURAL POPULATIONS.
IV. THE QUANTITATIVE GENETICS OF VIABILITY MUTANTS

John H. Gillespie

The current struggle over the interpretation of the observed genetic
variation in enzymes in natural populations has generally made little
direct use of the wealth of experimental results of the pre-electro-
phoresis era. Perhaps this is due, in part, to the contradictions which
continually cropped up when different laboratories did similar experiments. But while most of us have been concentrating on the electrophoretic data, a few workers have been duplicating and extending many of
the original experiments on the quantitative genetics of fitness traits.
Certainly no one has done a finer job in this than Terumi Mukai, and
it is because of his successes in settling long-standing disputes that
there is finally a firmly established phenomenology of the genetic variation in viability which can be confidently used for constructing models
of the genome.
    In this introduction I intend to give ten experimental observations
which I feel are critical in understanding the nature of the genetic
variation in *Drosophila*. These ten observations will be used to construct a model of the genome. The approach I am going to follow is
necessarily premature since we are not yet in a position to know much
about the genetic variation in total fitness. The level of precision
which has been attained in studies on viability is almost unthinkable
for fecundity or developmental time studies. For this reason I will be
rather casual in my definition of fitness, and will often confuse it

with viability. At this point in the history of our field this is the
best I feel that we can do.

The first three observations involve the experimental designs pioneered in the laboratories of Dobzhansky and Wallace (Wallace and Madden, 1953; Dobzhansky and Spassky, 1953; Dobzhansky and Spassky, 1954).
In this design the relative viabilities of chromosomes extracted from
nature are compared in the homozygous and heterozygous states using
inverted "analyzer" chromosomes with dominant markers and recessive
lethals. Three of the results from these are

1) $E(HOM) < E(HET)$. The average viability of lethal-free chromosomes
in the homozygous state is invariably less than that of the heterozygotes.

2) $VAR_g(HOM) > VAR_g(HET)$. The genetic variation between chromosomes
within the homozygotes is greater than the genetic variation between
chromosomes within the heterozygotes.

3) $VAR_e(HOM) > VAR_e(HET)$. The replication variance, excluding the
sampling variance, is greater for homozygotes than for heterozygotes.
This variance was called the microenvironmental variance by Dobzhansky
and Levene (1955). This phenomena is often called genetic homeostasis
and was first discovered by Dobzhansky and Wallace in 1953 although a
proper statistical description of it was not published until 1955.

Genetic homeostasis was further explored in the coupled papers,
Dobzhansky, Pavlovsky, Spassky, and Spassky, (1955), and Dobzhansky
and Levene (1955) in which they discovered the following.

4) When the laboratory environment is altered there are significant
genotype-environment interactions. This observation has not, to my
knowledge, been examined carefully since this original and important
discovery by Dobzhansky and his co-workers.

The next two observations are among the most controversial in the
field. They have to do with the heterozygous effects of lethal and
quasi-normal mutations. In reviewing the older literature one cannot
help but to feel that a great deal of the failure to agree came from
the fact that workers would set out to uncover a 5% or less effect
with only 20 to 50 chromosomes extracted from nature. With such designs
the power simply is not there. It is in this area where the opening
words of the abstract of the Mukai and Yamaguchi (1974) paper read
like a ray of hope: "Six hundred and ninety-one second chromosomes
were extracted from a Raleigh, North Caroline population....".

5) Spontaneously and naturally occurring lethal mutations are, on
the average, slightly deleterious in the heterozygous condition. For

naturally occurring mutations this was originally argued indirectly by
Dobzhansky and Wright (1941) and more recently by Crow and Temin (1964),
and was shown to be true directly by Hiraizumi and Crow (1960) and by
Mukai and Yamaguchi (1974). The confusion in such papers as that by
Dobzhansky and Spassky (1968) must be attributed to the small number
of chromosomes used to make the comparisons. For spontaneously occurring lethals there has always been agreement since the original work
by Stern, et.al. (1952). The recent paper by Yoshikawa and Mukai (1970)
supports Stern's finding.

6) Quasi-normal variants (Mildly-detrimental in the terminology of
Mukai and Crow) are nearly additive. This holds for spontaneously and
naturally occurring mutations. This observation is certainly the most
revolutionary since it seems to conflict with the basic tenants of those
who favor pervasive heterosis as well as those who favor lots of reces-
sive-deleterious alleles lurking around. The evidence here is less direct than we would like, but convincing nontheless. For naturally occurring variants the first hint of this came from the very clever arguments based on D:L ratios that were given by Greenberg and Crow (1960).
More direct evidence comes from the significant regression of heterozygote on the sum of homozygote viabilities demonstrated by Wills (1966)
and Mukai and Yamaguchi (1974). Finally there is the partitioning of
the genetic variation into additive and dominant components from a
partial diallel cross reported by Mukai et. al. (1974). For spontaneously occurring variants the results of Mukai et. al. (1972) also
argue persuasively for near additivity.

These two observations, taken together, suggest that there is an
inverse relationship between the homozygous and heterozygous effects
of mutant alleles. The next observation sits alone in that it involves
crosses with flies from different localities.

7) F_1 interpopulational hybrids are usually more viable than F_1's
from within population crosses. This surprising observation is due to
Vetukhiv (1953) and was confirmed by Brncic (1954).

8) The epistatic interactions between spontaneously or naturally occurring variants are synergistic. This is one of the few areas where
the laboratories of Crow, Dobzhansky, and Mukai obtained similar results. The relevant papers are Temin, et. al. (1969), Mukai (1969),
Spassky, et. al. (1965). A related and relevant observation by Kitagawa (1967) is that the heterozygous effects of lethals also show syn-
ergism.

Finally, we come to enzyme polymorphism.

9)   There is a lot of genetic variation in enzymes. See Lewontin (1974)
for the old view. The latest efforts by those searching for variants
within the variants shows that there is even more variation than previously expected. See Bernstein et. al. (1973) and Singh et. al. (1975)
for some of this.

10) The kinetic parameters of enzymes from heterozygotes appear to be,
in general, intermediate between those of the related homozygotes. This
has been documented in Gillespie and Langley (1974).

These ten observations represent a significant fraction of the experimental effort in population genetics. They concern, on the whole,
average properties of genotypes. It may never be possible to know the
nature of the contributions of individual loci to these average properties, but the properties, taken together, suggest a model for the
genome which will be developed in this paper.

A MODEL FOR THE GENOME

The essence of the model about to be described appeared first in 1934
in a paper on the physiological basis of dominance by Sewall Wright.
Wright's biochemical justification for his model is remarkably modern
in its overall feature, based, as it is, on the dynamic properties of
enzyme pathways. Rather than trying to justify the present form of the
model, *a priori*, as Wright did, I prefer to justify it *a posteriori* by
comparing its properties to the observations given in the introduction.
The motivation for the model comes from the need to transform an apparently additive phenotype, the kinetic properties of enzymes, into a
very non-linear one, the inverse relationship between the heterozygote
and homozygote effects of mutations, while retaining a mechanism for
maintaining genetic polymorphism.

For a locus which is the structural gene for an enzyme let us postulate the following:

I)   Each genotype is characterized by a random variable, $X(n,t)$, which
will be called the enzyme activity of the genotype. The enzyme activity
will generally fluctuate at random in both time and space, the arguments refer to the nth patch in the tth generation.

II)   The enzyme activity of a heterozygote in a specified environment
is exactly intermediate between the activities of the two associated
homozygotes in that environment. The justification for this assumption

has already been given in Gillespie and Langley (1974).

III) There exists a function, $\phi(x)$, which maps the activity of the enzyme onto the fitness of the genotype. It will be assumed that this *fitness function*, $\phi(x)$, is strictly monotonically increasing, is concave throughout its domain, and approaches a finite limit as $x\to\infty$. That is,

$$\phi'(x) > 0, \text{ and } \phi''(x) < 0, \text{ and } \lim_{x\to\infty} \phi(x) = K < \infty .$$

If , in a specified environment, genotypes A_1A_1 and A_2A_2 have enzyme activity x_1 and x_2, then I, II and III dictate the following:

genotypes:	A_1A_1	A_1A_2	A_2A_2
fitness:	$\phi(x_1)$	$\phi[(x_1 + x_2)/2]$	$\phi(x_2)$

If, in addition, the allele A_1 has frequency p in the population and if the population was in Hardy-Weinberg equilibrium before selection occurred, the change in p in this environment is

$$\Delta p(n,t) = pq \frac{p[\phi(x_1) - \phi((x_1 + x_2)/2)] - q[\phi(x_2) - \phi((x_1 + x_2)/2)]}{p^2\phi(x_1) + 2pq\phi((x_1 + x_2)/2) + q^2\phi(x_2)}$$

The arguments n and t have been suppressed from the $x_i(n,t)$ in this expression.

Before cataloging observations which seem to support the general nature of this model, lets assume a particular functional form for ϕ and see how its parameter may be estimated. The simplest mathematic form for ϕ would be

$$\phi(x) = (1+\alpha)x/(\alpha+x) \tag{1}$$

where ϕ has been scaled such that an activity of one corresponds to a fitness of one, [i.e. $\phi(1) = 1$]. Suppose, in a specific environment, we write the fitnesses of three genotypes as

A_1A_1	A_1A_2	A_2A_2
1	$1-hs$	$1-s$

Using (1) we can write the relationship between the heterozygous and homozygout effects as

$$h = \alpha/(2\alpha + s) \tag{2}$$

which is our desired inverse relationship. This expression can be re-
arranged to give α as a function of h and s:

$$\alpha = sh/(1 - 2h) \tag{3}$$

We have two independent methods to estimate α. The first uses the heterozygous effects of spontaneously occurring lethal mutations. In this
case s = 1 and .04 < h < .05 (see Crow and Temin (1964)) giving us a
range for α: .0435 < α < .0556. An independent method uses the heterozygous effects of mildly detrimental genes. Mukai et. al. (1972), estimated that for an average s value of .03 the average h value is .4.
Plugging these into (3) gives us α = .06. The agreement between this and
the previous estimate is remarkable, particularly so since they involve
very different regions of the ϕ curve. From this we see that the model
summarized in statements I to III is open to experimental examination
and even experimental justification through checks for internal consist-
ences.

There are many results in the genetics literature which give added
support to this model. The first group that I will mention concerns
the choice of an assymptotic rather than a unimodal from for ϕ. A unimodal curve would imply the existance of an optimum level of activity
giving similar deleterious effects if there is too much or too little
activity.
1) As far as I have been able to tell, there is only one known human
enzyme mutant which has deleterious clinical manifestations due to an
increase in the normal activity. This is a PRPP - synthetase high activity mutant found by Becker et. al. (1973). By contrast, many low
activity mutants with deleterious clinical manifestations are known.
2) In *Drosophila* there appears to be no dosage compensation for autosomal enzymes. The activity of an enzyme is proportional to the number
of copies of the gene (O'Brien and Gethmann, 1973; Stewart and Merriam,
1974). If there were an optimal level of activity we would expect the
deleterious effects of deletions to be matched by deleterious effects of
of duplications. In fact, as recently discovered by Lindsley and Sandler, et. al. (1972), the deleterious effects of hypoploidy far greater
than for hyperploidy. Another offshoot of this study is the demonstration that there is probably only one or at most a few loci in the en-

tire Drosophila genome which are haplo-lethal. If it turned out otherwise, it would force the abandonment of our model.

These observations together with certain *a priori* notions about
the dynamics of pathways which actually date back to Wright (1934) make
the general assymptotic form of φ more compelling than the unimodal
form. This is not to say that some enzymes will not be unimodel, or
that all enzymes might not cause a reduction in fitness if their activities are elevated high enough, but that in the domain of activities
that are expressed by most variants in most environments the assumpto-
tic form of φ is in better agreement with the data.

Another assumption is that φ(0) = 0. That is, a lack of activity of
the enzyme corresponds to lethality. Here we know that there are cases
where enzymes which lack activity are not lethal. However, the generality of this is open to question. For example, the recent work of Judd,
et. al. (1972) shows that there is essentially a one-to-one correspondence between chromomeres and complimentation groups in *Drosophila*. Of
the 16 complementation groups known in the zeste-white region, lethal
mutations have been found in all but 2 of these groups. For those enzymes where null homozygotes are viable, we can use

$$\phi(x) = (1-c)(1-\alpha)x/(\alpha+x) + c.$$

A final point concerns our postulate that the enzyme activities are
random variables deriving their randomness from the state of the environment. From the constractions of the model there would be two places
where randomness might be added, one way would be to shake the curve,
φ around at random, the other would be to shake the x's. Shaking φ,
however, will not give the genotype-environment interactions uncovered
by Dobzhansky and Levene (1955) since the relative orders of the genotypes would remain fixed. Shaking the x's will give the required genotype-environment interactions.

GENETIC POLYMORPHISM

One question we would certainly want to settle first is whether or not
the model, as described thus far, can explain the existence of genetic
polymorphism in enzymes. The answer is given in a recent paper (Gillespie, 1976a) which analyses various models of temporally and spacially

fluctuating environments. Let ΔM be the differences in the mean enzyme activities of two homozygotes and let σ^2 be their variances in
activities, then, if ΔM and σ^2 are small, polymorphism will occur if

$$|\Delta M/\sigma^2| < 1 + O(\alpha)$$

This important result illustrates two things. First, because of the
small value of α, the conditions for polymorphism are fairly insensitive to the assumptions made about the nature of the fluctuations. This
is convenient since it implies that we may be able to understand how
polymorphisms are maintained without a detailed understanding of the
environment. Second, the mean difference in activities between alleles
is usually of the same order or a smaller order than the variance in
activities. This will limit the contribution which polymorphic alleles
can make to the genetic variance in fitness. This point will be taken
up in the next section.

The above result can be extended to multiple alleles for certain
environmental structures. One interesting way to pose the question is
this: can an infinite number of alleles be packed on the axes of enzyme
activity or does a phenomenon similar to "limiting similarity" prevent
this? It turns out that an infinite number of alleles can be packed in
(Gillespie, 1976b). If δ is the difference in mean activities between
neighbouring homozygotes on the axis and if $\gamma = \delta/\sigma^2$, then the maximum
number of alleles which can be maintained is

$$n_{max} \sim 2/\sqrt{\gamma} \text{ as } \gamma \to 0$$

and the heterozygosity is

$$H \sim 1 - (1/3)\sqrt{\gamma/2} \text{ as } \gamma \to 0.$$

Thus the large numbers of alleles being uncovered by various new techniques are not at all fatal to this balancing-selection model.

HETEROZYGOTE-HOMOZYGOTE COMPARISONS

The various observations on the relative viabilities of chromosomes extracted from nature in the homozygous or heterozygous states also turn out to be compatible with the single-locus implications of the model. Clearly if genes combine additively across loci the single-locus behavior extends directly to chromosomal behavior.

In the analysis of Dobzhansky and Levene (1955) it was shown that the microenvironmental variance was similar in magnitude to the macro-environmental variance. This suggests that there is considerable variation in the environment from one vial to another, and, presumably, also within a vial. For this reason, in expressing the various moments for the genotypes, we need to carry out the averaging processes over both genotypes and environments. For concreteness, I will restrict the analysis in this section to a locus with only two alleles. In the expectations and variances to be given, a subscript of g refers to operations averaged over genotypes and a subscript e refers to the process carried over environments.

Consider first the mean fitnesses of the homozygotes and heterozygotes:

$$E(HOM) = p \ E_e \ \phi(x_1) + q \ E_e \ \phi(x_2)$$

$$E(HET) = p^2 E_e \ \phi(x_1) + 2pq \ E_e \ \phi((x_1 + x_2)/2) + q^2 E_e \phi(x_2).$$

How do these two compare? A little manipulation shows that

$$E(HET) - E(HOM) = 2pq \ E_e \ [\phi((x_1 + x_2)/2) - (\phi(x_1) + \phi(x_2))/2] > 0.$$

This shows that the concavity of ϕ implies that the average fitness of loci made homozygous will always be less than the average fitness of random genotypes pulled from the population. This is true for all alleles, whether they are held in equilibrium by mutation-selection balance, balancing selection, or whatever, providing the model is accurate.

The situation with the variance is analyzed similarly but with some approximation.

To simplify the analysis assumed that the environment is constant within each vial, but fluctuates from vial to vial. The total variance is then partitioned into environmental and genetic fractions by the methods pioneered by Wallace and Madden (1953) and Dobzhansky and Le-

vene (1955). The total variance may be written as the mean (over genotypes) of the variance (over environments) of fitness plus the variance
(over genotypes) of the mean (over environments) of fitness, or symbol-
lically

$$E_g \text{ VAR}_e \ \phi(x_{ij}) + \text{VAR}_g \ E_e \ \phi(x_{ij})$$

where x_{ij} refers to the activity of the ith genotype in the jth vial.
The two of our ten observations are that

$$E_g \text{VAR}_e(\text{HOM}) > E_g \text{ VAR}_e(\text{HET})$$

which is genetic homeostasis and

$$\text{VAR}_g \ E_e(\text{HOM}) > \text{VAR}_g \ E_e(\text{HET}).$$

Are these compatible with our model? Lets consider the case of genetic
homeostasis first. If we assume that the environmental variance across
environments of each genotype is small, we can approximate $\phi(x_{ij})$ a-
round EX_{ij} for the ith genotype and use this to show that $E_g\text{VAR}_e(\text{HOM}) >$
$E_g\text{VAR}_e(\text{HET})$ if and only if

$$[\phi'(\mu_1)^2 + \phi'(\mu_2)^2]/(1+\rho) - \phi' [(\mu_1 + \mu_2)/2]^2 > 0,$$

where $\mu_i = E_e x_i$ and ρ is the correlation of x_1 and x_2. This will always
be true because, while $\phi(x)$ is concave, $(\phi')^2$ is convex. In writing
down this inequality I assumed that both homozygote genotypes had the
same variance in activity. The remarkable aspect of this inequality is
that it implies that all loci contribute to genetic homeostasis, not
just those maintained by balancing selection.

Analysis of the second inequality is equally informative. The expression is for the variance in the mean differences of the genotypes.
Recalling from the conditions for polymorphism that the mean differences between genotypes must be small and of the same order as the variances, then we must conclude that the variance in the mean differences
must be of a smaller order of magnitude. For this reason it is unlikely that alleles held in by balancing selection will make a significant
contribution to the variance in the mean genotypic effects. If we assume that this variance is due primarily to alleles held in the population by mutation-selection balance we can use the same sort of appro-

ximations as exhibited in the Mukai and Yamagochi (1974) paper to show
that

$$VAR_g E_e (HOM) - VAR_g E_e (HET) > 0$$

if and only if

$$(1 - q/2)/(1 - 2q) > h^2$$

which is obviously true for our model.

The final result in this section concerns the behavior of F_1 interpopulational hybrids. If the allele frequencies in natural populations
are fluctuating due to some stochastic element, be it drift or a temporally fluctuating environment, then the difference in the mean viabilities between F_1 interpopulational and within populational hybrids can
be shown to equal

$$(\sigma_F^2/2) \ (1-R_F) \ E_e [\phi((x_1 + x_2)/2) - (\phi(x_1) + \phi(x_2))/2]$$

(Gillespie, 1976a), where σ_F^2 is the variance in allele frequency at
a locality (the variance calculated through time) and R_F is the correlation coefficient of the allele frequencies from two separate localities. Again, the convexity of ϕ gives us the desired result.

MULTI LOCUS CONSIDERATIONS

The natural extension of this model to multiple loci would be to make
ϕ a function of the activities at a large number of loci. When we be-
gin to work through the theory one parameter emerges as being critical
for understanding the behavior of the system, namely, $\partial^2 \phi/(\partial x_i \partial x_j)$,
which is a measure of the interaction between loci. If the sign of this
mixed partial derivative is negative we have the synergistic epistasis
which was described in the introduction. This is the sign we would
choose a priori for the multiple-argument analog to the one-dimensional
concavity, but it is more compelling to rely on the experimental evidence. It is important to note, in this context, that the model would
have been in serious trouble if the sign of the epistasis had been dif-

ferent for mildly detrimentals than for the heterozygous effects of
lethals. It is for this reason that the results of Kitagawa (1967)
are particularly valuable.

Without going into details here, it turns out that the behavior
of the multiple-locus model is analagous to that of the additive model
recently described by Gillespie and Langley (1976) if and only if the
epistasis is synergistic. The behavior is slightly different in that
there is disequilibrium at steady state, but the importance of correlations between alleles at different loci remains as does the excit-
ing result that adding loci in biologically reasonable ways can only
lead to an increase in the likelihood of polymorphism at any single
locus. The details of this analysis will be submitted shortly.

Summary.

    A stochastic model of the genome is presented and its properties are compared
to published observations on the quantitative genetics of viability in *Drosophila*.
In general, the agreement is quite remarkable, the convexity properties of the model
giving, for example, a mechanistic explanation for overdominance and genetic homeostasis. In addition, the model can account for the high levels of polymorphism presently being uncovered.

REFERENCES

Becker, M.A., Kostel, P.J., Meyer, L.J., and Seegmiller, J.E. 1973. Human phospho-
    ribosyl pyrophosphate synthetase: increased enzyme specific activity in a family
    with gout and excessive purine synthesis.
    Proc. Nat. Acad. Sci. U.S. 70: 2749-2752.

Bernstein, S., Throckmorton, L.H., and Hubby, J.L. 1973. Still more genetic varia-
    bility in natural populations.
    Proc. Nat. Acad. Sci. U.S. 70: 3928-3931.

Brncic, D. 1954. Heterosis and the integration of the genotype in geographic popul-
    ations of *Drosophila pseudoobscura*.
    Genetics 39: 77-88.

Crow, J.F., and Temin, R.G. 1964. Evidence for the partial dominance of recessive lethal genes in natural populations of *Drosophila*.
Amer. Natur. 98: 21-33.

Dobzhansky, Th., and Levene, H. 1955. Genetics of natural populations. XXIV. Developmental homeostasis in natural populations of *Drosophila pseudoobscura*.
Genetics 40: 797-808.

Dobzhansky, Th., Pavlovsky, O., Spassky, B., and Spassky, N. 1955. Genetics of natural populations. XXIII. Biological role of deleterious recessives in populations of *Drosophila pseudoobscura*.
Genetics 40: 781-796.

Dobzhansky, Th., and Spassky, B. 1953. Genetics of natural populations XXI. Concealed variability in two sympatric species of *Drosophila*.
Genetics 38: 471-484.

Dobzhansky, Th., and Spassky, B. 1954. Genetics of natural populations. XXII. A comparison of the concealed variability in *Drosophila prosaltans* with that in other species.
Genetics 39: 472-487.

Dobzhansky, Th., and Spassky, B. 1968. Genetics of natural populations. XL. Heterotic and deleterious effects of recessive lethals in populations of *Drosophila pseudoobscura*.
Genetics 59: 411-425.

Dobzhansky, Th., and Wallace, B. 1953. The genetics of homeostasis in *Drosophila*.
Proc. Nat. Acad. Sci. U.S. 39: 162-171.

Dobzhansky, Th., and Wright, S. 1941. Genetics of natural populations. V. Relations between mutation rate and accumulation of lethals in populations of *Drosophila pseudoobscura*.
Genetics 26: 23-52.

Gillespie, J.H. 1976a. A general model to account for enzyme variation in natural populations. II. Characterization of the fitness function.
Amer. Natur. (in press).

Gillespie, J.H. 1976b. A general model to account for enzyme variation in natural populations. III. Multiple alleles.
Evolution . (in press).

Gillespie, J.H., and Langley, C.H. 1974. A general model to account for enzyme variation in natural populations.
Genetics 76: 837-884.

Gillespie, J.H., and Langley, C.H. 1976. Multilocus behavior in random environments. I. Random Levene Models.
Genetics 82: 123-137.

Greenberg, R., and Crow, J.F. 1960. A comparison of the effect of lethal and detrimental chromosomes from *Drosophila* populations.
Genetics 45: 1153-16B.

Hiraizumi, Y., and Crow, J.F. 1960. Heterozygous effects on viability, fertility, rate of development, and longevity of Drosophila chromosomes that are lethal when homozygous.
Genetics 45: 1071-1083.

Judd, B.H., Shen, M.W., and Kaufman, T.C. 1972. The anatomy and function of a segment of the X-chromosome of *Drosophila melanogaster*.
Genetics 71: 139-156.

Lewontin, R.C. 1974. *The Genetic Basis of Evolutionary Change*. Columbia Univ. Press, N.Y.

Lindsley, D.L., and Sandler, L. *et.al.* 1972. Segmental aneuploidy and the genetic gross structure of the *Drosophila* genome. *Genetics* 71: 157-184.

Mukai, T. 1969. The genetic structure of natural populations of *Drosophila melanogaster*. VII. Synergistic interaction of spontaneous mutant polygenes controlling viability. *Genetics* 61: 749-761.

Mukai, T., Cardellino, R.A., Watanabe, T.K., and Crow, J.F. 1974. The genetic variance for viability and its components in a local population of *Drosophila melanogaster*. *Genetics* 78: 1195-1208.

Mukai, T., Chigusa, S.I., Mettler, L.E., and Crow, J.F. 1972. Mutation rate and dominance of genes affecting viability in *Drosophila melanogaster*. *Genetics* 72: 335-335.

Mukai, T., and Yamaguchi, O. 1974. Genetic structure of natural populations of *Drosophila melanogaster*. XI. Genetic variability in a local population. *Genetics* 76: 339-366.

O'Brien, S.J., and Gethmann, R.C. 1973. Segmental aneuploidy as a screen for structural genes in *Drosophila*: mitochondrial membrane enzymes. *Genetics* 75: 155-169.

Singh, R.S., Hubby, J.L., and Throckmorton, L.H. 1975. The study of genic variation by electrophoresis and heat denaturation technique at the octanol dehydrogenase locus in members of the *Drosophila virilis* group. *Genetics* 80: 637-650.

Spassky, B., Spassky, N., and Anderson, W.W. 1965. Genetics of natural populations. XXXVI. Epistatis interactions of the components of the genetic load in *Drosophila pseudoobscura*. *Genetics* 52: 653-664.

Stern, C., Garson, G., Kinst, M., Novitski, E., and Uphoff, D. 1952. The viability of heterozygotes for lethals. *Genetics* 37: 413-449.

Steward, B., and Merriam, J.R. 1974. Segmental aneuploidy and enzyme activity as a method for cytologenetic localization in *Drosophila melanogaster*. *Genetics* 76: 301-309.

Temin, R.G., Meyer, H.U., Dawson, P.S., and Crow, J.F., 1969. The influence of epistasis on homozygous viability depression in *Drosophila melanogaster*. *Genetics* 61: 497-519.

Vetukhiv, M. 1953. Viability of hybrids between local populations of *Drosophila pseudoobscura*. *Proc. Nat. Acad. Sci.* U.S. 39: 30-34.

Wallace, B., and Madden, C. 1953. The frequencies of sub- and supervitals in experimental populations of *Drosophila melanogaster*. *Genetics* 38: 456-470.

Wills, C. 1966. The mutational load in two natural populations of *Drosophila pseudoobscura*. *Genetics* 53: 281-294.

Wright, S. 1934. Physiological and evolutionary theories of dominance. *Amer. Natur.* 68: 24-53.

Yoshikawa, I., and Mukai, T. 1970. Heterozygous effects on viability of spontaneous lethal genes in *Drosophila melanogaster*. *Jap. J. Genet.* 45: 443-455.

Chapter 3.   Sex and Evolution

THE SEX HABIT IN PLANTS AND ANIMALS

J. Maynard Smith

This paper is concerned with the selective forces responsible for the evolution of breeding systems, and in particular with hermaphroditism as opposed to dioecy, with selfing as opposed to outcrossing, and with apomixis as opposed to sexual reproduction. It was written with plants in mind, but many of the arguments are relevant also to animals.

The problems are difficult because of the variety of factors which are relevant. These include:

  i)   The allocation of resources between male and female functions.
 ii)   The relative fitness of inbred and outbred offspring.
iii)   The probability of pollination.
 iv)   Gametophyte selection (e.g. between pollen tubes) and meiotic drive.
  v)   The mechanism of sex determination.
 vi)   Types of distribution, life history and ecological adaptation.
vii)   Long-term consequences for the population.

Since Darlington (1939), the last of these factors has been given prominence in many discussions of plant evolution. I shall concentrate on some of the shortterm selection pressures, since I believe that most features of plant breeding systems must be explained in these terms.

RESOURCE ALLOCATION: WHY ARE MOST PLANTS HERMAPHRODITE?

In this section I assume that a species has some type of self-incompatibility mechanism or barrier to self-fertilisation. If so, the sex habit will be determined primarily by selection acting on the number of
offspring produced by individuals of different types. It is assumed
that genes in a zygote can act as switches, directing development into
one or other type (male, female, hermaphrodite) or, in hermaphrodites,
can alter the relative allocation of resources to seeds or pollen.
(Botanically, "hermaphrodite" refers to a flower with both male and
female organs, or to a plant with such flowers; the argument of this
section applies equally to monoecious plants, with separate male and
female flowers on the same plant. That is, it applies to all plants
which are hermaphrodite in the zoological sense).

It is also assumed that the total production of seeds plus pollen
by an individual plant is constrained to lie within a "fitness set",
which cannot be altered by genetic change. It will be shown how the
form of the fitness set determines the sex habit.

Let m, f and h be the numbers of male, female and hermaphrodite
individuals in a population. It is supposed that a male can produce N
pollen grains, a female can produce n seeds, and a hermaphrodite can
produce αN pollen grains and βn seeds. The fitness set can then be
represented as a graph of β vs α, as in figures 1 and 2.

Suppose the population produces R offspring. Assuming self-incompatibility, and using "fitness" to measure the expected number of offspring produced by an individual,

$$\text{fitness of a male} = R/(m + \alpha h),$$

$$\text{fitness of a female} = R/(f + \beta h),$$

$$\text{fitness of a hermaphrodite} = R\left[\alpha/(m + \alpha h) + \beta/(f + \beta h)\right]. \quad (1)$$

If the situation is to be evolutionarily stable, two conditions
must be satisfied:

i) The fitnesses of any types actually present in the population
must be equal, and

ii) the values of α and β in hermaphrodites, if present, must be an
"evolutionarily stable stategy" or ESS (Maynard Smith & Price, 1973);
that is, the actual phenotype, $(\alpha^*\beta^*)$, of the hermaphrodites present
in the population must be as fit or fitter than any mutant phenotype
$(\alpha\beta)$ lying in the fitness set.

Case 1. h = 0. Population "dioecious".

By condition i), R/m = R/f, or m = f. This is the familiar conclusion that the sex ratio is 1:1 when the "costs" of male and female zygotes are equal (Fisher, 1930).

A dioecious population can be "invaded" by a hermaphrodite mutant
if α/m + β/f > 1/m; that is, if α + β > 1. It follows that a dioecious
population is stable only if the fitness set is concave (fig. 1a).

Case 2. m = f = 0, h = 1; population "hermaphrodite".

If (α*β*) is the phenotype of typical members of the population,
and (αβ) is a rare mutant, then (α*β*) is an ESS if the fitness of (αβ)
is less than or equal to that of (α*β*) for all points on the boundary
of the fitness set. For stability against small perturbations, this requires that (α/α* + β/β*) be at a local maximum when α = α*, β = β*.
This is equivalent to the requirement that the product α*β* be a maximum (figure 1b).

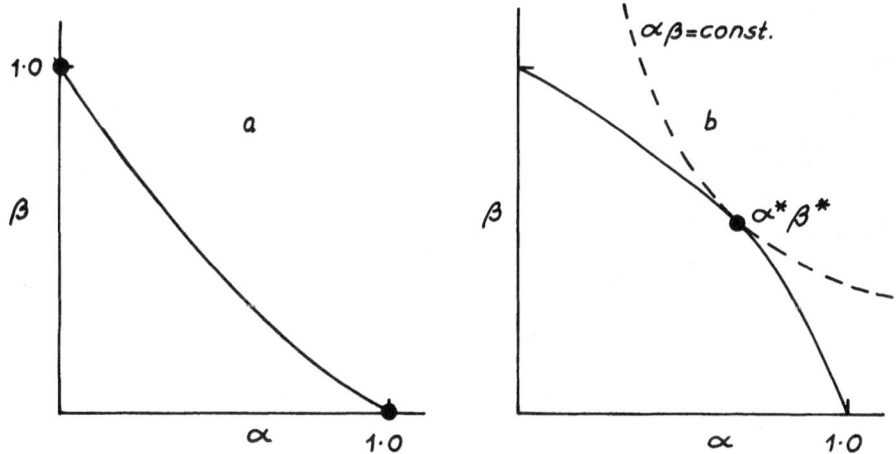

Figure 1. Fitness sets for which *a*, dioecy and *b*, hermaphroditism are the stable
strategies.

This corresponds to MacArthur's (1965) conclusion that selection will maximise the product of the number of males and females. If the fitness set is bounded by the line α + β = 1, as will be approximately true if pollen and seed production is limited by the same resources, then the ESS is for a hermaphrodite to divide its resources equally between pollen and seeds (Maynard Smith, 1971).

Case 3. h ≠ 0, f ≠ 0, m = 0. Population "gynodioecious".

The criterion α*.β* maximal provides a local stability criterion for for a hermaphrodite population. But can a population with h = 1 be invaded by males or females? It can be invaded by females if R/β* > 2R, or β* < 0.5. Provided, however, that α + β > 1 for some part of the fitness set, hermaphrodites will not be completely eliminated.

Figure 2 shows a fitness set for which the ESS is gynodioecy. Let (α*β*) be the stable phenotype of the hermaphrodites. Then the stable sex ratio, obtained by equating the fitness of females and hermaphro-

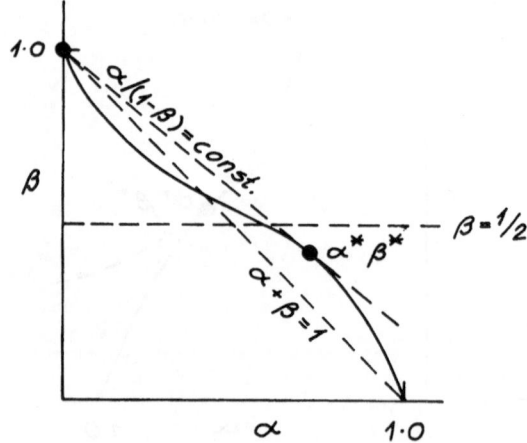

Figure 2.   Fitness set for which gynodioecy is the stable strategy.

dites, is given by f = h(1 - 2β*). This implies an excess of hermaphrodites, as is in fact observed in natural populations (Lloyd, 1974b).
The fitness of a rare mutant with a different resource allocation,
(αβ), is given by

$$V = R[α/(m + α*h) + β/(f + β*h)]$$
$$= \frac{R}{h}\left[\frac{α}{α*} + \frac{β}{(1-β*)}\right].$$

The condition for (α*β*) to be an ESS is that V should be at a
local maximum, which is equivalent to the requirement that α/(1-β)
should be at a maximum when α = α*, β = β*. The ESS can therefore be
found by drawing a tangent as in figure 2.

The conditions for androdioecy, h ≠ 0, f = 0, m ≠ 0, are similar.
All three types can coexist only in the artificial and unlikely case,
α + β = 1 exactly.

The conclusions can be summarised by saying that we would expect
to find monoecy or hermaphroditism in species for which the fitness
set is convex, dioecy if it is concave, and androdioecy or gynodioecy
for concave-convex sets.

Since the vast majority of species are hermaphrodite, is there a
general reason why fitness sets should be convex? I can offer two suggestions:

i) Pollen production occurs earlier in the season, for each species,
than seed maturation. The two processes therefore depend, in part, on
different "resources", and one would therefore expect a hermaphrodite
to do better than a linear combination of male and female.

ii) In insect-pollinated species, some energy must be expended by all
types on producing organs of attraction. A hermaphrodite plant may need
to allocate less total resource to such organs; this would be so if,
for example, a hermaphrodite flower produces as much pollen as a male
one and as much seed as a female one.

This would lead to the fitness set being convex. The conclusion
depends on assumptions, which merit further thought, on how a plant
will allocate its resources between organs of attraction and more directly reproductive structures. It does, however, suggest that dioecy
should be rarer in insect-pollinated than in wind-pollinated species.

Not all cases of dioecy will be explicable in terms of a concave
fitness set. The fitness values in i) assume self-incompatibility. In
self-compatible plants, dioecy may evolve because it prevents inbreeding depression (Baker, 1959). However, a long-lasting gynodioecious or

androdioecious system probably does imply a fitness set of the form illustrated in figure 2, since if it were an adaptation to prevent inbreeding, or to a wholly concave set, it would rapidly evolve into a
fully dioecious system. It would therefore be worth studying such species from this point of view. Thus, for a gynodioecious species, we
should seek for some reason why a plant would suffer a substantial
drop in seed production if it were to produce a few pollen grains, but
could produce a large quantity of pollen without incurring an equivalently increased loss in seed production.

Since writing this section I received from Dr. Eric Charnov an Ms
containing essentially the same mathematical argument and a very similar biological interpretation. Dr. Charnov has gone further in applying the argument to animals. Ghiselin (1969) has pointed out that hermaphroditism in animals tends to occur in species with low adult mobility, and in species in which the females brood the young. Charnov
suggests that low mobility will tend to be associated with a convex
fitness set because in such species there will be little sexual dimorphism (males will not evolve special locomotory or aggressive structures for seeking out and holding females); brooding will imply a convex set because female expenditure will occur at a different time from
male expenditure (just as seed ripening occurs later than pollen production).

THE EVOLUTION OF SELFING AND OUTCROSSING

Consider a population of self-incompatible hermaphrodites. On the average, if the population size is constant, each plant will contribute two
genomes to the next generation, one as pollen parent and one as seed
parent. A rare mutant which was capable of selfing would, on average,
contribute three genomes, two to its selfed offspring and one by outcrossing as a pollen parent. Clearly, unless offspring produced by
selfing are less fit than those produced by outcrossing, the selfing
habit will spread. How much less fit must inbred offspring be if self-
incompatibility is to be maintained?

Let selfing be caused by a recessive mutant m, such that m/m
plants always self and m/+ and +/+ never do. Counting at zygote formation, let the population contain a proportion P_o of m/m zygotes produced by outcrossing, and proportions P_1, P_2, ..., P_i m/m plants re-

sulting from 1, 2,..., i generations of selfing. The fitnesses of these plants are V_1, V_2,..., V_i respectively, relative to a fitness of unity for outcrossed plants. The frequency of gene m in outcrossed plants is p. $\bar{P} = \Sigma P_i V_i$ and p are small.

Writing p', P_i' etc. for the frequencies in the next generation:

$$p' = p + (1/2)\Sigma P_i V_i - p^2$$

$$P_o' = p^2$$

$$P_1' = P_o$$

$$P_2' = P_1 V_1$$

$$\ldots$$

$$P_i' = P_{i-1} V_{i-1} \, .$$

In the first equation, the term $(1/2)\Sigma P_i V_i$ represents the increase in the frequency of m due to pollen from m/m plants, and $-p^2$ the decrease due to the loss of homozygotes from the outcrossing population.

There will be some set of values of the V's for which there is an equilibrium; these are the critical values of the fitnesses such that the mutant frequency neither increases nor decreases. At this equilibrium,

$$p^2 = (1/2)\Sigma P_i V_i = (1/2)p^2(1 + V_1 + V_1 V_2 + \ldots + \Pi V_i + \ldots).$$

Writing $\alpha = 1 + V_1 + V_1 V_2 + \ldots + \Pi V_i + \ldots$, the condition for equilibrium is $\alpha = 2$. If $\alpha > 2$ the selfing habit will spread and if $\alpha < 2$ selfing will be eliminated. If $V_i = V$ for all i, then $\alpha = 1/(1-V)$, and the critical value of $V = 1/2$. That is, if all selfed generations have a fitness less than a half, selfing will be eliminated. It is easy to show that the same conclusion holds if selfing is due to a dominant gene.

The condition $\alpha > 2$ is for the initial increase of a selfing gene. It is difficult to treat the subsequent spread analytically. It is, however, easy to show that for a constant $V > 1/2$, selfing will increase to fixation. Conversely, if V_i is small for large i, selfing

cannot increase to fixation, as is obvious if one considers the case
$V_i = 0$ for some i. It may in fact often be the case that $\alpha > 2$ but V_i
is small for large i; in such cases there could be a balanced polymorphism for genes for self-sterility and self-compatibility, or the optimal genotype would be one allowing some proportion of selfed offspring. The data presented by Allard at this symposium confirm this
prediction.

What are plausible values for the V's? For an initially outcrossing
species, they may well be smaller than the critical ones. I do not know
of any data for a habitually outcrossing plant or animal hermaphrodite.
Figure 3 shows data for *Drosophila subobscura*, giving $\alpha = 1.71$. This
is a substantial overestimate, because the data are for brother-sister
mating and not selfing, and because the experiment was carried out in
optimal conditions with little competition. The data indicate that,
after falling to a very low value, V_i may rise again with large i as

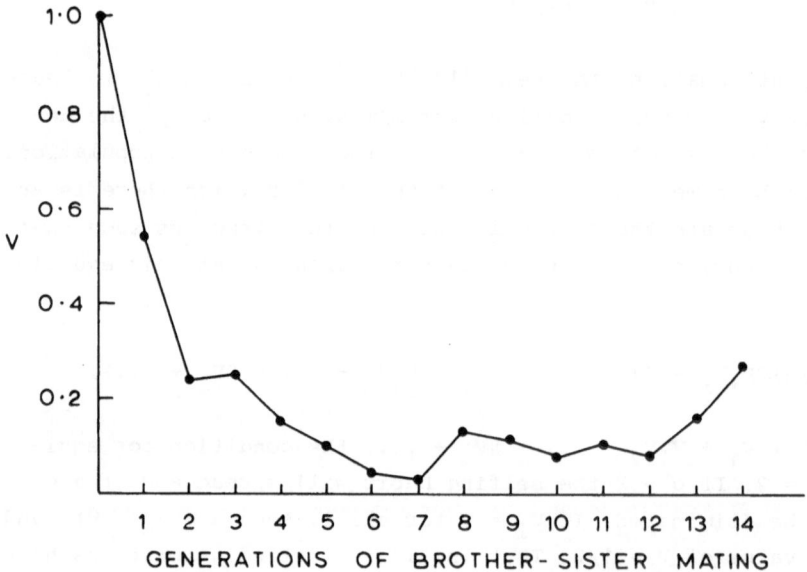

Figure 3. Mean number of offspring per day per pair in an inbred line of *Drosophila
subobscura* (data from Hollingsworth & Maynard Smith, 1955).

the least fit homozygotes are eliminated by selection. Sved (1971) estimated the fitness of *Drosophila melanogaster* homozygotes for the entire second chromosome as 0.15, confirming that V may fall very low.

One factor which would favour the establishment of selfing has been omitted: self-incompatible plants risk not being pollinated, whereas a selfing mutant would be certain of pollination. Mathematically, this is equivalent to a proportional increase in all the V values. This factor will be most important in sparsely distributed species, particularly if monocarpic. The association between selfing and the annual habit has been discussed from this point of view by Baker (1959).

So far, I have discussed the evolution of selfing from outcrossing. What of the reverse transition? As before, let V be the fitness of a plant produced by selfing. Consider a population of selfing plants, and a rare dominant mutant M such that M/+ plants do not self. The M/+ plants have 1/V times as many offspring as selfing plants, and transmit the gene M to half of them. Hence self-sterility will spread if V < 1/2; i.e. provided plants produced by outcrossing are twice as fit. If f is the chance of pollination of a self-sterile plant, this condition becomes V < f/2.

The same condition, V < f/2, can be shown to hold also for a self-compatible species with a high frequency of outcrossing, achieved, for example, by monoecy or by protandry. In practice, self-sterility is more likely to arise in such a species than in one in which inbreeding is the rule, for two reasons:

i) The frequency of lethal and deleterious recessives will be lower in a habitual selfer, and hence V will be larger. There is some observational evidence (e.g. Jinks & Mather, 1955) that hybrid vigour is less marked in habitually selfing species. There is, however, evidence that even in strongly inbreeding species, plants produced by outcrossing have a higher fitness, at least in the F_1; for example, Marshall & Allard (1970) found a higher frequency of adult heterozygotes in wild populations of *Avena sativa* than would be predicted from the frequency in seed.

ii) f is likely to be small in a selfing species, since there is little advantage in the production and dissemination of pollen.

The same arguments hold for the avoidance of selfing in hermaphroditic animals, and, qualitatively at least, for the avoidance of inbreeding in gonochoristic ones. Hermaphroditic gastropods show the whole range from self-sterility (e.g. *Helix*, *Cepaea*), through habitual outcrossing with occasional selfing (e.g. *Biomphalaria*) to habitual

selfing with occasional outcrossing (e.g. *Rumina*). In gonochoristic
animals, it is becoming apparent that some patterns of dispersion have
evolved because they reduce the likelihood of inbreeding (e.g. Bulmer,
1973; Itani, 1972).

Summarizing the conclusions of this section, if V_i is the fitness
of a plant produced by i generations of inbreeding (where $V_o = 1$), then
gene for selfing would enter a population of self-incompatible hermaphrodites if $\alpha > 2$, where

$$\alpha = 1 + V_1 + V_1 V_2 + \ldots + \Pi V_i + \ldots$$

For constant V, selfing will spread to the whole population if
$V > 1/2$. If $\alpha > 2$, but V_i is small for larger i, there could be polymorphism for self-compatibibility, or the establishment of a genotype permitting some proportion of selfed offspring. Starting from a population
of selfing plants, a gene for self-sterility would spread if $V < f/2$,
where f is the chance that a self-sterile plant would be pollinated, and,
as before, V is the relative fitness of a plant produced by selfing. In
practice, self-sterility is more likely to evolve in species with partial outcrossing (e.g. because or protandry or monoecy) than in a habitual selfer, because V will be smaller and f larger in such species.

SEX RATIO AND GENETIC MECHANISMS

In the two previous sections, it was assumed that resource allocation
and self-compatibility were under the control of autosomal genes. In
this section I discuss some complications which arise from sex-determ-
ining mechanisms and from gene action in the gametophyte. These problems have recently been discussed by Lloyd (1974a,b,c).

In dioecious species in which the male is the heterogametic sex it
is common to find an excess of females in natural populations (Lloyd,
1974b). The reason for this is not clear. Lewis (1942), Mulcahy (1967)
and Kaplan (1972) argue that an excess of females will maximise the
seed production of the population. Lloyd points out that although an
increased seed production may be a consequence of the unequal sex ratio,
it can hardly be the explanation of it, since selection maximises individual and not group fitness. If the argument were correct, one would
expect to find an excess of females in gynodioecious species also,

whereas the reverse is always the case; by similar reasoning, one would
expect to find an excess of females in gonochoristic animal species.

Correns (1928) showed that in *Silena alba* and *Rumex acetosa* the
excess of females is greater if larger amounts of pollen fall on the
stigma, and suggested that the explanation lies in the competitive
superiority of X-bearing pollen tubes. Lloyd (1974b) points out that
there are four genera in which there is evidence of selective fertilization, and that these are the only genera of dioecious angiosperms
in which there is visible differentiation between X and Y chromosomes
and in which the YY genotype is known to be inviable. Lloyd argues that
since the predominance of females is most marked in those genera in
which differentiation of the sex chromosomes is most advanced, the
explanation may be that genes hindering pollen tube growth have accumulated on the Y chromosome.

The explanation is a possible one, but it is difficult to see why
selection has not eliminated the genes concerned. A different possibility is suggested by Putwain & Harper (1972), who worked with *Rumex
acetosa*. They found that seed collected in the wild gave a 1:1 sex
ratio. The species also reproduces vegetatively, and after a season of
growth the sex ratio approached 2 females to 1 male. The allocation of
resources to leaves and reproductive organs differed in timing and extent between the sexes, and the authors suggest that the final sex
ratio is biassed because the sexes occupy different ecological niches
(the niches differ temporally and not spatially). The absence of a
sex difference in the seeds runs counter to the results of Correns
(1928) and Sprecher (1913); it may be that different amounts of pollen were present, although there is no evidence for this.

The excess of females in these dioecious species is therefore something of a puzzle. If Putwain & Harper's explanation holds for other
species, we would explain the association between an excess of females
and chromosomal differentiation on the grounds that ecological differentiation takes time to evolve. But the selective fertilization observed by Correns would remain unexplained. If on the other hand differential pollen tube growth is the main cause of the distorted ratios, it
is hard to suggest a selective explanation.

In dioecious plants and gonochoristic animals, the evolutionarily
stable sex ratio is close to 1:1 (Fisher, 1930). This is true only for
autosomal genes expressing themselves in the diploid phase. Selection
on genes expressed in the haploid phase (gametes or gametophytes), or
on cytoplasmic genes, will distort the 1:1 ratio. For almost all simple

genetic mechanisms of sex determination, the primary sex ratio resulting from a regular meiosis is 1:1. Consequently, selection on autosomal
genes acts to maintain the regularity of meiosis and to suppress the
effects of cytoplasmic genes and of genes in the gametophyte.

Does the same conclusion hold for gynodioecious species, in which
the evolutionarily stable sex ratio is not 1:1? A full answer to this
question calls for a lot of algebra, but in general it seems that the
conclusion does hold. For the two simplest types of genetic determination (M/+ female, +/+ hermaphrodite; +/+ female, M/+ and M/M hermaphrodite) Lloyd (1974c) has shown that, at equilibrium with a 1:1 meiotic
segregation, the genetic contributions of females and hermaphrodites
to the next generation are equal. It follows that there is no selection
on genes acting in the sporophyte tending to distort segregation. This
is not true, however, when +/+ is female, M/+ hermaphrodite and M/M
lethal. In this case, Lloyd shows that hermaphrodites contribute more
genes to the next generation than females. It can be shown that selection on autosomal genes in the hermaphrodite favours a distortion of
meiosis in favour of the production of an excess of M gametes.

APOMIXIS AND SIB COMPETITION

At first sight it seems that apomixis (or in animals, ameiotic parthenogenesis) confers the selective advantage of selfing without its disadvantages. The advantages arise because no resources need be wasted
on pollen, so that an apomictic mutant should have a twofold selective
advantage in a dioecious or hermaphrodite population (White, 1970;
aynard Smith, 1971). Initially, it is true, the genetic change which
suppresses meiosis and makes possible development without fertilization would probably not also suppress pollen production, so that the
twofold advantage might not accrue immediatly. At this stage, however,
it is possible that any pollen produced would spread the genes responsible for apomixis, although this would depend on the nature of the
genetic change involved. These advantages are not counterbalanced by
any loss of fitness caused by inbreeding.

For these reasons, we would expect apomictic varieties to be common,
as in fact they are. Why are they not universal? The group selection
explanation, that apomictic populations are eliminated because of their
slower evolutionary responses to changing environments, is not al-

together implausible. If sexual reproduction is to be maintained by
group selection, each apomictic mutant must be balanced by the extinction of an apomictic population. This could be the case, because
apomictic mutants (in animals, mutants capable of ameiotic parthenogenesis) may occur rather infrequently in evolution. It is important
to remember that the idea that recombination rates and chiasma frequency evolve by group selection cannot be defended on these grounds,
because within-population variation in these characters has been found
whenever it has been looked for.

The alternative view, that sexual reproduction and recombination confers a short term advantage by increasing variability, has also had its
adherents. The obvious difficulty with this view is that individual
selection will favour that breeding system which maximises the mean
fitness of the offspring rather than the variance of fitness. There is
however a way out of this difficulty once we recognize that there is
competition between sibs. The first clear statement of this idea known
to me is Baker (1959). He was considering the filling of a gap in a
forest, created by the death of a tree, ultimately by one new tree.
Each potential parent would contribute many competing seedlings. Therefore, he suggest, "If a parent tree had formed a high proportion of its
seeds by outcrossing, it is likely that there will be genetical differences between the competing offspring with an opportunity for the selection of that one which is best adapted to the prevailing conditions."

This idea was developed further by Williams & Mitton (1973) and
by Williams (1975). The essential features of most of the models in
Williams (1975) are that there is competition between sibs, and that
there is very intense selection. A combination of these two features can
give a short-term advantage to sex. To borrow an analogy from Williams,
an apomictic parent is like a man who buys a hundred tickets for raffle
with only one prize, and finds that they all have the same number; in
contrast, outcrossing parents buy fewer tickets, but all with different numbers.

I have attempted to articulate this model more precisely (Maynard
Smith, 1976). I imagine (figure 4) an environment divided into "patches", each large enough to support a single adult plant. R parents each
contribute N seeds to a patch; the survivor is that individual out of
the RN competitors which is best adapted to a particular patch. The
patches vary unpredictably in respect of a number of environmental
features (5 in the computer simulations), and there is a pair of alleles concerned with adaptation to each of those features. Thus R

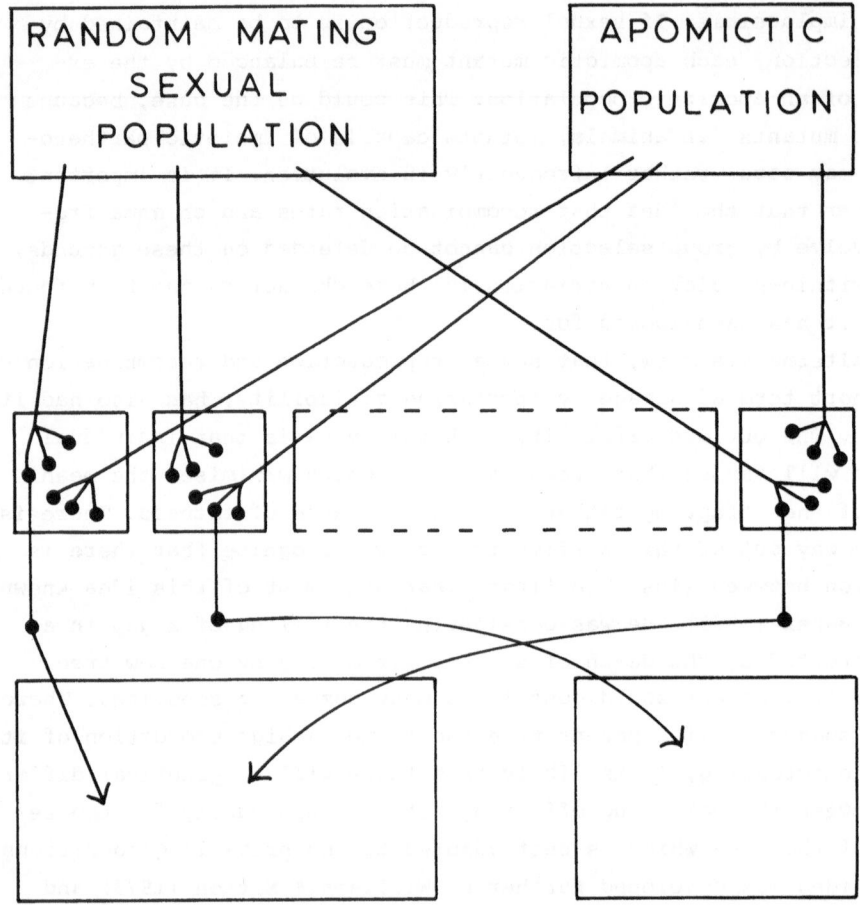

Figure 4. A model of sib competition, with R = 2 and N = 3.

measures the intensity of between-family selection and N of within-fam-
ily selection ("sib competition").

Both types of selection must be present before there is any selective advantage for sexual reproduction or recombination. It is possible
with this model for sexual reproduction to have a twofold advantage
over apomixis, sufficient to counterbalance the twofold disadvantage
of producing pollen, but this requires intense selection (RN of the
order of 40 or more). It is also possible to show that, for an entirely sexual population, a gene increasing recombination can be selectively favoured over an allele for reduced recombination.

It therefore seems clear that sib competition can lead to selection for sexual reproduction and recombination. I have two reservations about its importance in nature. First, it depends on a particular pattern of dispersal, in which offspring do not experience the same environment as their parents (if they did, apomixis would guarantee to them an optimal genotype), but in which sibs do compete with one another. I therefore have doubts about invoking the mechanism in plants with relatively poor powers of seed dispersal, since offspring will grow close to where their parents lived. This objection would disappear if there were large changes in the environment from generation to generation. At the other extreme, the argument can hardly apply to oysters, since dispersal is such that sib competition is unlikely.

The second reservation arises because in my computer simulation I assumed that the various environmental "features" of patches were independent (e.g. if a patch is "dry" rather than "wet", this does not alter the probability that it is "hot" rather than "cold"), and that only one pair of alleles was concerned with each feature. If one makes the more plausible assumptions that environmental features are correlated, and that many loci are concerned with adaptation to a single feature of the environment, this tends to cause selection for apomixis, or for tight linkage. Despite these reservations, sib competition is an important selective process favouring sexual reproduction and genetic recombination.

Summary.

The paper discusses the selective forces responsible for the evolution of breeding systems, and in particular of hermaphroditism as opposed to dioecy, of selfing as opposed to outcrossing, and of apomixis as opposed to sexual reproduction. Hermaphroditism is analyzed in terms of resource allocation to male and female functions, and it is shown that the evolutionarily stable strategy depends on the shape of the fitness set relating seed and pollen production. The conditions for self-compatibility to evolve in a self-sterile population are given in terms of the fitnesses V_i, after i generations of selfing. It is shown that polymorphism for the rate of selfing can easily arise. A model in which sexual reproduction is favoured over apomixis, or genes for high as opposed to low recombination within a sexual population, is described; it depends on competition between sibs in a unpredictable environment.

REFERENCES

Baker, H.G. 1959. Reproductive methods as factors in speciation in flowering plants. *C.S.H. Symp. Quant, Biol.* 24: 177-191.

Bulmer, M.G. 1973. Inbreeding in the great tit. *Heredity* 30: 313-325.

Correns, C. 1928. Bestimmung, Vererbung und Verteilung des Geschlechtes bei den höheren Pflanzen. *Handb. Vererbungsw.* 2: 1-138.

Darlington, C.D. 1939. *The Evolution of Genetic Systems.* Cambridge University Press.

Fisher, R.A. 1930. *The Genetical Theory of Natural Selection.* Oxford, Clarendon Press.

Ghiselin, M.T. 1969. The evolution of hermaphroditism among animals. *Quart. Rev. Biol.* 44: 189-208.

Hollingsworth, M.J., and Maynard Smith, J. 1955. The effects of inbreeding on rate of development and on fertility in *Drosophila subobscura.* *J. Genet.* 53: 295-314.

Itani, J. 1972. A preliminary essay on the relationship between social organisation and incest avoidance in non-human primates. In F.E, Poirier, ed. *Primate Socialization.* Random House.

Jinks, J.L., and Mather, K. 1955. Stability in development of heterozygotes and homozygotes. *Proc. Roy. Soc.* B 143: 561-578.

Kaplan, S.M. 1972. Seed production and sex ratio in anemophilous plants. *Heredity* 29: 359-62.

Lewis, D. 1972. The evolution of sex in flowering plants. *Cambr. Phil. Soc. Biol. Rev.* 17: 46-67.

Lloyd, D.G. 1974a. Theoretical sex ratios of dioecious and gynodioecious angiosperms. *Heredity* 32: 11-34.

Lloyd, D.G. 1974b. Female-predominant sex ratios in angiosperms. *Heredity* 32: 35-44.

Lloyd, D.G. 1974c. The genetic contribution of individual males and females in dioecious and gynodioecious angiosperms. *Heredity* 32: 45-52.

MacArthur, R.H. 1965. in *Theoretical and Mathematical Biology.* eds. T. Waterman & H. Morowitz. Blaisdell: New York.

Marshall, D.R., and Allard, R.W. 1970. Maintenance of isozyme polymorphism in natural populations of *Avena sativa.* *Genetics* 66: 393-399.

Maynard Smith, J. 1971. in *Group Selection.* ed. G.C. Williams. Chicago: Aldine-Atherton..

Maynard Smith, J. 1976. A short-term advantage for sex and recombination through sib competition. *J. theor. biol.* (in the press).

Maynard Smith, J., and Price, G.R. 1973. The logic of animal conflict. *Nature* 246: 15-18.

Mulcahy, D.L. 1967. Optimal sex ratios in *Silene alba.* *Heredity* 22: 411-423.

Putwain, P.D., and Harper, J.L. 1972. Studies in the dynamics of plant populations.
V. Mechanisms governing the sex ratio in *Rumex acetosa* and *R. acetosella*.
J. Ecol. 60: 113-129.

Sved, J.A. 1971. An estimate of heterosis in *Drosophila melanogaster*.
Genet. Res., Cambridge 18: 97-105.

White, M.J.D. 1970. Heterozygosity and genetic polymorphism in parthenogenetic
animals. in *Essays in Evolution and Genetics*. eds. M.K. Hecht and W.C. Steele.
New York: Appleton-Century-Crofts.

Williams, G.C. 1975. *Sex and Evolution*. Princeton University Press.

Williams, G.C., and Mitton. J.B. 1973. Why reproduce sexually?
J. theoret. Biol. 39: 545-554.

3. SEX AND EVOLUTION

EVOLUTION OF THE SEX RATIO IN THE WOOD LEMMING, *Myopus schisticolor*.

Bengt Olle Bengtsson

The wood lemming (*Myopus schisticolor* Lilljeborg) has probably the most
unequal sex ratio of all mammals. (The facts in this paragraph are
taken from Kalela and Oksala 1966, and Frank 1966.) In wild populations the frequency of males is normally between 0.20 and 0.30. This
unequal sex ratio is also found in animals kept in captivity. Kalela
and Oksala report that among 1073 new-born lemmings in a population
which started from 6 males and 11 females, the frequency of males was
only 0.248. The scarcity of males seems to be due to the existence,
both in natural and captive populations, of females which mainly or
exclusively produce female offspring.

    The situation has become even more complex with the surprising
finding by Fredga *et al*. (1976) that some female wood lemmings have
an XY chromosome constitution. These XY females seem to be fully fertile, unlike human XY females and other known cases in mammals where
the phenotypic sex is different from the chromosomal sex.

    The following hypothesis has been put forward to account for these
observations (Fredga *et al*. 1976; Fredga *et al*., in preparation).
A) There is an X linked factor which represses the male determining
effect of the Y chromosome. X chromosomes carrying this factor will
be denoted by X*. With this notation the genotype of the "sex reverted"
animals becomes X*Y.
B) The majority of eggs produced by the X*Y females contain the X*

chromosome. This implies that most or all (depending on the degree of
meiotic drive) of X*Y females' offsprings will be daughters, since
an embryo carrying an X* chromosome will always develop into a female.
The cause of the meiotic drive in X*Y animals may simply be that eggs
carrying a Y chromosome and no X chromosome have a much smaller chance
in being formed than normal eggs with an X and no Y chromosome. Fredga
et al. (1976) present some evidence that the Y chromosome is normally
absent from the cells undergoing the first meiotic division in XY females.

The correctness of this hypothesis will have to be determined by
analysis of the offsprings from different types of matings. A preliminary study of some pedigree data has given results that are consistent with the hypothesis.

We shall here consider what effects the postulated sex reversal
factor will have at the population level. It is, of course, of particular interest to see if the factor can explain the unequal sex
ratio observed in wood lemming populations. We shall also consider
what happens if one assumes that the sex reversal factor is linked to
a chromosome other than the X chromosome, and what effect different
modifiers to the system may have on the sex ratio.

MODEL

Consider an infinitely large, random mating population where all females mate even if there are fewer males than females in the population. We have to consider four different genotypes in the present situation. All males have the genotype XY, but the females can be of
three types: XX, XX* and X*Y. There can be no X*X* animals since no
males carry an X* chromosome.

It is assumed that the XX* females have normal fertility and segregate the two different X chromosomes in a Mendelian fashion.

Let the X*Y females produce Y carrying eggs with frequency m and
X* carrying eggs with frequency n, where m + n = 1. This implies that
a fraction $1/2 \times m$ of the X*Y females' embryos will have a YY chromosome constitution and will die. We introduce a fertility parameter,
f, for X*Y females such that the average fertility of these females
is $f(1 - (1/2)m)$, while the average fertility of XX and XX* females
is considered to be 1.

Two different ranges of values of the fertility parameter f are
of interest. These are considered separately. (1) If $0 \leq f \leq 1$, then
the X*Y females have a fertility which is even more decreased than
would be expected from their production of YY embryos. (2) If $1 < f <
(1 - (1/2)m)^{-1}$, the X*Y females compensate their fertility against the
loss of YY embryos, for example by giving better care to the surviving
young. We will not consider the possibility that the sex reversed animals have a fertility that is greater than or exactly equal to the
fertility of the other females, i.e. $f \geq (1 - (1/2)m)^{-1}$.

ANALYSIS

Assume that the frequencies of the genotypes XX, XX*, X*Y and XY are
a, b, c, and d, where $a + b + c + d = 1$. There are three different
types of matings which can occur in the population. Their relative
frequencies and the resulting offspring are given in Table 1.

The frequencies of the different genotypes in the next generation
will be

$$a' = ((1/2)a + (1/4)b) / (a + b + c) \cdot W \qquad (1a)$$

$$b' = ((1/4)b + (1/2)fnc) / (a + b + c) \cdot W \qquad (1b)$$

$$c' = ((1/4)b + (1/2)fnc) / (a + b + c) \cdot W \qquad (1c)$$

$$d' = ((1/2)a + (1/4)b + (1/2)fmc) / (a + b + c) \cdot W \quad (1d)$$

where

$$W = (a + b + fnc + (1/2)fmc) / (a + b + c).$$

It is clear from these equations that the frequency of X*Y females, c,
will always be identical to the frequency of XX* females, b. We can
thus put $c = b$ in the equations. These can be written

$$a' = (2a + b) / 4(a + b(1 + z))$$

$$b' = c' = b(1 + 2fn) / 4(a + b(1 + z))$$

Table 1. The possible types of matings and their offsprings

Mating type	Frequency	Offspring			
		XX	X*X	X*Y	XY
XX × XY	a/(a+b+c)	1/2			1/2
X*X × XY	b/(a+b+c)	1/4	1/4	1/4	1/4
X*Y × XY	c/(a+b+c)		fn1/2	fn1/2	fm1/2

$$d' = (2a + b(1 + 2fm)) / 4(a + b(1 + z))$$

where

$$z = f(n + (1/2)m) = f(1 - (1/2)m).$$

It is convenient to rewrite the equations in the following form

$$a' = (2 + u) / 4(1 + u(1 + z))$$

$$b' = c' = u(1 + 2fn) / 4(1 + u(1 + z))$$

$$d' = (2 + u(1 + 2fm)) / 4(1 + u(1 + z))$$

where

$$u = b/a. \text{ (u is, of course, always non-negative.)}$$

All the frequencies of the different genotypes are simple functions of u. We shall therefore now investigate the behaviour of u. From the equations it follows that

$$u' = b' / a' = u(1 + 2fn) / (2 + u).$$

Let Δu be equal to u' - u.

$\Delta u = u(1 + 2fn) / (2 + u) - u = u(2fn - 1 - u) / (2 + u).$

The following conclusions about the behaviour of u can be drawn.
If $2fn \leq 1$, u will decrease and go towards 0, irrespective of the ini-
tial value. If $2fn > 1$, u will always go towards the value $2fn - 1$.
This value is a globally stable equilibrium for u.

When $\hat{u} = 2fn - 1$, the different genotype frequencies are

$$\hat{a} = (2fn + 1) / (4 + (2fn - 1)(4 + 2f(1 + n))) \tag{2a}$$

$$\hat{b} = \hat{c} = (2fn - 1)(2fn + 1) / (4 + (2fn - 1)(4 + 2f(1 + n))) \tag{2b,2c}$$

$$\hat{d} = (2 + (2fn - 1)(2f(1 - n) + 1)/(4 + (2fn - 1)(4 + 2f(1 + n))) \tag{2d}$$

We can now interpret the results concerning u in terms of the genotype frequencies.

A) The factor causing the sex-reversal in X*Y females will be lost
form the population if and only if $2fn \leq 1$;

B) If $2fn > 1$, the factor will increase in the population and the
frequencies of the different genotypes will go to the equilibrium
given by $\hat{a}$, $\hat{b}$, $\hat{c}$ and $\hat{d}$ in (2a - 2d). This equilibrium is globally
stable. This means that even if the frequencies of the different geno-
types in a population are radically changed due to migration, drift
during a "bottle neck" phase, or some other evolutionary factor, the
population will always move back to the equilibrium where the frequency of males in the population is $\hat{d}$.

Some examples of the equilibrium frequencies for the different
genotypes when f and n take different values are given in Table 2.
In the limiting case when both f and n go towards 1, the frequencies
of the four genotypes all become equal, i.e. 0.25. If f and n are
both reasonable close to 1, the frequency of males in the population
at equilibrium is around 0.25 - 0.30.

From the condition for increase, $2fn > 1$, it is obvious that the
sex reversal factor can spread in the population even if the fertility of the sex reversed animals is much lower than the fertility of
the normal females. For example, when n = f = 0.75 the fertility of
the X*Y females is only 0.656, but the X* factor will increase nevertheless since 2fn is equal to 1.125. The important condition for increase is that the X*Y females contribute a sufficient number of X*
chromosomes to the next generation, and this depends not only on the

Table 2.   The frequencies of the different genotypes at
             equilibrium for different values of f and n

n	f	Equilibrium frequencies			
		XX	X*X	X*Y	XY
1.0	1.0	0.25	0.25	0.25	0.25
0.9	1.0	0.27	0.22	0.22	0.29
1.0	0.9	0.28	0.22	0.22	0.28
0.9	0.9	0.30	0.19	0.19	0.32
0.75	0.75	0.44	0.06	0.06	0.45

fertility of the X*Y animals but also on the drive for the X* chromosome.

The frequency of males decreases in the population when the sex
reversal factor increases. But it is interesting to notice that, except in the special case n = 1, the frequency of animals with an XY
chromosome constitution will increase in the population. This effect
is, however, rather small for most values of n and f.

OTHER TYPES OF INHERITANCE

What would have happened if the sex reversal factor was linked to a
chromosome other than the X chromosome? We shall consider three cases,
(1) that the factor is autosomal and dominant, (2) that it is autosomal and recessive, and (3) that it is Y linked. As in the X linked
case, it will be assumed that the XY females produce X and Y carrying
eggs in frequencies n and m, and that the fertility of these females
is f·(1 - (1/2)m).

Autosomal inheritance.

If the factor is dominant the genotype of the sex reversed females can be written AA*XY. (A*A*XY animals can not exist since males never carry the A* chromosome.) If the factor is recessive the sex reversed genotype is A*A*XY. One can show that such sex reversal factors will always be selected against and will disappear from the population under the condition on f which we consider, i.e. $f < (1 - (1/2)m)^{-1}$. The analysis is in both cases straightforward, but in the situation with a recessive factor rather laborious to do.

This result is not surprising. If the reversal factor is autosomal, and the sex reversed animals have lower fertility than other animals, it is to be expected that the factor will disappear from the population. This fact will not depend on whether there is a drive for the X or the Y chromosome in the sex reversed animals.

Y linked inheritance.

In this case the sex reversed animals' genotype can be written XY*. This Y* will thus function as a "silent" chromosome. It can be shown by local stability analysis that the Y* chromosome will increase in the population if and only if fm > 1. The condition for increase is in this case much more strict than the condition in the X linked case. A necessary, but not sufficient, condition is here that the fertility factor f is in the "compensating" region, i.e. that $1 < f < (1 - (1/2)m)^{-1}$.

If the condition for increase is fulfilled, and m is strictly smaller than 1, the population will go to an internal, globally stable equilibrium. If m is equal to 1, the frequency of males will continue to decrease, and the population will - according to the model - become extinct due to the lack of males.

The results concerning increase of sex reversal factors linked to different chromosomes are summarized in Table 3.

Table 3. The effect of different types of inheritance
for a sex reversal factor

Types of inheritance	Sex reversed animal	Condition for increase
Autos. Recessive	A*A*XY	-
Autos. Dominant	A*AXY	-
Y - linked	XY*	mf > 1
X - linked	X*Y	2nf > 1

DISCUSSION

We have shown that the hypothesis described in the introduction is
plausible from a population genetical point of view. It was primarily
put forward to explain the occurrence of fertile XY females and of females which only or predominantly produce daughters. But it turns out
that the hypothesis can also in a very satisfactory way explain the
unequal sex ratio found in both wild and captive wood lemming populations. If n and f are both close to one - which they are according
to the hypothesis - then all populations with the sex reversal factor
present will move to a stable equilibrium with about 25 to 30 per cent
males. These numbers correspond very well with the numbers observed
(Kalela and Oksala 1966).

No other type of inheritance of the sex reversion factor is as likely as the X linkage. A Y linked sex reversal factor may increase in a
population, but only if there is a drive for the Y chromosome in the
sex reversed females, and if the XY females compensate their fertility
against the loss of YY embryos. These assumptions are considered
highly unlikely.

A number of questions are raised by the analysis. The most important is, how the sex ratio in a population at equilibrium can be
changed by the appearance of new factors which in some way modifies
the sex reversal system existing.

The situation in the present case is, in this respect, similar
to the case with the sex ratio meiotic drive system in *Drosophila*
(for a mathematical study of this system, see Edwards 1961, and Thom-

son and Feldman 1975). Here there is an X linked factor which causes
meiotic drive for the X chromosome in the males, but there is no sex
reversal as in the present system. If the different possible genotypes
have different viabilities and fertilities, then the population may
move to a stable situation where there are fewer males than females
in the population. Thomson and Feldman studied what would happen if
new factors were introduced in the population which influenced the
strength of the drive and/or the effect of the meiotic drive factor
on the fertility. They showed that the fate of such a modifier depended on which chromosome the modifier was linked to. For example
an X linked modifier would increase if it caused even more drive for
the X chromosome (assuming that it had no effect on the fertility),
while a Y linked modifier would increase if it decreased the drive
for the X and increased the drive for the Y chromosome. The population could, thus, evolve both towards more equal and less equal sex
ratios, depending on the type of modifiers appearing.

The same is true in the wood lemming case. Here we have to consider two different types of modifiers, those which change the parameters f and n in the system, and those which change the system completely by taking away the sex reversal effect of the X* chromosome.

In the first case, it can be shown that a modifier which changes
the parameters of the X*Y females from f_1 and n_1 to f_2 and n_2, will
behave quite differently depending on which chromosome the modifier
is linked to. If it is X linked (which means that it is linked to the
X* chromosome), the modifier will increase if and only if $n_2 f_2 > n_1 f_1$.
Thus, if the effect on the fertility is the same, then an X* linked
modifier will increase only if it causes even more drive for the X*
chromosome than before. On the other hand, if the modifier is linked
to the Y chromosome, the condition for increase is $(1-n_2) f_2 > (1-n_1) f_1$.
Here the modifier will increase if it causes less drive for the X*
chromosome.

In the second case, where the modifier takes away the sex reversal
effect of the X* chromosome, we shall only consider the case where the
modifier is Y linked. This is equivalent to considering the introduction of a Y chromosome which is insensitive to the effect of the X*
chromosome. It is quite simple to show that this chromosome will increase in the population, irrespective of the values of f and n.

The sex ratio in the population can, thus, evolve both towards a
more unequal sex ratio (though in this case the frequency of males
can not fall below 0.25) and towards a more normal sex ratio, depending

on the modifiers that appear. We have here only considered modifiers
which in a direct way alter the sex reversal system. It is not clear
to the author how it is possible in this system to incorporate the considerations on parental expenditure and reproductive value suggested
by Fisher (1930; see also Bodmer and Edwards 1960).

It will be interesting to study species that are evolutionary
close to the wood lemming to see if they also have the sex reversal
polymorphism. This would indicate that the system has been maintained
for a long time and is thus evolutionary stable. There are indications
that at least one other species of lemmings, *Dicrostonyx torquator*
(Pallas), has a chromosome mechanism which may be similar to the system in the wood lemming (Fredga *et al.* 1976).

Summary.

   The frequency of males in wood lemming populations is normally around 0.20 -
0.30. This seems to be due to the fact that there are females who produce no, or
only a few, male offspring. It has been suggested that these females have an XY
chromosome constitution, where the X carries a male repressing factor. It is also
assumed that the XY females' eggs predominantly or exclusively carry the X chromosome. In these animals there is thus a meiotic drive favouring the X chromosome.
The effect of such a sex reversal factor at the population level is analysed. It
is shown that a population with this factor present, will move to a state where
the frequency of males is around 0.25 - 0.30, depending on the fertility and the
strength of the drive for the X chromosome in the XY females. If the sex reversal
factor is linked to another chromosome it will either never increase in the population (autosomal linkage) or increase in the population only under very special
circumstances (Y linkage). Modifiers which change the parameters of the system,
e.g. by strengthening the drive for the X chromosome, can alter the sex ratio in
the population both upwards and downwards depending on which chromosome they are
linked to.

Acknowledgement.

   I wish to thank K. Fredga, G. Thomson, M. Feldman, F. Christiansen and W. Bodmer
for their helpful comments and suggestions.

REFERENCES

Bodmer, W.F., and Edwards, A.W.F.   1960.   Natural selection and the sex ratio.
   Ann. Hum. Genet. 24: 239-244.

Edwards, A.W.F. 1961. The population genetics of the "sex-ratio" in *Drosophila pseudoobscura*.
Heredity **16**: 291-304.

Fisher, R.A. 1930. *The Genetical Theory of Natural Selection*. Oxford University Press.

Frank, F. 1966. Verschiebung des Geschlechtsverhältnisses in der Wuhlmaus-Gruppe (Microtidae).
Naturwissenschaften **53**: 90.

Fredga, K., Gropp, A., Winking, H., and Frank, F. 1976. Fertile XX and XY type females in the wood lemming *Myopus schisticolor*.
Nature **261**: 225-227.

Kalela, O., and Oksala, T. 1966. Sex ratio in the wood lemming, *Myopus schisticolor* (Lilljeb.), in nature and in captivity.
Ann. Univ. Turkuensis, Ser. A II **37**: 1-24.

Thomson, G.J., and Feldman, M.W. 1975. Population genetics of modifiers of meiotic drive: IV. On the evolution of sex-ratio distortion.
Theoret. Popul. Biol. **8**: 202-211.

3. SEX AND EVOLUTION

POPULATION GENETICS, DEMOGRAPHY AND THE SEX RATIO

Brian Charlesworth

Fisher (1930) was the first to propose a selective explanation for the
approximately one-to-one primary sex ratios of most organisms. A large
number of theoretical studies elaborating Fisher's theory have since
been published (e.g. Shaw and Mohler 1953; Shaw 1958; Kolman 1960; Bodmer and Edwards 1960; Edwards 1963; MacArthur 1965; Hamilton 1967; Emlen 1968; Crow and Kimura 1970; Leigh 1970; Spieth 1974; Eshel 1975;
Charnov 1975). Most of these authors have used discrete-generation population genetic models. For consideration of the human sex ratio, about
which we have the most information, discrete generation models are
clearly only a first approximation. Leigh (1970) used a continuous-time
model without age-structure, and Charnov (1975) presented a discrete
age-class model with overlapping generations but employed the intuitive
concept of the relative contributions of the sexes to future generations,
rather than an exact genetic model. Although such an intuitive approach
is valuable in suggesting ideas, a genetic model is clearly desirable.
In this paper I present a model for the evolutionary genetics of sex
ratio in a population with discrete age-classes. The biological question with which I am chiefly concerned is one raised by Fisher: at
what age does one expect the sex ratio to be equalised, when there is
differential mortality between the sexes and when parents can, at least
partially, replace offspring who die before maturity? As Edwards (1962)

has emphasised, Fisher's concept of "parental expenditure" is equivalent, in the human case, to differential mortality with replacement of
lost offspring. Instead of using parental expenditure directly, I use
mathematical models of reproductive behaviour which incorporate the
process of parental replacement of losses of offspring. These models
were originally developed for other purposes by human demographers,
notably Henry (1953, 1957, 1961, 1964a, 1964b, 1965) and Perrin and
Sheps (1964). I also consider possible evolutionary constraints on the
relations between the age of parents and the sex ratio of their offspring. Unless otherwise stated, I use the term "sex ratio" to denote
the fraction of newly-formed zygotes which are male, among the offspring
of a given class of individuals.

FORMULATION OF THE MODEL

In this section, the basic genetic and demographic models will be introduced. In order to simplify the mathematics, I use the technique of introducing a rare gene affecting sex ratio into a population with specified properties, and asking whether or not the new gene tends to increase in frequency. A population such that no rare gene affecting sex
ratio can spread is at an evolutionary equilibrium point with respect
to sex ratio. This criterion is a necessary but not sufficient one for
a stable genetic equilibrium, since the dynamics of a genetically polymorphic population are not considered. It corresponds to Maynard Smith's
(1972) concept of an "evolutionarily stable strategy". The use of this
technique will be illustrated with the simple case of no replacement of
lost offspring; the more difficult case is considered in the following
section.

I assume that the population is initially homozygous aa at an autosomal locus affecting sex ratio, and that a mutation to an allele A
occurs. (Similar results are obtained if an X-linked gene is studied).
The population is infinite in size and mating is sufficiently close to
random with respect to genotype that the rare allele A is effectively
represented only in Aa heterozygotes throughout the period in question.
The life history of an individual is divided into a number of discrete
age-classes; the probability of survival of a female from conception
to age x is l_x ($x = 1, 2, \ldots$) and the corresponding probability for

a male is l_x^*. These quantities are assumed to be constant over time and independent of genotype at the sex ratio locus. The population is assumed to be such that the fecundity of a mating is determined entirely by the age and genotype of the female partner. If there is differential mortality between the sexes and replacement, the fecundity of a female will be affected by the sex ratio of her offspring, and so cannot be treated as independent of her genotype at the sex ratio locus. Let m_x be the number of conceptions experienced at age x by an aa female in the original population, and write $k_x = l_x m_x$. It is assumed that m_x, and hence k_x, is constant over time. If sex ratio is unaffected by paternal age, it is known that a population with these properties will achieve a stable age-structure with constant finite growth rate λ_0 (Goodman 1969). It is not known whether this is always true when sex ratio is affected by paternal age, as is known to be the case in some human populations (Pollard 1969), but it seems to be a plausible assumption to make, at least as an approximation, particularly in view of the fact that age effects are very small. Let $\bar{r}_x$ be the sex ratio among zygotes produced by aa females aged x in a population at stable age distribution; similarly, let $\bar{r}_x^*$ be the sex ratio among offspring of males aged x. (When both parents' ages affect the sex ratio of their offspring, both $\bar{r}_x$ and $\bar{r}_x^*$ are affected by the age-distribution of the population and are independent of time only when the age distribution is stable). The aa population with constant age structure also has a male age-specific fecundity schedule, m_x^*, which is constant over time. This population satisfies the following equations

$$1 = \sum_{x=b}^{d} \lambda_0^{-x} k_x (1 - \bar{r}_x) = \sum_{x=b*}^{d*} \lambda_0^{-x} k_x^* \bar{r}_x^*, \qquad (1a)$$

$$1 = (1 - \bar{r}) \sum_{x=b}^{d} \lambda_0^{-x} k_x = \bar{r} \sum_{x=b*}^{d*} \lambda_0^{-x} k_x^* \qquad (1b)$$

where $\bar{r}$ is the overall sex ratio among new zygotes, b and $b*$ are the ages of first reproduction for females and males respectively, and d and $d*$ are the corresponding ages of final reproduction. In what follows the symbol Σ, omitting subscripts and superscripts, is used to denote summation over x from b to d for females or $b*$ to $d*$ for males, as appropriate.

We can now consider the fate of an A allele introduced into an aa population in stable age distribution. I assume that sex ratio is determined by parental age and genotype, so that Aa mothers aged x, mated with aa males according to the age distribution characteristic of the initial population, have fecundity M_x corresponding to a sex ratio of r_x among their offspring. Similarly, Aa males aged x, mated with aa females drawn from the initial population, have fecundity M_x^* and produce offspring with a sex ratio r_x^*. In general, $r_x \neq \bar{r}_x$ and $r_x^* \neq \bar{r}_x^*$. Let $K_x = l_x m_x$ and $K_x^* = l_x^* m_x^*$. As shown in Appendix A, the asymptotic rate of spread of allele A while still rare is given by s_o, the positive real root of a characteristic equation $f(s) = 0$. The detailed form of $f(s)$ is given in Appendix A. If $s_o > 1$, we conclude that A will increase in frequency, and so the initial population cannot be in evolutionary equilibrium. If $s_o \leq 1$ for all classes of modifiers of sex ratio, the initial population can be considered as being in evolutionary equilibrium. The nature of evolutionary equilibria for sex ratio can be investigated using the partial derivatives of s_o with respect to r_x and r_x^*, evaluated at $\bar{r}_x$ and $\bar{r}_x^*$. We have

$$s_o \approx 1 + \Sigma(r_x - \bar{r}_x)(\partial s_o/\partial r_x)_{\bar{r}_x} + \Sigma(r_x^* - \bar{r}_x^*)(\partial s_o/\partial r_x^*)_{\bar{r}_x^*} \qquad (2)$$

If all derivatives are zero, $s_o = 1$ to first order terms. This is a necessary, but not sufficient condition, for evolutionary equilibrium, and is the criterion which is used in the rest of this paper.

The use of this approach will now be illustrated with the simple case when there is no replacement of lost offspring, so that we have $K_x = k_x$ and $K_x^* = k_x^*$. Using the rule for the differentiation of an implicit function, eqn. (A.4) gives

$$(\partial s_o/\partial r_x) = (\lambda_o^{-x} k_x/C)[1/(2\bar{r}) - 1] \qquad (3a)$$

$$(\partial s_o/\partial r_x^*) = (\lambda_o^{-x} k_x^*/C)\{1 - 1/[2(1-\bar{r})]\} \qquad (3b)$$

where

$$C = 2(\partial f/\partial s_o)_{s_o=1} = [1/(2\bar{r})] \Sigma x\lambda_o^{-x} k_x [\bar{r}(1-\bar{r}_x) + (1-\bar{r})\bar{r}_x]$$

$$+ \{1/[2(1-\bar{r})]\} \Sigma x \lambda_o^{-x} k_x^* [(1-\bar{r})\bar{r}_x^* + \bar{r}(1-\bar{r}_x^*)]. \qquad (3c)$$

If and only if $\bar{r} = 1/2$, both these derivations are zero, and it can
be shown that all higher-order derivatives are zero as well. Genes affecting sex ratio are therefore neither selected for or against in a
population of sex ratio one-half. When $\bar{r} < 1/2$, eqns. (3) show that a
gene with a small effect on sex ratio and which either increases r_x or
r_x^* at each age, or which increases them at some ages but not others,
will be selected. The reverse is true when $\bar{r} > 1/2$. A sex ratio of 1/2
is therefore the only candidate for an evolutionary equilibrium in this
case, which corresponds to an equal cost of offspring of each sex in
the sense of Fisher (1930). This conclusion is in general agreement
with the results of discrete generation models (Spieth 1974), and with
the overlapping generations models of Leigh (1970) and Charnov (1975),
who did not explicitly allow for dependence of sex ratio on parental age.
There are no constraints on the distribution of sex ratio with respect
to parental age.

EVOLUTIONARY EQUILIBRIUM WITH REPLACEMENT

In this section, the equations for evolutionary equilibrium with replacement of lost offspring are derived. I then consider the effect of replacement on the overall population sex ratio, and discuss the expected
relations between sex ratio and parental age. I consider in detail only
the case of maternal control of the sex ratio.

The equilibrium equations.

Replacement together with differential mortality with respect to
sex means that the fecundities of females aged $x + 1$ and more will be
functions of the sex ratio of the offspring they produced at age x.
When evaluating $\partial s_o/\partial r_x$ using eqn. (A.4), we have to consider the partial derivatives of K_{x+y} ($y > 1$) with respect to r_x. (Male fecundities
are independent of r_x, since Aa males mate mostly with aa females). We
obtain

$$C(\partial s_o/\partial r_x)_{\bar{r}_x} = \lambda_o^{-x} k_x [1/(2\bar{r}) - 1] + \sum_{y=1}^{d-x} \lambda_o^{-(x+y)} [\partial K_{x+y}/\partial r_x]_{\bar{r}_x} \quad (4)$$

$$\{(1 - \bar{r}_{x+y}) + (1/2)\sum \lambda_o^{-z} k_z^*(1-\bar{r}_z^*) - [(1-\bar{r}_{x+y})/2]\sum \lambda_o^{-z} k_z^*\}$$

Write $\delta = 2\bar{r}-1$ as a measure of the deviation of the population sex ratio from 1/2. Using eqns. (1), we obtain

$$2\bar{r}\lambda_o^{-x} \, C(\partial s_o/\partial r_x)_{\bar{r}_x} = \sum_{y=1}^{d-x} \lambda_o^{-y} \, (\partial K_{x+y}/\partial r_x)_{\bar{r}_x} [\,(1-\bar{r}_{x+y})\delta + (1-r)\,] - k_x \delta$$

The values of $\bar{r}_x$ which correspond to an evolutionary equilibrium can, in principle, be obtained by setting eqn. (5) to zero for $x = b, b+1, \ldots d-1$ and solving for the r_x ($x = b+1, b+2, \ldots d$), but there are several difficulties associated with this. Instead, various approximations will be resorted to. The further analysis will be concerned almost entirely with the human case, although many of the basic principles should apply generally. The first problem is to represent mathematically the effect of variations in the sex ratio of offspring on fecundity at later ages. Vital statistics from a number of countries (Keyfitz and Flieger 1971) demonstrate that the mortality of males during their first year of post-natal life is about 1.2 times that of females. Similarly, males are disproportionately represented among stillbirths in later pregnancy (Ciocco 1940). The situation is less clear for early pregnancy wastage where data from chromosomal sexing of spontaneous abortions is variable and does not suggest an excess of males (Carr 1971, Pawlowitski 1972, Hamerton 1972). It may be fairly safely assumed that the total probability of dying before the end of the first year of post-natal life is higher for males than females.

Demographic studies of high fertility human populations have shown that foetal death or infant death during the first year or so of life, is accompanied by a reduction in the period of non-susceptibility to further conceptions following the conception of an offspring (Henry 1964a, b; Potter et al. 1965; Ginsberg 1973). In high fertility populations, the mean length of the non-susceptible period in the absence of loss of offspring is around two years at most. As far as human evolution is concerned, therefore, death of offspring after the first year of post-natal life was probably a relatively minor factor in replacement. We need to concentrate our attention on the effect of differential foetal and infant mortality on the gap between conceptions. The consequences of this effect are examined in the next two sections.

The equilibrium value of the population sex ratio.

In this section, r_x is treated as independent of x (which is empirically a good approximation), and Henry's simplest model of fecundity is used (Henry 1953, 1957). Let f_x be the probability that a female aged x is fertile (after the beginning of reproduction, f_x decreases with age as more and more females become sterile; once a female becomes sterile she is assumed never to reproduce again). The model assumes that a fertile female has a constant probability p of conceiving in a given age-interval, provided that she has emerged from the non-susceptible period following her previous conception. The mean length of this susceptible period is g, whose value depends on the probabilities of foetal wastage and infant death and the extent to which these shorten the non-susceptible period. As shown by Henry, this model implies that the probability that a fertile female conceives in an age-interval approaches, with increasing age, the constant value ϕ, given by $\phi = p/(1 + p\bar{g})$. Data on birth intervals suggest that this is a reasonable approximation to reality, except for very old or young mothers. We can therefore approximate m_x by ϕf_x. Setting $r = r_x$ for each x we can obtain

$$\partial m_x/\partial_r = -(\partial\bar{g}/\partial r)\phi m_x. \tag{6}$$

The derivative $\partial\bar{g}/\partial r$ can be determined as follows. Let D_i be the probability of loss of a daughter in the ith month after conception, and $D_i + d_i$ be the corresponding probability for a son. Loss of an offspring in the ith month is associated with a reduction in non-susceptible period of G_i months. (i is taken over all months where loss leads to reduction in non-susceptibility, say from 1 to 24). Let $d = \Sigma d_i$ be the total difference in probability of loss between males and females over this period, and let $G = \Sigma d_i G_i/d$. G is a measure of the mean reduction in non-susceptible period, weighted by the differential mortalities in the component intervals. We have

$$\partial\bar{g}/\partial r = - d\,G \tag{7}$$

Substitution of eqns. (6) and (7) into the characteristic equation, setting $\partial s_o/\partial r = 0$, yields the following equation for equilibrium

$$\delta \Sigma \lambda_o^{-x} k_x = d\,G\,\phi \Sigma \lambda_o^{-x} k_x [(1-\bar{r})\delta + (1-\bar{r})], \tag{8}$$

where $\delta = 2\bar{r} - 1$. Since the equilibrium sex ratio must be near 1/2, and d is known to be small, we can neglect the term in δ on the right-hand side of eqn. (8), and obtain

$$\delta \approx d \ G \ \phi/2 \qquad\qquad (9)$$

Data on human fertility (Henry 1953, 1957) suggest that a value of ϕ of about 0.04 per month is reasonable for high fertility populations. An upper limit for G is provided by estimates of the reduction in the non-susceptible period following foetal death; the data of Potter *et al.* (1965) and data quoted by Ginsberg (1973) suggest a value of 20 months. Taking into account the fact that much loss occurs in the post-natal period a value of G of 10 months is probably more realistic. δ may thus be placed between 0.2 and 0.4 times the difference between the total probability of loss of males and females over the first two years fol- lowing conception (d). Data on foetal wastage suggest that most of these losses are due to early foetal death and deaths in the first few months of infancy (Potter *et al.* 1965). This estimate of δ can be related to the secondary sex ratio at the end of the relevant period of infant life. Let this sex ratio be r'; we have

$$r'/(1-r') \ < \ (1-d) \ (1+\delta)/(1-\delta) \ \approx \ (1-d+2\delta) \qquad\qquad (10)$$

From the estimate of $\delta(2\delta < d)$ given above, one would therefore predict that the secondary sex ratio should be biased in favour of females by the end of the first year or so of infancy. This prediction will be exa- mined in the discussion section.

Relations between sex ratio and maternal age.

In this section I use a different simplified model of fecundity, where the age-class length chosen is adjusted to the period of non-sus- ceptibility in such a way that the latter covers at most one age-class subsequent to that in which conception occurred. To allow for scattering of dates of conceptions among females contained in a given age-class, we let there be a probability α_x that a female who conceives at age x and successfully rears the child past infancy is non-susceptible in age- class $x + 1$; there is a probability $1 - \alpha_x$ that she is susceptible. If the child dies before the end of infancy, which has probability $1 - v_x$, the female is non-susceptible in the next age-class with probability $\beta_x \ll \alpha_x$. If the length of an age-class is taken as 18 months, this representation should be reasonably accurate. We obtain the following basic equation for the fecundity of a fertile female aged x (c.f. Henry 1957)

$$m_x/f_x = \phi_x = p_x(1 - \phi_{x-1} h_{x-1}),$$
(11)

where p_x is the probability of her being fecundated while in age-class x and $h_x = \alpha_x v_x + \beta_x (1-v_x)$ is the overall probability that she is non-susceptible in age class x + 1. This gives

$$\partial \phi_{x+y}/\partial r_x = -(\partial h_x/\partial r_x) \phi_x \theta_{x,y} \quad (y=1,2,\ldots d-x),$$
(12)

where

$$\theta_{x,1} = p_{x+1}, \quad \theta_{x,y} = (-1)^{y-1} p_{x+1} \prod_{j=1}^{y-1} p_{x+j+1} h_{x+j} \quad (y>1).$$
(13)

Let $P_{x,y} = l_{x+y} f_{x+y}/l_x f_x$: this is the probability that a live fertile female aged x is still alive and fertile at age x + y. Substituting from eqn. (12) into eqn. (5), we obtain the equilibrium equation

$$\delta = -(\partial h_x/\partial r_x) \sum_{y=1}^{d-x} \lambda_0^{-y} P_{x,y} \theta_{x,y} [(1-\bar{r}_{x+y})\delta + (1-\bar{r})]$$
(14)

In order to get some insight into the dependence of $\bar{r}_y$ on y, we subtract from this equation the corresponding equation for x - 1. Writing $\Delta_x (u) = u_x - u_{x-1}$, where u is any function of x, and neglecting second-order terms, we obtain

$$(\partial h_x/\partial r_x) \sum_{y=1}^{d-x} \lambda_0^{-y} P_{x,y} \theta_{x,y} \Delta_x(\bar{r}_{x+y}) \approx (1/2\delta)\{(\partial h_x/\partial r_x)\Delta_x(T_x) +$$

$$T_x \Delta_x (\partial h_x/\partial r_x)\},$$
(15)

where

$$T_x = \sum_{y=1}^{d-x} \lambda_0^{-y} P_{x,y} \theta_{x,y} \quad (T_x > 0 \text{ when } \lambda_0 \geq 1).$$

The terms in braces on the right-hand side of eqn. (15) measure the difference between the extents to which females aged x and x-1 can replace

the additional losses of offspring caused by an increased fraction of
male zygotes. Because of increasing death and sterility risks with increasing age, one would intuitively expect this difference to be negative for most age-classes. This can be established more formally as
follows. Because of senescence we can expect $P_{x+1,1}/P_{x,1} < 1$ and $P_{x+1} \leq$
P_x for most ages. There is evidence (Henry 1964a, Potter et al. 1965)
that the non-susceptible period following a live birth increases in
length with age. This would create a tendency for $h_{x+1} > h_x$. There is
also a strong trend for foetal wastage and infant mortality to increase
with maternal age, which would at least partially counterbalance this
tendency. We can fairly safely assume, therefore, that for all except
very small x

$$(P_{x-1,1} \; P_x \; h_x)/(P_{x,1} \; P_{x+1} \; h_{x+1}) > 1 \tag{16}$$

As shown in Appendix B, this gives the result that, at least for x such
that d-x is even, we have $\Delta_x(T_x) < 0$. This disposes of the first part
of the right-hand side of eqn. (15). The other part involves $\Delta_x(\partial h_x/\partial r_x) =$
$-\Delta_x \; d_x(\alpha_x - \beta_x)$, where d_x is the difference between local probabilities
of death before the end of infancy for sons and daughters of mothers aged
x (d_x is assumed > 0). Now infant mortality and foetal wastage increase
with maternal age and $\alpha_x > \beta_x$, so that we may expect $\Delta_x(\partial h_x/\partial r_x) < 0$, since
the magnitude of d_x will tend to increase with increasing overall death-
rates. Hence the contribution of $\Delta_x(T_x)$ is, at least partially, counterbalanced by this term. But $\Delta_x(T_x)$ has terms arising both from increased
risk of sterility and increased risk of death with advancing age, while
$\Delta_x(\partial h_x/\partial r_x)$ reflects only one aspect of senescence. Moreover, it seems
plausible to assume that with increasing age there may be an increasing
inability of females to shorten the gap between successive conceptions
in response to loss of an offspring (see above), and this would to some
extent balance the effect of increased losses. It can be concluded,
therefore, that the term in braces is negative for most values of x.
Eqn. (14) can be shown to imply $\delta > 0$, so that we have for most x

$$\sum_{y=1}^{d-x} \lambda_0^{-y} \; P_{x,y} \; \theta_{x,y} \; \Delta_x(\bar{r}_{x+y}) < 0 \tag{17}$$

This in turn implies that, if sex ratio is an approximately linear
function of maternal age, we may expect a decrease in the proportion
of male offspring with increasing age of mothers for most ages. This

will be discussed further below.

DISCUSSION

The first conclusion suggested by this work, which is in agreement with the less complex models of sex ratio evolution referred to earlier, is that in species where replacement is impossible we expect a primary sex ratio of 1/2. Determinations of the chromosomal sex of mammalian blastocytes provide the most reliable data available on primary sex ratio; pooled results from three litter-producing species (where replacement of individual offspring losses is unlikely) give an overall sex ratio of 50.4%, with 95% confidence limits of ± 2.9% (Fechheimer and Beatty 1974). More data are clearly desirable.

The second conclusion, expressed in eqn. (9), is that the human primary sex ratio should be biased in favour of males when averaged over all parental ages, provided that the total probability of death before the end of infancy is higher for sons than daughters. This is in agreement with Fisher (1930) and studies of simple models of replacement (Crow and Kimura 1970). Data on chromosomal sexing of early human embryos seem to be rather variable between samples, but at present data on spontaneous and induced abortions seem to suggest a non-significant excess of *females* (Carr 1971, Hamerton 1972, Pawlowitski 1972). There is no suggestion of the gross excess of males reported in the earlier literature, which was presumably due to poor sexing of early foetuses. If it turns out to be the case that there is a high rate of preferential elimination of females in early pregnancy, such that the overall death-rate for daughters is higher than for sons, the above models will need some revision. The following argument suggests that this revision need not be drastic. Early pregnancy losses are probably almost completely replaced; hence we can work with m_x and M_x *etc.* as expressing the numbers of offspring alive at the end of the period of pregnancy when replacement is complete, and $\bar{r}$, r_x *etc.* as measuring sex ratio at this stage. With differential death of males and partial replacement later in pregnancy, for which there is good evidence, the sex ratio should be biased in favour of males at this stage, and the other relations derived will hold good for sex ratio at this stage rather than for the primary sex ratio.

A firm prediction of the present model is that the secondary sex
ratio should become biased in favour of females by the end of the first
year of postnatal life, approximately. (This is true even when sex ratio
is controlled by genes acting through the father or through the offspring themselves; details of the proof will be omitted here). This
conflicts with statements by some authors (Bodmer and Edwards 1960,
Edwards 1962), who predict a sex ratio of 1/2 at the beginning of ado-
lescence. This is based on the view that sex ratio will be equalised
at end of the period of parental expenditure, which is identified with
the end of parental care. As pointed out in the introduction to this
paper, the parental expenditure concept is logically equivalent to the
idea of replacement of offspring who are lost before a certain age; if
it is accepted that offspring older than about one year of age cannot
be replaced in populations which do not practice birth control, because
another baby is on the way in any case, there is no contradiction between the two viewpoints.

Data on human populations are in good agreement with the idea that
the sex ratio should be female-biased by the end of infancy, with the
infant mortality rates characteristic of pre-industrial societies. In
populations of European descent, for example, the overall sex ratio is
usually around 0.514 at birth. Data on contemporary underdeveloped
countries suggests that the male infant mortality rate is 1.2 times the
female (Keyfitz and Flieger 1971). To bring a sex ratio at birth of
0.514 down to 0.5 at one year of age requires a female infant mortality
rate of 0.21. This is within the range of mortality rates reported for
contemporary underdeveloped countries (Keyfitz and Flieger 1971) and
for many historical populations (Henry 1967). The results derived here
do not support the views of Beiles (1974) concerning the evolutionary
causes of a male-biased sex ratio.

We may finally consider data on the relations between sex ratio and
parental age. Data on the English population (James 1975), the U.S. population (Novitski and Kimball 1958) and the Australian population (Pollard 1969) show a negative correlation between maternal age and sex ratio. This appears to be due to a negative correlation between birth-order and sex ratio rather than maternal age itself (Novitski and Kimball
1958, Teitelbaum 1972). Since birth-order and maternal age are highly
correlated, there is a resulting relationship between sex ratio and maternal age. As described above, the evolutionary equilibrium for genes
affecting the sex ratio through the maternal genotype predicts a decrease in sex ratio with maternal age, in agreement with the observa-

tions. Intuitively one can view this result as follows. Early in life, females have a good chance of replacing lost offspring, so that the evolutionary equilibrium for young females is one which corresponds to population with a high level of replacement of losses *i.e.* one with a sex ratio biased heavily in favour of males, given that male offspring are lost more frequently than females. Later in life females have a higher chance of dying or becoming sterile before they are able to re-place a lost offspring, so that the sex ratio of their offspring corre-sponds to a population with a low level of replacement *i.e.* one with fewer males.

Alternative explanations for the relation between sex ratio and ma-ternal age can, of course, be put forward. Since there is an increase in foetal elimination and infant mortality rates with increasing mater-nal age, one could argue that there is a concomitant increase in the difference between male and female loss rates, which would account for the reduction in frequency of sons with maternal age. But this explana-tion would predict a relation with maternal age *per se*, rather than birth-order. On the selective hypothesis there is no difficulty in vi-sualising the selection of maternal sex ratio genes which respond to birth order, in such a way that the net effect is a correlation between maternal age and sex ratio. Similarly, James (1971) has proposed that there is an association between high coital rate and high sex ratio; this predicts a decrease in sex ratio as coital rate falls off with age. Again this theory would predict at least some effect of maternal age *per se*.

As far as paternal influence on the sex ratio is concerned, repla-cement will produce a male-biased primary sex ratio provided that there is a correlation between the ages of mates. It can also be shown that there will be a negative correlation between paternal age and sex ratio at evolutionary equilibrium. The correlation is expected to be weaker than with maternal genes. Novitski and Kimball (1958) and Moran *et al.* (1969) report such an effect, for U.S. data. Its existence has been questioned by Teitelbaum (1972), who failed to detect it using a diffe-rent statistical technique. Pollard (1969) reports an effect of paternal age using Australian data, and concludes that both parents must be young for there to be a high sex ratio.

The data on parental age and sex ratio are thus qualitatively in agreement with the selective theory. In view of our ignorance of the values of many of the variables affecting reproductive perforamnce in

this past, it is not possible to make any quantitative assessment of
the agreement. One cannot be certain that factors uncorrelated to selection are not involved, but at any rate the data do not disagree with
the theory.

APPENDIX A.   THE CHARACTERISTIC EQUATION FOR A RARE SEX RATIO GENE

Let ε_t and η_t be the frequencies of Aa individuals among female and
male zygotes respectively, produced by the population at time t. Let
B_t be the total number of zygotes at time t; we have (approximately)
$\bar{r}\, B_t$ males and $(1-r)\, B_t$ females among them. Neglecting higher order
terms in ε and η we have, for $t > d, d^*$

$$(1-\bar{r})B_t\, \varepsilon_t = [\,(1-\bar{r})/2]\Sigma B_{t-x}\, \varepsilon_{t-x}\, K_x(1-r_x) +$$

$$(\bar{r}/2)\Sigma B_{t-x}\, \eta_{t-x}\, K_x^*(1-r_x^*), \qquad\qquad (A.1a)$$

$$\bar{r}\, B_t\, \eta_t = [\,(1-\bar{r})/2]\Sigma B_{t-x}\, \varepsilon_{t-x}\, K_x\, r_x +$$

$$(r/2)\Sigma B_{t-x}\, \eta_{t-x}\, K_x^*\, r_x^* \qquad\qquad (A.1b)$$

But since the initial population is in stable age distribution, we have
$B_{t+1}/B_t = \lambda_0$ to terms of order ε and η, so that eqs. (A.1) become

$$\varepsilon_t = (1/2)\Sigma\lambda_0^{-x}\, K_x(1-r_x)\varepsilon_{t-x} +$$

$$[\bar{r}/(1-\bar{r})\,]\Sigma\lambda_0^{-x}\, K_x^*(1-r_x^*)\eta_{t-x}, \qquad\qquad (A.2a)$$

$$\eta_t = (1/2)\Sigma\lambda_0^{-x}\, [\,(1-\bar{r})/\bar{r}]K_x\, r_x\, \varepsilon_{t-x} + \Sigma\lambda_0^{-x}\, K_x^*\, r_x^*\, \eta_{t-x} \qquad (A.2b)$$

For $t < d, d^*$, we must add positive terms $g_1(t)$ and $g_2(t)$ to the right-
hand sides of (A.2a) and (A.2b) respectively, to account for contributions from individuals alive at time zero. Write the generating func-
tion for a function of t, U_t, as $G_s\{U\}$, where $G_s\{U\} = \sum_{t=0}^{\infty} s^t\, U_t$. Taking
g.f.'s in (A.2), we get

$$G_s\{\epsilon\}(1-G_s\{\lambda_o^{-x}\ K_x(1-r_x)\}/2) -$$

$$\{\bar{r}/[2(1-\bar{r})]\}G_s\{\eta\}G_s\{\lambda_o^{-x}\ K_x^*(1-r_x^*)\} = G_s\{g_1\} \qquad (A.3a)$$

$$- [(1-\bar{r})/(2\bar{r})]G_s\{\epsilon\}G_s\{\lambda_o^{-x}\ K_x\ r_x\} +$$

$$G_s\{\eta\}(1-G_s\{\lambda_o^{-x}\ K_x^*\ r_x^*\}/2 = G_s\{g_2\} \qquad (A.3b)$$

Using the method of Feller (1941) for inverting generating functions by expansion into partial fractions, it can be seen that, for large t, ϵ_t and η_t are proportional to s_o^t, where s_o is the largest zero of the determinant of eqns. (A.3), $f(s)$. We have

$$f(s) = 1-(1/2)G_s\{\lambda_o^{-x}\ [K_x(1-r_x) + K_x^*\ r_x^*]\} +$$

$$(1/4)G_s\{\lambda_o^{-x}\ K_x(1-r_x)\}G_s\{\lambda_o^{-x}\ K_x^*\} -$$

$$(1/4)G_s\{\lambda_o^{-x}\ K_x\}G_s\{\lambda_o^{-x}\ K_x^*\ (1-r_x^*)\}. \qquad (A.4)$$

This is the characteristic equation of the equivalent matrix representation of the process. The matrix can be shown to be primitive, given certain regularity conditions which will normally be satisfied in real situations. We can therefore apply the Perron-Frobenius theorem (Seneta 1973), which implies that s_o is unique, real and positive.

APPENDIX B.   PROPERTIES OF $\Delta_x(T_x)$

From eqns. (15) and (12), we can write

$$- \Delta_x(T_x) = (b_1-a_1) - (b_2-a_2) + (b_3-a_3) - ,\ldots, \qquad (B.1)$$

where

$$a_1 = \lambda_o^{-1}\ P_{x,1}\ P_{x+1}, \qquad b_1 = \lambda_o^{-1}\ P_{x-1,1}\ P_x,$$

$$a_k = \lambda_o^{-k} \; P_{x,k} \; P_{x+1} \; \prod_{j=1}^{k-1} P_{x+j+1} \; h_{x+j},$$

$$b_k = \lambda_o^{-k} \; P_{x-1,k} \; P_x \; \prod_{j=1}^{k-1} P_{x+j} \; h_{x+j-1}, \quad (k > 1)$$

From eqn. (16) we have

$$1 < \mu_k = b_k/a_k = P_{x-1,1} \; P_x \; h_x/P_{x+k-1,1} \; P_{x+k} \; h_{x+k-1}, \qquad \text{(B.2a)}$$

$$1 < \mu_1 = P_{x-1,1} \; P_x/P_{x,1} \; P_{x+1} \qquad\qquad\qquad \text{(B.2b)}$$

Let $A_k = (b_k - a_k) - (b_{k+1} - a_{k+1})$ $(k = 1,3,...)$. Then we have $\mu_{k+1} > \mu_k > 1$. Hence, for $A_k > 0$, we need

$$(\mu_{k+1}-1)/(\mu_k-1) < a_k/a_{k+1} \qquad\qquad\qquad \text{(B.3)}$$

Write $P_{x+k-1,1} \; P_{x+k} \; h_{x+k-1} = P_{x-1,1} \; P_x \; h_x(1-\xi_k)$, $(k > 1)$; $P_{x,1} \; P_{x+1} = P_{x-1} \; P_x \; \xi_1$; $\gamma_k = (\xi_{k+1}-\xi_k)/\xi_k$ $(\gamma_k > 0$ from eqn [16]). We thus have

$$(\mu_{k+1}-1)/(\mu_k-1) = \xi_{k+1}(1-\xi_k)/\xi_k(1-\xi_{k+1}) < (1+\gamma_k\xi_k)(1+\gamma_k) \qquad \text{(B.4)}$$

Now we must have $\xi_k < 1$; γ_k is a measure of the change in survival and reproductive parameters between ages $x + k - 1$ and $x + k$, and is therefore small, perhaps of the order of 1%-2%. But we have $a_k/a_{k+1} > 1/h_{x+k}$ (provided that $\lambda_o > 1$), on the assumptions about senescence. Since at least 25% of conceptions result in foetal wastage, with a resulting very short non-susceptible period, h_{x+k} can reasonably be assumed < 75%. There should be no difficulty, therefore, in satisfying $(1+\gamma_k)$ $(1+\gamma_k\xi_k) < a_k/a_{k+1}$, and hence (B.3). This implies $A_k > 0$.

If we choose x such that d-x is even, $-\Delta_x(T_x)$ will consist of a sum of A_k terms plus a term of form $+b_k$. $A_k > 0$ thus implies $\Delta_x(T_x) < 0$ in this case. Since b_k and a_k both decrease with increasing k, the contribution of the last term in $\Delta_x(T_x)$ is usually negligible except for large x, so that $\Delta_x(T_x) < 0$ for most x when $A_k > 0$ for each k.

Summary.

This paper examines the evolutionary genetics of modifiers of the sex ratio in age-structured populations. It is shown that, in the absence of parental replacement of offspring losses, a 1:1 sex ratio at conception is selectively favoured, in agreement with previous authors. Models of human reproductive behaviour which incorporate the phenomenon of replacement are studied. It is shown that, with higher death rates among sons than among daughters, there should evolve a male-biased sex ratio at the time of conception. By one year of age, a female-biased sex ratio is predicted, given the infant mortality rates characteristic of pre-industrial populations. The theory also predicts a negative correlation between parental ages and sex ratio. Data on human sex ratios are discussed in the light of these predictions.

Acknowledgements.

I am deeply grateful to E.L. Charnov for stimulating me to work on this problem, and to my wife for many discussions about sex ratio evolution.

REFERENCES

Beiles, A. 1974. A buffered interaction between sex ratio, age difference at marriage and population growth in humans and their significance for sex ratio evolution. *Heredity* 33: 265-278.

Bodmer, W.F. and A.W.F. Edwards. 1960. Natural selection and the sex ratio. *Ann. Hum. Genet.* 24: 239-244.

Carr, D.H. 1971. Chromosomes and abortion. *Adv. Hum. Genet.* 2: 201-257.

Charnov, E.L. 1975. Sex ratio selection in an age-structured population. *Evolution* 29: 366-368.

Ciocco, A. 1940. Sex differences in morbidity and mortality. *Quart. Rev. Biol.* 15: 59-73, 192-210.

Crow, J.F. and M. Kimura. 1970. An introduction to population genetics theory. Harper and Row, New York.

Edwards, A.W.F. 1962. Genetics and the human sex ratio. *Adv. Genet.* 11: 239-272.

Edwards, A.W.F. 1963. Natural selection and the sex ratio: the approach to equilibrium. *Amer. Nat.* 97: 397-400.

Emlen, J.M. 1968. A note on natural selection and the sex ratio. *Amer. Nat.* 102: 94-95.

Eshel, I. 1975. Selection on sex ratio and the evolution of sex-determination. *Heredity* 34: 351-361.

Fechheimer, N.S. and R.A. Beatty. 1974. Chromosomal abnormalities and sex ratio in rabbit blastocysts. *J. Reprod. Fert.* 37: 331-341.

Feller, W. 1941. On the integral equation of renewal theory. *Ann. Math. Stat.* 12: 243-267.

Fisher, R.A. 1930. The genetical theory of natural selection. Clarendon Press, Oxford.

Ginsberg, R.B.   1973.   The effects of lactation on the length of the post-partum ano-
    vulatory period: an application of a bivariate stochastic model.
    Theor. Popul. Biol. 7: 276-309.

Goodman, L.A.   1969.   The analysis of population growth when the birth and death rates
    depend upon several factors.
    Biometrics 25: 659-681.

Hamerton, J.L.   1971.   Human cytogenetics II. Clinical cytogenetics.
    Academic Press, New York.

Hamilton, W.D.   1967.   Extraordinary sex ratios.
    Science 155: 477-488.

Henry, L.   1953.   Fondements théoriques des mesures de la fécondité naturelle.
    Revue Inst. Int. Stat. 21: 135-151.

Henry, L.   1957.   Fécondité et famille. Modèles mathématiques I.
    Population 12: 413-444.

Henry, L.   1961.   Fécondité et famille. Modèles mathématiques II.
    Population 16: 27-48, 261-282.

Henry, L.   1964a.   Mesure du temps mort en fécondité naturelle.
    Population 19: 485-514.

Henry, L.   1964b.   Mortalité intra-uterine et fécondabilité.
    Population 19: 899-940.

Henry, L.   1965.   French statistical research in natural fertility.
    In *Public health and population change* (M.C. Sheps and J.C. Ridley, eds.).
    Pittsburgh University Press, Pittsburgh.

Henry, L.   1967.   Manuel de démographic historique.
    Droz, Geneva.

James, W.H.   1971.   Cycle day of insemination, coital rate and sex ratio.
    Lancet 1: 112-114.

James, W.H.   1975.   Sex ratio in twin births.
    Ann. Hum. Biol. 2: 365-378.

Kolman, W.   1960.   The mechanism of natural selection for the sex ratio.
    Amer. Nat. 94: 373-377.

Keyfitz, N. and W. Flieger.   1971.   Population. Facts and methods of demography.
    W.H. Freeman, San Francisco.

Leigh, E.G.   1970.   Sex ratio and differential mortality between the sexes.
    Amer. Nat. 104: 205-210.

MacArthur, R.H.   1965.   Ecological consequences of natural selection. Pp. 388-397.
    In *Theoretical and mathematical biology* (T.H. Waterman and H.J. Morowitz, eds.).
    Blaisdell, Mass.

Maynard Smith, J.   1972.   On evolution.
    Edinburgh University Press, Edinburgh.

Moran, P.A.P., E. Novitski and C. Novitski.   1969.   Paternal age and the secondary sex
    ratio in humans.
    Ann. Hum. Genet. 32: 315-317.

Novitski, E. and A.W. Kimball.   1958.   Birth order, parental ages and sex of offspring.
    Amer. J. Hum. Genet. 21: 123-131.

Pawlowitski, I.H.   1972.   Frequency of chromosome abnormalities in abortions.
    Humangenetik 16: 131-136.

Perrin, E.B. and M.C. Sheps. 1964. Human reproduction: a stochastic process.
    Biometrics 20: 28-45.

Pollard, G.N. 1969. Factors influencing the sex ratio at birth in Australia.
    J. Biosoc. Sci. 1: 125-144.

Potter, R.G., M.L. New, J.B. Wyon and J.E. Gordon. 1965. Applications of field stu-
    dies to research on the physiology of human reproduction. Pp. 377-399.
    In *Public health and population change* (M.C. Sheps and J.C. Ridley, eds.).
    Pittsburgh University Press, Pittsburgh.

Seneta, E. 1973. Non-negative matrices.
    Allen and Unwin, London.

Shaw, R.F. 1958. The theoretical genetics of the sex ratio.
    Genetics 43: 149-163.

Shaw, R.F. and J.D. Mohler. 1953. The selective significance of the sex ratio.
    Amer. Nat. 87: 337-342.

Spieth, P.T. 1974. Theoretical considerations of unequal sex ratios.
    Amer. Nat. 108: 837-849.

Teitelbaum, M.S. 1972. Factors associated with the sex ratio in human populations.
    Pp. 90-109. In *The structure of human populations* (G.A. Harrison and A.J. Boyce,
    eds.).
    Clarendon Press, Oxford.

3. SEX AND EVOLUTION

A COMPARATIVE STUDY ON ENZYME POLYMORPHISMS IN SYMPATRIC DIPLOID AND
POLYPLOID POPULATIONS OF *LUMBRICILLUS LINEATUS* (O.F.M.) ENCHYTRAEIDAE
OLIGOCHAETA

Bent Christensen, Jens Jelnes and Uffe Berg

The Oligochaete worm *Lumbricillus lineatus* is a littoral species often
abundantly present in wrack beds. Amphimictic diploid and parthenogenetic tri-, tetra and pentaploid populations have been found. Due to
peculiarities in the reproductive behaviour of the parthenogenetic
polyploids, two different chromosome "races" have been found to occur
together in all the localities studied. The aim of the present study
is to compare such two closely related forms occurring in the same environment but subjected to different genetical conditions.

MATERIAL AND METHODS

Breeding biology.

Amphimictic diploids. Reproduction as in other monoecious organisms.

Parthenogenetic triploids. The oogenesis represents a number of
unique features such as asynaptic meiosis and reductional distribution of the univalents (Christensen, 1960). The triploids never produce mature sperm but they must be fertilized by sperm from one of the
other chromosomal types in order to produce viable eggs. However, the

sperm does not enter the egg, but eggs which have not been activated
by sperm die at an early stage.

Parthenogenetic tetraploids. The oogenesis resembles that of the
triploid, but - in contrast to what is found here - the tetraploids
produce morphologically normal spermatozoa. However, the first spermato-
cyte division is highly irregular with a varying number of uni- and
multivalents in addition to normal bivalents, and the nuclei produced
are unbalanced. It is not known whether tetraploid eggs have to be activated by sperm in order to develop normally.

The areas investigated, cf. Fig. 1.

Area I. Stations along the inner Danish coastline. The salinity
ranges from 8 - 30 per mill with small variations at a given station.
Usually an abundance of decaying seaweeds due to the sheltered condition of the sampling sites chosen. Diploids and triploids were found
together at all stations.

Area II. Three stations along the west coast of Jutland. Salinity
appr. 30 per mill, and only small variations. Due to the exposed conditions in this area decaying seaweed is of minor importance as a food
supply and the worms were often found in coarse sand or gravel just a-
bove zero level, especially in the immediate surroundings of smaller
or larger stones overgrown with green algae. Diploids and triploids
were found together at all three stations.

Area III. Two stations along the Hvide Sande canal which connects
the oligohaline (less than 5 per mill salinity) Ringkøbing Fjord with
the saline North Sea. One station is located immediately inside and
the other station outside (towards the North Sea) a dam built in 1931
in order to regulate the water level in Ringkøbing Fjord through sluices
in the dam. This means that the salinity at the collection sites varies
according to the position of the sluices. The worms from the inner station were found in decaying seaweed as along the inner Danish coastline
(Area I) whereas the situation at the outer station resembles that a-
long the west coast of Jutland (Area II).

Triploids and tetraploids were found together at both stations.
The entire absence of diploids might indicate that the abnormal spermatozoa produced by the tetraploid are used to activate the parthenogenetic eggs of both the triploid and the tetraploid.

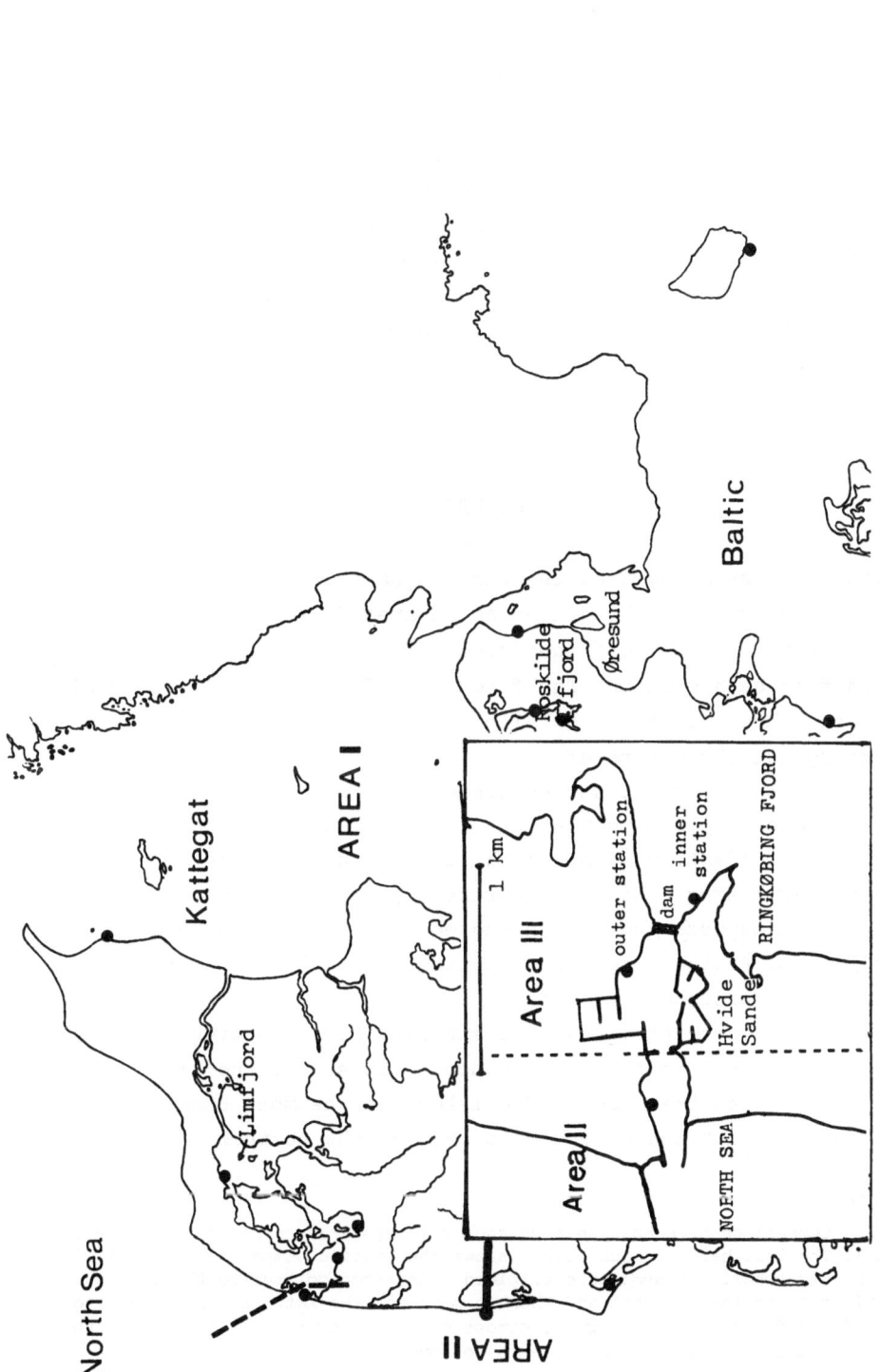

Figure 1. Map showing the areas studied. Area I: Stations along the Limfjord and into the Baltic. Area II: Three stations along the west coast of Jutland. Area III (inset): Two stations along the Hvide Sande canal located immediately inside and outside a dam (with sluices) that separates Ringkøbing Fjord from the North Sea.

The markers used.

The phosphoglucomutase and phosphoglucoseisomerase loci were used
as genetical markers. Electrophoresis was performed in the buffer of
Detter et al. (1968) using the methods of Jelnes (1974). Detection of
the enzyme activity after slicing the gel horizontally followed the
methods of Jelnes (1971 and 1974). The nomenclature used is as follows:
The two loci are abbreviated *Pgm* and *Pgi* respectively. Within both loci
the most common allele is called *1.00* and other alleles are termed according to their anodic position relative to that. Diallelic polyploids
are given according to the two alleles present and no attempt is made
to indicate the dosage.

RESULTS

Nine stations extending from the western part of the Limfjord and into
the Baltic exemplifies the situation in Area I. The histogrammes in
Fig. 2 show the phenotype frequencies in the diploid and triploid populations from each station. The frequency of genotypes in the diploid
populations show that the *1.00* allele is strongly dominant within both
loci throughout the entire area.

A comparison of the observed frequencies of identical phenotypes
in diploid and triploid populations from the same station reveals a
high degree of identity within both loci in the majority of stations.

In summary: Area I is characterized by a strong dominance of a
particular homozygote and a high degree of identity between diploids
and triploids. Furthermore the situation is stable as no change has
been observed throughout the years.

In Area II three stations were studied. From the frequency distribution of genotypes in the diploid populations, cf. the histogrammes
in Fig. 3, is seen that the *1.00* alleles are the most common also in
this area. But with the exception of Pgm in the northernmost station

---

Figure 2. Comparison of phenotype frequencies in sympatric diploids and triploids
in Area I. The left half of each column shows the genetic composition of the diploid, the right half its sympatric triploid. The three columns to the left show
the stations along the Limfjord and then in sequence the Kattegat, Roskilde fjord
(2 sts), Øresund stations, the two rightmost columns show the stations along the
Baltic. N: sample sizes, given above each column. (see opposite page).

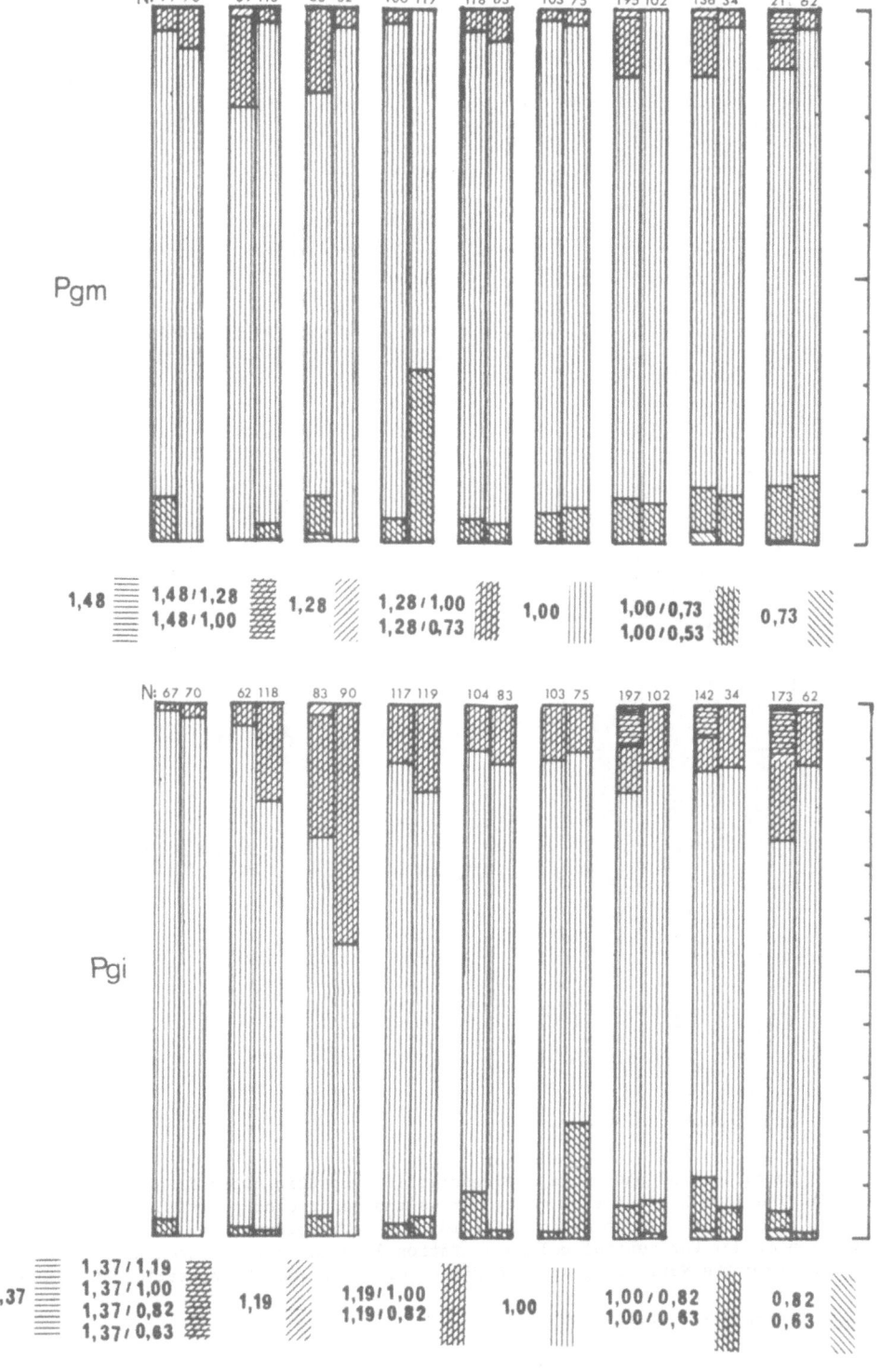

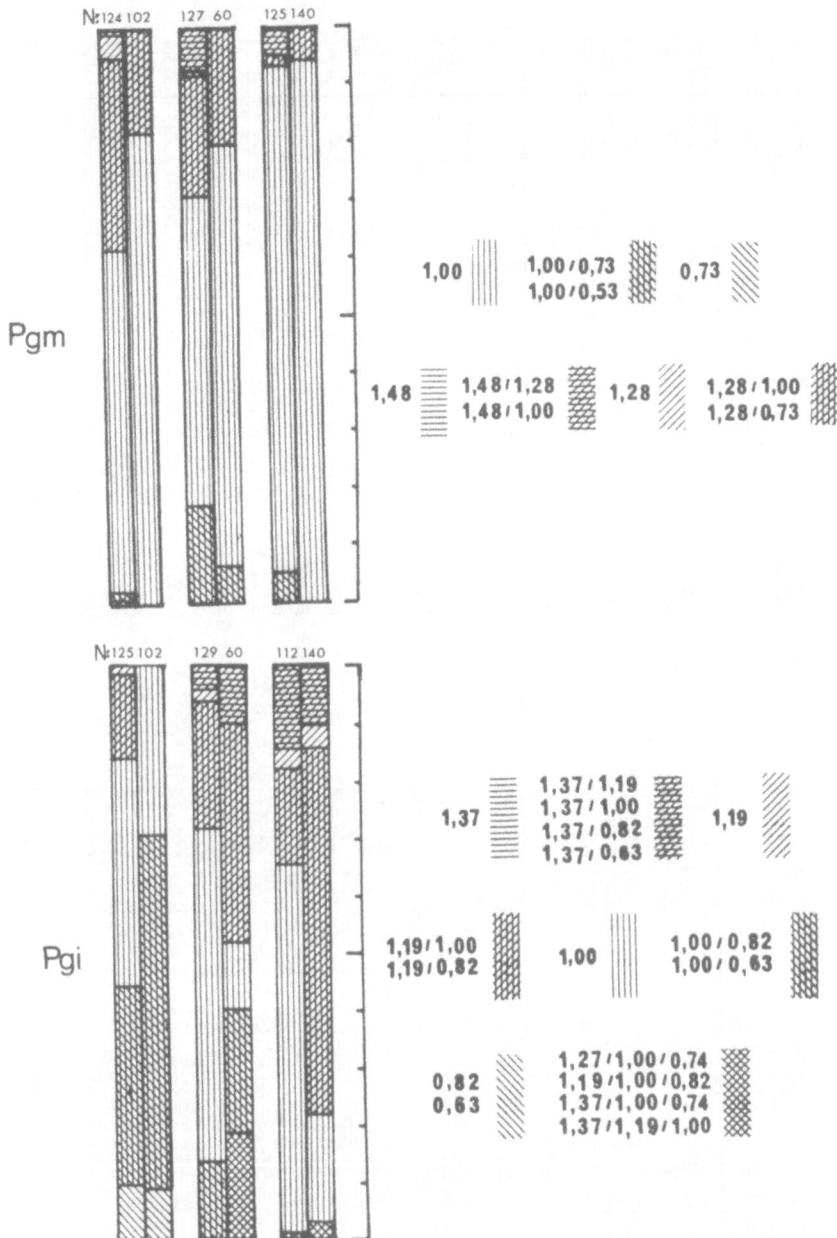

Figure 3. Comparison of phenotype frequencies in sympatric diploids and triploids in area II. The left half of each column shows the genetic composition of the diploid, the right half its sympatric triploid. The left column shows the southern-most station at Ho, the central column a station located near the opening of Hvide Sande canal into the North Sea, and the right column the northern station at Thyborøn.
N: sample sizes, given above each column.

(Tyborøn) the predominance is less pronounced than in Area I and a
higher number of heterozygotes is produced. Comparison of phenotype
frequencies in diploids and triploids within the same station shows a
high degree of identity in *Pgm* phenotypes at the northernmost station,
whereas this locus is less similar at the other stations. Here the major homozygote phenotype reaches a higher frequency in the triploid.
The *Pgi* phenotypes differ strongly between the diploids and triploids
within all three stations. The difference is due to a much higher frequency of heterozygotic phenotypes in the triploid.

In summary: There is a greater genetic variation in this area compared to Area I and the frequency of identical *Pgi* phenotypes is different in sympatric diploid and triploid populations. The situation
has been stable for two years.

Area III differs from the two former by the presence of parthenogenetic tri- and tetraploids and in the entire absence of diploid individuals. Another difference is that the situation is highly dynamic
and this brought us to study this particular area in more detail. Since
autumn 1972 we have made observations each autumn and spring (with the
exception of the autumn 1974) on the station outside the dam. (cf. Fig.
1) and since the autumn '73 on the inner side. The results are presented in Tables 1-4. Among the triploids from the outer station the majority of individuals belong to four different types, namely *Pgm 1.00*
Pgi 1.19/1.00, *Pgm 1.28/1.00 Pgi 1.00/0.82*, *Pgm 1.28/1.00 Pgi 1.27/*
1.00/0.74 and *Pgm 1.00/0.73 Pgi 1.19/1.00/082*. From Table 1 is seen
that the frequency of these types varies in a regular manner: the two
former are common in spring, the two latter in the autumn. This is
shown schematically in the histogrammes Fig. 4. In the autumn 1975 no
triploids were found.

Among the tetraploids the majority of individuals belong to three
different types: *Pgm 1.28/1.00 Pgi 1.19/1.00*, *Pgm 1.28/1.00 Pgi 1.00/*
0.82 and *Pgm 1.54/1.00 Pgi 1.19/1.00*. From Table 2 and the histogrammes in Fig. 4 is seen that the frequency of *Pgm 1.28/1.00 Pgi 1.19/1.00*
is decreasing and this type may now have disappeared from the locality.
Pgm 1.28/1.00 Pgi 1.00/0.82 shows the opposite trend and is now the
major type. *Pgm 1.54/1.00 Pgi 1.19/1.00* has become second in frequency.

On the inner side of the dam the vast majority of triploid individuals belong to the two types *Pgm 1.00 Pgi 1.19/1.00* and *Pgm 1.28/1.00*
Pgi 1.00/0.82, and there is no indication of seasonal or long-term
changes. Among the tetraploids *Pgm 1.28/1.00 Pgi 1.19/1.00* is the major
type, and it may have increased its frequency.

Table 1. Percentages of genetic variants among triploids at the outer station in Area III. Sample sizes are given below

Pgm				*Pgi*		
		1.00	1.19/1.00	1.00/0.82	1.27/1.00/ 0.74	1.19/1.00/ 0.82
1.00	aut.72	2.8	12.7			
	spr.73	8.0	62.0			
	aut.73		2.3	2.3	2.3	
	spr.74	5.4	40.5	2.7		
	spr.75	15.2	39.4			
	aut.75					
1.28/1.00	aut.72	1.4			74.7	
	spr.73	2.0		8.0	4.0	
	aut.73			4.6	68.2	
	spr.74			35.1	10.8	
	spr.75	3.0		24.2	12.1	
	aut.75					
1.54/1.00	aut.72					
	spr.73		14.0			
	aut.73					
	spr.74		5.4			
	spr.75		6.1			
	aut.75					
1.00/0.73	aut.72		1.4			7.0
	spr.73			2.0		
	aut.73					20.5
	spr.74					
	spr.75					
	aut.75					

Sample sizes: aut.72:71; spr.73:50; aut.73:44; spr.74:37; spr.75:33; aut.75:no triploids were found.

Table 2. Percentages of genetic variants among tetraploids at the outer station in Area III. Sample sizes are given below

Pgm		Pgi		
		1.00	1.19/1.00	1.00/0.82
1.00	aut.72		21.1	
	spr.73	3.5	3.5	
	aut.73	3.5	1.8	
	spr.74	1.4	2.7	
	spr.75	9.7	3.2	3.2
	aut.75		3.4	3.4
1.28/1.00	aut.72	15.8	47.4	15.8
	spr.73	5.3	43.9	29.8
	aut.73	5.3	10.5	71.9
	spr.74	1.4	13.7	80.8
	spr.75	21.0	6.5	32.3
	aut.75			76.3
1.54/1.00	aut.72			
	spr.73			7.0
	aut.73			7.0
	spr.74			
	spr.75		22.6	1.6
	aut.75		16.9	

Sample sizes: aut.72:19; spr.73:57;aut.73:57; spr.74:73; spr.75:62; aut.75:54.

Table 3. Percentages of genetic variants among triploids at the
            inner station in Area III. Sample sizes are given below

| *Pgm* | | *Pgi* | | | |
		1.00	1.19/1.00	1.00/0.82	1.27/1.00/0.74
1.00	aut.73		82.4		
	spr.74		30.2	2.3	
	spr.75	1.9	70.4		
	aut.75		+		
1.28/1.00	aut.73			17.6	
	spr.74		2.3	65.1	
	spr.75			24.1	3.7
	aut.75			+	

Sample sizes: aut.73:68; spr.74:43; spr.75:54; aut.75:only few individuals were
                found.

In summary: In Area III only triploids and tetraploids were recorded and within both a few types were strongly dominating. On the
outer side of the dam seasonal changes in frequency were observed a-
mong the triploids and a long-term trend among the tetraploids.

DISCUSSION

The data presented above raise several interesting questions such as
the origin of the polyploids and the evolution within polyploid, parthenogenetic organisms, subjects that have been extensively studied
by the Helsinki group (Suomalainen and Saura, 1973, and Lokki et al.,
1975).

However, in the present context we will focus upon those aspects
of the study that are particularly relevant to the theme of this symposium, selection in natural population. We do not pretend to have
measured the selective forces and we only want to discuss the presence

Table 4. Percentages of genetic variants among tetraploids at the inner station in Area III. Sample sizes are given below

Pgm		Pgi		
		1.00	1.19/1.00	1.00/0.82
1.00	aut.73	41.2	2.9	
	spr.74	13.8	3.4	
	spr.75	5.6	2.8	
	aut.75	+		
1.28/1.00	aut.73	2.9	47.1	
	spr.74		44.8	
	spr.75		88.9	2.8
	aut.75		+	
1.54/1.00	aut.73	5.9		
	spr.74	13.8	24.1	
	spr.75			
	aut.75			

Sample sizes: aut.73:34; spr.74:29; spr.75:37; aut.75: only few individuals were found.

of and the possible way in which natural selection might operate.

If a directional change and regular seasonal fluctuations in the genetic composition of populations are taken as evidence for selection, this force operates at the outer station in Area III, cf. Fig. 4.

The seasonal changes in the genetical composition of the triploid population show that *Pgm 1.00 Pgi 1.19/1.00* and *Pgm 1.28/1.00 Pgi 1.00/ 0.82* are favoured during the spring whereas in the autumn a change has occurred towards a strong predominance of *Pgm 1.28/1.00 Pgi 1.27/1.00/ 0.74*. The reason for this could be seasonal changes in the environment-al conditions caused by waterflow through the sluices in the dam. The direction of this flow mainly depends upon the water level in Ringkø-bing Fjord and this is generally high in the spring causing an outflow of less saline water to the collection site outside the dam. The salin-ity conditions here therefore approach those on the inner side of the

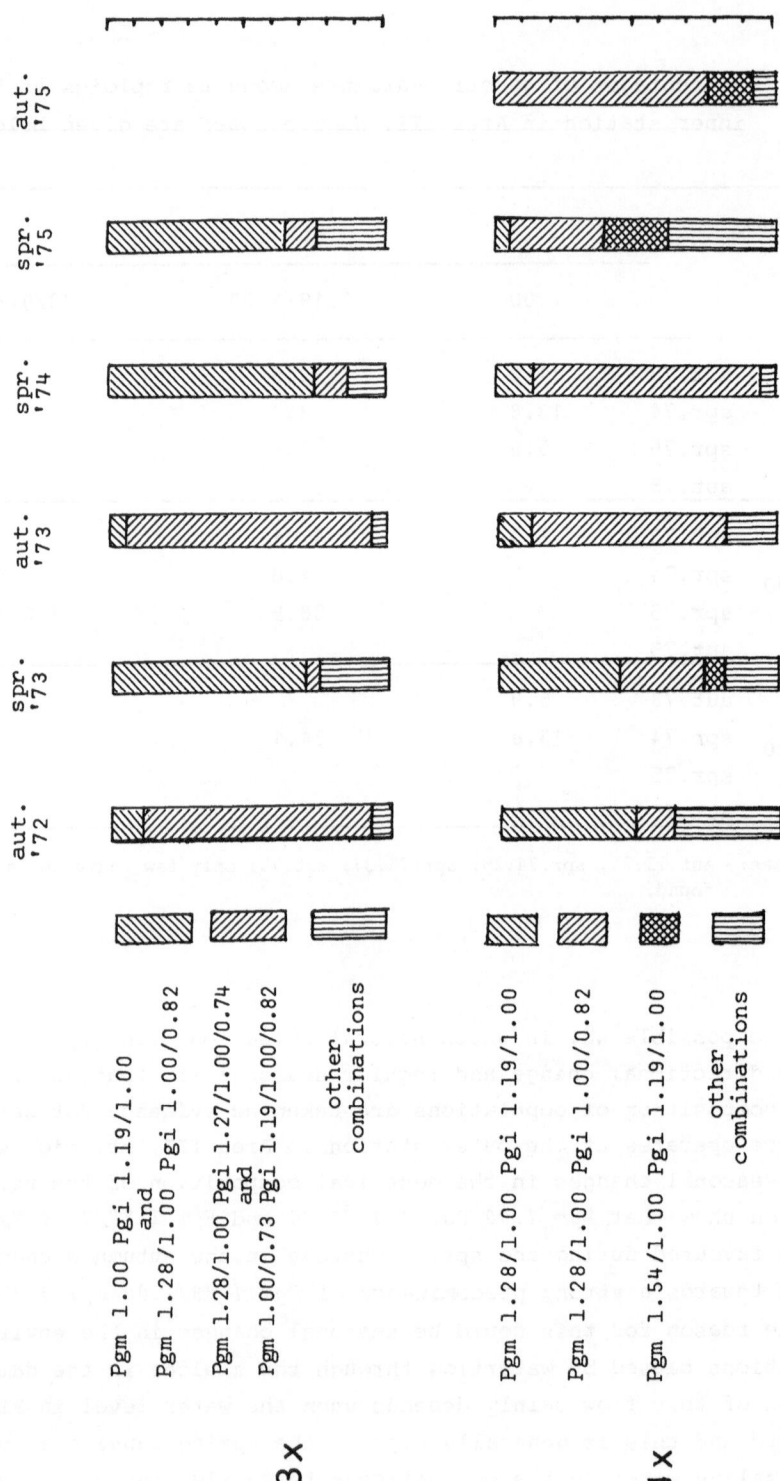

dam at this time of the year whereas later in the summer and autumn
they are more like the North Sea conditions. The fact that the major
types in the spring samples outside the dam are identical to those
that are strongly dominating on the inner side supports the causal relationship outlined. It would thus seem as though the triploid has not
succeeded in evolving a genotype adapted to the full range of environmental fluctuations at this locality.

In Area II along the west coast of Jutland amphimictic diploids
and parthenogenetic triploids occur together. The diploids in this area tend to produce a higher number of Pgi heterozygotes than in Area
I. And from Fig. 3 is seen that the difference in *Pgi* phenotypes between the diploids and triploids in Area II is mainly due to a still
higher frequency among the triploid of the most common heterozygote
phenotype in the diploid population. The diploids and triploids in
Area II thus differ in the same direction from the stations in Area
I with respect to *Pgi* and the mechanism responsible for that must be
natural selection. The higher frequency in the triploid of the most
common heterozygote in the diploid might indicate that the *Pgi* polymorphism in Area II is based upon heterozygote superiority. However,
the presence of homozygotes and other heterozygotes shows that some
other mechanism without unconditioned superiority of any phenotype
maintains the *Pgi* variation in this area.

In Area I the phenotype frequencies are very similar in sympatric
diploids and triploids. Could this high degree of identity be due to
a frequent origin of triploid individuals from the diploid form due
to some failure in meiosis or fertilization? A possibility that would
represent a great challenge to the value of the entire investigation.

The following data presented here can be used as arguments against
this possibility.

1)   In Area I there is in the majority of stations a general tendency towards a lower frequency of heterozygotes in triploids compared to sympatric diploids, cf. Fig. 2. In the extreme case of constant
prereduction and an abortive second meiotic division resulting in the
fertilization of a diploid female pronucleus, the expected distribution of triploid phenotypes is identical with that of the diploid. Oth-

Figure 4.   Schematical representation of the genetical changes observed among tri-
             and tetraploids at the outer station in Area III. (see opposite page).

er - and more likely - failures produce a higher frequency of heterozygotes in triploids, and the opposite trend is observed.

2)   The same failure in the reproduction of the diploid must occur
in Area II and here it would tend to produce a higher *Pgi* heterozygosity and a lower *Pgm* heterozygosity in the triploids compared with the
ancestral diploid, and how this could come about is difficult to see.

In addition to the high degree of identity there is in Area I a
strong dominance of a particular allele and the small variations in
allele frequencies show no obvious geographical trend. This might suggest another possibility. Is the variation perhaps mutational? If this
is the case, the relative frequencies of alternative alleles must show
the same pattern of variation in diploid and triploid populations on
the assumption of identical mutation rates. But due to the different
degree of ploidy the absolute numbers of the various alleles are different in the two forms. This means that in the present case one would
expect a general tendency for the absolute number of rare alleles to
be higher in triploid "populations" compared to diploid containing the
same number of individuals. This high number of rare alleles is, however, found in the same number of individuals and consequently, the
frequency of heterozygotes must be higher among the triploids compared to the diploids. As already mentioned above, the actual observations
do not bear out this expectation, to the contrary, the opposite trend
is found in the stations in Area I. An accumulation of selectively neutral mutations is thus unlikely to account for the variation observed,
and we are - mostly by exclusion - left with natural selection as a
possible agent.

If this holds true, and if the different degrees of ploidy make
no difference in the response of identical phenotypes to environmental
factors, then the identical distribution of phenotypes in the two forms
suggests a situation where natural selection favours a certain distribution of phenotypes in this particular environment. A heterogeneous
environment where different phenotypes exploit different resources may
form the ecological basis for such a model. Although the selective differentials might be stronger in Area II and in particular in Area III
the constant presence of several genetic variants among the parthenogenetic forms suggests a similar mechanism here.

Summary.

Polymorphism in the phosphoglucoseisomerase and phosphoglucomutase loci are studied in amphimictic diploid and parthenogenetic tri- and tetraploid *Lumbricillus lineatus*. Stations within three different areas were studied. In Area I diploids and triploids were found together; the degree of heterozygosity was low within both loci and sympatric diploids and triploids were similar with respect to frequency of identical phenotypes. In Area II where diploids and triploids were found there is a higher degree of heterozygosity in *Pgi* and the frequency of identical phenotypes differs strongly in the two forms whereas the *Pgm* locus is more like the situation in Area I. In Area III parthenogenetic tri- and tetraploids were found together; both seasonal and long-term changes in the genetic composition were observed. — It is concluded that selection is responsible for the genetic variation, and it is suggested that the selection favours a certain distribution of phenotypes in a given environment without unconditioned superiority of any phenotype.

REFERENCES

Christensen, B. 1960. A comparative cytological investigation of the reproductive cycle of an amphimictic diploid and a parthenogenetic triploid form of *Lumbricillus lineatus* (O.F.M.) (Oligochaeta, Enchytraeidae).
Chromosoma (Berl.) 11: 365-379.

Detter, J.C., Ways, P.O., Giblett, E.R., Baughan, M.A., Hopkinson, D.A., Povey, S., and Harris, H. 1968. Inherited variations in human phosphohexose isomerase. *Ann. Hum. Genet.* 31: 329-338.

Jelnes, J. 1971. The genetics of three isoenzyme systems in *Ephestia kuehniella* A. *Hereditas* 69: 138-140.

Jelnes, J. 1974. Genetics of three isoenzyme systems in *Aricia artaxerxes* F. (Lep., Rhophalocera).
Hereditas 76: 79-82.

Lokki, J., Suomalainen, E., Saura, A., and Lankinen, P. 1975. Genetic polymorphism and evolution in parthenogenetic animals. II. Diploid and polyploid *Solenobia triquetrella* (Lepedoptera: Psychidae).
Genetics 79: 513-525.

Suomalainen, E., and Saura, A. 1973. Genetic polymorphism and evolution in parthenogenetic animals. I. Polyploid curculionidae.
Genetics 74: 489-508.

3. SEX AND EVOLUTION

SELECTION AND GENETIC DIFFERENTIATION IN PARTHENOGENETIC POPULATIONS

Anssi Saura, Juhani Lokki and Esko Suomalainen

The evolutionary potential of a species is a function of the amount of
its genetic variation. According to Fisher's fundamental theorem of
natural selection the rate of increase in fitness of a species at any
time is equal to its genetic variance in fitness at that time. The notion that parthenogenesis leads to decreased variability has been stated
already by Petrunkévitch (1905). Later authors have claimed that parthenogenesis is "a blind alley of evolution" (cf. e.g. Darlington 1932;
White 1945; Fisher 1958). Parthenogenetic populations should become
genetically uniform. If they are also polyploid, new (mostly recessive)
mutations have increased difficulties in expressing themselves. Furthermore, parthenogenetic populations incorporate new, beneficial mutations at a slower rate than comparable bisexual populations (Muller
1932; Crow and Kimura 1965).

Single-gene variation in natural populations has for the past ten
years been measured by the method of gel electrophoresis of enzymes.
This method has now been applied to the study of genetic variability of
a wide range of organisms (cf. Powell 1975). The method has several advantages. When applied to parthenogenetic forms, it is subject to many
sources of error as compared with bisexually reproducing organisms.
First, the allele relationships within and between loci can not be
studied without first establishing these relationships in the bisexual
forms of preferably the same species. The results even then represent

enzyme phenotypes rather than genotypes, since nonfunctioning alleles
can be detected only in a homozygous condition. Polyploid organisms
have more than two doses of each chromosomal gene. The allele dose of
a polyploid heterozygote in each enzyme zone is difficult to quantify
with certainty, unless each allele is visible as a distinctive zone.
Yet we prefer to use the words allele and genotype in the following,
since the limitations mentioned hold true for results obtained with e-
lectrophoresis in general.

The parthenogenetic beetles studied by us have several advantages
as experimental material; they have an apomictic parthenogenesis; in
addition to a diploid bisexual race there may be several races with different degrees of polyploidy. Many of these races are geographically
widespread. When recombination needs not be considered, the effects of
migration may also be established. We compare here the genic polymorphism in a wingless, sluggish beetle (*Otiorrhynchus scaber*) with corresponding flying forms (mainly *Adoxus obscurus*). The unit of selection in an apomictic animal is the entire individual. Cytological phenomena in parthenogenetic beetles have been extensively studied by
the author Suomalainen (cf. Suomalainen 1969; Suomalainen *et al.* 1976).

MATERIAL AND METHODS

Beetles (*Otiorrhynchus scaber* and *Adoxus obscurus*) have been caught by
suddenly jerking their food plants. The beetles fall in catalepsia and
they may then be collected in a net. Most of the material is from northern Europe, but much material has also been collected in central Europe.
The bisexual *Adoxus* sample, used as a reference, is from Canada.

The insects have been brought to the laboratory alive and deep
frozen. They have thereafter been assayed by starch gel electrophoresis. The enzyme assay and electrophoresis procedures have been described earlier by Suomalainen and Saura (1973). The following enzyme systems were assayed: acid phosphatase (Acph), adenylate kinase (Adk),
amylase (Amy), aldehyde oxidase (Ao), α-glycerophosphate dehydrogenase
(α-Gpdh), esterase (Est), hexokinase (Hk), leucine aminopeptidase (Lap),
malate dehydrogenase (Mdh), malic enzyme (Me), 6-phosphogluconate dehydrogenase (6-Pgdh), phosphoglucomutase (Pgm), superoxide dismutase (Su)
and triosephosphate isomerase (Tpi). The allele relationships of enzyme
variants have been deduced by comparing enzyme patterns of parthenoge-

netic forms with bisexual forms of the same species. Enzyme patterns were furthermore compared with corresponding *Drosophila* patterns, which were ascertained by crossing flies.

RESULTS

The enzymes identified in this study are designated by the abbreviations given in the preceding section. The patterns observed in the variable enzymes of the diploid bisexual samples conform to those expected for diploid organisms on the assumption of a Hardy-Weinberg equilibrium. We therefore assume that each variable enzyme is controlled by a different locus. The invariant zones are also assumed to be controlled by invariable loci. The enzyme variants found in the parthenogenetic individuals are given allele and locus symbols by comparing their enzyme phenotypes with those observed for the diploid bisexual race.

When several enzymes appear in an assay system, a hyphenated number is added to the symbol of the enzyme and the corresponding locus. The zone of activity with least anodal migration is called 1, the next is 2, etc. At each locus one allele has been arbitrarily named 1.00. All other alleles are designated by reference to that standard, adding to or subtracting from 1.00 the number of millimeters by which the migration of the enzyme coded by each allele differs from that standard.

We have studied the genetic constitution of each individual with regard to about twenty enzyme loci. A parthenogenetic individual is either homozygous or heterozygous at each locus. In some cases the heterozygosity may be rather complex so that apparently three or four alleles are present at a locus in a single tetraploid individual. When the activity of only two alleles can be noticed in a polyploid individual, there is in general reason to suspect that the set of bands staining stronger results from the activity of more than one dose of an allele. When the results obtained for each locus for each parthenogenetic individual are combined we obtain the genotype of this particular individual over the loci studied. In the remainder of this paper this will be called the overall genotype characterizing each individual. These genotypes may also be called clones or biotypes but we prefer the word overall genotype, since it represents a qualitative description of its allelic composition.

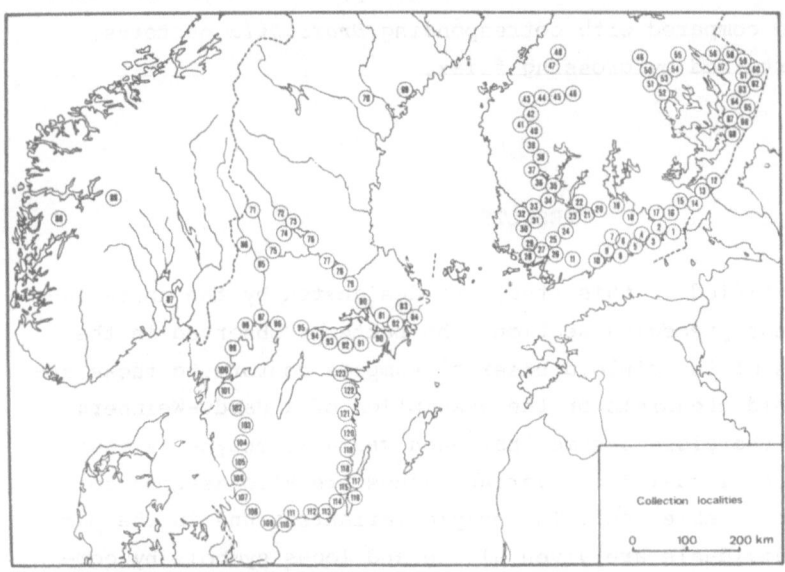

Figure 1. The collection localities of the tetraploid northern European *O. scaber* samples (Saura *et al.* 1976a).

Otiorrhynchus scaber. We have collected a total of 482 tetraploid individuals from Scandinavia. The collection localities are shown in Figure 1. In general four weevils were collected from each locality. Nine loci were monomorphic and all weevils had an identical allele configuration at these loci. Three of these monomorphic loci represented cases of permanent heterozygosity.

A total of 75 different overall genotypes have been recognized in the material. A total of 27 of them have been found to occur in more than one population. Figure 2 shows the number of populations in which a genotype occurring in more than one population and designated with a Roman number has been found as well as the number of individuals having that overall genotype in the total material.

Half of the total number of weevils studied belong to three genotypes, I, VII and XVI. The geographic distribution of these three genotypes is illustrated in Figures 3-5. The remaining genotypes are rarer; fourteen of them have been found only in two or three populations. The distribution of the less frequent genotypes is given in detail in Saura *et al.* 1976a.

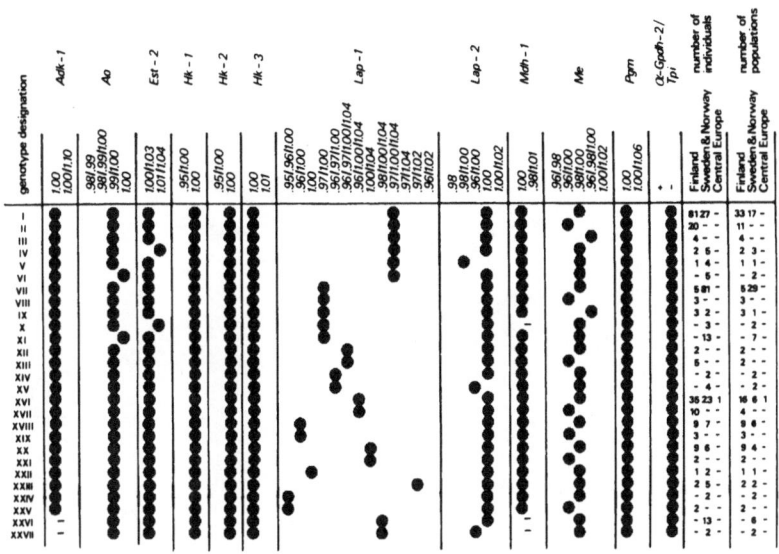

Figure 2. A compilation of the genotypes occurring in more than one population, which gives the symbol of the genotype, the overall genotype with regard to variable loci and the number of individuals having the overall genotypes in different major areas as well as the number of populations, in which the genotype has been found within each major area. Dashes indicate missing data. Modified from Saura *et al.* 1976a.

Adoxus obscurus. Triploid parthenogenetic *Adoxus obscurus* beetles were collected at 52 localities in Scandinavia (Figure 6). In order to establish the variability within a population over 30 beetles were assayed for two localities. Figure 7 shows the variation at loci, which are polymorphic in the triploid parthenogenetic beetles. The data for monomorphic loci are not included. The genotypes occurring in more than one population are each identified with a Roman number.

An inspection of Figure 7 shows that with the exception of five localities Type I occurs throughout the area studied. Altogether 261 out of the 328 beetles assayed (i.e. 80%) belong to this type. Type II has been found in 13 individuals, which originate from six populations. Five of these localities are in Sweden and one in Finland. Type III occurs in 34 individuals but only in three localities. These localities are in eastern central Finland. Type IV has been encountered in 11 individuals, which originate from four populations. These

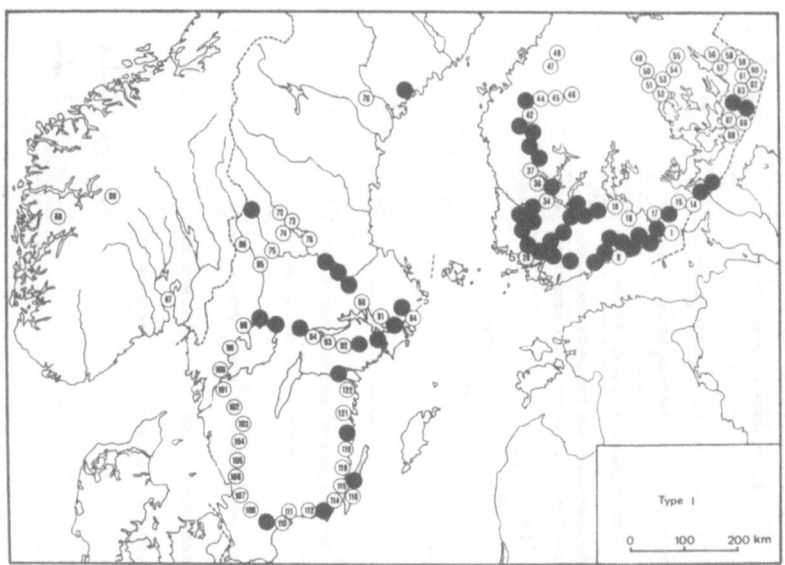

Figure 3.   The distribution of certain genotypes in the northern European localities.
A black circle indicates a locality, where the genotype identified with a Roman number has been found (Saura *et al.* 1976a).

localities are the ones from southeastern Finland. Only nine individuals represent allele combinations, which do not belong to the four
types mentioned above. The distribution of the less frequent types
(II through IV) is shown in Figure 8.

DISCUSSION

Many of the *Otiorrhynchus scaber* genotypes have widespread and overlapping distributions. The parallel occurrence of many different genotypes and the amount of variation within a single population has been
previously established (Suomalainen and Saura 1973). The overlapping
distributions can be taken as indicating either that the original
colonizing parthenogenetic population has been polymorphic or that
the distributions reflect successive phases of colonization. Presu-

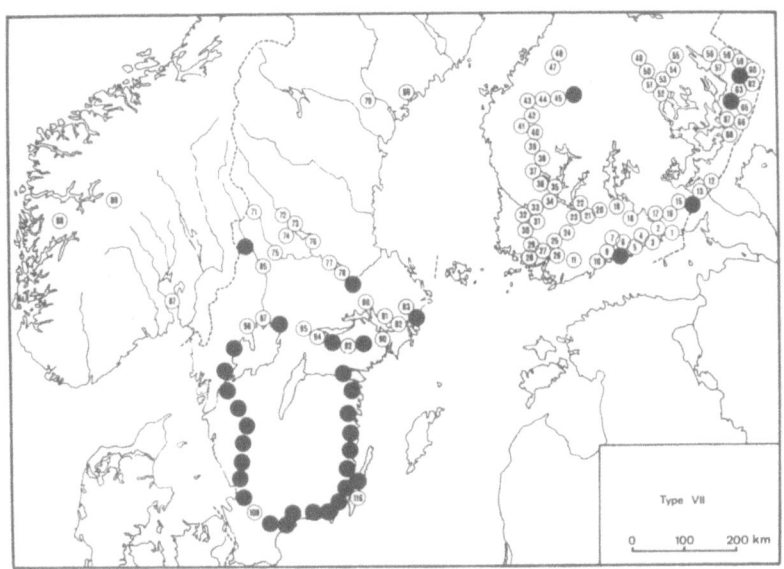

Figure 4.   See Figure 3.

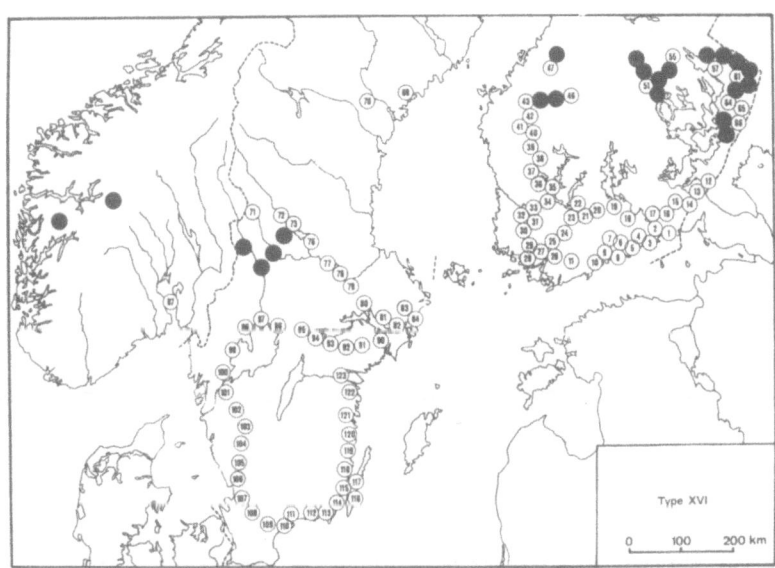

Figure 5.   See Figure 3.

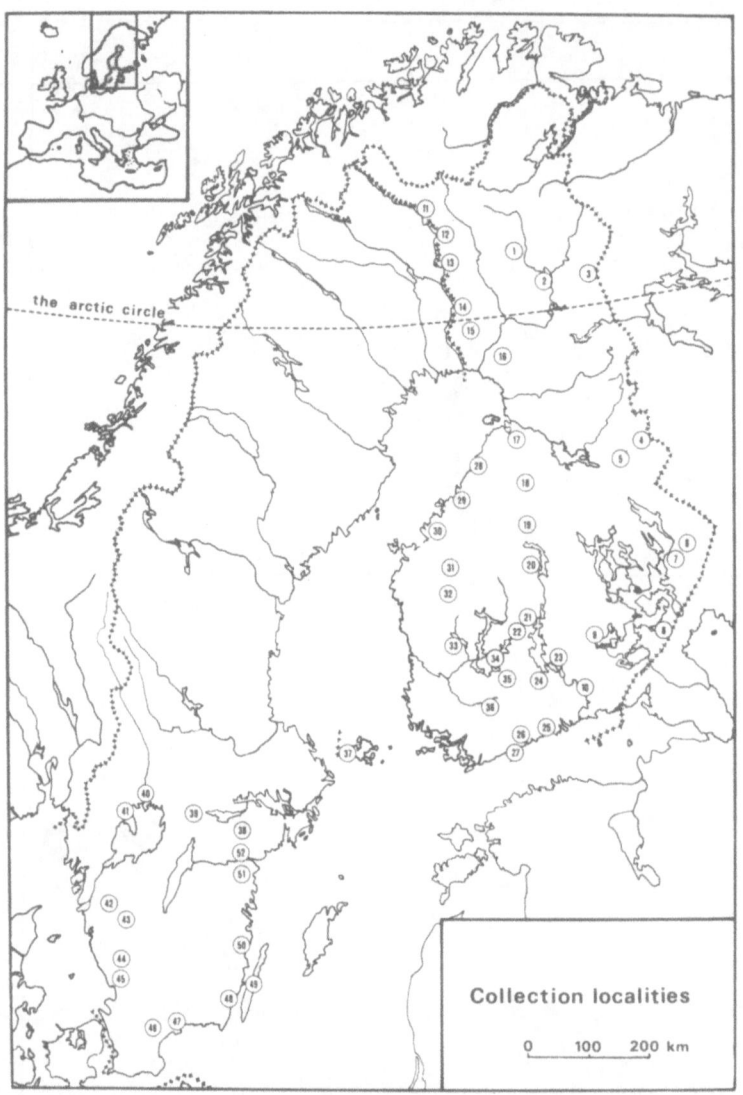

Figure 6.   The collection localities of parthenogenetic *Adoxus obscurus* samples in
northern Europe (Lokki *et al.* 1976a).

mable these two processes have acted together. Their end result is
the simultaneous occurrence of several genotypes in a single population. Our method of taking a small sample only from each population
does not, of course, indicate the total amount of variation within
the population. It indicates that the areas of occurrence of the geno-

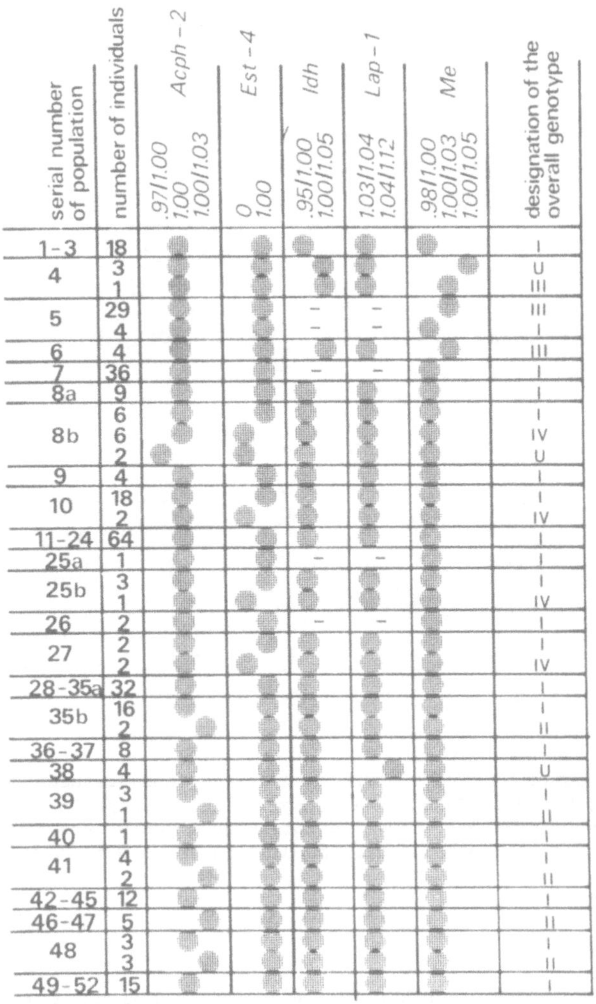

Figure 7. Serial number of population, number of individuals studied of each genotype within the population and the overall genotype of each individual in triploid
parthenogenetic populations of *A. obscurus*. Invariable loci are not included. The
overall genotypes are identified with a Roman number when they occur in more than
one population. U denotes unique overall genotype.

types can be defined and that the relative abundance of these genotypes differs from region to region.

Half of the total number of weevils studied belonged to three genotypes, I, VII, and XVI. These three genotypes are, accordingly, the

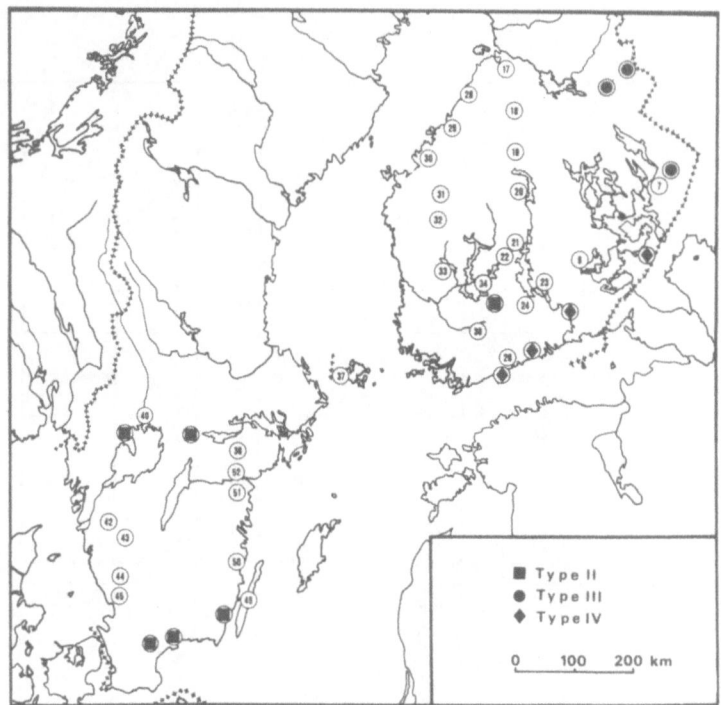

Figure 8. The distribution of the less frequent genotypes in the area studied. Type I has been found in all populations, except in 4, 6, 38, 46 and 47. In the northern-most area (not shown in this Figure) only Type I has been encountered. The single genotype found only in population 38 is not shown.

most important ones, when we consider the geographic distribution of different types. The distributions of the three types do not appreci-ably overlap (cf. Figures 3, 4 and 5). The distribution of these types is remarkably well in accord with the vegetation zones of north-western Europe illustrated in Figure 9. The zones have been drawn following Ahti *et al.* (1968), but the map corresponds also in almost every detail with the biotic zonation of Sjörs (1963), even though different authors apply different names to the zones. Here we use the names proposed by Ahti *et al.*

Genotype XVI occurs in the central boreal zone in Finland and Swe-den, and, curiously enough, in an approximately corresponding area on the coast of Norway. A single individual belonging to this type has been found in Zürichberg in Switzerland. It may either have a similar

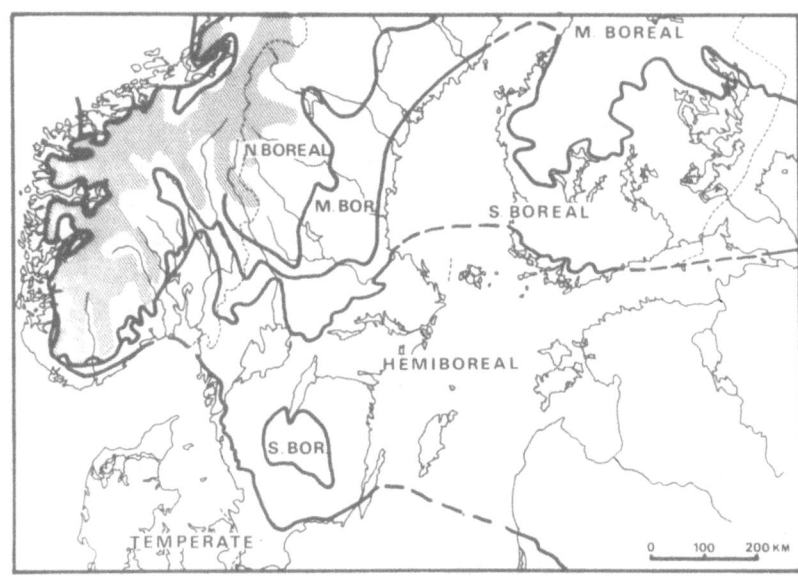

Figure 9. The biotic zones of northwestern Europe according to Ahti *et al*. (1968).
The solid lines indicate zonal boundaries and the shading alpine zone.

phenotype only, or it reflects a distribution of Type XVI through Europe at habitats in increasing altitudes towards the south.

Type I has a southern boreal distribution in Sweden and Finland, but it also occurs in some hemiboreal populations. The distribution of the major type VII is hemiboreal. The distributions of the less frequent types are not presented here. The more common of them show also clear-cut regularities correlated with biotic factors (cf. Saura *et al*. 1976a). A limited distribution may indicate that a form has recently arisen through mutation. It may also give evidence of its migration history.

The geographic distribution of different genotypes is, accordingly, correlated with the biotic zones of northwestern Europe. The differentiating characteristics are mainly allele configurations at two loci, *Lap* and *Me*. The *Me* allele *0.96* occurs in the most continental parts of our area of study; *Lap-1* allele *0.96* is found in the middle boreal zone; allele *0.97* is southern boreal in a combination $Lap\text{-}1^{0.97/1.00/1.04}$; and hemiboreal in a combination $Lap\text{-}1^{0.97/1.00}$.

It is, of course, true that a correlation between a genetic character and some environmental factor can always be demonstrated. It is
also clear that similar phenotypes have probably originated in different parts of the distribution area of the tetraploid race of *O. scaber*.
However, the correlations between the major biotic factors and the
major genotypes of tetraploid *O. scaber* are so numerous and evident
that we may conclude that different genotypes are adapted to different environmental conditions.

We have previously, (Lokki *et al.* 1975) demonstrated that the distribution of different genotypes of a flightless parthenogenetic moth
(*Solenobia*) can be explained by their different origin and migration
histories. *O. scaber* is likewise a flightless insect. The active migration of this species is hardly more effective than that of *Solenobia*. The tetraploid race of *O. scaber* must have colonized northern
Europe recently, i.e. not earlier than about ten thousand years ago
at the close of the Ice Age. The species lives, at least as an adult,
mainly on the Norway spruce (*Picea abies*), which is an ubiquitous
native tree in northern Europe. Spruce trees have been to some extent moved by man. This does not, however, explain the distribution
of *O. scaber* genotypes.

The tetraploid genotypes of *O. scaber* represent balanced and apparently succesful genetic complexes. As such they are good examples
of permanent heterozygosity. Their adaptability is at least in part
attributable to heterosis. The differences between and within populations are attributable to mutation. These mutations have occurred and
established themselves in the populations since the origin of parthenogenetic reproduction. This notion was first put forward on morphological evidence by Suomalainen (1961).

The fact that parthenogenetic populations are polymorphic presents
a case where a single (and probably limited) resource is exploited
by coexisting separate genotypes. According to the mutual exclusion
principle, separate genotypes with identical niches can not coexist -
a point stressed by White (1970). Different ecological niches alternating, e.g. mosaic-fashion in space, would present a case optimally
exploited by a polymorphic population. The niche differences are presumably not seasonal, since the development of *O. scaber* takes two
years in northern Europe. A two-year cycle subdivided into regularly
recurring temporal niches is improbable. The geographically differentiated distribution of several genotypes indicates that some of the
niche parameters are geographically separate. The species consists of

an array of genotypes, comparable to geographic races. The polymorphism is attributable to environmental changes, to each of which a population responds by a rapid multiplication of the best existing genotypes.

In comparison with two flightless parthenogenetic insects, *Solenobia triquetrella* (Lokki *et al.* 1975) and *Otiorrhynchus scaber*, parthenogenetic *Adoxus obscurus* is genetically highly monomorphic. *S. triquetrella* populations differ drastically from each other, and all evidence indicates that they are polyphyletic. In contrast to them, *O. scaber* genotypes differ from each other slightly, so that the differences can be explained by single mutations. *A. obscurus* genotypes represent very clearly a monophyletic lineage.

The major food plant of Scandinavian *Adoxus*, the fireweed, is a circumpolar plant species. It is a characteristic element of the secondary succession, which appears almost everywhere where the soil nitrogen content is elevated (e.g. following fires). *Adoxus* must be able to follow its vagrant host plant. When we consider the distribution of the minor genotypes III and IV, it becomes apparent that the geographic distributions do not reflect migration barriers. The reasons for this phenomenon may be assumed to be adaptive differences at the genotypes (e.g. Types III and IV may well favor a more continental climate in comparison with Type I). The success of Type I can be explained by the assumption that it has superior fitness in Scandinavia. Whenever local genotypes originate by mutation they are in general not capable of competing with Type I and are therefore eliminated within this area. Population 38 is an example of local differentiation in our material. The population structure of an actively migrating parthenogenetic insect is very different from that of the flightless weevils. *Adoxus* lives on a weedy host plant species, which probably also exhibits very little genetic differentiation along geographical transects. This is in our opinion the explanation to the low number of different *Adoxus* genotypes in northwestern Europe.

The populations of flying parthenogenetic weevil, *Polydrosus mollis*, have differentiated in central and northern Europe into numerous genotypes, which are geographically discrete (Lokki *et al.* 1976b). *P. mollis* lives on beech (*Fagus*), oak (*Quercus*) and hazel (*Corylus*). In spite of the fact that these trees and bushes are common in central Europe, they have been removed from most natural biotopes by human activity. *Polydrosus mollis* does not form continuous populations in

central Europe, but occurs here and there. In northern Europe these
trees and bushes are comparatively rare and they can be found only on
the best soils. The plant communities inhabited by *P. mollis* are stable
(in fact they are climax communities in central Europe) in comparison
with the unstable environment of *Adoxus obscurus*. The permanence of
the environment does not impose continuous migration on *P. mollis* populations, as is the case in *Adoxus*. Therefore *P. mollis* populations
are isolated from each other quite effectively. The ecological requirements of *P. mollis* are stringent at least in northern Europe. Migration does not level off differences arising by mutation in local populations of this flying weevil. Environmental diversity has presumably promoted the establishment of this polymorphism, much of which
is probably adaptive.

The parthenogenetic Curculionids are aneuploid to a certain extent (Suomalainen 1940). This aneuploidy has been offered as a partial
solution to parallel allele configurations observed in different lineages of tetraploid *Otiorrhynchus scaber*. Mutation is, however, the
mechanism generating new alleles in an apomictically parthenogenetic
lineage. A certain proportion of mutations lead to enzyme phenotypes
resembling a homozygous condition. If we ignore back mutations, mutations leading to nonfunctioning alleles may happen rather freely in
an organism without recombination. In fact, we have some evidence of
homozygosity for silent alleles in *Adoxus*. Mutations leading to functioning heterozygosity should be more numerous than ones leading to
homozygosity. There is, indeed, much evidence for permanent heterozygosity in our material. Parthenogenetic populations should accumulate
mutations and have elevated degrees of heterozygosity in comparison
with bisexual populations. Table 1 shows the degrees of heterozygosity per locus per individual in polyploid parthenogenetic insect populations.

Solenobia has an automictic parthenogenesis, and the origin of
polyploidy is different from that in beetles with apomictic parthenogenesis. When there is no recombination, the automixis of *Solenobia*
is functionally equivalent to apomixis.

An inspection of Table 1 shows that the heterozygosity has, indeed,
increased in polyploid parthenogenetic beetles with the exception of
Otiorrhynchus scaber. The degree of heterozygosity of diploid bisexual
O. scaber is, however, exceptionally high. The causes underlying this
high diploid heterozygosity are poorly understood (Suomalainen and

Saura, Lokki and Suomalainen

Table 1. The observed degrees of heterozygosity per locus per individual in different diploid and polyploid populations of parthenogenetic insects.

Species	Bisexual race	Parthenogenetic races		
		2n	3n	4n
Adoxus obscurus (Lokki *et al.* 1976a)	0.18		0.34	
Polydrosus mollis (Lokki *et al.* 1976b)	0.14	0.36	0.37	
Otiorrhynchus scaber (Suomalainen and Saura 1973; Saura *et al.* 1976a)	0.31		0.25	0.38
O. salicis (Saura *et al.* 1976b)	0.12		0.24	
O. singularis (Suomalainen and Saura 1973)			0.37	
Strophosomus melanogrammus (Suomalainen and Saura 1973)			0.30	
Solenobia triquetrella (Lokki *et al.* 1975)	0.23	0.23		0.23

Saura 1973). As for *Solenobia*, heterozygosity values for each of the three types and two degrees of polyploidy are equal. This, of course, fits well with the postulated origin of parthenogenesis in *Solenobia* (Seiler 1961). As for weevils with two different degrees of polyploidy, the results for *P. mollis* do not support the contention that triploidy has originated by a chance fertilization of a diploid parthenogenetic female. In addition to the circumstance that the degrees of heterozygosity are virtually identical in diploid and triploid parthenogenetic races, their enzyme phenotypes are also very similar (cf. Lokki *et al.* 1976b).

Due to the accumulation of new mutations, animals with apomictic
parthenogenesis should become completely heterozygous (a view expressed e.g. by Darlington 1937; Suomalainen 1950 and White 1945). When
there is free recombination, parthenogenesis should again result in
complete homozygosity, unless opposed by selection (this view is mathematically treated by Asher 1970; Asher and Nace 1971 and Templeton and
Rothman 1973). Our studies have also shown influences of selection, e.g.
the original genotype is maintained at many loci.

When we consider mutations occurring in a parthenogenetic lineage,
we must consider their effects on the fitness of the mutant individuals. If we exclude dominant lethals, a majority of mutations lead to
the formation of either completely identical enzyme molecules, or molecules, which, even though they are slightly altered, function approximately as well as the original enzyme. A certain proportion of these
molecules function better, leading to a higher fitness, while those,
which function unsatisfactorily may lower the fitness of the individual concerned. A relatively high proportion of mutations (e.g. frame
shifts, duplications, deletions, nonsense mutations, mutations affecting a sensitive point of the molecule) produce either no enzyme at all
or defective enzyme molecules. Because of the functional reserve of
enzyme function, which a diploid organism has, nonfunctioning alleles
do not need to lower its fitness (see e.g. Harris 1975 for examples).
We must, of course, not assume that mutations are always one-way e-
vents; there are back mutations, but the back mutation rate is negligible.

We may briefly consider the amounts of heterozygosity in a diploid
and polyploid parthenogenetic lineage as a result of recurrent mutation as well as the accumulation of nonfunctioning alleles. Nonfunctioning is defined here as a recessive character; nonfunctioning alleles do not produce appreciably any functioning protein. The effects
of selection are not considered in calculating the amounts of heterozygosity in different levels of ploidy.

On the basis of the assumptions mentioned above we may calculate
the degrees of heterozygosity in apomictically parthenogenetic lineages. According to a deterministic model, the number of heterozygous
loci (H) as a function of generations in a diploid lineage is (both
alleles are functioning)

$$H = Ne^{-vt} - ne^{-(u+v)t},$$

where u is the mutation rate to functioning and v to nonfunctioning alleles. N is the total number of loci per individual and n is the number of loci with two similar alleles, when t = 0.

When at least two functioning alleles are present, the number of heterozygous loci is the following in a triploid case

$$H = (n_2+3n_1)[e^{-(u+\frac{3}{2}v)t} - e^{-(u+v)t}] - n_1e^{-\frac{3}{2}(u+v)t} + N(3e^{-vt} - 2e^{-\frac{3}{2}vt}),$$

where n_1 and n_2 represent different allele combinations, when t = 0 (cf. Lokki 1976b).

In a tetraploid case we have

$$H = N[3e^{-2vt} - 8e^{-\frac{3}{2}vt} + 6e^{-vt}] + (n_2+4n_1)[e^{-(\frac{3}{2}u+2v)t} - e^{-\frac{3}{2}(u+v)t}] +$$

$$-(n_4+2n_3+3n_2+6n_1)[e^{-(u+2v)t} + e^{-(u+v)t} - 2e^{-(u+\frac{3}{2}v)t}] +$$

$$-n_1e^{-2(u+v)t},$$

where n_1, n_2 etc. again represent different allele combinations when t = 0 (cf. Lokki 1976b). These functions are illustrated in Figures 10 and 11.

In order to study the theoretical variance of heterozygosity, a stochastic model has been constructed in the diploid case (Lokki 1976a). Then we have the following probability generating function

$$P(x,y,t) = [(x-1)e^{-vt}+1]^{N-n}[(x-1)e^{-vt}+(y-x)e^{-(u+v)t}+1]^{n},$$

the power series expansion of which gives the probabilities for any zygosity condition in a parthenogenetic lineage. The standard deviations of heterozygosity thus obtained are presented in Figure 10.

When we compare the polyploid cases with the diploid case we note that the amount of functioning heterozygosity increases very markedly within the same period of time. A transition from triploidy to tetraploidy does not yield much more functioning heterozygosity than that already obtained by a transition to triploidy. Triploidy is, in fact, the most common degree of ploidy in parthenogenetic weevils. Of the 54 parthenogenetic races and species of these insects studied this far, two are diploid, 31 are triploid, 14 are tetraploid, five are pentaploid and two are hexaploid (Suomalainen *et al.* 1976).

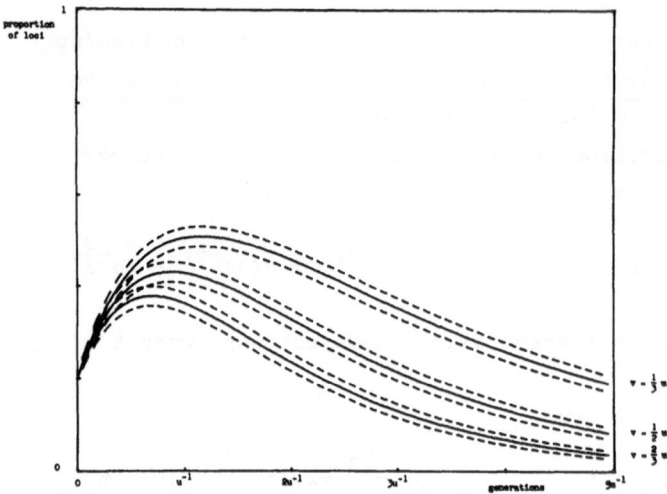

Figure 10. The amount of functioning heterozygosity (with two functioning alleles)
in a diploid parthenogenetic lineage. The dots give the theoretical standard deviations multiplied with two. This gives a confidence level of 95%. Initial heterozygosity is 0.2. u is the mutation rate to a functioning allele and v is the mutation
rate to a nonfunctioning allele (Lokki 1976a)

Even though polyploidization gives a parthenogenetic population
possibilities for an increased genetic homeostasis, the accumulation
of nonfunctioning alleles will later produce diverse physiological
disadvantages. Such disturbances can be easily visualized in metabo-
lically important and branching biochemical pathways. In them substrate concentrations and reversibility potentials are crucial for
the functioning of the metabolism as a whole. A disturbance in the mutual quantities of genes coding for enzymes may later prove to be
an unbearable physiological stress. This will be the cost that the
parthenogenetic population must pay for its increased heterozygosity
with concomitant adaptive advantages. The whole nature of the accumulation process in apomictically parthenogenetic forms is very different from the effects of deleterious alleles in populations with
recombination.

The advantages conferred by an acquisition of polyploidy to a
parthenogenetic lineage are twofold. First it gives the population
more genetic adaptability by increasing the functioning heterozygosity.
Another advantage is that polyploidy has a buffering effect against
mutations producing nonfunctioning alleles. This also gives a poly-

a)

b)

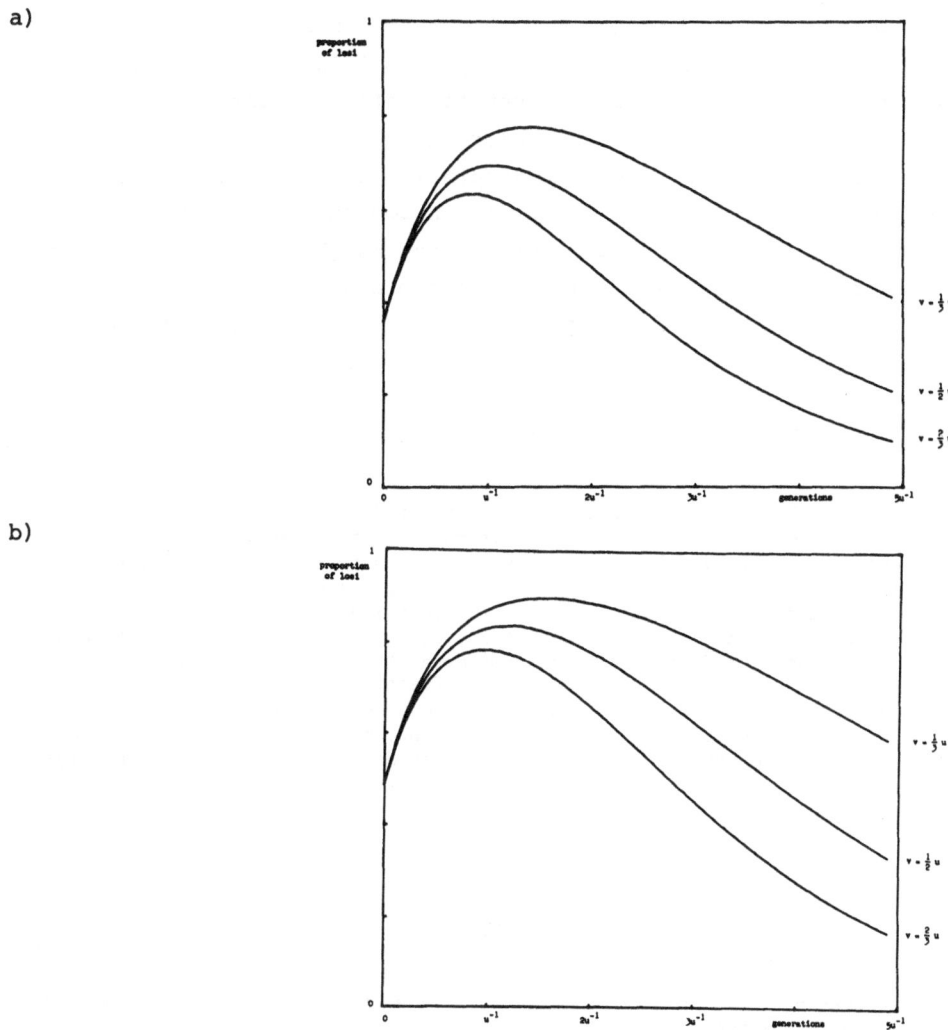

Figure 11.   The amount of functioning heterozygosity a) in a triploid and b) in a tetraploid parthenogenetic lineage. The initial heterozygosity in diploids is 0.2; u and v as in Figure 10 (Lokki 1976b).

ploid population an increased life expectancy in relation to a diploid
parthenogenetic population. As shown by Suomalainen (1940) apomictically parthenogenetic populations can tolerate considerable aneuploidy.
These considerations support the observation that parthenogenetic animals are capable of evolution and adaptation to the environment. Yet
they do not have a long-term advantage in competition with animals
reproducing by the union of gametes - even though they may be succesful for very long periods of time. This explains, in part, the absence of speciation and further divergence of parthenogenetic forms.

Summary.

   Genic polymorphism at enzyme loci has been studied in different insects with
apomictic parthenogenesis. The populations of the wingless weevil, *Otiorrhynchus
scaber*, have diverged into numerous genotypes in northern Europe. This diversity
is due solely to mutations, which have occurred after the establishment of the
parthenogenetic mode of reproduction. The geographical distributions of these genotypes are correlated with major biotic factors. The flying beetle, *Adoxus obscurus*,
is genetically highly uniform within the same area. This is taken to indicate that
nonmobile parthenogenetic forms may adapt very effectively to their environment. In
the absence of recombination, this is accomplished solely by mutations and selection. The effects of mutations on the life expectancy and heterozygosity of an apomictically reproducing monophyletic lineage is considered in connection with polyploidy.

REFERENCES

Ahti, T., Hämet-Ahti, L., and Jalas, J. 1968. Vegetation zones and their sections
   in northwestern Europe.
   Ann. Bot. Fenn. 5: 169-211.

Asher, J.H. 1970. Parthenogenesis and genetic variability. II. One-locus models
   for various diploid populations.
   Genetics 66: 369-391.

Asher, J.H., and Nace, G.W. 1971. The genetic structure and evolutionary fate of
   parthenogenetic Amphibian populations as determined by Markovian analysis.
   Am. Zool. 11: 381-398.

Crow, J.F., and Kimura, M. 1965. Evolution in sexual and asexual populations.
   Amer. Natur. 99: 439-450.

Darlington, C.D. 1932. *Recent advances in cytology*. 1st Edition. Churchill, London.

Darlington, C.D. 1937. *Recent advances in cytology*. 2nd Edition. Blakiston's Sons
   and Co. Inc., Philadelphia.

Fisher, R.A. 1958. *The genetical theory of natural selection*. 2nd Edition. Dover
   Publications, New York.

Harris, H. 1975. *The principles of human biochemical genetics*. North-Holland/American Elsevier, Amsterdam.

Lokki, J. 1976a. Genetic polymorphism and evolution in parthenogenetic animals. VII. The amount of heterozygosity in diploid populations. *Hereditas* 83: 57-64.

Lokki, J. 1976b. Genetic polymorphism and evolution in parthenogenetic animals. VIII. Heterozygosity in relation to polyploidy. *Hereditas* 83: 65-72.

Lokki, J., Saura, A., Lankinen, P., and Suomalainen, E. 1976a. Genetic polymorphism and evolution in parthenogenetic animals. V. Triploid *Adoxus obscurus* (Coleoptera: Chrysomelidae). *Genet. Res.* 28: 27-36.

Lokki, J., Saura, A., Lankinen, P., and Suomalainen, E. 1976b. Genetic polymorphism and evolution in parthenogenetic animals. VI. Diploid and triploid *Polydrosus mollis* (Coleoptera: Curculionidae). *Hereditas* 82: 209-216.

Lokki, J., Suomalainen, E., Saura, A., and Lankinen, P. 1975. Genetic polymorphism and evolution in parthenogenetic animals. II. Diploid and polyploid *Solenobia triquetrella* (Lepidoptera: Psychidae). *Genetics* 79: 513-525.

Muller, H.J. 1932. Some genetic aspects of sex. *Amer. Natur.* 66: 118-138.

Petrunkévitch, A. 1905. Natural and artificial parthenogenesis. *Amer. Natur.* 39: 65-76.

Powell, J.R. 1975. Protein variation in natural populations of animals. *Evol. Biol.* 8: 79-119.

Saura, A., Lokki, J., Lankinen, P., and Suomalainen, E. 1976a. Genetic polymorphism and evolution in parthenogenetic animals. III. Tetraploid *Otiorrhynchus scaber* (Coleoptera: Curculionidae). *Hereditas* 82: 79-100.

Saura, A., Lokki, J., Lankinen, P., and Suomalainen, E. 1976b. Genetic polymorphism and evolution in parthenogenetic animals. IV. Triploid *Otiorrhynchus salicis* (Coleoptera: Curculionidae). *Ent. Scand.* 7: 1-6.

Seiler, J. 1961. Untersuchungen über die Entstehung der Parthenogenese bei *Solenobia triquetrella* F.R. (Lepidoptera: Psychidae). III. *Z. Vererbungsl.* 92: 261-316.

Sjörs, H. 1963. Amphi-Atlantic zonation. Nemoral to Arctic, pp. 109-125. In *North Atlantic Biota and their History*. Edited by A. Löve and D. Löve. Pergamon Press, Oxford.

Suomalainen, E. 1940. Beiträge zur Zytologie der Parthenogenetischen Insekten. I. Coleoptera. *Ann. Acad. Sci. Fenn. Ser.* A LIV 7: 1-145.

Suomalainen, E. 1950. Parthenogenesis in animals. *Advan. Genet.* 3: 193-253.

Suomalainen, E. 1961. On morphological differences and evolution of different polyploid parthenogenetic weevil populations. *Hereditas* 47: 309-341.

Suomalainen, E. 1969. Evolution in parthenogenetic Curculionidae. *Evol. Biol.* 3: 261-296.

Suomalainen, E., and Saura, A. 1973. Genetic polymorphism and evolution in par-
    thenogenetic animals. I. Polyploid Curculionidae.
    Genetics 74: 489-508.

Suomalainen, E., Saura, A., and Lokki, J. 1976. Evolution in parthenogenetic in-
    sects.
    Evol. Biol. 9: 209-257.

Templeton, A.R., and Rothman, E.D. 1973. The population genetics of parthenogene-
    tic strains of *Drosophila mercatorum*. I. One locus model and statistics.
    Theoret. Appl. Genet. 43: 204-212.

White, M.J.D. 1945. *Animal cytology and evolution*. 1st Edition. Cambridge University
    Press, Cambridge.

White, M.J.D. 1970. Heterozygosity and genetic polymorphism in parthenogenetic
    animals. pp. 237-262.
    In *Essays in Evolution and Genetics in Honor of Theodosius Dobzhansky*. Edited
    by M.K. Hecht and W.C. Steere. Appleton-Century-Crofts, New York.

Chapter 4. Ecology and Evolution

CACTUS-BREEDING *DROSOPHILA* - A SYSTEM FOR THE MEASUREMENT OF NATURAL
SELECTION

J.S.F. Barker

"Only by studying the ecology of natural and artificial populations can
we come to understand the forces or factors that act on gene frequencies within them" (Clarke, 1975). The need for the joint consideration
of population ecology and population genetics is not a new idea, but as
Clarke said, the point is "so obvious and elementary that ..... it has
often been overlooked". In the study of evolution, we would hope to understand, at the intra-specific level, the factors determining both the
numbers of organisms and the kinds (genotypes) of organisms and the nature of any interactions between numbers and kinds (Birch, 1960). At
the community level, we ask similar questions, but have to include consideration of higher-order interactions between numbers and kinds of
different species.

In recent years, a start has been made to the development of theory
incorporating both genetic and ecological parameters. Most attention
has been devoted to density-dependent selection (Anderson, 1971; Roughgarden, 1971; Clarke, 1973), evolution of the niche (Levins, 1968;
Roughgarden, 1972), and the evolution of life-history strategies (Lewontin, 1965; Cohen, 1967; King and Anderson, 1971; Mertz, 1971; Charlesworth, 1973). Pianka (1974) has discussed these topics, and the theory has been reviewed and extended by Christiansen and Fenchel (1977).

In experimental analyses of natural populations, there are few examples where genetic and ecological parameters have been measured in

the one study. In early studies, Fisher and Ford (1947), and Ford and Sheppard (1969) estimated the frequency of the *medionigra* gene and population numbers in an isolated colony of the moth *Panaxia dominula*, and have shown selection coefficients to vary from year to year, while Gershenson (1945) showed that there was a positive correlation between population numbers and the frequency of melanic forms in the hamster *Cricetus cricetus*. More recently, correlations between genetic changes and changes in population numbers have been demonstrated in small mammals (see reviews of Krebs *et al.*, 1973; Smith *et al.*, 1975). In these cases, density-dependent selection for, or a dispersal of, different genotypes is hypothesised to cause changes in population numbers. The emphasis in all these studies has been on population dynamics of individual populations and temporal changes in genetic composition; they do not encompass a full integration of genetics and ecology.

Recent studies where allozyme frequencies have been estimated in a number of different populations of one species present an alternative approach. Here, ecology does impinge on genetic interpretation, but it is often *a posteriori*. That is, the study is primarily genetic in defining the magnitude of genetic variation, but then environmental factors are considered in attempting to interpret observed patterns of variation. The geographical distributions of gene frequencies or average heterozygosity may suggest selective factors of possible importance, but the ecological content of the interpretation is usually rather fortuitous. Of course, once possible selective forces are defined in this way, further study may lead to a closer integration of genetic and ecological parameters, although this may be precluded, or at least delayed, because of lack of knowledge of the ecology of the species.

In the context of the topic of this Symposium, our question relates to the measurement of natural selection. This implies a number of steps in the investigative process. Suppose we are considering a single polymorphic locus, then the question resolves to the following:

(1) Is selection acting directly on the locus of interest?

(2) Or, are any differences in gene frequency between populations or any changes within a population due to linkage disequilibrium?

(3) If selection is indeed acting on the locus of interest, what is the nature of the selection process in terms of the life-history and ecology of the natural population?

(4) Whether or not selection is acting, how important are migration or drift in influencing patterns of gene frequency in space or in time? By phrasing the problem in this way, one recognises the ecological im-

plications of what is primarily a genetic question, and realises implicitly the need for ecological information if a full understanding is
to be obtained.

With this recognition, one can consider the total program of possible experimentation, with the first problem being the choice of an
organism. Studies in this area probably have not been initiated in this
way very often. This is not to say that simply following one's curios-
ity may not prove valuable and informative, but that even where a particular species is chosen because of interest in some aspect of its
biology, consideration still should be given to its possible advantages
and limitations, so that an optimum program may be defined.

Therefore, assuming that such a beast might exist, what are the requirements of an ideal organism, and what might we need to know, or at
least, be able to find out concerning its biology? In terms of studies
of allozyme variation in a sexually-reproducing, outbreeding species,
these requirements and associated implications can be summarized as
including:

(1)   Species distribution - preferably widespread or at least existing
in a range of environments in definable populations, so that different
populations are subject to different sets of environmental conditions.
Implies knowledge of -
    (i)   factors controlling distribution
          Environmental - climatic, barriers to dispersal
          Biotic - interspecific competition
    (ii) niche requirements

(2)   Population structure - preferably with individual populations
varying in size, and with at least some isolated from the rest of the
species.
Implies estimation of -
    (i)   population size
    (ii) population dynamics - changes in number in time
    (iii) dispersal and of migration between populations

(3)   Reproductive biology and breeding structure -
          Breeding season(s)
          Age at reproduction
          Length of reproductive life
          Total fecundity

Identification of offspring of individual parents
Are populations panmictic?

(4) Suitability for sampling of natural populations -
Can large numbers be collected without disturbing population structure?
Will samples be unbiassed for gene and genotype frequencies?

(5) Suitability for laboratory (or at least controlled) breeding -
Mode of inheritance of variants detected
Linkage relationships
Study of laboratory populations

(6) Suitability for field manipulation -
Artificial colonies
Perturbation of gene frequencies

(7) Suitability for non-destructive assay of individual genotype.

This listing is not comprehensive, and the sections are not mutually
exclusive, but clearly a species fulfilling these requirements would
provide the opportunity for evaluating the forces operating to maintain genetic variation and for the measurement of natural selection.

CACTUS-BREEDING *DROSOPHILA*

The *mulleri* subgroup of the *repleta* group of *Drosophila* comprises at
least 28 species (Wasserman, 1962; Richardson *et al.*, 1975), most of
which are known to be cactophilic; that is, breeding and feeding in
rotting arms, cladodes or fruit of one or more of a variety of cactus
species. The phylogeny of these species has been reasonably well defined on morphological, cytological and reproductive criteria (Wasserman, 1962). Some of the species have been used in studies of isozyme
variability (Johnson *et al.*, 1968; Richardson *et al.*, 1975), and isozyme variation in four species of the subgroup has been used in a
study of genetic differentiation and speciation (Zouros, 1973).

Although these species do not fulfil all the requirements elabo-

rated above, sufficient is known of the ecology, breeding site, nutritional requirements and field behaviour of some of them to make them of
particular value, as has also been recognized by Heed and his collaborators (Rockwood-Sluss *et al.*, 1973; Johnston and Heed, 1975). All the
species are native to the Americas, but one (*D. buzzatii*) which is presumed to have originated in Argentina, is known to have spread to many
other countries where its "hosts" (various *Opuntia* species) have become established within historical times, either cultivated or as weeds
(Carson and Wasserman, 1965). The countries which *D. buzzatii* has invaded include Australia (Mather, 1957), where the history of *Opuntia*
colonization, spread, control and present distribution is well-document-
ed (Mann, 1970). Thus *D. buzzatii* was chosen for intensive study, and
we have been fortunate to be able to extend the work to the related
cactophilic species, *D. aldrichi*, which we have found also living in
Opuntia cactus in Australia in some parts of the *D. buzzatii* species
distribution (Mulley and Barker, 1976).

Distribution and population structure of *D. buzzatii* and *D. aldrichi*
in Australia.

Nine species of cactus became major pests of agricultural and grazing land in Australia. The infestation was out of control by 1870, due
to the absence of natural enemies. When the biological control program
started in 1920 about 24.3×10^6 hectares were affected, with about
half of this area covered by prickly pear so dense that the land was
useless (Fig. 1), and it was estimated that the pear was spreading at
the rate of about 400,000 hectares per year. *Cactoblastis cactorum*
(Berg) (Lepidoptera:Pyralidae) was introduced from South America as
a potential control agent in 1925, and by 1940, complete control had
been achieved. Thus the *Opuntia* now occur in an island distribution,
both within the limits of the original infestation where island size
ranges from just one or two plants to a fairly dense infestation over
a few hundred hectares, and elsewhere in southern N.S.W., Victoria and
South Australia. In these latter areas, some patches of pest pear remain as *C. cactorum* does not establish viable populations, but most
islands comprise from one to about 30 plants of the cultivated species,
O. ficus-indica.

Available evidence (Barker and Mulley, 1976) suggests that *D. buzzatii* is specific to the cactus niche, and we suspect that it exists

throughout the entire *Opuntia* distribution in Australia. Assuming complete specificity of *D. aldrichi* to the cactus niche, it has the same
potential distribution, but it is restricted within this by currently
unknown factors, and is concentrated in the northern part of the *Opun-*
tia distribution. The one population known outside this area (locality 5 in the Hunter Valley of N.S.W.) is separated by some 900 km from
the main distribution area. However, *D. aldrichi* has not been found in
localities 3 and 4 (respectively 40 km and 17 km from locality 5),
where extensive collecting has been done on a number of occasions.
Clearly we have an interesting problem in defining just what factors
determine the ecological boundary of *D. aldrichi*, particularly as variation in these factors within the distribution may impose strong selective pressures.

Although the *D. aldrichi* distribution and population structure are
not yet well defined, the distribution of *D. buzzatii* certainly fulfils many of the requirements previously specified - *i.e.* an island
distribution with considerable variation in island population size,
variation in inter-island distances, a number of peripheral isolate
populations, variation among islands in ecology (both in *Opuntia* species and other flora), with the whole distribution extending over a
wide geographical and climatic range, and with other populations outside Australia that have been isolated from the native South American
populations for varying lengths of time.

PRESENT STATUS OF THE STUDY

Our aim has been and still is the experimental study of allozyme polymorphism, with the ultimate aim of understanding the selective or other
forces acting on the polymorphic loci. At the start of the study in
1971, the primary aim was to define the extent of genic variation, to
describe the macrogeographic variation in gene frequencies and heterozygosity and to determine associations of any spatial variation with
environmental variables. Once this phase was well advanced, and following the discovery of both *buzzatii* and *aldrichi* at locality 5, this
population was chosen for a study of temporal variation in gene frequencies and heterozygosity, together with a description of microgeographic variation, population subdivision and breeding structure
throughout the year. Given the known breeding site of the species, and

the importance of yeasts in the nutrition of *Drosophila*, the most recent phase has involved analyses of yeast species distribution, and the
possible selective influence of yeast species on *buzzatii*.

Macrogeographic Variation.

     Results of collections made in localities 1 to 35 (Fig. 1), with
some populations sampled on more than one occasion, have been present-

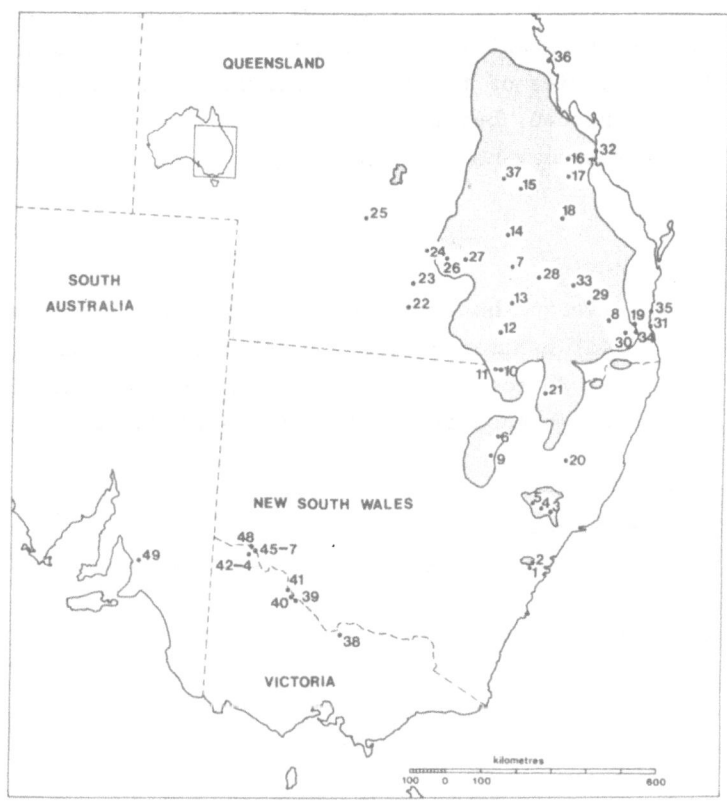

Fig. 1. Distribution of the main *Opuntia* infestations in 1920 (shaded areas), and
the localities from which *D. buzzatii* populations have been sampled.

ed by Barker and Mulley (1976). Assays of recent collections in localities 38 to 49 are not yet completed, but average gene frequencies appear to be similar to those elsewhere in the distribution. Of 29 presumptive loci surveyed, only six are consistently variable (*viz*. - *Es-
terase-1*, *Esterase-2*, *Alcohol dehydrogenase-1*, *Pyranosidase*, *Aldehyde
oxidase* and *Phosphoglucomutase*), with one other (*Alcohol dehydrogena-
se-2*) sporadically variable at low frequency throughout the total distribution, giving estimates of the percentage of polymorphic loci of
19.2 ± 0.35 (1% criterion) and 15.3 ± 0.33 (5% criterion). Average heterozygosity was 0.065 ± 0.031. These estimates of genetic variation
are lower than those reported for most other species of *Drosophila*, but
are not significantly different from estimates available for other cactophilic species of the *mulleri* subgroup (Table 1). Clearly more information is needed, but the average heterozygosity in species of the *mulleri* subgroup may be less than in other *Drosophila* species.

Table 1.   Average heterozygosity (H̄) in species of the *mulleri* subgroup -
             (a) estimated from the eight loci common to our study and that
             of Zouros (1973), (b) estimated from the 12 common loci assay-
             ed in Australian populations of *buzzatii* and *aldrichi*

Species and race	H̄ ± standard error
(a)	
mojavensis A	.023 ± .022
B I	.133 ± .077
B II	.086 ± .057
arizonensis	.147 ± .079
mulleri	.171 ± .093
aldrichi	.197 ± .109
buzzatii (Australia)	.196 ± .100
(b)	
aldrichi	.125 ± .052
buzzatii	.157 ± .069

If this is so, the low level of genetic variation in these species may be a function of their ecology, possibly in terms of the niche-variation hypothesis (Van Valen, 1965) or that of environmental grain (Levins, 1968; Gillespie, 1974). As *buzzatii* is apparently specific to the cactus niche, one might consider that it would have a narrow feeding and breeding niche, or that immature stages would experience their environment as relatively fine-grained. Certainly the immature stages always occur in the same apparently uniform environment - in rot-pockets inside the thick, leathery covering of *Opuntia* cladodes, where they are protected, at least to some extent, from direct effects of external environmental variation in temperature and humidity. However, rot development and the rotting process through to final drying out are partly dependent on these environmental factors, and no doubt there are differences between rot-pockets or even between different parts of the one rot-pocket in temperature, pH, microflora and chemical composition. In this regard, the flies apparently are more perceptive than we are, because among rot-pockets that we see as similar, preliminary observations indicate that 30-40% are not being utilized as breeding sites. As yet, little is known of the environmental heterogeneity experienced by the species, so that these and other possible hypotheses cannot be distinguished.

For all six variable loci, there were significant differences in gene frequency among the 50 samples from localities 1 to 35, as tested either by heterogeneity X^2 or by the standardised gene frequency deviation (Christiansen *et al.*, 1976). These tests confound temporal as well as spatial variation, as collections were made in different months over 2½ years. However, most collections were made in three separate trips, and within each of these sets of samples, there were highly significant differences in gene frequency among populations for all six loci.

Such differences among populations could be due of course to selection and/or drift. Multivariate analyses of genotype-environment associations indicate that selection is acting, although we do not yet have evidence to prove that selection is acting directly on the allozyme loci. Genotype-environment associations have been studied in a number of species, the stated aim of these studies being to identify environmental forces having a selective effect. Although a number of multivariate techniques have been used, only one or two have been used in any one study, and different workers apparently regard different

techniques as superior for this purpose. Thus principal components were
used by Johnson *et al* (1969), Tomaszewski *et al* (1973) and Rockwood-
Sluss *et al* (1973), likelihood ratio tests and regression analysis by
Kojima *et al* (1972), principal components and canonical correlation by
Johnson and Schaffer (1973), principal components with varimax rotation
and stepwise multiple regression by Bryant (1974*a,b*), factor analysis
by Taylor and Mitton (1974), canonical correlation after correction for
geographical location by Schaffer and Johnson (1974), principal components and multiple regression by McKechnie *et al* (1975), and stepwise
multiple regression by Nevo and Bar (1976). Each technique has certain
limitations and disadvantages, but as computer programs for most of
them are available in packages such as BMD or SPSS, they may be used
simply because they are available. We do not believe that any one procedure can be considered as ideal in testing for genotype-environment
associations, and have preferred therefore to use a battery of procedures. In this way, one might hope not only to determine significant associations, but also to test the "robustness" of any associations detected.

Quantitative climatological data used in the analyses (Table 2) were
derived from the tables of Austclimdata (Kieg and McAlpine, 1969), which
provide 30 year normals on a weekly basis. The climatological components
have been defined in terms of both means and variabilities, and the aim
was to assess associations of both gene frequencies and heterozygosity
with both sets of environmental components. Other variables used also
are listed in Table 2. We have used latitude and distance inland as coordinates of geographical location. Major geographical features and
climatic zones generally run parallel to the coast, so that populations
at similar distances inland, but at different latitudes, are more likely
to be in similar environments than are populations on the same longi-
tude but at different latitudes.

As a first step, the method of canonical correlation, which finds
the patterns in the two sets of variables showing the greatest possible
correlation, was applied to the data for the four combinations of gene
frequencies or heterozygosities with the measures of environmental
means or environmental variabilities (Table 3). In all cases, the first
canonical correlation was highly significant, indicating that significant associations do exist. Particular associations between genetic
and environmental variables have been detected using other multivariate
procedures, although our analyses are not yet complete. However, rather
than discuss these in any detail, some of the results will be used to

Table 2.   Environmental components used in analyses of genotype-envi-
            ronment associations

Environmental means

RAIN	Average annual precipitation (inches)
AVTEM	Average daily temperature
SUMMAX	Average daily maximum temperature in
	December, January, February
WINMIN	Average daily minimum temperature in
	June, July, August
RELHUM	Average daily (3.00 PM) relative humidity
LAT[a]	Latitude
DINLD[a]	Distance inland (shortest distance to coast)
DMARG[a]	Distance from margin of species distribution
MONTH	Month of collection
SPOP[a]	Number of *Opuntia* species growing in locality

Environmental variabilities

SDWT, CVWT	Standard deviation and coefficient of variation of
	average weekly temperature
AVTRY,SDTRY,CVTRY	Mean, standard deviation and coefficient of
	variation of daily temperature range (year)
AVTRS,SDTRS,CVTRS	As above (Summer)
AVTRW,SDTRW,CVTRW	" (Winter)
SDRAIN,CVRAIN	Standard deviation and coefficient of variation
	of weekly precipitation
SDRHUM,CVRHUM	Standard deviation and coefficient of variation
	of weekly averages of relative humidity

---

a: Also included in the set of environmental variabilities in the multivariate analyses

Table 3. Canonical correlations between gene frequencies or expected
heterozygosities at the six variable loci and environmental
means or environmental variabilities

Variable sets	First canonical correlation
Gene frequencies - Environmental means	0.88***
Gene frequencies - Environmental variabilities	0.90**
Heterozygosity - Environmental means	0.76**
Heterozygosity - Environmental variabilities	0.80***

** P < 0.01,   *** P < 0.001

emphasise the need for caution in interpreting such analyses.

Firstly, a significant association between some combination of environmental variables and some genetic variable is not in itself sufficient evidence for selection. As pointed out by Schaffer and Johnson
(1974), most of the environmental patterns will show a relationship
with geographical location, so that any similarity of heterozygosity or
gene frequency in neighbouring locations could be explained either as
due to selective effects of the similar environments or as due to migration and genetic drift. The two hypotheses are confounded. Correc-
tions for geographical location can be used to get patterns of genetic
and environmental variables free of location effects, but if environmental variables linearly related to the geographical coordinates do
have selective effects, the correction for location will remove some
effects of selection. Therefore significant environmental variables
may be overlooked. Geographically distant populations with similar patterns of genetic and environmental variables, and with genetically different populations located between them, will provide the strongest e-
vidence for selection.

Secondly, our results indicate that different environmental variables may be judged important by different multivariate analyses, and
careful study of the results is necessary to assess meaningful associations. For example, $Est-1^a$ gene frequency showed significant positive
associations with DINLD and LAT, and negative association with RELHUM
in a stepwise multiple regression. But for principal components of environmental means, SUMMAX (+) and RAIN (-) had higher loadings than
RELHUM (-) and DINLD (+), while the loading for LAT was negative. That
is, DINLD and RELHUM show as important in both analyses, although two
other variables have higher loadings in the principal component anal-
lyses. However, the results for LAT with a positive relationship for
one analysis and a negative one for the other, would lead to different
interpretations if only one or the other analysis had been done. Yet
the difference is a function only of the statistical methodology - a
partial regression coefficient in one case considering LAT in isola-
tion from other variables, and as part of a pattern of variation in
the second case. After correcting for geographical location, however,
$Est-1$ gene frequencies do not show association with any environmental
variables. As a second example, expected heterozygosity (H) at $Est-2$
provides an interesting contrast. Stepwise multiple regression showed
significant association with DINLD (-,P < 0.001), AVTRY (+,P <0.05),
RAIN (-,P < 0.05) and RELHUM (-,P < 0.05), but regressions of H on

the first four principal components of the environmental means and on the first four principal components of the environmental variabilities were not significant. Similarly, no associations were detectable by factor analysis. But after correcting for geographical location, H shows strong association with a number of environmental variables.

As a final example, *Adh* gene frequencies seem to provide evidence for selection. Gene frequencies at most localities were not significantly different, and only two groups of populations deviated from the overall average. At localities 1-5, the frequency of $Adh-1^b$ was significantly lower, and at localities 31 and 35 significantly higher than the overall average. Localities 1-5 are in the same geographical region, but are probably genetically isolated from each other (Barker and Mulley, 1976, and below). Localities 31 and 35 are strikingly similar environmentally in that the *Opuntia* is growing right on the coast within a few metres of salt water. Locality 32 also is similar, except that the *Opuntia* is growing on a sandy beach rather than on mud-flats as at localities 31 and 35. For collections made at the same time (August, 1974) the standardised gene frequency deviations were 3.888 (locality 31), 1.813 (locality 32) and 5.723 (locality 35). Localities 31 and 35 are only 25 km distant from each other, while locality 32 is 520 km distant. Thus migration could be postulated to account for the similarity in *Adh* frequencies at localities 31 and 35, but this is unlikely as they were significantly different for *Est-2* and *Pyr* gene frequencies. However, the similar *Adh* gene frequencies at locality 32 could hardly be due to migration. Given that the environments of these three localities are so different from all other localities, but are very similar to each other, selection could well be implicated for this locus.

Although we have demonstrated some significant genotype-environment associations after correcting for geographical location, this does not negate the possibility of some gene-flow between spatially isolated populations. However, direct evidence from observed gene frequencies indicates that at least some populations probably are genetically as well as spatially isolated from each other, and knowledge of the history of the *Opuntia* distribution would suggest that they have been so isolated at least since 1940.

We have argued that *buzzatii* (and presumably *aldrichi* also) entered Australia in one or more of the 1230 crates of *Opuntia* imported during the biological control program (Barker and Mulley, 1976). When the potential of *Cactoblastis cactorum* as a control agent of *Opuntia* was recognised in 1925, a massive program for distribution of *Cactoblastis* throughout the infestation commenced. The first field liberation of over 2 1/4 million eggs was made in February - March 1926. The following summer nearly 8 million eggs were released, and by November 1929, 389 million eggs had been distributed from rearing cages at the experiment station. During this period and through the 1930's, rotting pear containing *Cactoblastis* also was distributed extensively. Therefore, at this time the cactophilic *Drosophila* species would have found an almost unlimited environment, and must have rapidly produced enormous populations throughout the *Opuntia* infestation. Then as the pear was controlled and its distribution receded by 1940 to the islands as found today, so the *Drosophila* population would have contracted to spatially isolated populations.

Among the populations we have studied, there are three sets that are geographically contiguous which were sampled at the same time, *viz.* localities 1 and 2 (5.5 km apart); 3, 4 and 5 (22 km from 3 to 4, and 17 km from 4 to 5), and 10 and 11 (12 km apart). The populations in each set deviate from the overall gene frequencies at each locus in the same direction (as expected with either selection or gene-flow), but localities 1 and 2 were significantly different in gene frequencies at *Est-1* and *Est-2*, and localities 10 and 11 were significantly different at *Est-1*, *Est-2*, and *Ao*. Locality 4, which is situated <u>between</u> localities 3 and 5, had significantly lower *Est-1*^{a} gene frequency than both localities 3 and 5, and was significantly different from locality 5 at *Est-2* and *Pgm*.

Temporal and microgeographic variation.

The study of temporal variation, started at locality 5 in February 1974, is combined with a study of microgeographic variation. Ten collecting sites have been defined in two parallel linear transects, and flies are being collected monthly at each site. Full analysis of the results has not been completed, and as we want to try to relate any temporal changes in gene frequencies and heterozygosity to environmental variables, we plan to continue this study for some years.

Nevertheless, results obtained to date have led us to postulate some important aspects of the ecology and behaviour of the species. In the populations investigated in the spatial study and in the temporal study, observed heterozygote frequencies at individual loci were gen-

erally less than expected. At locality 5, significant differences in
gene frequencies among sites have been detected for all variable loci,
but usually only in the period November to March which coincides with
the time of maximum *Cactoblastis* activity when new rots are continually developing. These new rots in the fleshy cladodes may remain as
suitable breeding sites for four to six weeks, so that following colonization, at least two generations could develop in each rot. Thus we
postulate that in these summer months, flies emerging in a particular
rot tend to remain there, the founder event and subsequent inbreeding
leading to genetic differences among rots and an observed heterozygote
deficiency. From late summer to autumn, these rots dry out, and the
population becomes dependent on the much smaller number of old, long-
persistent rots in the basal "stems" of the *Opuntia*. The flies inhabiting these rots through the winter will include not only any that developed there the previous summer, but also ones from surrounding areas that have been forced to leave the drying cladode rots. That is,
the partly-isolated sub-population structure would be broken down, and
gene frequencies would not be expected to be different among sites during these months.

Over the first 18 months of the study, gene frequencies at all
variable loci have shown significant changes in time. Clear cyclical
changes are not obvious at all loci, and should not be expected with
such a short period of study. In addition, drought conditions prevail-
ed through the summer of 1974-75, so that the patterns of environmental change were not the same in both years. Nevertheless, some of the
changes in gene frequency suggest cyclical patterns, so that continuing study may detect real associations between temporal environmental changes and changes in the genetic composition of the population.
For example, at *Est-1*, the *a* allele shows significantly high frequencies in summer - autumn, and significantly low frequencies in winter -
spring, while a reverse pattern is exhibited by the *c* allele. Similarly, the changes in frequency of the *a* and *b* alleles of *Est-2* show an
inverse relationship, with *Est-2*^{a} at significantly high frequencies in
summer - autumn, and significantly low frequencies in winter - spring.

Yeast species distribution and selection at allozyme loci.

Seven species of yeasts have been isolated from rotting *Opuntia* (Table 4). Y2 to Y6 apparently occur throughout the *Opuntia* distribution, although not all have been found at all localities that have been assayed. Y7 and Y8 have been isolated only from fruit and cladodes of *O. ficus-indica* at localities 38 - 49 and at some other localities in this region where no *buzzatii* were found. In these *O. ficus-indica* samples, Y3 was most common, being isolated from 14 of 24 rots. Rot samples from locality 5 have been assayed on three occasions. In March and April, 1975, rotting cladodes were collected and returned to the laboratory, samples taken for yeast assay, and all emerging *Drosophila* collected (Table 5). Although the sample size was very small, Y3 again was most common. As noted previously, not all rots produced emerging *Drosophila*, but four of six such rots were devoid of yeast, suggesting that they were not suitable breeding sites. On the other hand, two rots (14H and 15E) were used as breeding sites although no yeasts were isolated. In October, 1975, 36 rots were assayed for yeast species. Yeasts

Table 4.   Yeast species isolated from rotting *Opuntia*

Code No.	Name[b]
Y2[a]	*Candida krusei* (Cast.) Berkhout.
Y3	*Pichia membranaefaciens* (Hansen)
Y4	*Candida sp.*
Y5	*Candida sp.*
Y6	*Pichia sp.*
Y7	⎫ Not yet identified
Y8	⎭

a: Y1 kept as code for *Saccharomyces cerevisiae*

b: Identified by Centraalbureau voor Schimmelcultures, The Netherlands

Table 5.   Microflora assay and *Drosophila* emergences for rots collected at locality 5 in March and Arpil, 1975

Rot No.		Yeasts isolated	Other Fungi[a]	Bacteria[a]	No. of emergences	
					buzzatii	*aldrichi*
March:	14A		+	+	-	-
	14B	Y4,Y6	+	+	5	-
	14C		+	+	-	-
	14D	Y3	+	+	61	36
	14E		+	+	-	-
	14F		-	+	-	-
	14G	Y4,Y6	+	+	-	1
	14H		+	+	8	25
	14I	Y3	+	+	-	-
	14J	Y3	-	-	1	-
April:	15A	Y3,Y5	+	+	2	-
	15B	Y3,Y5	+	+	56	4
	15C	Y6	+	-	-	-
	15D	Y3,Y5	+	+	5	122
	15E		+	+	-	26

a: Presence (+) or absence (-)

were isolated from only 23 of the rots, but two species were isolated
from seven of them, so that the numbers of each species were Y3 - 5,
Y4 - 6, Y5 - 17, and Y6 - 2. Clearly, Y5 was most common, so that the
prevalence of these yeast species may vary through the year, thus contributing to temporal environmental heterogeneity.

In both laboratory and field experiments, Y3 was found to be most
attractive to *buzzatii*. In the laboratory experiment, six cages were
set up each with 150 adult *buzzatii*, three cages with males, and three
with females. Discs of agar (3 cm diameter) were smeared with an even
film of 1:1 suspension of one of the yeasts Y2 to Y6. Five discs (each

coated with a different yeast) were assigned at random to nine possible positions on the base of each cage, and the cages placed under even lighting at $25^{\circ}C$ and 70% relative humidity. The cages were left undisturbed for about 45 minutes, and then without moving the cages or disturbing the flies in any way, the numbers of flies on each disc were recorded every 30 minutes over a four hour period to give nine replicate sets of observations. Analysis of variance of these data showed yeasts to be the only significant effect ($P < 0.001$), with Y3 clearly more attractive than the other four yeasts tested (Table 6).

The field experiment was done at locality 5. Each of the five yeasts was grown on 200 ml of sterile, mashed *Opuntia* cladodes in two litre plastic ice cream containers at $25^{\circ}C$ for three days, with four containers for each yeast. In the field, the containers were placed at four different sites. In addition to the five wild yeasts, one container with *Saccharomyces cerevisiae* and banana mash (our normal bait for field collections) was included at each site, with the six containers placed in a random sequence in a circle of approximately three metre diameter. Flies attracted to the baits were collected periodically, and the containers were put into a new random sequence after each collection. The numbers of *buzzatii* of each sex collected from each yeast at each site were used as data for an analysis of variance. The main effect for yeasts again was highly significant ($P < 0.001$), with Y3 most attractive, although the order for the other yeasts differed from the laboratory experiment (Table 7). In addition, the site x yeast interaction was significant ($P < 0.01$), an effect which might reflect differential response due to heterogeneity in the yeasts naturally occurring at each site at the time of the experiment.

Table 6. Mean number of flies on each yeast at each replicate scoring, and analysis of differences among yeasts using Tukey's w-procedure (underlined means not significantly different)

Yeast	Y5	Y2	Y6	Y4	Y3
Number	0.91	2.63	5.65	6.63	15.50

Table 7.   Mean number of each sex of *D. buzzatii* collected from each
             yeast at each site, and analysis of differences among yeasts
             using Tukey's w-procedure (underlined means not significant-
             ly different)

Yeast	Y4	Y2	Y5	Y6	Y1[a]	Y3
Number	3.00	7.50	7.75	8.12	12.87	14.25

a: *Saccharomyces cerevisiae*

The flies collected on these yeasts were returned to the laboratory, and assayed for *Est-1*, *Est-2*, *Pyr*, *Ao* and *Pgm*. For each locus, heterogeneity X^2 analyses of the numbers of each genotype and of each allele showed significant effects of yeasts for *Est-2* (Table 8) but not
for any other locus. Clearly no firm conclusions can be drawn from this
one experiment, but if these results are repeatable, preferences for
different yeasts and differences in relative fitness of genotypes when
utilizing different yeasts could prove to be a potent force maintaining genetic variability.

Two laboratory experiments provide some support for this contention. The first experiment determined the relative nutritional value
of Y3, Y5 and Y6, measured in terms of immature stage viability and
developmental time, and the fitness of alternative alleles at five polymorphic loci for development on these three yeasts. The flies used
originated from 75 isofemale lines from locality 5, maintained as separate lines in the laboratory for four generations, then mixed in a
population cage and maintained for a further six generations. Eggs were
collected on agar discs smeared with a mixture of equal proportions of
the three yeasts. Newly hatched larvae were collected using the techniques of Podger and Barker (1966), with 100 larvae placed in each of
11 vials with one of the three test yeasts incorporated into and on
the medium.

Both sex and yeast had significant effects on immature stage viability. Viability was significantly less on Y6 (Table 9), and assuming
an equal sex ratio among eggs laid, the viability of females was sig-

Table 8.   *Esterase-2* gene frequencies in flies collected on different
            yeasts

Yeast	Males					Females				
	$2N^a$	a	b	c	d	$2N^a$	a	b	c	d
Y1	116	.24	.32	.14	.30	66	.29	.32	.21	.18
Y2	50	.24	.32	.08	.36	56	.38	.30	.05	.27
Y3	100	.46	.18	.12	.24	114	.33	.31	.16	.20
Y4	28	.68	.14	.07	.11	16	.25	.44	.19	.13
Y5	64	.27	.25	.11	.38	40	.35	.30	.10	.25
Y6	70	.40	.29	.09	.23	40	.15	.40	.13	.33

$$\text{Het } X^2_{(15)} = 35.601** \qquad \text{Het } X^2_{(15)} = 16.654$$

a: No. of genes assayed

** $p < 0.01$

Table 9.   Mean viability (%) and developmental times (days) of
            D. buzzatii developing on different yeasts (Experiment 1) or
            deposited as eggs onto different yeasts (Experiment 2)

Experiment	Character	Yeast		
		Y3	Y5	Y6
1	Viability	36.5	40.4	28.4
	Developmental time	11.68	11.95	11.80
2	Viability	53.6	68.8	56.9
	Developmental time	11.64	11.53	11.74

nificantly less than that of males on all yeasts. Developmental time
was significantly longer for males than for females, but there were no
significant effects of yeast.

The second experiment was done in a similar way, except that ovi-
positing females were given a choice of the three yeasts for egg lay-
ing, and the newly emerged larvae were all raised on the same medium
containing equal proportions of the three yeasts. Therefore differences
among yeasts reflect differences in female choice and in the genotypes
of eggs deposited on the different yeasts. Again there were significant
effects of yeast and of sex, but on both immature stage viability and
developmental time. Viability was significantly higher for individuals
deposited on Y5 (Table 9) while again the viability of females was significantly less than that of males. Developmental time was significantly shorter for Y5, and again significantly longer for males than for
females.

Thus Y3, which was most common in rots in the field in some periods
of the year, was also most attractive to *buzzatii*, while immature stage
viability and developmental time in both experiments were low to intermediate for this yeast. In contrast, Y5, which was the most common yeast
at other periods of the year, was poorly attractive to *buzzatii*, yet individuals developing from eggs laid on Y5 (experiment 2) or actually developing on Y5 (experiment 1) showed highest viability.

In both experiments, samples of the parents and emerging progeny
were assayed for allozyme genotypes at five polymorphic loci. In experiment 1, differences among yeasts approached significance for *Est-1*
genotypes, with progeny on Y5 and Y6 showing a decreased frequency of
a/a homozygotes, as compared with the parents which had been maintained on *Saccharomyces cerevisiae*. For *Pyr*, although gene frequencies were
not different among progeny on the three yeasts, all progeny were significantly different from the parents with an increased frequency of
the *a* allele, and of *a/a* homozygotes.

In experiment 2, there were significant differences among progeny
from the three yeasts for *Est-2* and *Pgm*, and for these two loci, some
significant differences between parents and progeny. For *Est-2* (Table
10), progeny from Y3 were not different from the parents, those from
Y5 were lower in $Est-2^c$ frequency, while those from Y6 were lower in
$Est-2^c$ and $Est-2^d$ frequency. For *Pgm* (Table 11), progeny from Y6 were
not different from the parents, while those from Y3 and Y5 were both
lower in Pgm^a frequency, again implying differential selective effects
of the three yeasts.

Table 10. *Esterase-2* gene frequencies in parents, and in
progeny deposited as eggs onto different yeasts

	2N	a	b	c	d	e
Parents	390	.339	.356	.108	.190	.008
Progeny						
Y3	598	.366	.314	.077	.238	.005
Y5	620	.387	.345	.050	.208	.010
Y6	610	.390	.393	.067	.144	.005

Progeny - Among yeasts: $\chi^2_{(8)}$ = 24.919, P < 0.01

Table 11. *Phosphoglucomutase* gene frequencies in parents, and in
progeny deposited as eggs onto different yeasts

	2N	a	b	c
Parents	398	.058	.942	.000
Progeny				
Y3	604	.031	.969	.000
Y5	638	.028	.972	.000
Y6	606	.059	.937	.003

Progeny - Among yeasts: $\chi^2_{(2)}$ = 9.595 (excluding Pgm^c)
P < 0.01

While these results are not definitive, and more detailed experimentation needs to be done, they do serve to emphasise our main point
that knowledge of the ecology of a species is vital in attempting to
understand the evolutionary forces acting on it.

FUTURE DIRECTION OF THE STUDY

Our main objective now is to continue the studies in progress, and to
initiate studies of rot ecology and field manipulation of populations.

Although *buzzatii* and *aldrichi* are utilizing the same environment
in that adults of both species have been collected from individual rots
in the field, and immature stages of both species have been found simultaneously in individual rots, the species show different seasonal
distribution patterns. At locality 5, both species are present in all
months of the year, but *buzzatii* is most abundant in late spring and
early summer, while *aldrichi* is most abundant in late summer and early
autumn. This difference must reflect differential adaptation to environmental conditions, but the effect may be indirect and mediated
through environmental effects on the *Opuntia* and on rot development.
Thus studies on the rot environment and on rot development and utilization by the two species are necessary.

Comparisons of genetic variation in *buzzatii* and *aldrichi* also
indicate a need for more detailed information on the rot environment.
Although average heterozygosity is apparently the same in both species,
there are some intriguing differences in the loci that are polymorphic.
For example, *Mdh* and *Idh* are polymorphic in *aldrichi* but not in *buzzatii*, while *Adh* is polymorphic in *buzzatii* but not in *aldrichi*. For the
variable loci of both species, studies of the biochemical and physiological properties of the allozymes need to be related to the ecological conditions of the rot environment.

Although we have argued that at least some of the populations investigated are genetically as well as spatially isolated, direct information on migration potential can be obtained. To this end, we are
planning to establish new *Opuntia* patches in the vicinity of locality
5. *Cactoblastis* will be introduced to ensure rot development, and then
these new sites will be used in one of two ways. Some will be sampled
periodically to determine whether *buzzatii* and/or *aldrichi* have successfully colonized them. In others, *buzzatii* and *aldrichi* derived

from locality 5 will be deliberately released, but these initial populations will have defined gene frequencies different from those at
locality 5. Then if no migration into the former sites occurs, changes
in gene frequency in the latter sites towards those of locality 5 would
be indicative of selection.

Clearly we do not yet have any unequivocal answers to the questions
posed in the introduction to this paper. However, the study has pro-
gressed far enough to confirm our belief that the cactus-breeding *Drosophila* provide a valuable system for studies in population biology,
and to show that these questions can, at least potentially, be answer-
ed.

Summary.

A full understanding of natural selection must encompass information on the selection process in terms of the life-history and ecology of natural populations. This
implies the choice of a suitable organism for study such that ecological, as well
as genetical parameters, can be estimated. The requirements for such an organism
are detailed. Although the cactus-breeding *Drosophila* do not meet all the requirements, sufficient is known of the ecology, breeding site, nutritional requirements,
and field behaviour of some of them to make them of particular value.

Thus allozyme polymorphism in Australian populations of *D. buzzatii* and *D. aldrichi* is being studied. Three phases of the study are in progress - (1). Magnitude
of genic variation, macro-geographic variation in gene frequencies and heterozygosity, and associations of spatial variation and environmental variables, (2). In
one population, temporal variation in gene frequencies and heterozygosity, together
with a description of microgeographic variation, and breeding structure throughout
the year, (3). Yeast species distribution, and selective influence of yeast species.

Results available to date are presented, and future aspects of the study are outlined.

Acknowledgements.

I am grateful to Mr. P.D. East, Dr. J.W. James, Mr. J.C. Mulley, Mr. G.L. Toll
and Mr. P.R. Widders for allowing me to quote unpublished work, and to the Australian Research Grants Committee for financial support. I am indebted to Dr. J. Bundgaard for the opportunity to spend a Study Leave as Guest Professor in the Department of Ecology and Genetics, Aarhus University, during which time this paper was
prepared.

REFERENCES

Anderson, W.W. 1971. Genetic equilibrium and population growth under density-regu-
lated selection.
Amer. Natur. 105: 489-498.

Barker, J.S.F., and Mulley, J.C. 1976. Isozyme variation in natural populations of
Drosophila buzzatii.
Evolution 30: 213-233.

Birch, L.C. 1960. The genetic factor in population ecology.
Amer. Natur. 94: 5-24.

Bryant, E. 1974a. On the adaptive significance of enzyme polymorphisms in relation
to environmental variability.
Amer. Natur. 108: 1-19.

Bryant, E. 1974b. An addendum on the statistical relationship between enzyme polymorphisms and environmental variability.
Amer. Natur. 108: 698-701.

Carson, H.L., and Wassermann, M. 1965. A widespread chromosomal polymorphism in a
widespread species, *Drosophila buzzatii.*
Amer. Natur. 99: 111-115.

Charlesworth, B. 1973. Selection in populations with overlapping generations. V.
Natural selection and life histories.
Amer. Natur. 107: 303-311.

Christiansen, F.B., and Fenchel, T.M. 1977. *Theories of Populations in Biological
Communities.* Springer Verlag, Berlin.

Christiansen, F.B., Frydenberg, O., Hjorth, J.P., and Simonsen, V. 1976. Genetics
of *Zoarces* populations IX. Geographic variation at the three phosphoglucomutase
loci.
Hereditas 83: 245-256.

Clarke, B. 1973. Mutation and population size.
Heredity 31: 367-379.

Clarke, B. 1975. The contribution of ecological genetics to evolutionary theory:
Detecting the direct effects of natural selection on particular polymorphic loci.
Genetics 79: 101-113.

Cohen, D. 1967. Optimizing reproduction in a randomly varying environment.
J. Theoret. Biol. 16: 1-14.

Fisher, R.A., and Ford, E.B. 1947. The spread of a gene in natural conditions in
a colony of the moth *Panaxia dominula L.*
Heredity 1: 143-174.

Ford, E.B., and Sheppard, P.M. 1969. The *medionigra* polymorphism of *Panaxia dominula.*
Heredity 24: 561-569.

Gershenson, S. 1945. Evolutionary studies on the distribution and dynamics of
melanism in the hamster (*Cricetus cricetus L.*). I. Distribution of black hamsters
in the Ukrainian and Bashkirian Soviet Socialist Republics (U.S.S.R.).
Genetics 30: 207-232.

Johnson, F.M., Richardson, R.H., and Kambyssellis, M.P. 1968. Isozyme variability
in species of the genus *Drosophila*. III. Qualitative comparison of the esterases of
of *D. aldrichi* and *D. mulleri.*
Biochem. Genet. 1: 239-247.

Johnson, F.M. and Schaffer, H.E. 1973. Isozyme variability in species of the genus *Drosophila*. VII. Genotype-environment relationships in populations of *D. melanogaster* from the eastern United States. *Biochem. Genet.* 10: 149-163.

Johnson, F.M., Schaffer, H.E., Gillespy, J.E., and Rockwood, E.S. 1969. Isozyme genotype-environment relationships in natural populations of the Harvester Ant, *Pogonomyrmex barbatus*, from Texas. *Biochem. Genet.* 3: 429-450.

Johnston, J.S., and Heed, W.B. 1975. Dispersal of *Drosophila*: The effect of baiting on the behaviour and distribution of natural populations. *Amer. Natur.* 109: 209-216.

Kieg, G., and McAlpine, J.R. 1969. Austclimdata. A magnetic tape with estimated mean weekly climatic data for the Australian continent. C.S.I.R.O. Division of Land Research, Tech. Memo. 69/14.

King, C.E., and Anderson, W.W. 1971. Age-specific selection. II. The interaction between r and K during population growth. *Amer. Natur.* 105: 137-156.

Kojima, K., Smouse, P., Yang, S., Nair, P.S., and Brncic, D. 1972. Isozyme frequency patterns in *Drosophila pavani* associated with geographical and seasonal variables. *Genetics* 72: 721-731.

Krebs, C.J., Gaines, M.S., Keller, B.L., Myers, J.H., and Tamarin, R.H. 1973. Population cycles in small rodents. *Science* 179: 35-41.

Levins, R. 1968. Toward an evolutionary theory of the niche. *In*: E.T. Drake (ed.): *Evolution and Environment*. pp. 325-340. Yale University Press, New Haven, Conn.

Lewontin, R.C. 1965. Selection for colonizing ability. *In*: H.G. Baker and G.L. Stebbins (eds.): *The Genetics of Colonizing Species*. pp. 77-94. Academic Press, New York.

McKechnie, S.W., Ehrlich, P.R., and White, R.R. 1975. Population genetics of Euphydryas butterflies. I. Genetic variation and the neutrality hypothesis. *Genetics* 81: 571-594.

Mann, J. 1970. *Cacti naturalised in Australia and their control*. S.G. Reid, Government Printer, Brisbane.

Mather, W.B. 1957. Genetic relationships of four Drosophila species from Australia. *Genetics of Drosophila*, Univ. Texas Publ. 5721: 221-225.

Mertz, D.B. 1971. The mathematical demography of the California Condor population. *Amer. Natur.* 105: 437-453.

Mulley, J.C., and Barker, J.S.F. 1976. The occurrence and distribution of *Drosophila aldrichi* in Australia. *Drosophila Inf. Serv.* 52: (in press).

Nevo, E., and Bar, Z. 1976. Natural selection of genetic polymorphisms along climatic gradients. *In*: S. Karlin and E. Nevo (eds.): *Population Genetics and Ecology*. pp. 159-184. Academic Press, New York.

Pianka, E.R. 1974. *Evolutionary Ecology*. Harper & Row, New York.

Podger, R.N., and Barker, J.S.F. 1966. Collection of large numbers of larvae of homogeneous age and development. *Drosophila Inf. Serv.* 41: 195.

Richardson, R.H., Richardson, M.E., and Smouse, P.E. 1975. Evolution of electrophoretic mobility in the *Drosophila mulleri* complex. *In*: C.L. Markert (ed.): *Isozymes* IV. *Genetics and Evolution*. pp. 533-545. Academic Press, New York.

Rockwood-Sluss, E.S., Johnston, J.S., and Heed, W.B. 1973. Allozyme genotype-environment relationships. I. Variation in natural populations of *Drosophila pachea*.
  Genetics 73: 135-146.

Roughgarden, J. 1971. Density-dependent natural selection.
  Ecology 52: 453-468.

Roughgarden, J. 1972. Evolution of niche width.
  Amer. Natur. 106: 683-718.

Schaffer, H.E., and Johnson, F.M. 1974. Isozyme allelic frequencies related to selection and gene-flow hypotheses.
  Genetics 77: 163-168.

Smith, M.H., Garten, C.T., Jr., and Ramsey, P.R. 1975. Genic heterozygosity and population dynamics in small mammals. *In*: C.L. Markert (ed.): *Isozymes* IV. *Genetics and Evolution*. pp. 85-102. Academic Press, New York.

Taylor, C.E., and Mitton, J.B. 1974. Multivariate analysis of genetic variation.
  Genetics 76: 575-585.

Tomaszewski, E.K., Schaffer, H.E., and Johnson, F.M. 1973. Isozyme genotype-environment associations in natural populations of the Harvester Ant, *Pogonomyrmex badius*.
  Genetics 75: 405-421.

Wasserman, M. 1962. Cytological studies of the repleta group of the genus *Drosophila*: The mulleri subgroup.
  Studies in Genetics, II, Univ. Texas Publ. 6205: 85-117.

Zouros, E. 1973. Genic differentiation associated with the early stages of speciation in the *mulleri* subgroup of Drosophila.
  Evolution 27: 601-621.

4. ECOLOGY AND EVOLUTION

TEST OF THE HYPOTHESIS THAT MIGRATION BALANCES SELECTION
IN DIFFERENTIATED SUBPOPULATIONS OF *SPIRORBIS BOREALIS*

Roger W. Doyle and Laura J. Richards

Spirorbis borealis offers a clear-cut example of an important type of
variable selection pressure: zygote dispersal, followed by disruptive
selection as a pre-adult, followed by mating within subpopulations. It
is possible to make quantitative estimates of the intensity of selection on a behavioural polymorphism in this species. Theoretical conditions for protecting polymorphisms against fixation by natural selection have been proposed by Christiansen (1974), Strobeck (1974) and
others. The question considered here is whether migration between subpopulations is sufficient to counterbalance the strong selection a-
gainst one of the phenotypes, using any of the simple, conventional
models of inheritance. The work has been done on a *Spirorbis* population in Terence Bay, approximately 20 km south of Halifax, Nova Scotia.

Adult *Spirorbis* are small, tube-dwelling annelids which live per-
manently attached to the fronds of brown algae. Before metamorphosis
to the adult occurs there is a brief dispersal phase, during which
the larvae swim freely in the plankton looking for a place to settle.
The two principal substrates, *Fucus vesiculosis* and *Ascophyllum nodosum*, vary in their relative suitability from place to place in the
intertidal and near sub-tidal zones. In some areas animals which prefer to settle on *Fucus* have the highest survival while in other areas
survival is greatest among animals which settle readily on *Ascophyllum*.

When tested in the laboratory the preferences for the different types
of substrates have been found to vary in parallel with this difference
in substrate suitability (Doyle 1975).

DIRECTION OF THE SELECTION

Spirorbis borealis is hermaphroditic with both sets of gametes matur-
ing at about the same time. Release of gametes is thought to be arranged so that cross-fertilization is much more prevalent than self-
fertilization except in isolated individuals (Gee and Williams 1965;
Doyle 1974). Sperm are released into the water and offspring are a mixture of selfed, full- and half-sibs with the latter category predomi-
nating. Larvae are released about two weeks after fertilization and
spend from 15 minutes to 12 hours searching for suitable substrate.
The algae occupied by *Spirorbis* are fairly well intermixed in most of
the *Spirorbis* zone, although *Ascophyllum* has a tendency to be found
somewhat higher on the shore line. In tide pools and other protected
places, however, *Ascophyllum* is often the only fucoid alga present
in the zone. *Spirorbis* has a maximum post-settlement life expectancy
of 1 1/2 years.

The life-expectancies of the newly-settled larvae are very different on the two algae (Figure 1). This difference in survival comes a-
bout through the very poor adhesion of *Spirorbis* to the surface of
Ascophyllum fronds. Mortality of newly-settled larvae is greater on
Ascophyllum then on *Fucus* and mortality on both algae is greater in
highly turbulent - that is, wave-swept - parts of the coastal environment. The difference in suitability of the substrates also affects
adults, as can be seen in Figure 2. The relative suitability of *Ascophyllum* decreases with turbulence, resulting in a strong association
between the occupancy of the two substrates and a measure of the intensity of wave action (Table 1). The turbulence measurements in this
table represent the loss of weight of small plaster-of-Paris blocks
attached directly to algal fronds during 3-day intervals. The weight
losses are presented as standard deviations from the mean erosion of
all blocks set out during the same period. The association between the
presence of *Spirorbis* on *Fucus* or *Ascophyllum* and the amount of erosion of the plaster blocks is significant at p = p.004 by the Wilcox-
on two-sample test. Presumably the association would have been great-

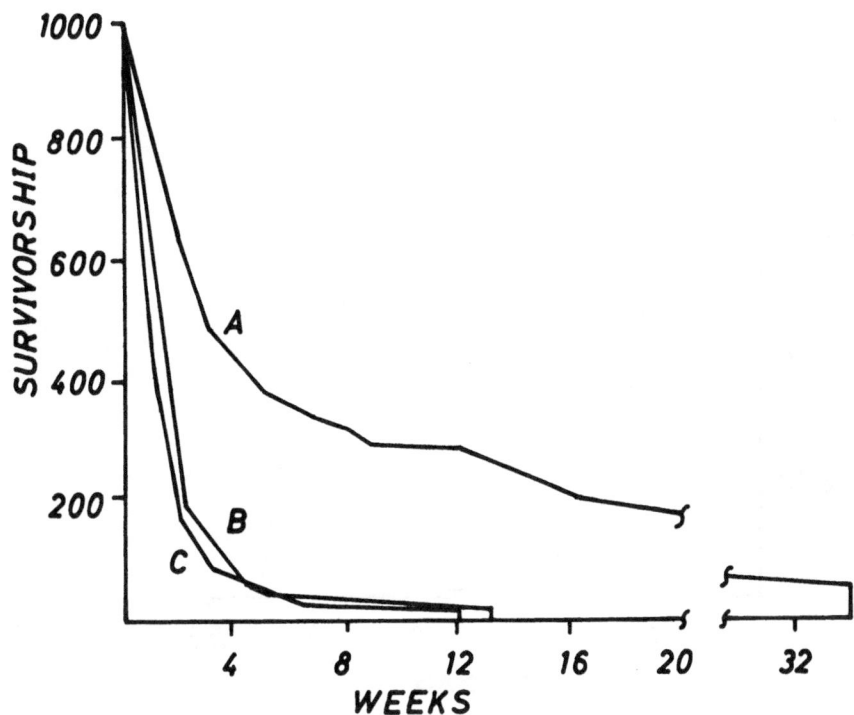

Figure 1. Survivorship curves for newly-metamorphosed *Spirorbis*. A: on *Fucus* in a moderately turbulent location. B: on *Ascophyllum* in a protected tidal pool. C: on *Fucus* transplanted to a highly turbulent location.

er had erosion rates been measured in bad weather, but as it happens they were not. Direct observations of the distribution of *Spirorbis* on the shore soon convinces one that in turbulent areas there is no chance of the animals surviving to reproductive maturity on *Ascophyllum*.

On the other hand it is easy to find embayments and tide pools, frequently very small and low on the shore, which are protected enough to allow *Spirorbis* to colonize *Ascophyllum* in profusion. If *Fucus* is found at all in these areas it is very crowded with *Spirorbis* soon after the start of the reproductive season. *Ascophyllum* is effectively the only substrate available and *Spirorbis* larvae which refuse to

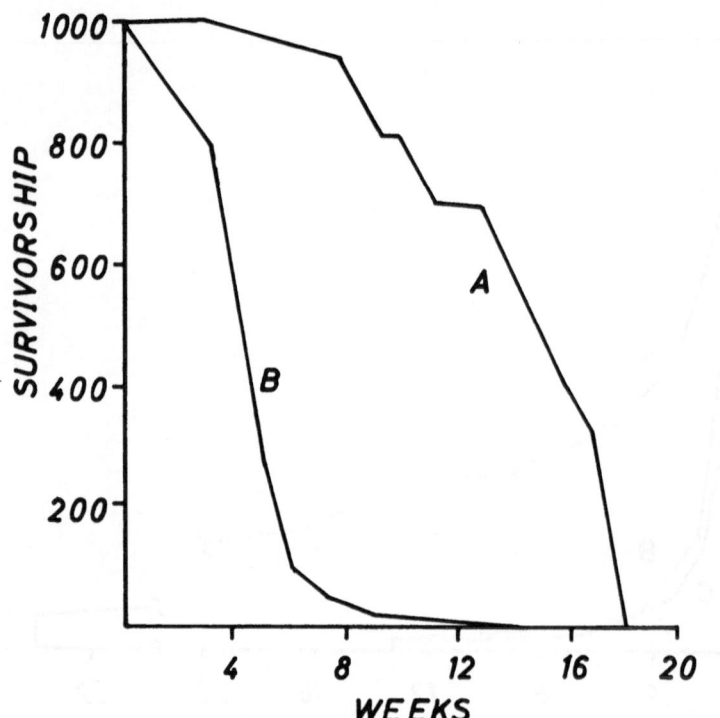

Figure 2.   Survivorship curves for adult *Spirorbis*. A: on *Fucus* in a moderately turbulent location. B: on *Ascophyllum* transplanted to the same location as A.

settle on it have a low relative fitness. The shore-line must consequently be considered "coarse-grained" with respect to the direction
of selection for willingness to settle on *Ascophyllum*. The heterogeneity of substrate suitabilities is to some extent paralleled by a
heterogeneity of substrate preferences among subpopulations of animals
collected in different areas (Table 2).

Table 1.   Turbulence measurements associated with
              substrates occupied by *Spirorbis borealis*

	Substrates				
Occupied *Fucus* empty *Ascophyllum*				Occupied *Fucus* occupied *Ascophyllum*	
erosion	rank	erosion	rank	erosion	rank
-1.62	1	0.28	42	-1.12	3
-1.15	2	0.61	43	-1.12	4
-1.05	6	0.62	44	-1.09	5
-0.93	10	0.72	45	-0.98	7
-0.89	12	0.75	46	-0.97	8
-0.87	13	0.81	47	-0.95	9
-0.84	14	0.88	48	-0.91	11
-0.70	15	1.15	49	-0.65	18
-0.68	16	1.25	50	-0.63	19
-0.67	17	1.68	51	-0.60	20
-0.59	21	1.86	52	-0.50	24
-0.57	22			-0.49	25
-0.53	23	n = 36		-0.36	30
-0.46	26			-0.33	31
-0.45	27			-0.22	35
-0.43	28			0.27	41
-0.38	29				
-0.31	32			n = 16	
-0.30	33				
-0.25	34				
-0.18	36				
-0.10	37				
-0.09	38				
-0.08	39				
0.20	40				

Table 2.   Larval settlement frequencies

Subpopulation	Sample	Number settled		
		Fucus	*Asc.*	total
turbulent (on *Fucus*)	1	340	21	361
	2	542	43	585
	3	1625	33	1658
	4	1101	110	1211
Weighted proportions		0.95	0.05	
protected (on *Ascophyllum*)	1	422	109	531
	2	528	226	754
	3	1625	88	1713
	4	1252	497	1749
Weighted proportions		0.81	0.19	

DIFFERENTIATION OF SUBPOPULATIONS

Freshly collected batches of *Ascophyllum* and *Fucus* were collected from adjacent turbulent and exposed areas on the shore of Terence Bay, Nova Scotia. The algae, which bore many thousands of mature *Spirorbis*, were placed in separate tanks of running seawater. Overflow from the tanks was fed into screened boxes containing numerous pieces of both types of algae, on which settlement of larvae released over the course of a week took place. The test algae substrates were removed at two-day intervals (called "samples" in Table 2) and replaced with fresh algae. The results of the experiment can be summarized by noting that the mean indidence of *Ascophyllum* preference is 19% in the offspring of parents collected in the protected region and only 5% in the offspring of parents collected in the turbulent region. This association between parental substrate and offspring substrate is a type of statistical interaction which elsewhere has been termed "habitat loyalty" (Doyle 1976).

Table 3.   Parent/offspring habitat loyalty.

Parental substrate	Larval settlement (pooled data)	
	Fucuc	*Asc.*
Fucus	.95	.05
Asc.	.81	.19

	Larval settlement (unpooled data)	
Sample	Weight	Trace
1	892	1.15
2	1339	1.23
3	3371	1.03
4	2960	1.19

Trace (pooled data) = 1.14.
Trace (unpooled data) = 1.13 ± .08, $p < .05$

Habitat loyalty is conveniently quantified as the trace of the square matrix of relative settlement frequencies, as in Table 3. The null hypothesis of no "loyal" interaction between parent and offspring substrate preferences is represented by a trace of 1.00. Settlement frequencies were found to interact significantly with sample according to a contingency table analysis on Table 2, so the value of the trace was calculated separately for each sample. The weighted value of the trace is 1.13 ± .08 which is significant at the .05 level. Thus, the turbulent and protected environments have differentiated subpopulations of *Spirorbis* larvae which show "loyalty" to the substrate on which their parents managed to survive. Note, however, that there is a strong overall preference for *Fucus*.

OPTIMUM PHENOTYPES IN THE SUBPOPULATIONS

It has been shown elsewhere (Doyle 1975) that there is reason to believe that when a larva is presented with a choice of substrates its
fitness is maximal if it either (a) settles on the first substrate it
encounters without exhibiting any preference or (b) swims around until it finds the substrate with the highest survival value, ignoring
all others. Which behaviour is optimal in any part of the environment
depends on the relative survival-value of the substrates and on their
abundance in relation to the probability of death in the plankton. The
survivorship curves indicate that in the turbulent areas phenotype (b)
has a relative fitness of 1; no matter how much *Ascophyllum* the larvae
encounter they are better off continuing to look for a *Fucus* frond.

In protected areas a phenotype willing to settle on *Ascophyllum*
has a relative fitness greater than zero, and possibly much greater
than that of a *Fucus*-only phenotype. The optimal preference phenotype
is probably type (a), *i.e.*, one in which both substrates are equally
acceptable. It can be shown that natural selection is never stabilizing in the matter of substrate choice. There are no situations in
which, say, a "choose-*Fucus*-30%-of-the-time" phenotype is optimal.
Acceptance of this theoretical point is necessary before selection intensities, which are expressed in terms of the fitness of an optimum
phenotype, can be calculated.

The best estimate of the optimal substrate preferences of larvae
in the two types of subpopulation is given in Table 4. Selection is disruptive in that the incidence of *Ascophyllum* settlement is above optimum for the turbulent subpopulations and below optimum for the protected populations. The overall habitat loyalty is predicted to be 1.50.
Further discussion of selection intensities will be directed mainly to
the turbulent subpopulation where the optimum phenotype is known with
more certainty.

SELECTION INTENSITY IN TURBULENT AREAS

The incidence of *Ascophyllum* preference is estimated to be .05 (Table
2). Since the relative fitness of this trait is zero the intensity of
selection, *i.e.*, the amount of selective death in the population, is
also .05. The selection intensity in the protected areas is not known

Table 4. Optimal substrate-preference phenotypes

Parental habitat	Larval preferences	
	Fucus	*Asc.*
Fucus	1.0	0
Asc.	0.5	0.5

with any certainty, but the proportion of selective deaths among animals migrating from the protected to the turbulent areas is 0.19.

The selection regime in the turbulent areas is analogous to artificial "truncation" selection in which part of the population is selected against absolutely, with the minor difference in that under artificial selection the 0-fitness individuals are generally in the majority. In truncation selection the intensity of selection is usually called the selection differential or "reach" and is calculated as the difference in mean of the population before and after selection. For a normally distributed variable the selection differential is $-f(t)/F(t)$, where $F(t)$ is the proportion surviving. With this convention the individuals selected against are on the right-hand side of the distribution and the selection differential is measured from the mean before selection to the mean after selection. The calculated value of the selection differential within the turbulent population is -0.11σ and within the migrants from the protected to the turbulent areas it is -0.34σ. Among the animals in the protected subpopulation which do not migrate the direction of selection is, of course, reversed.

The above calculation of selection differentials depends on the assumption that numerous genetic and non-genetic influences on substrate preference combine additively and independently to give a behavioural variable which is normally distributed in the population. Individual *Spirorbis* with a value exceeding a certain threshold choose *Ascophyllum* when given a choice, and those below the threshold choose *Fucus* or do not settle at all. The actual choice of substrate - an all-or-none expression of the phenotype - changes at a threshold value

which has the same absolute value in both populations. This is a conventional model in quantitative genetics (Falconer 1965).

MIGRATION-SELECTION EQUILIBRIUM

We now consider whether the selection against *Ascophyllum* preference
in the turbulent subpopulation can be counterbalanced by immigration
from the protected subpopulations. A necessary condition for migrational equilibrium is easily derived from the quantitative genetic model.

The preference data in Table 2 reflect the difference between the
mean phenotypes of the two populations. By definition, they also reflect the difference in the mean genotypic values, since the parental
substrate has no direct influence on the preferences of the offspring.
To calculate migrational equilibrium it is necessary to suppose that
random mating occurs among the migrants and the local survivors of
selection, and that the genotypic differences combine additively in
the quantitative genetic sense. This by no means implies the absence
of dominance or other gene-gene interactions; it merely implies that
in the offspring there is a positive correlation between the phenotypic values of parents and offspring.

The mean phenotype of the selected individuals in the turbulent
population is $-f(t)/F(t)$ where $f(t)$ is

$$(1/\sqrt{2\pi})\ \exp(-t^2/2),$$

and $F(t)$ is

$$\int_{-\infty}^{t} f(t)\,dt.$$

For the population of migrants from the protected to the turbulent locations, t is replaced by t'. Selection in the turbulent population reduces the incidence of *Ascophyllum* preference and in so doing shifts
the population mean by an amount $-f(t)/F(t)$. If this shift is to be
counterbalanced in the next generation by random mating with the survivors of larvae immigrating from the protected subpopulations, the
mean of these immigrants must be less negative than the mean of the
turbulent population before selection. The deviation of this mean from
the threshold before selection is $-t$, and the mean of the surviving

immigrants deviates from the threshold by $-f(t')/F(t') - t'$. Thus a necessary, but not sufficient, condition for migrational equilibrium is simply

$$f(t')/F(t') < t - t'. \qquad (1)$$

If this condition is met it is possible to calculate the value of a parameter called "migration rate" which can stabilize the mean of the turbulent population against selection in each generation. At equilibirum, the weighted breeding values of the survivors from the two sources must sum to zero. Let m represent the fractional contribution of immigrants to the breeding population in the turbulent area, let ρ represent the unknown parent-offspring covariance in this population, and let g be a constant whose value depends on the breeding system and which is probably close to 2.0 in *Spirorbis*. Then at equilibrium the sum of the breeding values is:

$$m\rho g\{-f(t')/F(t') + (t - t')\} + (1 - m)\rho g\{-f(t)/F(t)\} = 0. \qquad (2)$$

or, substituting Q' and Q for the quantities in curly brackets,

$$mQ'\rho g + (1 - m)Q\rho g = 0.$$

The factor ρg drops out of the solution, giving

$$m = Q/(Q - Q') \qquad (3)$$

If the conditions of the model are met in the real world, then m should represent the real immigration rate at equilibrium. It can easily be shown that when the phenotypic value of the population is in migrational equilibrium then gene frequencies are also in equilibrium.

The incidences of *Ascophyllum* preference in the turbulent and protected subpopulations are .05 and .19, respectively. The corresponding values for t and t' are 1.65 and 0.88. Substituting in equation (1) we find that .33 < .77. Therefore it is possible to calculate a stabilizing migration rate. Substituting in equation (3) and solving for m, the migration rate is m = -.107/(-.435-.107) = .197.

Thus, an immigration rate of 20% in each generation would just counterbalance the directional selection in the exposed subpopulation. Unfortunately, this figure seems impossibly large to anyone familiar with the ecology of the area.

GENERAL COMMENT ON THE RESULT

Life-histories involving the sequence: dispersal→selection→mating are
common among sessile marine invertebrates (Scheltema 1971). Disruptive
selection with 100% differences in the relative fitnesses of substrate-
preferences must also be common and easily recognized. As pointed out
earlier there is no intermediate optimum for substrate preference at
the level of the individual, although phenotypes which grow less
"choosy" as time in the plankton increases are adaptive (Doyle 1975)
and are widely observed (Knight-Jones 1953).

Theoretical models explaining the maintenance of genetic variability in subdivided populations, like those of Strobeck (1974) and
Christiansen (1974) invariably assume the simplest and best-behaved
genetic architecture and breeding systems. We, in contrast, have tried
to identify simple, well-behaved ecological situations leading to
measureable selection pressure, *then* look at the genetics. The result
is an interesting dilemma: the simple quantitative genetic model gives
conclusions which are obviously wrong. Although they are not described
here the obvious one-locus and two-locus models do not work either.

We have biased our sample of the genome by choosing a part which
is under measureable selection and in which genetic differences (between populations in this case) can be measured with reasonable sample
sizes. In deciding to study selection that can be detected by ordinary
ecological techniques we have, naturally, fastened on phenotypic variation which is very important to the animal. The simple genetic models
on which theory has been based do not appear to be very useful when
applied to this ecologically significant variation. Between-population
differences may not combine additively in a straight-foreward way, for
example, in *Spirorbis borealis*. It is interesting to speculate that
this may generally be true of traits under strong disruptive selection
in nature.

Summary.

This paper tests the adequacy of simple models of migration-selection equilibrium in explaining the frequencies of habitat-preference phenotypes in a natural population.

Spirorbis borealis is a tubicolous annelid which attaches to the brown algae *Fucus vesiculosis* and *Ascophyllum nodosum* in the littoral zone. Except in areas protected from wave action, survival is much lower on the latter substrate. Larvae from adults in exposed areas show a reduced incidence of settlement on *Ascophyllum* in the laboratory, compared with the incidence among offspring of animals in protected areas. A necessary condition for maintaining selection-migration equilibrium is derived for a quantitative trait under truncation selection. The migration rate required to prevent selective elimination of *Ascophyllum*-preference phenotypes in the wave-swept areas is unreasonably large. This is also true of simple one-locus models of the genetics of the preference behaviour. The simplifying assumptions required to construct the mathematical models of selection-migration equilibrium evidently falsify the situation.

REFERENCES

Christiansen, F.B. 1974. Sufficient conditions for protected polymorphism in a subdivided population.
Amer. Natur. 108: 157-166.

Doyle, Roger W. 1974. Choosing between darkness and light: the ecological genetics of photic behaviour in the planktonic larvae of *Spirorbis borealis*.
Mar. Biol. 25: 311-317.

Doyle, Roger, W. 1975. Settlement of planktonic larvae: a theory of habitat selection in varying environments.
Amer. Natur. 109: 113-126.

Doyle, Roger W. 1976. Analysis of habitat loyalty and habitat preference in the settlement behaviour of planktonic marine larvae.
Amer. Natur. 110: 719-730.

Falconer, D.S. 1960. An introduction to quantitative genetics. Ronald Press, New York. 365 pp.

Gee, J.M., and Williams, G.B. 1965. Self- and cross-fertilization *Spirorbis borealis* and *S. pagenstecheri*.
Jour. Mar. Biol. Assoc. U.K. 45: 275-285.

Knight-Jones, E.W. 1953. Decreased discrimination during settling after prolonged planktonic life in larvae of *Spirorbis borealis* (Serpulidae).
Jour. Mar. Biol. Assoc. U.K. 32: 337-345.

Scheltema, R.S. 1971. Larval dispersal as a means of genetic exchange between geographically separated populations of benthic marine gastropods.
Biol. Bull. 140: 284-322.

Strobeck, Curtis. 1974. Sufficient conditions for polymorphism with N niches and M mating groups.
Amer. Natur. 108: 152-156.

4. ECOLOGY AND EVOLUTION

THE SELECTION REGIME OF *PHILAENUS SPUMARIUS* (L.) (HOMOPTERA)

Olli Halkka and Erkki Mikkola

In visual polymorphisms, as a rule there exist geographical and local
discontinuities in the frequencies of the genes responsible for colour
variability. Many of these discontinuities cannot be satisfactorily explained on the assumption of discontinuities in the intensity of visual
selection. For types of selection other than visual, it is often jus-
tifiable to suppose at least linkage disequilibrium (see Frydenberg
1963), if not supergene linkage (see Turner 1967), between the "visual" genes and the genes subject to other types of selection.

The visual polymorphisms in a species are often older than its
other ecological polymorphisms. Many visual polymorphisms are older
than the time needed for species formation. Thus in the genus *Philaenus* two species, *P. spumarius* (L.) and *P. signatus* Melichar, have 11
colour morphs in common (Halkka and Lallukka 1969). No complex polymorphisms are known from other, closely related species belonging to
the family Aphrophoridae. It appears unlikely that convergent evolution
could have been taken place only in the two *Philaenus* species with a
probably common ancestor, and not in the lineages leading to the other
species.

Such persistent maintenance of visual polymorphism is almost certainly due in part to persistent visual selection. But the geographical and local variability in gene frequencies in *Philaenus* populations
cannot be explained solely on the basis of this mode of selection

(Halkka 1964). In the present paper, an attempt is made to estimate the relative significance of the various components, visual and non-visual, of the selection regime of *Philaenus*.

MATERIAL

In the years 1969-75, altogether 40737 *Philaenus* nymphs were isolated on their food plants in tiny plastic boxes or "miniature cages" (for the method, see Halkka *et al*. 1970). From these minicages, 27872 adults emerged and were classed by sex and colour morph.

The minicages were used on 12 small rocky islands, situated in the Tvärminne archipelago on the Finnish side of the Gulf of Finland. Maps of the study area have been published by Halkka *et al*. (1970). The habitats of *Philaenus* in this archipelago, the "miniature meadows", are distinct entities with sharp borders. The meadows are like tiny (5 m^2 and over) islands in the "seas" formed by the predominantly bare granite surfaces of the true islands. They might be called "islands of second degree". In such meadows, it is often possible to isolate well over 95% of the *Philaenus* nymphs. With the minicage method, successive generations of the univoltine *Philaenus* can be investigated with equally high efficiency.

The minicage method is very tedious, however, and unsuitable for large populations (2000 individuals and more). From such populations, samples were taken with sweep nets (see Halkka *et al*. 1970). Altogether, about 20 populations in the Tvärmine archipelago have been investigated by this method. In both the minicage method and the sweep net method, all the individuals were released as soon as their sex and colour phenotype had been noted.

Populations living on the Finnish mainland and in other European countries have been examined exclusively by the sweep net method.

MODES OF SELECTION IN *PHILAENUS*

Visual selection.

   Although the ecology of *Phileanus* has been dealt with in hundreds
of publications, very little attention has been paid to predators,
either invertebrate (see Harper and Whittaker 1976) or vertebrate (see
Halkka and Kohila, in press). Harper and Whittaker (op. cit.) released
radioactive *Philaenus* nymphs or adults in the field and, after a short
interval, measured the radioactivity in potential invertebrate predators. Levels of radioactivity significantly higher than the background
were recorded in a number of Coleopterans and Araneid and Phalangid
spiders. But, according to Harper and Whittaker (1976), the rate of
predation was too low to be of any significance in the maintenance of
colour polymorphism in the prey.

   The best guide to potential vertebrate predators of *Philaenus*
is the rather extensive literature dealing with predator - prey relations in meadow habitats typical of the spittlebug. Food lists are
available for many bird species. Halkka and Kohila (in press) found
86 papers listing the food of birds and other predators living in meadow habitats occupied by *Philaenus*. Altogether 10 species of birds and
one Amphibian (*Rana temporaria*) have been reported to prey on *Philaenus*. But with each predator species the intensity of predation was
found to be either uniformly very low or only sporadically intense.
Hence, it appears premature to give any detailed estimate of the significance of visual selection in the maintenance of the colour polymorphism. The available evidence indicates, paradoxically, that the intensity of selection by predators is too low to account for the presence of the great variability observed, with at least 11 colour morphs
present in continental populations of *Philaenus spumarius*.

Climatic selection.

   Colour morph frequencies in European *Philaenus* populations display
distinct clines and area effects (Halkka 1964, Halkka and Mikkola 1965,
Halkka, Raatikainen and Vilbaste 1975). Most of the clines observed
run south - north direction and closely parallel to temperature gradients. For many of the area effects, it is possible to offer an explanation based on the climatic peculiarities (such as sharp differ-

ences in humidity) of the region in question. (For the term "area effect", see Cain and Currey 1963).

The clines and area effects seen in colour morphs may be caused by
some factors that change concomitantly with climate. But although alternative explanations cannot be dismissed, geographical variability
in climate appears to be the most probable cause of the geographical
variability in colour allele frequencies in *Philaenus*. For this reason,
selection by climate is here taken as one of the three main components
of the selection regime of this species.

The clines in allele frequency can only with great difficulty be
explained by supposing differential predation of the different morphs
along the cline. The clines, therefore, provide strong evidence for
non-visual selection and give substance to hypotheses postulating linkage of the colour alleles with "ecophysiological" alleles. (For the
genetics of the colour polymorphism, see Halkka *et al.* 1973).

Site-specific (niche-specific) selection.

i) *Food plants as subniches*. Fortunately, there have been many
studies on the ecology of *Philaenus*, both in Europe (see Halkka *et al.*
1967, Witsack 1973) and in N. America (see Weaver and King 1954). The
following points appear to be of genetic significance: 1) The species
is univoltine, 2) it overwinters in the egg, passes through the nymphal stage in about 5 weeks and has maximum adult life-span of about
3 months, 3) the period between copulation and oviposition lasts several weeks, and the eggs, 100-200 in total, are laid in small groups
over a period of 2-4 weeks, 4) both nymphs and adults tap the xylem of
various herbs and (occasionally) grasses, more than 95% of the nearly
4000 recorded food-plant species being dicotyledons.

In view of its extreme polyphagy, *Philaenus* appears at first sight
to be an almost impossible species in which to study ecological genetics. But the spittlebug is far from evenly distributed among its
thousands of food plants. Certain plants, most of them tall and robust
perennial herbs, stand out as favourite sources of xylem. Of 40737 nymph-
food plant combinations recorded in Finland in 1969-75, almost one-fourth
(9008 or 22.1%) pertained to a single plant species, the meadowsweet,
Filipendula ulmaria.

In the island populations investigated during the present study,
the nymphs feed on 65 different species of plant. The 10 most favoured
food-plants harbour about 87.7% of the nymphs. The total niche of the

species can be divided into 65 subniches, labelled according to the names
of the food-plants. The subniches so defined can be characterized in
terms of the ecology of the plant species. Among the variables, certain
peculiarities of microclimate, as well as a specific combination of invertebrate and vertebrate predators, are associated with each food-plant.
ii) *Subniche-specific variability of survival rate.* Survival rates of
nymphs isolated in minicages at the 3rd to 5th instar vary greatly according to the food-plant species (Table 1). As subniches, the plant
species included in the table vary in carrying capacity from year to
year. In other words, there is subniche-specific oscillation in the selection coefficient, with dissimilar amplitudes, minima and maxima for
the different subniches (Table 1).

   The minima, or bad-year carrying capacities, are much more variable
than the maxima. In fact, the variability of the minima in Table 1
mainly reflects the drought tolerance of the food-plants. In the very
dry summer of 1973, many dead or dying nymphs were isolated in minicages on plants that were withering. Measured in terms of survival of
late-instar nymphs, environmental heterogeneity is greatest when conditions are at their worst.

   On the most favoured food-plants, the survival percentages are
fairly high during the late nymphal instars (Table 1). This means that
many of the losses curbing a rapid increase of population size in this
fecund animal must occur in the early nymphal instars. The small fraction of nymphs surviving to adulthood (1-2%) is highly selected, and
without doubt differentially on different food-plants (Table 1).
Two essential features of the evolutionary strategy of *Philaenus* seem
to be high fecundity and high survival of eggs. When overwintered eggs
resulting from outdoor cultures are provided in spring with luxurious
greenhouse plants, a single pair may give rise to 96 adult individuals
(the record progeny from crossing experiments made in Finland). Witsack (1973; Central Europe) has shown that in non-conditioned outdoor
cultures a single spittlebug female may lay as many as 200 eggs. Thus,
in *Philaenus*, egg-to-adult viability is an important component of fitness. Further, the good winter survival of the eggs indicates that the
decisive factor is the fitness of the nymph.
iii) *Distribution of the colour morphs on the food-plants.* Table 2
shows the distribution of the morphs on the seven most important food-
plants. The table has been compiled by pooling frequency data from several island populations of *Philaenus* gathered in 6 successive years.
In the table, FU = *Filipendula ulmaria,* LV = *Lysimachia vulgaris,* SV =

Table 1. Survival percentages of *Philaenus* nymphs on the ten most important food-plants of the species in 1969-74. The last column gives the values for the worst year (for most plants 1973) and the best year (mostly 1969).

	adults	dead nymphs	total number of individuals	survival percentage mean	survival percentage range
Filipendula ulmaria	5482	3126	8608	63.7	20.0 - 95.0
Lysimachia vulgaris	5235	1347	6582	79.5	55.0 - 98.0
solidago virgaurea	3540	2658	6198	57.1	8.0 97.0
Lytr-um salicaria	3735	1089	4824	77.4	47.0 - 99.0
Chrysanthemum vulgare	1052	405	1457	72.2	26.0 -100.0
Potentilla palustris	1096	349	1445	75.8	63.0 - 97.0
Tripleurospermum maritimum	732	295	1027	71.3	11.0 - 99.0
Achillea millefolium	690	185	875	78.9	10.0 -100.0
Rymex acetosa	259	405	664	39.0	7.0 - 80.0
Angelica silvestris	204	450	654	31.2	11.0 - 89.0
Other food-plants	3278	1415	4693	69.8	0.0 -100.0
Group totals and mean survival percentage	26323	11724	38047	69.2	

Table 2. Niche width ard niche overlap of phenotypes of female *Philaenus* as shown by percentages bred from nymphs feeding on the different platn species. On account of the high frequency of *typ* (p^t/p^t), most non-*typ* phenotypes are heterozygous for p^t, and hence the table shows the distribution of genotypes (alleles).

	FU	LV	SV	LS	PP	CV	TM	other food-plants	number of plant species utilized	total number of females	frequency of morph (%)
typ (p^t)	23.8	21.3	13.7	13.2	4.2	4.2	3.1	16.5	44	10235	80.5
tri (p^T)	9.4	26.4	12.7	25.0	4.2	1.4	5.7	15.1	22	212	1.7
mar (p^M)	27.5	17.6	12.3	13.7	4.3	3.6	2.6	18.2	30	839	6.6
lat (p^L)	33.9	9.8	15.5	17.4	2.9	2.2	2.5	18.2	27	407	3.2
fla (p^F) (p^C)	21.7	13.5	6.8	27.0	4.6	3.0	5.1	14.3	18	237	1.9
ice (p^C)	28.2	16.8	19.7	6.0	6.0	4.3	2.6	16.5	23	351	2.8
lop (p^O)	12.4	28.4	20.0	7.0	4.9	6.5	2.1	18.6	22	429	3.4
totals and food plant utilization (%)	3037	2632	1764	1708	535	525	396	2123		12710	
	23.9	21.0	13.9	13.4	4.2	4.1	3.1	16.7			100.1

Solidago virgaurea, LS = *Lythrum salicaria*, PP = *Potentilla palustris*, CV = *Chrysanthemum vulgare*, TM = *Tripleurospermum maritimum*. The full names of the colour morphs included in the table are to be found in previous papers on *Philaenus* (see e.g., Halkka *et al.* 1973).

For the set-up comprising the absolute values corresponding to Table 2, a chi-square value was calculated and found to be significant (χ^2 = 273.5, d.f. 42, P < 0.001). The distribution of the colour morphs on the food-plants thus deviates clearly from that predicted from the null hypothesis. But the biological significance of this deviation is not easy to deduce. As shown by Cochran (1954) and others, pooling of frequencies from different populations may result in false correlations. For this reason, it would be a mistake to conclude that, in Table 2, a positive or negative deviation from the mean percentage of food-plant utilization (bottom row) always indicates preference or repugnance for the food-plant concerned. In some of the meadows and islands, owing to the founder principle, some of the morphs and some of the food-plants may have attained exceptionally high frequencies, independently of each other.

Significant χ^2 values may be obtained for many of the colour morph - food-plant combinations in 2 x 2 setting during a single generation of a population inhabiting an isolated meadow or island. In each 2 x 2 setting, the number of colour morph x on food plant y and on all other food plants is compared with the number of all other colour morphs on y and on all other food plants (for examples, see Table 3). It is legitimate to sum chi-squares for successive generations and calculate a composite chi-square value, with degrees of freedom equal to the number of generations (Cochran 1954). In an instance like the present, such composite chi-squares are meaningful only on the condition that all the partial chi-squares contributing significantly to the end result deviate in the same direction from the values predicted from the null hypothesis. (This holds for Table 3).

The composite χ^2 values are often statistically significant when separate values are summed for 3 or 4 successive generations (years), but not for longer periods. In one instance, a colour morph preferred a food-plant for 5 or 6 years. On the island of Mellanspiken, the morph *lat* (*lateralis*) favoured *Filipendula ulmaria* in 1970-72 and 1974-75 (data for 1973 are lacking) (χ^2 = 11.47, d.f. 5, P ∼ 0.05).

We do not yet understand why chi-square significance breaks down in practically every population when the summing procedure is extended over four or more generations. There are many possible explanations.

Table 3. Occurrence of the colour morph *mar* on *Filipendula ulmaria*
on the island Stora Bönholmsgrundet (x^2 = 10.24, d.f. = 3,
P < 0.05)

	mar	other morphs
1969		
Filipendula ulmaria	8	65
other food-plants	11	243
1970		
Filipendula ulmaria	5	58
other food-plants	33	668
1971		
Filipendula ulmaria	7	84
other food-plants	24	712

For example, the environments provided by the meadows and the food-plants may not remain constant in time (see Table 1). In many of the meadows, the ground slopes gently, and the meadow has a wet and a dry end. The humidity gradient among the slope may be very different in wet and dry years. From what we know about correlations between colour morph frequencies and macroclimate (see p. 000), it appears possible that the morphs have different microclimatic optima. If a morph tended to prefer the same microclimate in successive generations, it would appear at different levels on a slope in wet and dry years. But in a succession of years with a similar microclimate, the morph would remain in the plant stand that occupied the zone supplying its favoured microclimate. The results of chi-square calculations in a 2 x 2 setting might then suggest that the morph favours a given plant species when in reality the plant merely happened to occupy the microclimatic optimum. If this were so, chi-square significance would break down when the microclimatic optimum shifted away from the plant stand.

Thus the plant - colour morph combinations may depend on the weather of the study year or on the varying qualities of the sites in which the populations live. To test for such possible dependence, an analysis of variance model was constructed for estimating the effect of year, meadow and food-plant, and the pair-wise interactions of these. In this model

$$y_{jst} = m+b_j+c_s+d_t+(bc)_{js}+(bd)_{jt}+(cd)_{st}+e_{jst}$$

where y_{jst} is the relative frequency of the morph studied among the spittlebugs on plant type j in meadow s in year t and e_{jst} is the error term. With this model, there is no need to measure plant species coverages; however, the model only applies to cases in which there are no zero frequencies, i.e. to the two most common colour morphs *typ* (*typicus*) and *mar* (*marginellus*). The coefficients are subject to the usual restriction that their sum over any appearing index is zero (Table 4). This means that e.g. the linear coefficients b, c, d represent deviations from the general mean m.

From the morph - food-plant combinations, sets of data in balanced design were selected with the imposed restriction of only one observation per cell and consequent omission of 3-factor interactions. The model is thus orthogonal, permitting partition of total variance into components due to plant, meadow or year, and their pairwise interactions.

Although some of the single-factor or interaction coefficients are high, no general trend in the distribution of effects is evident. The model was applied to seven orthogonal set-ups with the result that, in five instances, the only factor found to affect the frequency of either *typ* or *mar* was the meadow (Table 4). For the food-plant, sub-significant values (P = 0.084 or 0.088) were obtained in two set-ups. Thus the analysis did not establish any effect of the food-plant on the frequencies of the two commonest colour morphs. Obvious weaknesses of the model, in particular the inequality of the variances in the relative frequencies, may have impaired its power of detecting weak but real effects.

Table 4. Summary of analysis of variance results for the seven data sets. Subdivision of the multiple correlation squared (R^2) (= percentage of variance explained) into its components. The columns show the components of R^2 due to different variables and their pairwise interactions. Rows: 1 *typ* on *Potentilla palustris* (PP), *Lythrum salicaria* (LS), *Lysimachia vulgaris* (LV), *Filipendula ulmaria* (FU) and other food plants (OFP) in the islands Östra Mellanspiken (ÖM; meadow B) and Porskobben (P) in 1972-75. 2a *typ* on FU and OFP in the islands Allgrundet (A) and ÖM in 1972 and 1974, 2b *mar* in the same setup, 3a *typ* on FU and OFP in ÖM, A (two meadows) and Bönholmsgrundet (B) in 1969-71, 3b *mar* in the same setup, 4a *typ* on LV and OFP in B and P in 1969-71, 4b *mar* in the same setup.

Data set	R^2	year	meadow	plant	year x meadow	year x plant	meadow x plant
1	.7672	.0417	.0207	.2069	.0958	.2203	.1817
2a	.9926* [a]	.0013	.9443** [b]	.0147	.0198	.0005	.0118
2b	.9928*	.0002	.8911**	.0024	.0541	.0408	.0040
3a	.9149	.0146	.6838**	.0139	.0478	.0856	.0690
3b	.9681**	.0117	.8483***	.0205	.0371	.0229	.0274
4a	.9644	.1516	.5925*	.1739	.0093	.0249	.0120
4b	.8401	.1845	.2773	.2011	.0976	.0694	.0100

a: indicates the significance level of the whole model and
b: the significance level of this set of coefficients in the whole model.

iv) *Meadow-specific morph frequencies.* In five instances the significance of the "meadow effect" is a trivial result, as direct observations have shown that in many of the meadows the colour morph frequencies are site-specific and unique. Each meadow differs from all the others in the distribution and coverages of the prime food-plants of *Philaenus* and is a unique mosaic composed of microclimatically dis-

Table 5. Estimates of coefficients and analysis of variance for data
set. (For abbreviations, see the legend to Table 4)

	meadow			plant				
	Ö.M. B	Porsk.	PP	LS	LV	FU	others	
meadow				-.095	.056	-.014	.068	-.014
Ö.M. B	.018	Interactions		.083	-.064	.039	-.050	-.007
Porsk.	-.018			-.083	.064	-.039	.050	.007
year								
1972	-.043	.063	-.063	-.151	-.003	.094	.009	.051
1973	.029	.004	-.004	.001	.035	-.001	-.053	.017
1974	.004	-.024	.024	.045	-.029	.028	.028	-.071
1975	.009	-.042	.042	.103	-.001	-.121	.015	.003

Analysis of variance

Source of variation	Sum of squares	DF	mean square	F	P
year	.02820	3	.00940	.718	.560
meadow	.01401	1	.01401	1.071	.321
plant	.13971	4	.03492	2.667	.084
year x meadow	.06471	3	.02157	1.647	.230
year x plant	.14878	12	.01239	.947	.537
meadow x plant	.12270	4	.03067	2.343	.113
residual	.15714	12	.01309		
total	.67524			1.465	.246

similar elements.

The frequencies of the morphs thus seem to correlate significantly with the ecological associations composed of the food-plants (and
their concomitant variables) but have not been found to correlate for
long periods with any one of the component plant species.

A number of physical and biotic factors seem to influence the colour morph - food-plant associations in different ways in different
growth seasons.

The food-plant - colour morph associations have been worked out
for the 3rd to 5th nymphal stages. The adults are much more mobile
than the nymphs and probably feed on many food-plant species during
their lifetime. In *Philaenus*, the initial male majority changes to a
female majority during the preoviposition period. Thus the inseminated
female is the most effective propagule of the spittlebug. Before oviposition, the females may spend several weeks wandering far from their
original sites and may colonize new territories. Do the females, under
such circumstances, show fidelity, i.e. a tendency to oviposit on a
plant of the same species as that on which they fed as nymphs?

Some light is thrown on this question by the "clustering" observed
on many islands, including Bönholmsgrundet. On this island, the *marginellus* colour morph occurred in three consecutive years (1969-71) in
clusters in the *Filipendula* stands (Table 3). This clustering may reflect a preference of the egg-laying female for *Filipendula*, which
would suggest fidelity or homing.

Many of the populations included in the present study display
what might be called "allele protection". This occurs in two forms,
retention of many alleles in a very small population and retention of
a very rare allele in a large population. To give an instance of the
first form, five alleles (p^t, p^T, p^M, p^L and p^F) were retained through
the years 1969-73 in a population (Ostspiken) that changed in size as
follows: 8, 22, 41, 187, 225.

Retention of a very rare allele in a large population is exempli-
fied by the population of Gulkobben. In this population, the frequency
of p^T (the rare top dominant allele) in the 7 consecutive years (1969-
75) was 0.005, 0.005, 0.006, 0.002, 0.003 and 0.004. During these years,
the minimum size of the sweep net sample was 811, and the maximum 2109.

To study allele protection in relation to population size, the percentage of *typ* females (p^t/p^t) was plotted on semilog paper against
the total number of females (Fig. 1). These values gave a significant

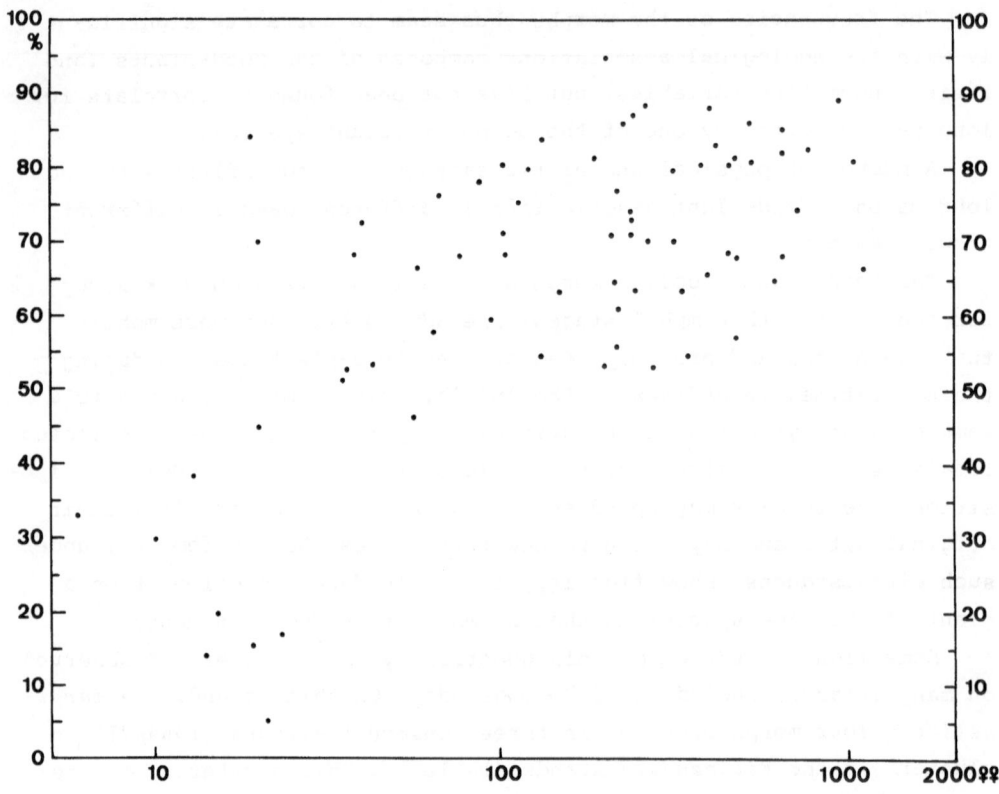

Figure 1. A correlation diagram showing the frequency of the *typ* phenotype (p^t/p^t) in the females plotted against the logarithm of the number of individuals of this sex.

positive correlation, r = 0.67.

The values in Figure 1 are for 6 successive years (1969-74) from each of 12 populations. This means that the variability can be split into two components, temporal and spatial. In analysis of covariance, the spatial (island-specific) component turns out to be much more explanatory than the temporal component (loss in multiple correlation squares 0.398 and 0.028 respectively).

DISCUSSION

The nymphs of *Philaenus* in their frothy "cuckoo-spit" masses change
position only about once a week (Halkka *et al.* 1967). They seldom
crawl more than 0.6 m from the previous sucking site, and may choose
a new site on the same plant. The adults move over a much wider range,
as they jump vigorously from plant to plant in the habitat - meadows
rich in species. In Levin's terminology, the environment is coarse-
grained for the nymph and fine-grained for the adult (Levins and Mac-
Arthur 1966). This mode of distribution of individuals at different
stages of the life-cycle corresponds nicely to the basic set-up for
Levene's (1953) well-known model of selection in a spatially heterogeneous environment.

Recently, certain specific conditions requisite for "multi-niche
polymorphism" or "protected polymorphism" (Prout 1968) have been postulated by Bulmer (1972), Christiansen (1974), Strobeck (1975) and others as refinements of Levene's (1953) model (see also Thoday 1972).
Under the conditions specified by Christiansen (1974), homing towards
the neighbourhood of the birthplace enhances the possibility of polymorphism (see also Levene 1953). The data from Bönholmsgrundet suggest
that in *Philaenus* protected polymorphism may be contingent on fidelity
to a certain set of food-plant species. This may, or perhaps more often may not, involve return to the birthplace. Such fidelity is a
means of securing optimal distribution of particular genotypes in
particular niches, without necessarily increasing the rate of inbreeding. Niche fidelity of this type also tends to maintain the integrity
of a polymorphic species, rather than leading to sympatric speciation
(cf. Maynard Smith, 1966).

The miniature meadows investigated for the present study are mo-
saics in two respects: plant stands and variability in microclimate. It
seems possible that both these factors affect the frequencies of the
colour morphs of *Philaenus*. The two factors are, of course, interdependent and vary temporally with the general character of the weather
of each growing season.

In *Philaenus* populations, the maintenance of rare alleles may depend on the continued presence of combinations of those plant species

that provide the favoured subniches of this allele. Allele losses may
be partially due to factors specifically affecting the rare niches
that support the rare alleles. Problems related to this principle were
discussed by Clarke (1972), but he was mainly concerned with a model
in which reduction of population size promotes homozygosity. In *Philaenus*, overwhelmingly commonest homozygote p^t/p^t seems to be less abundant in small than in large populations.

The Maximin strategy proposed by Templeton and Rothman (1974),
which complements the theories of Levins (1968), contains elements
which appear relevant in explaining changes in homozygosity in *Philaenus*. During periods of minimum fitness for the whole population,
droughts for example, the individual colour morphs might each be expected to show maximum fitness in a particular combination of subniches.

In the spittlebug, concentration of a large part of preadult mortality in the nymphal phase allows rapid expansion in good growth seasons. From 1971 to 1972, the size of one of the populations (Allgrundet)
increased from 468 to 4661.

The principle of protected polymorphism allows extremely fine adjustment of allele frequencies in less good years. In stable meadows
with practically unchanging food-plant coverages, the frequencies of
the colour alleles may remain almost the same for many generations.
During the study period, 1969-75, the best examples of stable gene
frequencies were the populations (both sampled with sweep nets) of
Lillhamn (χ^2 = 12.27, d.f. 12, P about 0.40)˙ and Rovholmen (χ^2 = 12.62,
d.f. 18, P about 0.80).

After the early nymphal stages, the next precarious period in the
life-cycle of *Philaenus* is the rather long interval (4-6 weeks) between
copulation and oviposition. Many of the colour alleles are expressed
in the females only, and this may reflect how greatly the population
depends for its continuance on the survival of the females during
this period. Apostatic selection (Owen and Wiegert 1962) and protection of the females by warning colouration (Thompson 1973) have been
claimed to operate during the preoviposition phase. If these types of
visual selection really are at work (cf. section "Visual selection"),
they may contribute to the longevity of the females. Visual selection,
perhaps more effective in large, dense populations, may tend to pro-
tect morphs other than *typ*. Possibly a balance exists between visual
and niche-specific types of selection.

The hypothetical structure here outlined, although worked out entirely from results of field studies, appears to fulfil a number of

predictions based on theory. The theory suggests that in natural populations the alleles (genotypes) are specialized for different subniches, and this is indicated by the differential distribution of the
colour genes in the different minimeadows. It further suggests that
natural subniches, (food-plant species) differ in their carrying capacity, and this, too, is observable in the populations investigated. In
Philaenus, each subniche contributes a unique component to the total
selection regime (Table 1).

Summary.

    The selection regime of the Homopteran *Philaenus spumarius* has been found to
comprise at least three components: 1) visual, 2) climatic and 3) niche-specific.
These, by their interaction, maintain a complex colour polymorphism. The genes (alleles) determining colour seem to be linked with genes governing climatic and niche-
specific polymorphisms. As a result of this linkage, colour variability may persist
in populations in which the intensity of visual selection is too low to maintain all
morphs.
    The various food plants of *Philaenus* constitute subniches of the total niche of
each population. Their influence is particularly important at the nymphal stage. In
this univoltine species the survival rate of 3rd to 5th instar nymphs varies from
year to year between limits which are specific for each subniche. This means that
the niche-specific component of the selection regime is divisible into subcomponents.
Differences in survival rates between subniches are most pronounced in years when
survival in general is poor. The selection coefficients of the different subniches
have different mean values and different amplitudes.
    In many of the populations investigated with regard to the temporal constancy
of the allele (or superallele) frequencies, the frequency fluctuates over a very
small amplitude. The precise maintenance of focal frequencies is probably contingent
on the multitude of subniche-specific selection pressures, which can be arranged in
pairs with opposite signs. In such a complex selection regime yearly changes in the
intensity of the individual selection pressures are likely to cancel each other out.

Acknowledgements.

    Investigations involving the isolation of tens of thousands of individuals in
natural populations necessitate teamwork. The names of the persons who have assisted in the fieldwork are to be found in other papers on the spittlebug; here I express my thanks to them collectively. For much of the planning and methodology of
the basic ecological investigations, the present author is deeply indebted to Professor Mikko Raatikainen. For devoted help in various phases of the work, thanks are
due to Liisa Halkka, Riitta Hovinen, Tarja Kohila, Juhani Lokki and Risto Väisänen,
and, specifically for checking the English, to Jean Margaret Perttunen. The statistical calculations were performed at the Computing Centre of the University of Helsinki. Expenses have been defrayed by grants from the University of Helsinki and
from the National Research Council for Sciences (Academy of Finland). This paper is
report no. 541 from Tvärminne Zoological Station.

REFERENCES

Bulmer, M.G. 1972. Multiple niche polymorphism.
  Amer. Natur. 106: 254-257.

Cain, A.J., and Currey, J.D. 1963. Area effects in *Cepacea*.
  Phil. Trans. Roy. Soc. Lond. Ser. B. 246: 1-81.

Christiansen, F.B. 1974. Sufficient conditions for protected polymorphism in a subdivided population.
  Amer. Natur. 108: 157-166.

Clarke, B. 1972. Density-dependent selection.
  Amer. Natur. 106: 1-13.

Cochran, W.G. 1952. The χ^2 test of goodness of fit.
  Ann. Math. Statist. 23: 315-345.

Cochran, W.G. 1954. Some methods for strengthening the common χ^2 tests.
  Biometrics 10: 417-451.

Frydenberg, O. 1963. Population studies of a lethal mutant in D. melanogaster. I.
  Behaviour in populations with discrete generations.
  Heredity 50: 89-116.

Halkka, O. 1964. Geographical, spatial and temporal variability in the polymorphism of *Philaenus spumarius*.
  Heredity 19: 383-401.

Halkka, O., Halkka, L., Raatikainen, M., and Hovinen, R. 1973. The genetic basis
  of balanced polymorphism in *Philaenus* (Homoptera).
  Hereditas 74: 69-80.

Halkka, O., and Kohila, T. 1976. Persistence of visual polymorphism, despite a
  low rate of predation, in *Philaenus spumarius* (L.) (Homoptera).
  Ann. Zool. Fenn. (in the press).

Halkka, O., and Lallukka, R. 1969. The origin of balanced polymorphism in the
  spittlebugs (*Philaenus*, Homoptera).
  Ann. Zool. Fenn. 6: 431-434.

Halkka, O., and Mikkola, E. 1965. Characterization of clines and isolates in a
  case of balanced polymorphism.
  Hereditas 54: 140-148.

Halkka, O., Raatikainen, M., and Halkka, L. 1976. Conditions requisite for stability of polymorphic balance in *Philaenus spumarius* (L.) (Homoptera).
  Genetica 46: 67-76.

Halkka, O., Raatikainen, M., Halkka, L., and Lallukka, R. 1970. The founder principle, genetic drift and selection in isolated populations of *Philaenus spumarius* (L.) (Homoptera).
  Ann. Zool. Fenn. 7: 221-238.

Halkka, O., Raatikainen, M., Vasarainen, A., and Heinonen, L. 1967. Ecology and
  ecological genetics of *Philaenus spumarius* (L.) (Homoptera).
  Ann. Zool. Fenn. 4: 1-18.

Halkka, O., Raarikainen, M., and Vilbaste, J. 1975. Clines in the colour polymorphism of *Philaenus spumarius* in eastern Central Europe.
  Heredity 35: 303-309

Harper, G., and Whittaker, J.B. 1976. The role of natural enemies in the colour
  polymorphism of *Philaenus spumarius* (L.).
  J. Anim. Ecol. 45: 91-104.

Levene, H. 1953. Genetic equilibrium when more than one ecological niche is available.
Amer. Natur. **87**: 331-333.

Levins, R. 1968. Evolution in changing environments. Princeton University Press, Princeton, N.J. 120 pp.

Levins, R., and MacArthur, R. 1966. The maintenance of genetic polymorphism in a spatially heterogeneous environment: variations on a theme by Howard Levene.
Amer. Natur. **100**: 585-589.

Maynard Smith, J. 1966. Sympatric speciation.
Amer. Natur. **100**: 637-650.

Owen, D.F., and Wiegert, R.G. 1962. Balanced polymorphism in the meadow spittlebug, *Philaenus spumarius*.
Amer. Natur. **96**: 353-359.

Prout, T. 1968. Sufficient conditions for multiple niche polymorphism.
Amer. Natur. **102**: 493-496.

Sokal, R.R., and Rohlf, F.J. 1969. Biometry, San Francisco, 776 pp.

Strobeck, C. 1975. Selection in a fine grained environment.
Amer. Natur. **109**: 419-425.

Templeton, A.R., and Rothman, E.D. 1974. Evolution in heterogeneous environments.
Amer. Natur. **108**: 409-428.

Thoday, J.M. 1972. Disruptive selection.
Proc. R. Soc. London B **182**: 109-143.

Thompson, V. 1973. Spittlebug polymorphic for warning coloration.
Nature **242**: 126-128.

Turner, J.R.G. 1967. On supergenes. I. The evolution of supergenes.
Amer. Natur. **101**: 195-221.

Weaver, C.R., and King, D.R. 1954. Meadow spittlebug.
Ohio Agric. exp. Sta. Res. Bull. **741**: 1-100.

Witsack, W. 1973. Experimentell - ökologische Untersuchungen über Dormanz - Formen von Zikaden (Homoptera, Auchenorrhyncha) 2. Zur Ovarial - Parapause und obligatorischen Embryonal - Diapause von *Philaenus spumarius* (L.) (Aprhophoridae).
Zool. Jahrb. Syst. **100**: 517-562.

4. ECOLOGY AND EVOLUTION

EFFECTS OF A VIRUS ON COMPETITION AND SELECTION IN BARLEY

Jens Sandfaer

Viruses occur widely in the plant kingdom and it seems probable that
they influence selection in plant populations in many cases. There are,
however, few cases in which the effects of a virus on selection in a
plant population have been studied in detail.

   At Risø we have studied the competition between some barley varieties. The competition in different variety mixtures was investigated
and, with a single exception, we found that the varieties in the mixtures competed exclusively for the same growth factors according to the
concepts developed by de Wit (1960) for plant competition; no other
form for interaction occurred between the varieties in these mixtures.
In a single mixture, however, we found that the two varieties not only
competed for the same growth factors, but that there was also another
form of interaction indicating that one of the components inhibited the
growth of the other. A detailed study of this mixture led us, by acci-
dent, to study the influence of a virus on competition and selection in
barley. In the following, I will review the results of these investigations (Sandfaer 1970a and b, 1973, and 1974).

BARLEY VARIETIES AND THE BSM-VIRUS

In our experiments we worked primarily with two two-rowed spring barley
varieties, Tystofte Prentice and Svalöf Freja. Prentice is an old Danish
variety, about 70 years old, whereas Freja is a more recent variety
first marketed in Denmark in 1943. Prentice is a late variety with long
straw. The grain yield of Prentice is about 14 per cent lower than that
of Freja, but its yield of straw is high. With regard to the number of
seeds per unit area, which is the most important character from a fitness point of view, Prentice has about 11 per cent fewer seeds than
Freja. From these data it was expected that in a mixture of the two varieties Freja would increase in frequency and Prentice would decrease.
We found, however, that the opposite was the case. In mixtures of the
two varieties the frequency of Prentice increased.

A closer study of the mixtures revealed that a sterility interaction occurred between the two varieties. Grown in pure stand, Freja had
about 3 per cent sterile flowers but when this variety was grown in an
1:1 mixture with Prentice its sterility increased to about 18 per cent.
Prentice grown in pure stand had about 16 per cent sterile flowers, and
when grown in mixture with Freja this percentage decreased to about 12
per cent. It took some time and required a considerable number of experiments before we found the solution to this rather unexpected sterility
interaction. The key discovery was that Prentice carries the barley
stripe mosaic virus (BSMV).

The BSM-virus is a seed-born virus and it is the only seed-born virus known in the grass family. The spread of the virus from diseased to
healthy plants is accomplished through leaf-to-leaf contact especially
during periods of air turbulence. Transfer of the virus through pollen
has been demonstrated experimentally, but this is not of significance
because barley is almost entirely self-pollinating. The most common leaf
symptoms are bleached, yellow or light-green stripes of various length,
light-green to yellow mottling, and sometimes almost entire yellowing
of the leaves. These symptoms, however, vary considerably depending -
among other factors - on the strain of the virus and the barley variety.
Prentice has almost certainly carried the virus for many years; however,
leaf symptoms are normally absent which probably explains why the virus
infection was not detected earlier. Infected plants are known to have
more sterile flowers than uninfected plants and, as will be described
later, the virus also induces triploids and aneuploids in the progeny
of infected plants. Only a certain percentage of the seeds produced by

a virus-infected plant are virus-infected, but in the field the virus
spreads from diseased to healthy plants through leaf-to-leaf contact.
Therefore, in each generation, a variety carrying BSMV consists of a
mixture of healthy and diseased plants, the latter being infected either
as seeds or during growth.

The occurrence of BSMV in Prentice and not in Freja, which was found
to be virus-free, explains the sterility interaction found in the 1:1
mixture of the two varieties. In the mixture, the virus is transferred
from Prentice to Freja causing the high sterility of Freja. The high
sterility of Prentice in pure stand was found to be due to the presence
of the virus. We also observed some decrease in the sterility of Prentice when grown in mixture. This is also expected: it can be predicted
that the spread of the virus from virus-infected to virus-free Prentice
plants is reduced when the virus-infected Prentice plants are diluted
with virus-free Freja plants.

THE EFFECT OF VIRUS-INDUCED STERILITY ON COMPETITION AND SELECTION

The effect of BSMV on the competitive relationships between Freja and
Prentice was investigated in an experiment comprising nine mixtures
containing different proportions of the two varieties. These nine mixtures were duplicated giving a total of 18 mixtures that were grown for
five consecutive years together with the two varieties in pure stands.
The results from the first and fifth years of the experiment are shown
in the replacement diagrams (Figure 1). To the left in each diagram is
shown the yield of Freja in pure stand. Moving to the right are mixtures
with a decreasing percentage of Freja. The dots around the falling curve
indicate the yield of Freja in the different mixtures. The straight line
indicates the yield expected from Freja in pure stand assuming that
there is no competition - no interaction - between the two varieties.
The observed yields are below the straight line, especially in the fifth
year. This agrees with expectations: it takes some time for Freja to become fully infected with BSMV. The observed yields for Prentice grown
in the different mixtures are rather close to those expected. The results show that in the mixtures the yield of Freja was suppressed with
no corresponding increase in the yield of Prentice.

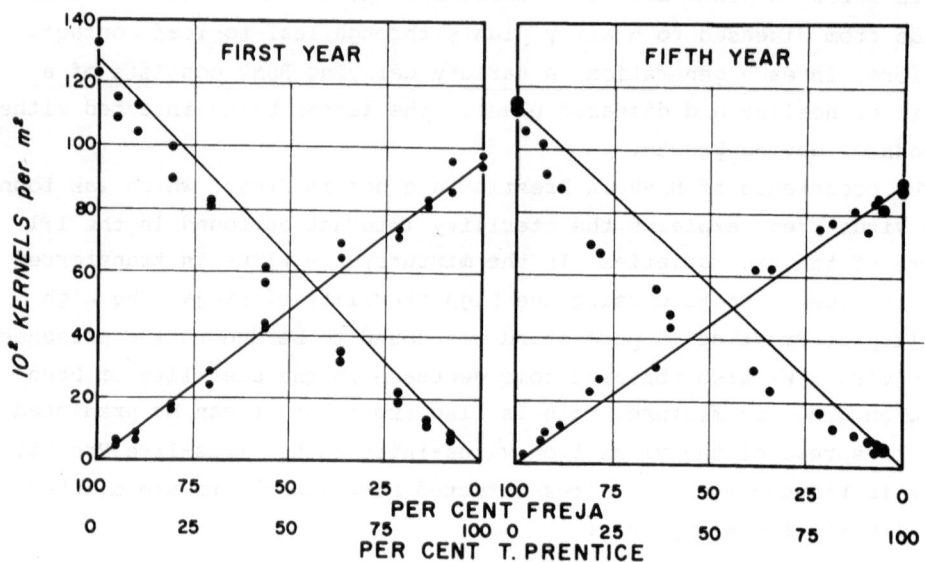

Figure 1.   Replacement series graphs for mixtures of Prentice and Freja.

The sterility in Freja grown in mixtures with different proportions
of the two varieties is shown in Figure 2 for the second and fifth years
of the experiment mentioned above. The sterility was not counted in the
first year of the experiment; at that time we did not realize that a
sterility interaction occurred in these mixtures. In the *second year*
of the experiment about three per cent sterile flowers were found in
Freja in pure stand, and in mixtures with low frequencies of Prentice
there was only a slight increase in the sterility of Freja. In these
mixtures there were only few Prentice plants from which the virus could
be transmitted to Freja. With increasing frequency of Prentice in the
mixtures, the sterility in Freja increased. This was the case until the
1:1 mixture where the sterility in Freja reached a plateau with about
20 per cent sterile flowers. The virus-induced sterility in Freja consequently depended upon the relative frequencies of the two varieties
in the mixture. In the *fifth year* there was increased sterility in Freja, also including mixtures with only low frequencies of Prentice. This
was also expected. Each year some Freja plants became virus-infected

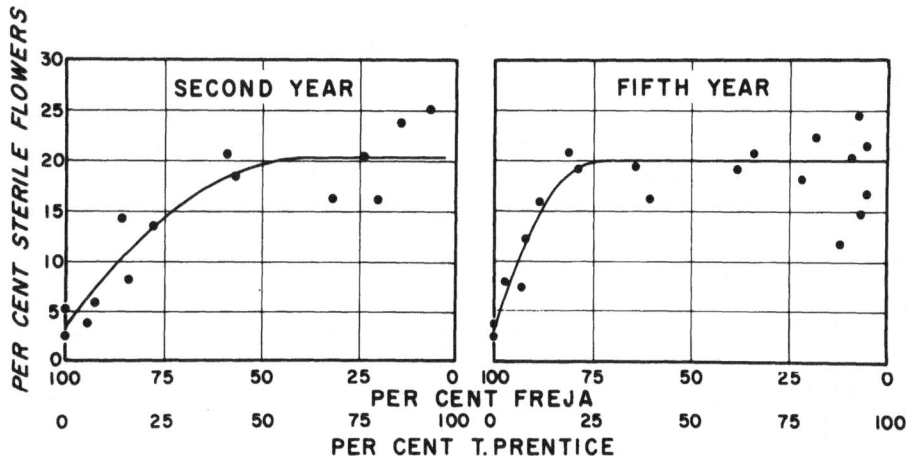

Figure 2. Percentage of sterile flowers in Freja grown in pure stand and in mixtures with different composition.

and in the next generation these plants also contributed to the spread of the virus. Had the experiment continued, Freja would probably have finally shown about 20 per cent sterile flowers in all mixtures.

Selection in these mixtures is illustrated in the ratio diagrams (Figure 3) for the first and the fifth years of the same experiment. The input ratio of the two varieties is on the abscissa, the ordinate axis shows the output ratio. The straight lines are based on the yield data from pure stands of the two varieties. In the *first year* the observational points from mixtures with a high frequency of Freja were scattered around the straight line, whereas the observations from mixtures with a low frequency of Freja (a high frequency of Prentice) were all below the line. The slope of the regression line calculated from the data from the mixtures was significantly greater than one (P < 0.05). The results from the first year show that selection in the mixtures was frequency-dependent. This is in accordance with expectations based on the frequency-dependent sterility interaction found in these mixtures (Figure 2). The results show that the minority component was at a selective disadvantage: in mixtures with a high frequency of Freja, this variety had a selective advantage, and in mixtures with a high frequency of Prentice, this variety had a selective advantage. In the *fifth year* the observations were below the line at all frequencies and the

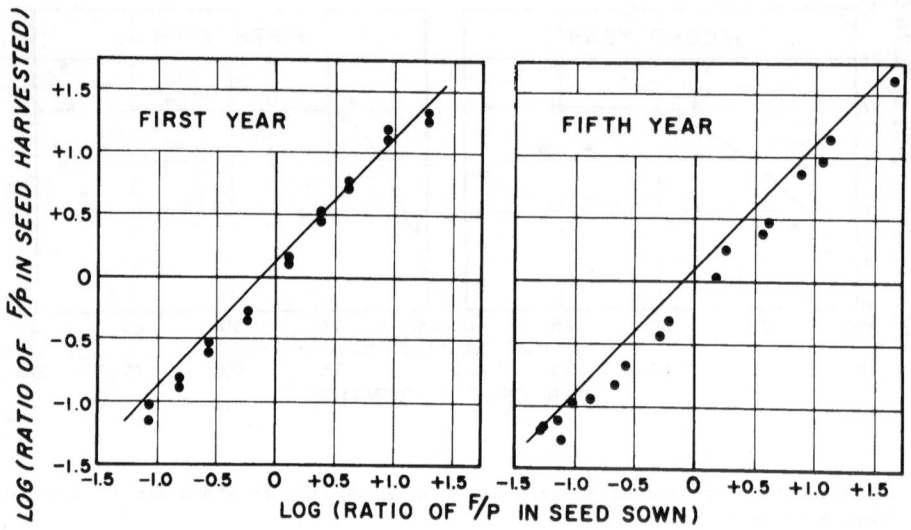

Figure 3.   Ratio diagrams for mixtures of Prentice and Freja.

slope of the regression line was not significantly different from one.
Freja had a selective disadvantage at all frequencies and revealed no
tendency to a frequency-dependent selection. The results show that there
was a change in the type of selection in these mixtures during the experiment. This is in accordance with the change in the sterility interaction in the mixtures resulting from the build-up of the virus infection in the Freja component of the mixtures.

THE EFFECT OF VIRUS-INDUCED TRIPLOIDS AND ANEUPLOIDS ON COMPETITION
                           AND SELECTION

Not all the BSMV-induced, so-called sterile flowers are really sterile,
i.e. without a viable embryo. In some cases an embryo starts to develop,
but development ceases at an early stage, resulting in a very light,
shrivelled grain which can easily be erroneously classified as s ste-

rile flower. It was found that some of these shrivelled grains were able to germinate and that a considerable fraction of the plants from these seeds were triploids and aneuploids.

This finding was investigated in more detail in an experiment in which eleven barley varieties were inoculated with BSMV and the frequency of triploids and aneuploids was determined in the progeny from these virus-infected plants, as well as in the progeny from non-inoculated, virus-free plants of the same varieties (Table 1). In the *virus-free material*, averaged over all eleven varieties, a low frequency of 0.04 per cent triploids was found together with some very few aneuploids. There was, however, considerable variation between varieties in the frequency of triploids. In the *virus-infected material* the frequency of triploids and aneuploids, averaged over varieties, was 1.76 and 0.38 per cent, respectively. From these data it is obvious that BSMV increased the fre-

Table 1. The frequency of triploids and aneuploids in BSMV-free and BSMV-infected material of eleven barley varieties

| Variety | Virus-free material | | Virus-infected material | |
| | Percentage of | | Percentage of | |
	triploids	aneuploids	triploids	aneuploids
Amsel	0	0	0.88	0.52
Atlas	0.03	0.02	1.07	0.27
Bomi	0.23	0	0.45	0.23
Bonus	0.02	0	1.18	0.23
Domen	0.01	0	0.19	0.04
Emir	0	0	0.86	0.14
Freja	0.03	0	1.97	0.70
Kenia	0.01	0.01	3.37	0.66
M. Baldric	0.04	0.01	3.77	0.52
Proctor	0.02	0	1.68	0.20
T. Prentice	0	0	3.14	0.45
Average	0.04	0.00	1.76	0.38

quency of triploids as well as of aneuploids. All varieties responded
to virus infection with an increase in the frequencies of triploids and
aneuploids, but with considerable variation in frequencies among varieties.

The BSM-virus comprises many strains characterized by their ability to induce leaf symptoms. In the previous experiments we used only a
single strain of the virus. In a further experiment, however, we compared the effect of six different strains of BSMV on the frequencies of
triploids and aneuploids in three barley varieties. The six strains differed in their ability to induce leaf symptoms ranging from so-called
mild to severe strains. Our data show that all six strains are able to
induce triploids and aneuploids in the three varieties analyzed. The
data do not indicate any clear connection between the ability of a virus
strain to induce leaf symptoms and its ability to induce triploids and
aneuploids. About the same frequency of triploids was found after inoculation with the mild strains as after inoculation with the severe
strains.

Barley triploids and aneuploids have a great percentage of sterile
flowers and their offspring consist primarily of aneuploids. It might,
therefore, be assumed that the occurrence of BSMV in a barley variety
not only decreases the number of seeds for the next generation, and in
this way reduces its selective value, but also causes a genetic deterioration of the population. This is probably not the case. The triploids
and aneuploids originate from very thin, shrivelled seeds. In a normal
seed-cleaning procedure nearly all of these seeds will be discarded.
Therefore, the frequency of triploid and aneuploid seeds sown in the
field will be low as a rule. The triploid and aneuploid plants usually
germinate a few days later and are very weak in the initial growth
stages. Their selective values are so low that it seems unlikely that
a significant number of these plants would be able to survive to maturity under the strongly competitive conditions normally present in a
barley field.

DISCUSSION AND CONCLUSIONS

The results presented demonstrate that BSMV influences selection in barley populations through induced flower sterility. In the present case frequency-dependent selection was found in the first years of the experiment but after some years the selection changed to frequency-independent. This could be explained from the changes observed in the sterility interaction in the mixtures, which changes were again found to be due to the spread of the virus in the mixtures.

The influence of a virus on competition between two grass species was reported by van Berg and Elberse (1962). In a study of the competition between *Lolium perenne* and *Anthoxanthum odoratum*, they found that a depression of the yield of *Lolium* in the mixtures was caused by a virus that was carried by *Anthoxanthum* and transferred to *Lolium* during the experiment. The virus was supposed to be harmful to *Lolium* only.

Our results also demonstrate that BSMV in barley has genetic effects through the induction of triploids and aneuploids. It is concluded that the selective values of triploids and aneuploids in all probabilities are so low that they constitute no significant risk of genetic deterioration in the populations.

Genetic effects of BSMV have also been demonstrated in maize. Sprague *et al.* (1963) and Sprague and McKinney (1966 and 1970) found in maize an effect of BSMV called Aberrant Ratio (AR). AR describes a phenomenon associated with virus infection in which F_2 and backcross ratios exhibit consistent and significant departures from Mendelian expectations at several heterozygous marker loci.

Different viruses have been shown in animal tissue cultures to affect the chromosomes causing chromosomal aberrations, deformation of chromosomes, and polyploidy (review in Nichols 1970). In our experiments several cases of chromosome fragments and chromosome damage were observed, indicating that the genetic effect of BSMV in barley is not restricted to causing the formation of triploids and aneuploids.

It is very difficult to establish when Prentice became infected by the virus. Prentice is an old variety that was in common use in Denmark from 1904 to about 1925. Perhaps Prentice carried the virus from its origin. Another obvious possibility is that the infection occurred at a later date during the long period in which Prentice was cultivated. There is, however, some support - to be presented in the following - for the hypothesis that BSMV was more common in barley in Denmark about

the turn of the century, and that Prentice has carried the virus for a
long period.

For many years Prentice has been characterized as a variety with
many sterile flowers. We now know that this is due to BSMV. We have
established a virus-free line of Prentice and the sterility in this line
is as low as in other virus-free varieties. This suggests that the virus
infection of Prentice is not of recent date.

Tests for BSMV have been carried out on a number of barley samples
from the variety collection at the Agricultural University, Copenhagen.
Six varieties were found to be virus-infected. They are all old varieties. Newer varieties had no virus even though several of them had been
maintained in the collection for a considerable number of years. This
suggests that the virus was previously more common in Denmark than it
is today. No virus infection has been detected in commercial samples of
the varieties commonly grown at present.

About the turn of the century, flower sterility in barley in Denmark
was recognized and studied by the Danish geneticist, Professor Wilhelm
Johannsen, as part of his classic studies on pure lines. In a lecture
to the Royal Danish Agricultural Society in 1898, he explained his reason for working with this character: He said that "... in the last
couple of years there has been an alarming increase in the flower sterility of my barley material which has caused me to investigate this
matter". (Johannsen 1899).

Johannsen carried out selection experiments for high and low frequency of sterile flowers in different barley varieties. The frequency
of sterile flowers differed between varieties. Goldthorpe and the Prentice varieties were found to have high flower sterility (Johannsen
1899). In all the varieties investigated, Johannsen was able to isolate
lines with a high frequency of sterile flowers (Johannsen 1903). In
lines selected from Goldthorpe he observed a very peculiar frequency
distribution of plants with different frequencies of sterile flowers;
distributions with two and even three maxima occurred. Johannsen analyzed a case with a double-peaked distribution in more detail. He found
that plants from the peak with a low frequency of sterile flowers were
true-breeding, whereas plants from the peak with a high frequency of
sterile flowers in the next generation produced a double-peaked curve.
This picture was also reproduced in the following generations (Johannsen 1905 and 1913). A possible explanation of the results could be that
the material was infected with BSMV. Selected plants with low sterility
were virus-free and therefore true-breeding with low sterility, whereas

selected plants with high sterility were virus-infected. As mentioned before, only a fraction of the progeny of a virus-infected plant are virus-infected and, therefore, in the next generation is it again possible to select virus-free plants with low sterility.

Johannsen (1899 and 1913) also observed that not only really sterile flowers occurred, but also flowers which produced thin, shrivelled seeds. The same observation was, as mentioned earlier, also made in our experiments after virus infection.

The observations mentioned above are not conclusive, but they yield some support to the hypothesis that BSMV was previously more common in barley in Denmark, and that about the turn of the century it tended to be a problem in barley breeding. Even though Johannsen's experiments did not identify the cause of the increased flower sterility, they yielded a sound basis for the elimination of the problem. Professor Johannsen recommended that plant breeders should practice strong selection, not only for fertile spikes, but also for fertile plants. Such selection, together with vigorous seed cleaning, and the discarding of all small shrivelled kernels (among which most of the virus-infected kernels are found, Inouye 1962) should reduce the frequency of BSMV. It is therefore not unlikely that, without knowing of the existence of the virus, plant breeders have eliminated it from their barley breeding material.

Summary.

Barley stripe mosaic virus (BSMV) influences competition and selection in barley through the induction of sterile flowers and the formation of triploid and aneuploid seeds. In variety mixtures a frequency-dependent selection was found in the first years of the experiment, but after some years the selection changed to frequency-independent. This could be explained from the spread of the virus in the mixtures. The frequency-dependent selection resulted from a frequency-dependent, virus-induced sterility interaction between the varieties in the mixtures.

BSMV caused a pronounced increase in the frequency of triploid and aneuploid seeds. The selective values of triploids and aneuploids are, however, so low that in all probability they constitute no significant risk of genetic deterioration of the population.

REFERENCES

Berg, J.P. van den and W.T. Elberse. 1962. Competition between *Lolium perenne* L. and *Anthoxanthum odoratum* L. at two levels of phosphate and potash. *J. Ecol.* 50: 87-95.

Inouye, T. 1962. Studies on barley stripe mosaic in Japan. *Ber. Ohara Inst. Landwirtsch. Biol. Okayama Univ.* 11: 412-496.

Johannsen, W. 1899. Nogle studier over variation og forædling med særligt henblik på Goldthorpe-byg. *Tidsskr. Landbr. Planteavl* 5: 63-86.

Johannsen, W. 1903. Om arvelighed i samfund og i rene linier. *Overs. Kgl. Danske Vidensk. Selsk. Forh. No.* 3: 235-294.

Johannsen, W. 1905. Arvelighedslærens elementer. Gyldendal, Copenhagen.

Johannsen, W. 1913. Elemente der exakten Erblichkeitslehre. Gustav Fisher, Jena. 2. Ausgabe.

McKinney, H.H. and L.W. Greeley. 1965. Biological characteristics of barley stripe-mosaic virus strains and their evolution. *Tech. Bull. No.* 1324, *Agr. Res. Serv. U.S.D.A.*: 1-84.

Nichols, W.W. 1970. Virus-induced chromosome abnormalities. *Ann. Rev. Microbiol.* 24: 479-500.

Sandfaer, J. 1970a. Barley stripe mosaic virus as the cause of sterility interaction between barley varieties. *Hereditas* 64: 150-152.

Sandfaer, J. 1970b. An analysis of the competition between some barley varieties. *Risø Report No.* 230: 1-114.

Sandfaer, J. 1973. Barley stripe mosaic virus and the frequency of triploids and aneuploids in barley. *Genetics* 73: 597-603.

Sandfaer, J. 1974. Triploids and aneuploids in barley varieties infected with different strains of barley stripe mosaic virus (BSMV). *Hereditas* 78: 326.

Sprague, G.F. and H.H. McKinney. 1966. Aberrant ratio: An anomaly in maize associated with virus infection. *Genetics* 54: 1287-1296.

Sprague, G.F. and H.H. McKinney. 1970. Further evidence on the genetic behaviour of AR in maize. *Genetics* 67: 533-542.

Sprague, G.F., H.H. McKinney and L. Greeley. 1963. Virus as a mutagenic agent in maize. *Science* 141: 1052-1053.

Wit, C.T. De. 1960. On competition. *Versl. Landbouwk. Onderz. Ned.* 66 (8): 1-82.

4. ECOLOGY AND EVOLUTION

SELECTION AND INTERSPECIFIC COMPETITION

Tom M. Fenchel and Freddy B. Christiansen

Other organisms constitute a component of the environment to which animals respond by natural selection. Among the different kinds of interspecific interactions, competition for common resources is attributed an important role in the evolution of the single species and for the structure of biological communities. The rapidly growing evidence for the importance of interspecific competition include the relationship between species numbers and environmental complexity, resource sharing among coexisting species, and character displacement, *viz.*, that where the distribution pattern of two congeners overlap, they differ more with respect to morphological, physiological, or behavioral characters (assumed or demonstrated to be related to resource exploitation) than where the same species occur allopatrically (e.g., Christiansen and Fenchel, 1977; Fenchel, 1975, a,b; Fenchel and Kofoed, 1976; Huey *et al.*, 1974; Hutchinson, 1959; MacArthur, 1972; Pianka, 1973, 1974).

A number of theoretical contributions (MacArthur and Levins, 1967; MacArthur and Wilson, 1967; Roughgarden, 1972) have considered the evolution of competing species. These models do not, however, consider outbreeding, Mendelian populations. The purpose of the present paper is to develop some general models which describe the evolution of competing species and thus of character divergence and niche width in Mendelian populations, and which include intraspecific competition and

a restricted resource spectrum. This approach is similar to that of
Roughgarden (1976).

THE NICHE AND EXPLOITATIVE COMPETITION

The niche relationships between an organism and its environment will
always be determined by a large number of parameters; these may be
classified into three main dimensions; time, habitat, and resource
(Pianka, 1973). We will in the following only consider a one-dimensional resource niche and the exploitative competition between two or more
species along such a resource axis. The niche concept used in this paper was originally developed by Levins (1968) and MacArthur and Levins
(1967). For further discussion on habitat and time niches and multidi-
mensional niche relationships see Christiansen and Fenchel (1977). However, as also shown by some examples in this contribution, the description of the competitive interactions along one resource axis is not always an unrealistic simplification of nature.

The niche of an organism with respect to a resource quality, ρ,
(e.g., food particle size) may be described as a utilization function,
U, which describes the rate at which the individuals of the population
exploit the resource (Fig 1A). The utilization function is in the following assumed to be Gaussian; this is often not unrealistic; also, other bell-shaped $U(\rho)$ curves will probably not alter the qualitative predictions of the models (Christiansen and Fenchel, 1977).

The utilization function may be resolved into two components:

$$U(\rho) = \omega f(\rho) \quad , \tag{1}$$

where f is a probability distribution, describing the preference with
respect to resource quality, so that

$$\int f(\rho) d\rho = 1 \quad .$$

The parameter ω measures the rate of consumption per individual so that
$1/\omega$ is proportional to the carrying capacity, K, of the species. The
definition (1) has a number of shortcomings. Thus it implies that the
species exploit the resources independently of their abundance, i.e.,
f is independent on the functional form of the resource distribution

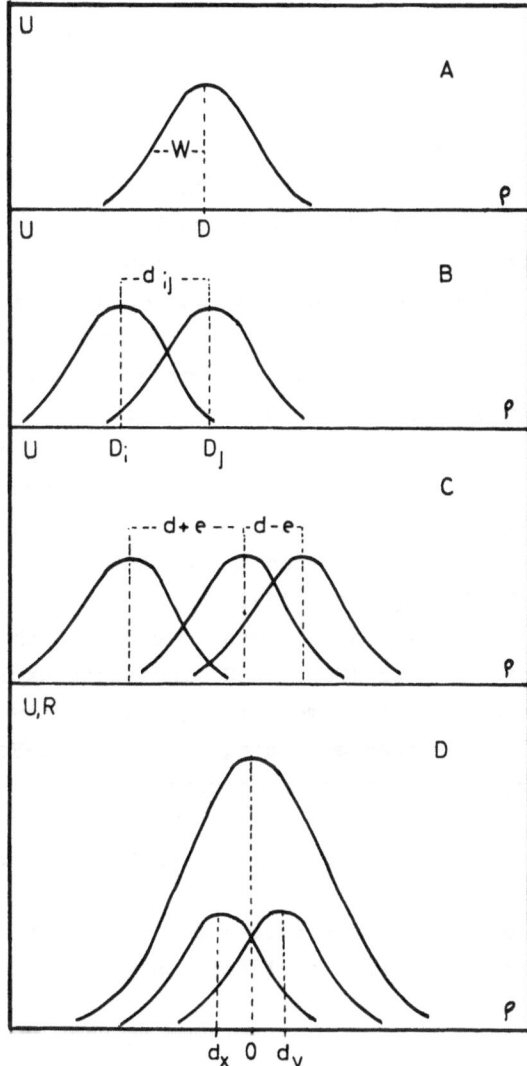

Figure 1.   The ecological niche and different niche interrelations along a single resource axis. For further explanation, see the text.

function R(ρ) and ω is independent of resource availability (Christiansen and Fenchel, 1977). The total resource utilized per time unit and individual is assumed to be

$$\int U(\rho)\ R(\rho)\ d\rho \quad .$$

We may characterize the niche by its width, W, which is the standard deviation of f(ρ); thus, e.g., a small W indicates that the animal in question is a specialist. The niche is further characterized by its modal position on the resource axis, D.

The niche width may be ressolved into three components which are: 1) individual variance in resource utilization (within phenotype component, V_w), 2) variation between individuals (between phenotype components, V_b), and 3) variation due to age structure (V_a). All three components may be influenced genetically and environmentally and V_b and V_a may have a genetic component. Given the genetical variance, the genetic components of V_b and of V_a probably usually respond more rapidly to selection than V_w (Christiansen and Fenchel, 1977; Fenchel, 1975 a,b; Roughgarden, 1972).

The partial or total overlap of the utilization curves of two or more species (Fig. 1B-C) indicates exploitative competition and the degree of overlap measures the intensity of the competition. If the populations grow according to the Lotka-Volterra equations, i.e., the growth of species i is given by

$$dx_i/dt = r_i x_i\ (K_i - \underset{ij}{\Sigma}\ \alpha_{ij} - x_i)/K_i \quad , \tag{2}$$

then various considerations show that a meaningful interpretation of the competition coefficients α_{ij}, in terms of the utilization functions U_i and U_j is

$$\alpha_{ij} = \left(\ \int U_i U_j\ d\rho\ \right) \Big/ \left(\ \int U_i^2\ d\rho\ \right). \tag{3}$$

In terms of the niche parameters d_{ij} (= $D_i - D_j$) and W (assuming W_i = W_j = W), it can be shown that definition (3) yields

$$\alpha_{ij} = \exp\left[-\ d_{ij}^2/(4W^2)\right] \tag{4}$$

(Christiansen and Fenchel, 1977; MacArthur and Levins, 1967).

Equations (2) imply that there is a limit to how similar species

can be with respect to resource utilization and still coexist in a homogeneous environment. If we assume identical carrying capacities and
niche widths, it can be shown that the limiting similarity (i.e., the
closest niche packing which does not imply competitive exclusion) is
approximately

$$d/w > 1, \tag{5}$$

a result which is fairly robust with respect to the details of a number assumptions (number of competing species, the exact shape of $U(\rho)$).
The theory of limiting similarity is discussed in details in Christiansen and Fenchel (1977). MacArthur (1972), and May (1974).

We have so far considered a homogeneous, unlimited resource spectrum. This is probably always an unrealistic assumption, usually the
resource spectrum will be limited at its borders. We will therefore also consider cases where the resource spectrum itself has a Gaussian
distribution with a variance given by σ^2 (Fig. 1D). In such a system,
the carrying capacity of species i will be a function of D_i. This can
be described by the function

$$K(D) = \exp\left\{-D^2/\left[2(\sigma^2 + w^2)\right]\right\} \tag{6}$$

which is the carrying capacity, K, measured relative to the carrying
capacity, had the niche position been in the center of the resource
spectrum, so that $D = 0$ (Christiansen and Fenchel, 1977).

THE EVOLUTION OF COMPETING SPECIES

In order to study the effect of selection on a system of competitors
along one resource axis we will use a basic set up consisting of three
species; x, y, and z (Fig. 1C) which grow according to a discrete generation version of (2), such that

$$
\begin{aligned}
x' - x &= Rx\,(1 - x - \alpha_x y - \beta z) \\
y' - y &= Ry\,(1 - \alpha_x x - y - \alpha_z z) \\
z' - z &= Rz\,(1 - \beta x - \alpha_z y - z)
\end{aligned} \tag{7}
$$

where the left sides are the population increments in one generation, $R_x = R_y = R_z = R$ are the finite growth rates (equality used only to make stability conditions for a three species equilibrium simple) and R is small and positive, $\alpha_{xy} = \alpha_{yx} = \alpha_x$, $\alpha_{yz} = \alpha_{zy} = \alpha_z$, $\alpha_{xz} = \alpha_{zx} = \beta$, and where $K_x' = K_y' = K_z' = 1$. We also assume that in species y there is a genetic variation which influences the competition with x and z and that it is bound to a single autosomal locus with two alleles, A and a with the frequencies p and q (= 1 - p) respectively. The structure of the population is given by the following table:

Genotype:	AA	Aa	aa	Σ
Number:	y_1	y_2	y_3	y
Frequency:	p^2	2pq	q^2	1
Competition coefficients	$\begin{cases} \alpha_{x1} \\ \alpha_{z1} \end{cases}$	$\begin{matrix} \alpha_{x2} \\ \alpha_{z2} \end{matrix}$	$\begin{matrix} \alpha_{x3} \\ \alpha_{z3} \end{matrix}$	

In terms of the parameters of (7) we have that

$$\alpha_x = p^2\alpha_{x1} + 2pq\alpha_{x2} + q^2\alpha_{x3} \qquad (8a)$$

$$\alpha_z = p^2\alpha_{z1} + 2pq\alpha_{z2} + q^2\alpha_{z3} \qquad (8b)$$

and each of the genotypes changes through a generation according to

$$y_i' - y_i = Ry_i (1 - \alpha_{xi}x - y - \alpha_{zi}z). \qquad (9)$$

The change in gene frequency, finally, is given by

$$p' - p = \left[(y_1' + y_2'/2)/y' \right] - p$$

$$= \left[(y_1' + y_2'/2) - py' \right] / y'. \qquad (10)$$

Using this basic set up, we will in the following sections explore the effect of selection under different assumptions with respect to species numbers, shape of the resource spectrum, the components of niche width, and the presence of intraspecific competition. In most cases details of formal proofs of the models will not be given in the following;

the reader is referred to Christiansen and Fenchel (1977) for a more
detailed treatment.

The evolution of a two species system.

We will first consider a two species system (like (7) where z is
absent) in a homogeneous resource spectrum. When the expressions for
y_i' and y' from (9) and (7) are substituted into (10) we have

$$p' - p = -R[\ y_1 x \alpha_{x1} + y_2 x \alpha_{x2}/2 - pyx\alpha_x]/y'. \qquad (11)$$

When α_x from (8a) and the genotype frequencies from the table are substituted into (11) we get

$$p' - p = (-Rpqxy/y')\ \left[p(\alpha_{x1} - \alpha_{x2}) + q(\alpha_{x2} - \alpha_{x3}) \right] \qquad (12)$$

Now, substituting $(1 - p)$ for q in (8a) and differentiating with respect to p yields

$$\delta\alpha_x/\delta p = 2\,[p(\alpha_{x1} - \alpha_{x2}) + q(\alpha_{x2} - \alpha_{x3})]\ . \qquad (13)$$

Comparing (13) and (12) reveals that $p' - p$ has the opposite sign of
$\delta\alpha_x/\delta p$ so that the change in gene frequency will always be so that α_x
decreases. If the genes A and a affect the niche position while W_y is
kept fixed, this means that d_{xy} will increase without bounds, while if
d_{xy} is kept fixed then W_y will decrease. We have thus demonstrated character displacement. Note, however, that we have ignored intraspecific
competition and assumed an infinite resource spectrum; the predictions
that d_{xy} will grow infinitely or that W will totally collapse are not
realistic.

We will here shortly review the evolution of a two species system
in a limited resource spectrum. Assume a Gaussian resource spectrum with
the variance σ^2 and with a modal position of 0 along the resource axis (Fig. 1D). Further assume the presence of two competing species, x
and y, with a fixed niche width W and with a carrying capacity of 1,
had their niche positions been 0. Assume further that there is genetical
variance which allow the species to move their actual niche positions,
d_x and d_y. From (4) and (6) we have that

$$\alpha = \exp[- (d_x - d_y)/(4W^2)] \qquad\qquad (14)$$

and

$$K_i = \exp[- d_i^2 /2 \, (\sigma^2 + W^2)] \qquad\qquad (15)$$

We finally assume that x and y grow according to (2) which yields the the equilibrium population size of x as

$$\hat{x} = K_x - \hat{y}\alpha \; .$$

Now if a genotype is present which changes the niche position from d_x to d_x' and thus K_x to K_x' and α to α', then it will increase in the population if

$$(K_x' - K_x) - \hat{y}(\alpha' - \alpha) > 0 \; . \qquad\qquad (16)$$

Substituting (14) and (15) into (16) (and after long calculations) the shifts in niche positions (given as d/W) by the two species from some initial positions and the globally stable two species equilibirum for different values of the resource spectrum width, σ, can be calculated (Fig. 2). It can be shown that the expression $W^2/\sigma^2 (< 1)$ equals the equilibrium value of α; thus competition becomes intenser as the niche widths approach the width of the resource spectrum. If $W/\sigma > 1$, then the evolutionary stable coexistence is not possible.

The evolution of a three species system.

We will now consider a three species system (Fig. 1,C) where the center species is subject to "diffuse competition". The populations are assumed to grow according to our general system (7) and we will first ignore intraspecific competition.

Figure 2. The evolutionary shift in niche modes for two competing species with fixed niche widths, W, in a Gaussian resource spectrum, the width of which is given by $W^2/\sigma^2 = 1/3$. The solid lines are the stable points for d_x/W and the interrupted lines the stable points for d_y/W. Below: The globally stable niche positions for two competitors with fixed niche widths W in Gaussian resource spectra with variances measured as W^2/σ^2 (see opposite page).

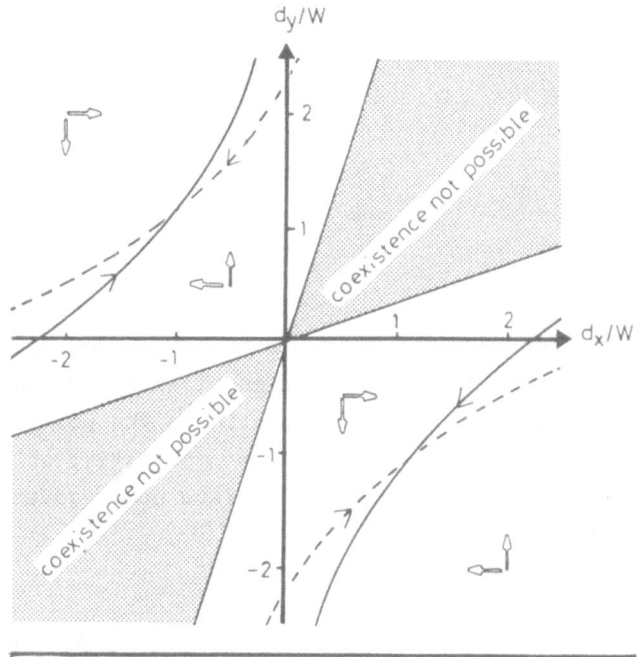

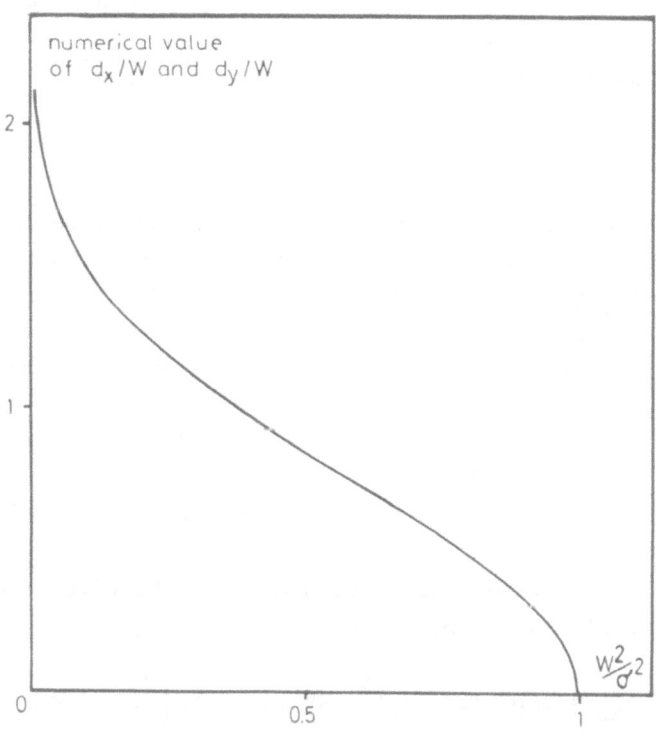

By a method similar to the one used for arriving at (12) we find
that

$$p' - p = (-Rpqy/y') [p(\alpha_1^* - \alpha_2^*) + q(\alpha_2^* - \alpha_3^*)] \qquad (17)$$

where

$$\alpha_i^* = x\alpha_{ix} + z\alpha_{iz} \qquad . \qquad (18)$$

Equation (17), however, is much more complicated than (12) since it is
a function of the population sizes and we cannot offer a general solution. We will instead investigate the possibility for the increase of
a rare gene A in a population consisting only of the genotype aa, so
that we can ignore the very rare homozygotes AA and thus terms of the
order p^2.

Equation (17) then yields

$$p' - p = (-Ry/y')p(\alpha_2^* - \alpha_3^*) \qquad (19)$$

which show that a rare gene will increase if

$$\alpha_2^* < \alpha_3^* \qquad (20)$$

i.e., if the heterozygote experiences less competition.

We will now consider the example shown in Fig. 1C. Each species has
a fixed niche width, W, but the modal distances between y and x and between y and z are (d + e) and (d - e) respectively.

The competition coefficients of this system is given by

$$\alpha_{x3} = \exp[-(d + e_3)^2/(4W^2)]$$

$$\qquad (21)$$

$$\alpha_{z3} = \exp[-(d - e_3)^2/(4W^2)] \qquad .$$

Introducing a gene which affects the size of e, the rare heterozygote
will have competition coefficients given by

$$\alpha_{x2} = \exp[-(d + e_2)^2/(4W^2)] = \exp\left\{-[(d + e_3) + (e_2 - e_3)]^2/(4W^2)\right\}$$

$$\qquad (22)$$

$$\alpha_{z2} = \exp[-(d - e_2)^2/(4W^2)] = \exp\left\{-[(d - e_3) - (e_2 - e_3)]^2/(4W^2)\right\} \qquad .$$

By Taylor expansion around $(e_2 - e_3)$ combined with (21) we find

$$\alpha_{x2} = \alpha_{x3}[1 - (\overset{\bullet}{e}_2 - e_3)(d + e_3)/(2w^2)]$$

$$\alpha_{z2} = \alpha_{z3}[1 + (e_2 - e_3)(d + e_3)/(2w^2)]$$

(23)

Calculations on the condition (20), using (21) and (23), show that for all $\alpha < 0.54$ (which is the limiting similarity which if exceeded, means the exclusion of y) a rare allele will increase if $e_2 < e_3$, or in other words; the niche of the middle species will migrate towards the midpoint position between the niches of the two competitors.

The effect of intraspecific competition on niche width.

As already discussed, the niche width of a population may be increased either by increasing individual variance in resource utilization or by increasing variation between individuals. We will first consider the last case, i.e., polymorphism at a locus which affects the position of the utilization functions of genotypes belonging to a species, y, which is sandwiched in between two competitors.

We have already found that in diffuse competition, the intermediate species will tend toward the midpoint between the competitors. If the heterozygote is closest to the midpoint, then polymorphism and consequently an increase in variance will be maintained.

Polymorphism will also relax intraspecific competition. If γ_{ij} is the competition coefficient describing the effect of genotype j on genotype i then we have that

$$\gamma_{ij} = \exp(-\varepsilon^2/4w^2) \approx 1 - \varepsilon^2/(4w^2)$$

(24)

when ε is a small displacement of the niche of genotype j relative to that of genotype i. It can be seen from (24) that the decrease of intraspecific competition is only of the order ε^2. From (23) it is seen that the interspecific competition is by a similar displacement relaxed by the order of $\varepsilon (= e_2 - e_3)$. Thus, as long as the modal position of y is not symmetric relative to that of x and z, stronger selection forces will act on the character divergence and niche increase will not take place. When y has reached the midpoint, however, $\alpha_x = \alpha_z$ and the selection pressure due to interspecific competition becomes independent of the first degree of ε, and we may expect that selection will tend to increase niche width.

Allowing the polymorphism of y of our system (7) to relax intra-specific competition requires changes in the equation (9) which de-scribes the growth of the genotype i, which now becomes

$$y_i' - y_i = Ry_1 (1 - \alpha_{xi}x - \sum_{j=1}^{3} \gamma_{ij}y_i - \alpha_{zi}z)$$

and the growth of the total population is given by

$$y' - y = Ry(1 - \alpha_x x - \sum_{i=1}^{3} \sum_{j=1}^{3} \gamma_{ij}y_i y_j/y - \alpha_z z) \; .$$

The change in gene frequency can then be found as

$$p' - p = (-Rpqy/y') \Big\{ [p(\alpha_1^* - \alpha_2^*) + q(\alpha_2^* - \alpha_3^*)]$$

$$+ [p(\bar{\gamma}_1 - \bar{\gamma}_2) + q(\bar{\gamma}_2 - \bar{\gamma}_3)] \; y \Big\} \qquad (25)$$

where α_i^* is given by (18) and

$$\bar{\gamma}_i = p^2 \gamma_{i1} + 2pq\gamma_{i2} + q^2 \gamma_{i3} \; . \qquad (26)$$

Assuming again that A is rare so that we can neglect terms of the or-der p^2, the condition for the increase of A becomes from (25) and (26)

$$(\alpha_2^* - \alpha_3^*) + y(\gamma_{23} - \gamma_{33}) < 0. \qquad (27)$$

Now assume that the niche position of the dominating genotype is at the midpoint between x and z so that $\hat{x} = \hat{z}$ and $\alpha_{x3} = \alpha_{z3} = \alpha$. The com-petition coefficients of the new genotype are given by

$$\alpha_{x2} = \exp[-(d + \varepsilon)^2/(4w^2)]$$

$$\alpha_{z2} = \exp[-(d - \varepsilon)^2/(4w^2)] \; .$$

After a Taylor expansion and ignoring cubic terms, the condition (27) for the increase of A becomes

$$(d/W)^2 < (1 - 2\alpha + \alpha^4)/[\alpha(1 - \alpha)]$$

which is fulfilled for $\alpha < 0.08$ or $d/W > 3.2$, that is for a loose packing of the niches. Thus only when the niche distance is large will the middle species acquire new genes which increases its utilization function.

A similar analysis may be performed for either species in the model for two species competing for a Gaussian resource. At any of the points on the curve of stable euqilibria in Fig. 2 we may ask whether niche expansion by polymorphism will occur. It turns out that polymorphism will be maintained at an equilibrium where the niche position of the species is sufficiently close to zero, $viz.$ at Fig. 2 from the arrowheads and closer. The most extreme equilibrium point where polymorphism will be maintained (the arrowhead) will always be further from the resource center than the two species global equilibrium, but for W^2/σ^2 greater than about 0.5, the two points are very close. The two species global equilibrium will therefore always produce polymorphism, but the amount is likely to be small for $W^2/\sigma^2 > 1/2$.

Finally, the case where an increase in the niche width of species y due to a genetic variation which increases the individual variance with respect to resource utilization will be mentioned. Allowing for the effects on intra- and interspecific competition and for the carrying capacity which for genotypes with an increased niche width is larger, the result is qualitatively identical to the one arrived at above, $viz.$, increase in niche width will only take place for a loose packing ($\alpha < 0.15$ or $d/W > 2.8$); otherwise selection will tend to decrease individual variance with respect to resource utilization.

The models on intraspecific competition in this section should not lead to too far reaching conclusions about the evolution of niche width. The main interest of these considerations in the present context is that they show that intraspecific competition will not affect the qualitative predictions on character displacement significantly.

GENERAL DISCUSSION AND EVIDENCE FROM NATURE

The theory of the previous sections allows to make some general predictions which can be tested in the field. We will in the following
give a general discussion of the theory in the light of evidence from
nature.

Colonization of a habitat with a Gaussian resource spectrum.

Before discussing field evidence, it will be fruitful to summarize
the evolution of species which colonize a habitat with a restricted resource spectrum in the light of the theory. Roughgarden (1976) has independently and based on another line of thought arrived at results similar to those outlined below.

Consider a habitat with an unexploited resource with a Gaussian
distribution. A species which can exploit this resource colonizes the
habitat. When established, selection will tend to displace the mode of
the niche of the species toward the center of the resource distribution center (similar to the movement of the middle species subject to
diffuse competition). Having reached the middle, weaker selection forces (which are proportional to the quadratics of the displacement as
compared to the movement towards the middle which is driven by selection forces proportional to the displacement) will tend to increase the
niche variance by including new genotypes or (probably at a slower
rate) increasing the variance in individual resource utilization.

Unless the species has filled the resource spectrum totally
($W^2/\sigma^2 = 1$, in Fig. 2), then a second species can invade. If the first
species has achieved a sufficient genetical variance, we will observe
a rapid change in niche position and it may end up having a marginal
niche position. Given sufficiently genetical variance in both species,
they will end up having symmetrical niche positions around the mode of
the resource spectrum as shown in Fig. 2. A third species may invade
and if the criterion of limiting similarity is not violated, all three
species will coexist. If the third species has an intermediate niche
position relatively to the two first species, the latter will show a
niche displacement towards the borders of the resource spectrum and
the new species will move its niche towards the midpoint between those
of its competitors.

This picture of the evolution of competing species allows compar-

ison with nature. Since we have only discussed one niche dimension it
is implicit that only closely related species, usually congeners, can
be used for testing the described theory. One may study systems which
consist of two or more coexisting congeners which are ecologically i-
dentical save for a differential utilization of resource qualities.
Other useful study systems are constituted by congeners which differ only with respect to a habitat dimension (e.g., salinity, humidity) and
therefore usually do not coexist but may overlap in distribution along
environmental gradients or in patchy environments (Fenchel, 1975 a).
Archipelagos, where sympatric and allopatric populations of congeners
on different islands can be studied are classical objects for such
studies (e.g., Lack, 1947; MacArthur and Wilson, 1967).

Niche distances in communities of competitors.

The theory predicts that in multispecies systems (>2) of coevolved
competitors, the niche distances between neighbours along the resource
axis should be equal and the niche widths should be equal (and not vi-
olate the limiting similarity principle of $d/W \gtrsim 1$). This prediction is
examplified in a large number of cases where coexisting congeners have
been studied thoroughly (e.g., Fig. 3; for other examples, see Mac-
Arthur, 1972): Many studies have relied on character measures (e.g.,
whole body lengths, measures of trophic organs) rather than resource
utilization functions as measures of niche distance. In such studies,
the within phenotype component of niche width is not measured and in
many cases the exact relation (or indeed the correlation) between some
metric measure and resource utilization is not established. However,
the generalization of Hutchinson (1959) that coexisting congeners frequently have size ratios of about 1.3 (linear measures) may be taken
as an example of equal niche spacing.

Character displacement.

This term, originally coined by Brown and Wilson (1956), describes
what the theory predicts; a shift in niche position where two closely
related species overlap in distribution. An example is shown in Fig. 4.
Although many examples are known from the literature, most of these are
not sufficiently analyzed and could well be interpreted in other ways
(Grant, 1972). Thus, most of the examples, as implied in term itself,

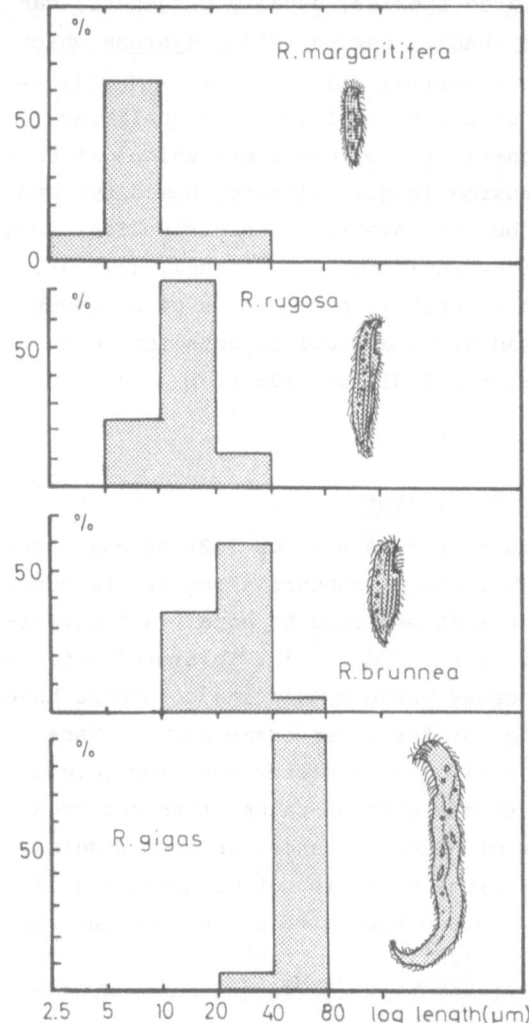

Figure 3. The frequency distribution of ingested diatoms in four coexisting species of the ciliate protozoan genus *Remanella*. (Redrawn from Christiansen and Fenchel, 1977; data after Fenchel, 1968).

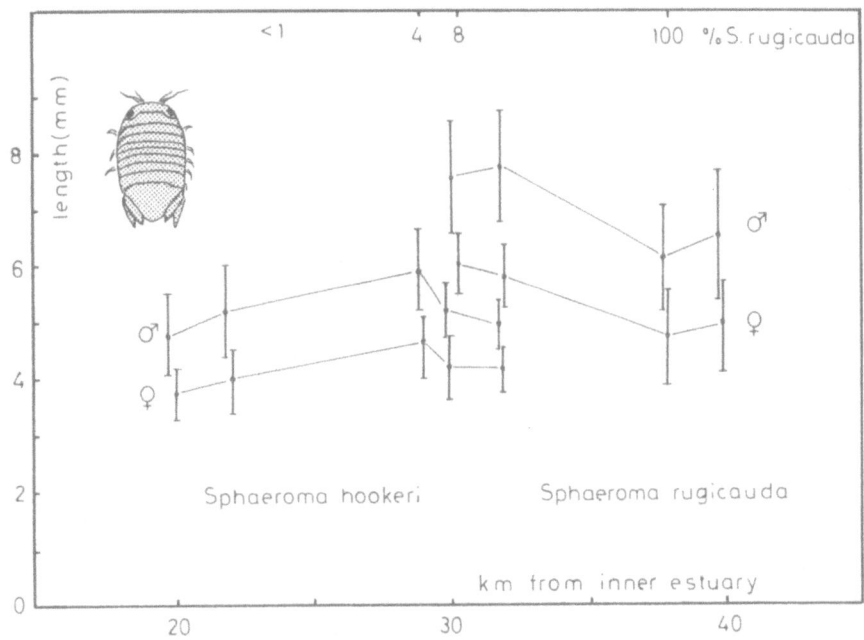

Figure 4. The body lengths (with standard deviations) of two congeneric species of isopods along a Danish estuary (Mariager Fjord) with overlapping distribution along a few km long zone. (Jens-Ole Frier, unpublished data).

only include character measures and not the ecological niche. Further-more, most of these examples have not been reproduced, *viz.*, only sin-gle cases of coexistence and of allopatric distribution are given and the findings may in some cases represent clines which have evolved due to some other, unknown environmental factor.

More recently, a number of critical studies of character displace-ment have been made (e.g., Huey and Pianka, 1974; Huey *et al.*, 1974). Cody (1974) gives a number of interesting examples of character dis-placement with respect to breeding periods and behaviour and to "cross-overs" in character displacement. Fenchel (1975 a,b) studied character displacement in three species of mud snails (*Hydrobia* spp.) in a pat-chy, estuarine environment. This example is reproducable; i.e., many

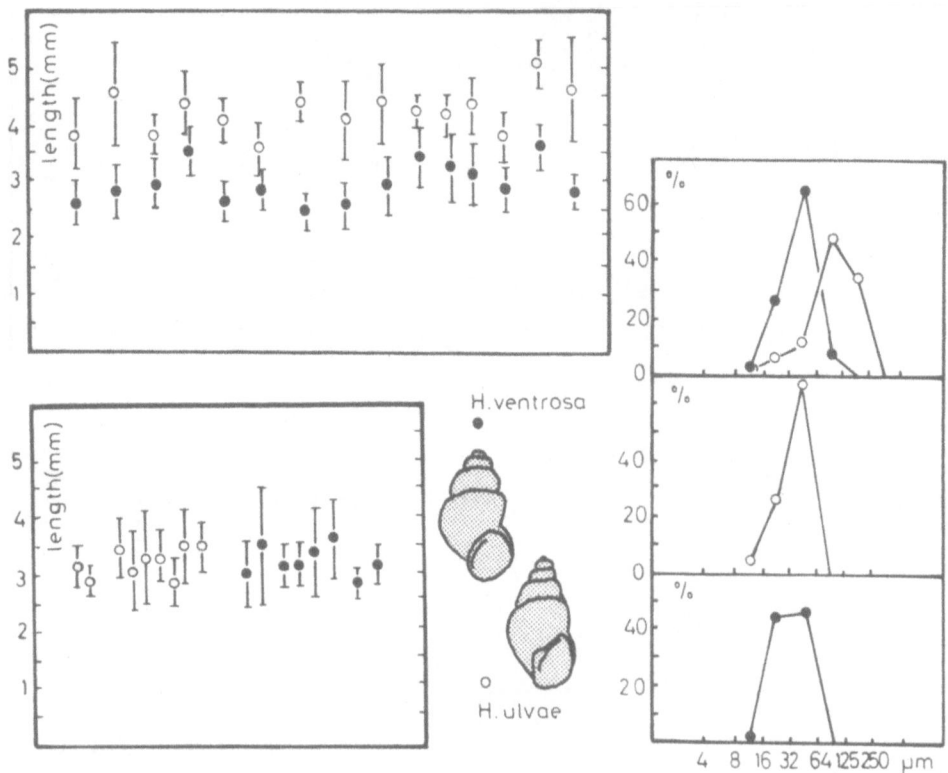

Figure 5. Left: the shell lengths of two species of mud snails (with standard deviations) in 15 localities with coexistence and 17 localities with allopatric occurrence. Right: the average size distribution (volume %) of ingested food particles
of coexisting and allopatric populations of the two *Hydrobia* spp. (After Fenchel,
1975b).

isolated localities with coexistence as well as with allopatric populations could be studied and compared. For example, wherever two of the
species (*H. ulvae* and *H. ventrosa*) are found coexisting, their size
ratio is between 1.2 and 1.9 whereas allopatric populations of both
species have about the same, intermediate sized individuals (Fig. 5).
The system is also interesting because recent hydrographic and man

made changes in the estuary set a limit to the time scale involved in
the evolution of character displacement in the different subpopulations.
None of these could be more than 100 - 150 years old and some are as
recent as 10 years. It could also be shown that the size difference does
lead to a partitioning in the food particle sizes so that the observed
character displacement leads to a niche displacement of about 1 (Fig.
5). Fenchel and Kofoed (1976) could show that the species do in fact
show exploitative competition for food (mainly sediment living diatoms),
that the niche overlap is unity for two species with populations with
similarly sized individuals and that competition is relaxed by the size
difference. It could also be shown that snails of different sizes (at
natural population densities) affect the availability of diatoms of
different size ranges in the sediment (Fig. 6).

There are reasons to believe that character displacement, contrary
to the suggestion by MacArthur and Levins (1967) and MacArthur (1972),
is a very common phenomenon. We believe that it will prove important
for studying selection in natural environments since the factors to
which the populations respond are well defined. The most useful approach will be, however, not as much to add new examples to the ones
known as to analyze the individual cases carefully.

Niche width and competition.

The theory predicts that niche width should decrease with increasing competition intensity. Thus, we would expect that in character displacement we would also observe a decrease in variance together with
the niche mode displacement. Indeed, some of the described cases (including the classical study of Lack on the Darwin finches and the example shown on Fig. 4) do suggest this to be the case. An interesting
case has recently been described by Johnson (1973). The numerous species of *Drosophila* of the Hawaiian archipelago are not shared equally
among the different islands. It could be shown that enzyme polymorphism
was inversely correlated with species richness, giving support to the
prediction that competition decreases genetical variance.

We cannot, however, expect to find a complete accordance with the
theory in all observed cases. The crude models imply that the necessary genetical variance always is present and that there are no physiological constraints on any evolutionary trends. This must in particular
be kept in mind when systems which have evolved only recently, such as

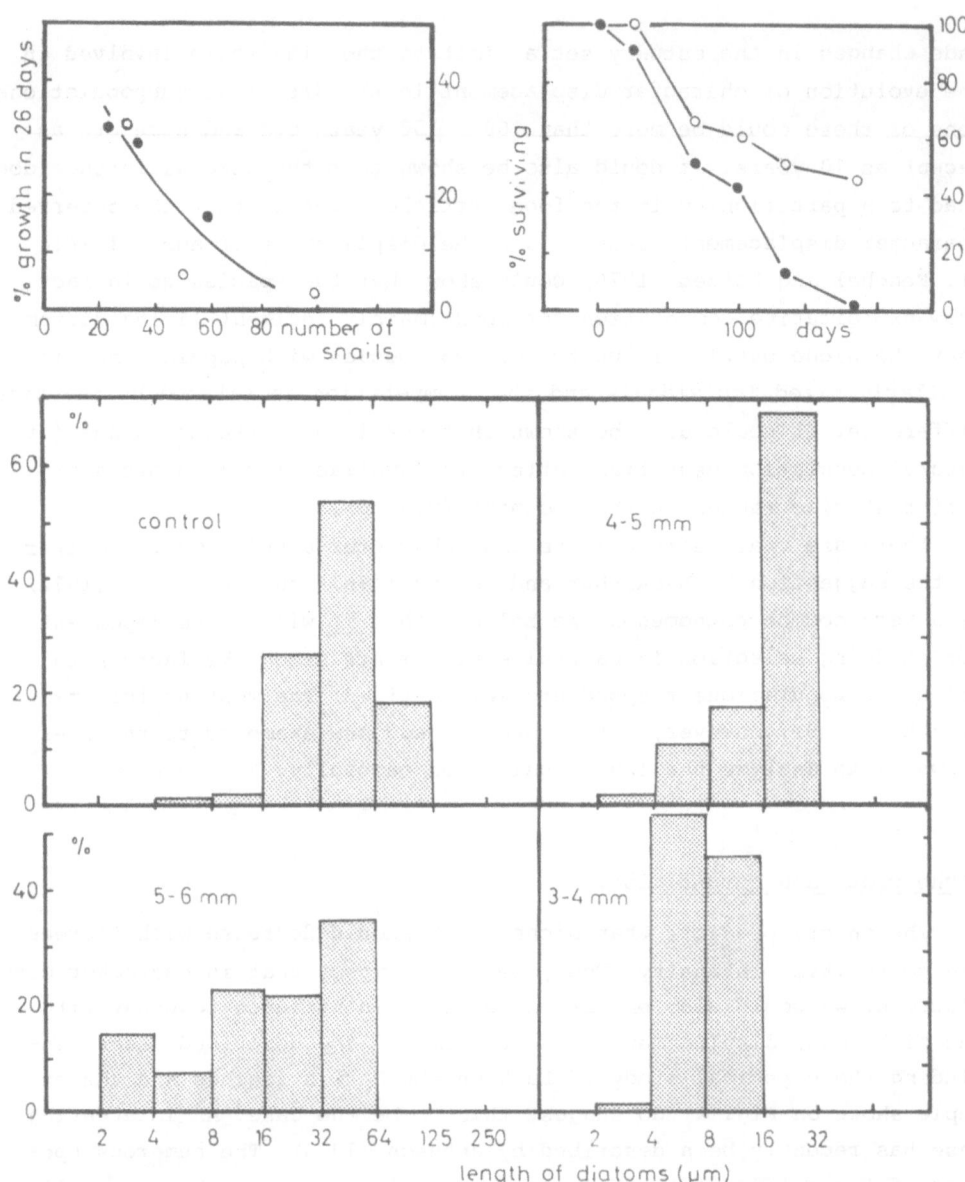

Figure 6. Above, left: the average growth of *Hydrobia ulvae* as function of own density (open circles) and as function of the density of *H. ventrosa* of the same size
range (closed circles) in experimental containers with natural sediment. Right: the
survival of small *H. ventrosa* in the presence of *H. ulvae* of similar sizes (closed
circles) and in the presence of larger sized *H. ulvae* in experimental containers with
natural sediment. Below: the size distribution (volume %) of diatoms in illuminated,
natural sediment after a week without snails and after grazing of *H. ulvae* of different size classes and with natural population densities. (After Fenchel and Kofoed,
1976).

the snail systems, are under discussion. Here the variances in the utilization functions are about equal in allopatric and sympatric populations. An analysis has shown, however, that by far the greatest component of the niche width is the within phenotype variance in resource
utilization (Fenchel, 1975 b). Thus, while there is probably often a
large genetical variance available with respect to average body size,
which can rapidly respond to selection, the within phenotype component
will probably respond slowly to selection since changes will involve
a number of complex morphological and physiological changes. It has
been shown, however, that the age structure of niche width has been
affected in the coexisting populations of both species where breeding
periods are short and non-overlapping. In contrast, allopatric populations produce eggs throughout the summer. This probably contributes
to a decrease in niche overlap (Fenchel, *op.cit.*).

Summary

    Some theoretical models which describe the evolution of Mendelian species popul-
ations involved in interspecific, exploitative competition along a single resource-
quality dimension are discussed. The effects of selection include shifts in niche
modes and in niche widths as the result of inter- as well as intraspecific competition; these effects are described for infinite and limited resource spectra. The
models are finally discussed in the light of field evidence exemplifying niche differences between coexisting congeners, character displacement, and the evolution of
niche width.

Acknowledgement.

    We are grateful to cand. scient. Jens-Ole Frier for permission to use the un-
published results shown in Fig. 4.

                                REFERENCES

Brown, J., and Wilson, E.O. 1956. Character displacement.
    Syst. Zool. 5: 49-64.
Cody, M.L. 1974. Competition and the structure of Bird Communities. Princeton Uni-
    versity Press, Princeton.

Christiansen, F.B., and Fenchel, T. 1977. Theories of Populations in Biological Communities. Springer Verlag, Heidelberg.

Fenchel, T. 1968. The ecology of marine microbenthos II. The food of marine benthic ciliates.
Ophelia 5: 73-121.

Fenchel, T. 1975a. Factors determining the distribution patterns of mud snails (Hydrobiidae).
Oecologia (Berl.) 20: 1-17.

Fenchel, T. 1975b. Character displacement and coexistence in mud snails (Hydrobiidae).
Oecologia (Berl.) 20: 19-32.

Fenchel, T., and Kofoed, L.H. 1976. Evidence for exploitative interspecific competition in mud snails (Hydrobiidae).
Oikos 27: 367-376.

Grant, P.R. 1972. Convergent and divergent character displacement.
Biol. J. Linnean Soc. 4: 39-68.

Huey, R.B., and Pianka, E.R. 1974. Ecological character displacement in a lizard.
Amer. Zool. 14: 1127-1136.

Huey, R.B., Pianka, E.R., Egan, M.E., and Coons, L.W. 1974. Ecological shifts in sympatry: Kalahari fossorial lizards (*Typhosaurus*).
Ecology 55: 304-316.

Hutchinson, G.E. 1959. Homage to Santa Rosalia or why are there so many kinds of animals?
Amer. Natur. 95: 145-159.

Johnson, G.B. 1973. Relationship of enzyme polymorphism to species diversity.
Nature. (London) 242: 193-94.

Lack, D. 1947. Darwin's finches. Cambridge University Press. Cambridge.

Levins, R. 1968. Toward an evolutionary theory of the niche. *In*: E.T.Drake (ed.): Evolution and Environment. pp. 325-340. Yale University Press. New Haven, Conn.

MacArthur, R.H. 1972. Geographical Ecology. Harper and Row, New York.

MacArthur, R.H., and Levins, R. 1967. The limiting similarity, convergence and divergence of coexisting species.
Amer. Natur. 101: 377 -385.

MacArthur, R.H., and Wilson, E.O. 1967. The Theory of Island Biography. Princeton University Press, Princeton.

May, R.M. 1974. On the theory of niche overlap.
Theoret. Popul. Biol. 5: 247-332.

Pianka, E.R. 1973. The structure of lizard communities.
Ann. Rev. Ecol. Syst. 4: 53-74.

Pianka, E.R. 1974. Niche overlap and diffuse competition.
Proc. Nat. Acad. Sci. USA 71: 2141-45.

Roughgarden, J. 1972. Evolution of niche width.
Amer. Natur. 106: 683-718.

Roughgarden. J. 1976. Resource partitioning among competing species. A coevolutionary approach.
Theoret. Popul. Biol. 9: 388-424.

4. ECOLOGY AND EVOLUTION

COEVOLUTION IN ECOLOGICAL SYSTEMS: RESULTS FROM "LOOP ANALYSIS" FOR
PURELY DENSITY-DEPENDENT COEVOLUTION

Jonathan Roughgarden

Much of the theory of population ecology is concerned with predicting
equilibrium population size on the basis of assumptions about the interactions between populations. Familiar population interactions are interspecific competition, predation, symbiosis including parasitism and mu-
tualism, and others. Recent years have witnessed the proliferation of
equations which model these interactions in ways especially suited for
certain species. Most population dynamic models predict that the interacting populations will attain stable equilibrium abundance provided
the parameters in the model satisfy certain requirements which are special to each model. Many models also allow for other possibilities including cycling of various forms. Nonetheless almost all models contain
stable coexistence at an equilibrium point as *one* of the possibilities.

Whenever species are coexisting at a stable equilibrium point it is
clear that their equilibrium abundance are functions of the parameters
in the model. For example the equilibrium abundance of each competitor
in a Lotka-Volterra competition system is a function of all the competition coefficients, α_{ij}, and all the carrying capacities, K_i. However,
the parameters in a population dynamic model are themselves subject to
evolutionary modification by natural selection. Therefore, natural selection also indirectly controls the equilibrium population sizes because natural selection directly controls the parameters. In this con-

text the basic question an ecologist may ask of coevolutionary theory
is: to what value does natural selection set each parameter in an ecological model and what are the equilibrium population sizes attained
as a result? To answer this question is to understnd how the combined
effect of natural selection in each of the interacting populations shapes
the *final* configuration of the whole ecological community.

There are two separate clauses in the question posed above. The first
clause concerns the parameter values themselves.In some circumstances
this clause is the focus of interest. For example, the competition coefficients, α_{ij}, are often interpreted in terms of the similarity of
resource use by two competitors. A coevolutionary theory might predict
the value of α_{ij} and thereby predict the degree of similarity of two
competitors. In this case the degree of similarity itself is the primary feature of interest; the population sizes of the competitors being
of second interest. In other circumstances the parameters in the model
are of secondary interest to the equilibrium population sizes which they
cause. This is especially true if it can be shown that the effect of
evolution on the parameters of a model leads to a lowering of the equilibrium population size of a species. Showing this raises the possibility of natural selection causing the extinction of a population. Another
circumstance where the population sizes are of primary interest is where
the amount of standing crop at different trophic levels (the energy py-
ramid) is to be explained. In this case data on the summed abundance of
several interacting populations are available, but the details of the
interactions are largely unknown. Thus an ecological system is a set of
populations together with their interactions. A coevolutionary theory
of ecology must predict both the equilibrium sizes of the populations
and their interactions from evolutionary first principles.

For the question above to be meaningful it must be true that there
exists a stable equilibrium point to talk about. That is, we view natural selection in each of the populations as influencing the parameters in the model provided that the parameters are restricted to values
where a stable equilibrium point exists. We will be able to detect when
selection is *tending* to destroy an initially stable equilibrium (see
example 2) but the evolutionary analysis itself is restricted to the
region of parameter space in which there exists a stable equilibrium
point.

In any theory for the evolution of the parameters in ecological models one must specify which species exert *evolutionary control* over the

parameters. Consider a Lotka-Volterra system again. We typically expect
evolution within species-i to control its own carrying capacity, K_i.
But it is also possible that some other species pollutes the environment thereby causing K_i to be under the control of another species. Less
far fetched, the parameter α_{ij} may be under the joint evolutionary control of both species-i and species-j. The assignment of the evolutionary
control of the parameters is the most critical assumption in the theory
of coevolution.

The simplest assignment of evolutionary control is to assume that
each species controls only the parameters in its own equation for population growth. This assumption rules out the possibility of species-i
controlling the carrying capacity of species-j, and also implies that
no parameter is under the joint control of two or more species. For
example α_{ij} is assumed to be under the sole control of species-i. This
assumption is equivalent to assuming that the fitness of genotype A_iA_j
in species-s, $W_{s,ij}$, is a function only of the various population sizes
and not of the gene frequencies in the various species. This assumption
means that the coevolution is purely density-dependent; there is no interspecific frequency-dependence.

Provided certain assumptions are met, this paper predicts the values to which natural selection will set the parameters of an ecological
model and predicts the consequences of this evolution for the equilibrium population sizes. These predictions are derived by combining explicit
population genetic formulae for natural selection at one locus with two
alleles in each species with the ecological model under examination.
Simple criteria are developed which predict whether evolution within
a species leads to its own net increase or decline, and whether evolution within a species leads to the increase or decline of any other species. The key assumptions are that the ecological model possesses a
stable equilibrium point for the range of parameter values being considered, and that the assignment of evolutionary control leads to purely density-dependent coevolution.

THE GRAPH OF A COMMUNITY

An ecological system consists of the populations and their interactions.
The following paragraphs present a useful graphical representation of
an ecological system.

   To develop a coevolutionary model one begins with a population dynamic model defined in discrete time with the time interval being one generation. The original population dynamic model is of the form

$$\Delta N_i = [W_i(N_1 \ldots N_S) - 1]N_i \quad \text{for } i = 1 \ldots S \tag{1}$$

where N_i is the abundance of species-i at time t and W_i is the average
number of offspring produced per individual in species-i. If $W_i = 1$
there is direct replacement by each individual and $\Delta N_i = 0$. By definition
$W_i \geq 0$. By assumption this population dynamic model has a positive locally stable equilibrium. This means that the eigenvalues of the gradient
matrix evaluated at equilibrium lie in a unit circle in the left half
of the complex plane (the circle with radius 1 and center at -1). The
stability criteria from continuous time models are necessary but not
sufficient for the discrete time models discussed here. At a positive
locally stable equilibrium point we define the quantities a_{ij} as $\partial(\Delta N_i)/$
∂N_j evaluated at $N_i = \hat{N}_i$. That is, a_{ij} are the elements of the gradient
matrix evaluated at equilibrium. They may be computed explicitly as
$a_{ij} = \hat{N}_i \, \partial W_i/\partial N_j$ where $\partial W_i/\partial N_j$ is evaluated at $N_i = \hat{N}_i$ for all i. With
these quantities Levins (1974) has posed a graphical representation of
an ecological system at equilibrium which proves very helpful. Each
species abundance at equilibrium is placed at a node and lines are drawn
between nodes to indicate the population interactions. For example, with
three species we have

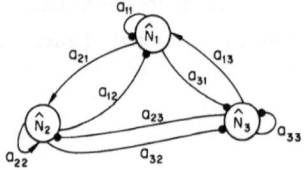

A line terminating with an arrow indicates a positive effect, *e.g.* a_{21}
indicates that N_1 has a positive effect on N_2, and a_{22} indicates that

N_2 grows autocatalytically. A line terminating with a dot indicates a
negative effect. This graph is also called a Coates graph in the elec-
trical engineering literature (Chen 1971).

There are several important measures defined on this graph and on
certain subgraphs formed from it.

1) The *feedback* (Levins 1974) in the community is defined as F ≡
$(-1)^{S+1}$ det(a_{ij}) where S is the number of species in the community. To
see this more explicit, note that as $a_{ij} = \hat{N}_i \, \partial W_i/\partial N_j$ we get a simple
expression for the determinant so that

$$F = (-1)^{S+1} \ (\prod_{j=1}^{S} \hat{N}_j) \ \det(\partial W_i/\partial N_j) \tag{2}$$

If S = 1, then F reduces to F = $\hat{N} \, \partial W/\partial N$. *Thus when S = 1 the sign of F
is identical to the sign of the density dependedence in the population.*
2) A *subcommunity* (of the order S - 1) is a system obtained by deleting
one species from the system. G_i denotes the graph of the subcommunity
obtained by deleting species-i. For example G_1 from the preceding example
is

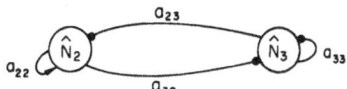

The *feedback of the subcommunity* is defined, just as above, by $F_i \equiv$
$(-1)^S$ det$_{ii}(a_{ij})$ where the symbol det$_{ii}(a_{ij})$ means the determinant of
the matrix of rank S - 1 obtained by deleting the ith column and i^{th}
row from the matrix (a_{ij}). In terms of the W's we have

$$F_i = (-1)^S \ (\prod_{j \neq i} \hat{N}_j) \ \det(\partial W_i/\partial N_j) \tag{3}$$

*If S = 2 then the sign of F_i is identical to the sign of the density
dependence in the undeleted species.*
3) In this paper we also introduce the *connecting web from species-j
to species-i.* It is represented by the graph, G_{ij}, obtained by deleting
all the lines leading into j and all the lines leading out of i. G_{12}
in the example is

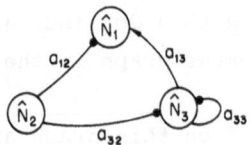

Species-2 is seen to influence species-1 directly through a_{12} and also indirectly via species-3. We introduce a definition of the *feedback of a connecting web*, G_{ij}, as $F_{ij} \equiv (-1)^{S+i+j+1} \det_{ji}(a_{ij})$ where $\det_{ji}(a_{ij})$ means the determinant of the matrix of rank $S - 1$ obtained by deleting the jth row and i^{th} column from (a_{ij}). In terms of the W's we have

$$F_{ij} = (-1)^{S} (-1)^{i+j+1} (\prod_{k \neq j} \hat{N}_k) \det_{ji}(\partial W_i / \partial N_j) \qquad (4)$$

If $S = 2$ then the sign of F_{ij} is identical to the sign of $\partial W_i / \partial N_j$; i.e. it is negative if j competes with or preys upon i and positive if i preys upon j, *etc.* By saying j exerts a negative (positive) effect upon i we mean $F_{ij} < (>) 0$.

These feedback measures completely specify whether the evolution within some species increases or decreases its own equilibrium population size, and increases or decreases the equilibrium population sizes of other species.

A key lemma is

Lemma 1.   For any community at a stable equilibrium point $F < 0$; $\Sigma F_i < 0$.

That is, a necessary condition for local stability is that the feedback of the entire community be negative, and that the sum of the feedback from all subcommunities be negative. This result is an immediate consequence of the conditions on the coefficients of the characteristic equation which are necessary for the eigenvalues to have negative real parts.

THE COEVOLUTIONARY MODEL AND THE ASSOCIATED PURE ECOLOGICAL MODEL

A *coevolutionary model* is obtained by fusing the standard formulae for natural selection at one locus with two alleles with a population dynamic model, yielding for i = 1,...,S

$$\Delta N_i = (W_i - 1)N_i; \quad \Delta P_i = [p_i(1-p_i)/(2W_i)]\partial W_i/\partial p_i \qquad (5)$$

where W_i is the mean fitness in species-i,

$$W_i = p_i^2 W_{i,11} + 2p_i(1-p_i)W_{i,12} + (1-p_i)^2 W_{i,22},$$

and p_i is the allele frequency in species-i. By assumption $W_{i,jk} = W_{i,jk}(N, \ldots, N_S) \geq 0$, that is, the fitnesses are non-negative and pure-ly density-dependent. The assumption of pure density dependence implies that $\partial W_i/\partial p_j \equiv 0$ for $j \neq i$. A coevolutionary equilibrium satisfies, for every i, one of the following conditions

$$W_i = 1, \ \partial W_i/\partial p_i = 0; \qquad (i)$$

$$W_{i,11} = 1, \ \hat{P}_i = 1; \qquad (ii) \qquad\qquad (6)$$

$$W_{i,22} = 1, \ \hat{P}_i = 0. \qquad (iii)$$

In the first condition species-i is at a polymorphic or "interior" equi-librium and in the other two species-i is at a fixation or "boundary" equilibrium.

To analyze the coevolutionary model above, we refer continually to another model which is called the *pure ecological model*. The pure eco-logical model is a population dynamic model of the form (1) which is de-fined by taking the gene frequency in every species as a constant while allowing the population sizes to vary. Since the p's are constants in this model the equilibrium population sizes can be regarded as functions of the p's. Whenever this pure ecological model leads to a stable equi-librium, the resulting equilibrium community can be represented by the community graph. For example with two populations we might have

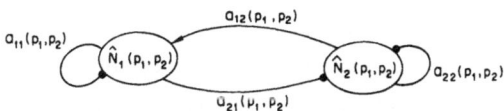

The $\hat{N}_i(p_1, \ldots, p_S)$ are defined as functions of the p's *implicitly* by the system of equations

$$W_i(P_i, N_1, \ldots, N_S) = 1, \quad i = 1, \ldots, S \tag{7}$$

and the a's are defined by

$$a_{ij}(P_1, \ldots, P_S) = \hat{N}_i(P_1, \ldots, P_S) \, \partial W_i(P_i, \hat{N}_1, \ldots, \hat{N}_S)/\partial N_j. \tag{8}$$

Furthermore, for every set of p's we can evaluate the feedback in the various subgraphs of the equilibrium community, and we can obtain a complete picture of the equilibrium ecological system which results. For this discussion to be well posed, we *assume that the pure ecological model has a unique positive locally stable equilibrium point for any set of fixed p's.* This assumption together with that stipulating purely density-dependent fitnesses are the two principal assumptions in this paper.

The set of equilibrium communities generated by the pure ecological model for all p's comprise the set of communities which are *evolutionarily possible.* The question to be solved is: which of the evolutionarily possible communities will result from coevolution? That is, if the gene frequencies are not in fact held constant but instead evolve to some values, say $(\hat{p}_1, \ldots, \hat{p}_S)$, then the coevolutionary equilibrium community which results is identical to the community which would result from the pure ecological model with the p's as $(\hat{p}_1, \ldots, \hat{p}_S)$.

ANALYSIS AND RESULTS

We first establish the central lemma

Lemma 2. Assume that the fitnesses are purely density-dependent and that the associated pure ecological model has a unique locally stable equilibrium for any set of fixed p's. Then an equilibrium point in the coevolutionary model $(\hat{p}_1, \ldots, \hat{p}_S, \hat{N}_1, \ldots, \hat{N}_S)$ is locally stable if and only if the mean fitness in each species is a local maximum with respect to the gene frequency in that species; i.e. for each i, W_i is a local maximum with respect to p_i.

This applies regardless of whether all species are polymorphic, all fixed, or a mixture with some polymorphic and the rest fixed. This

lemma is shown separately for completely polymorphic equilibria, completely fixed equilibria and mixed equilibria. The local stability of
the equilibrium is established by showing that the eigenvalues, λ_i, of
the following matrix satisfy $|\lambda_i + 1| < 1$, *i.e.* lie in the unit circle
centered on -1 in the complex plane. The matrix may be partitioned into
blocks as

$$\begin{pmatrix} B & C \\ D & A \end{pmatrix}$$

where $b_{ij} = \partial(\Delta p_i)/\partial p_j$, $c_{ij} = \partial(\Delta p_i)/\partial N_j$, $d_{ij} = \partial(\Delta N_i)/\partial p_j$, and $a_{ij} = \partial(\Delta N_i)/\partial N_j$. From the assumption that the fitnesses are only density
dependent we have that $\partial W_i/\partial p_j = 0$ for $j \neq i$ so therefore B is a diagonal
matrix, $b_{ij} = 0$ for $i \neq j$. For the same reason the matrix D is also diagonal with $d_{ij} = 0$ for $i \neq j$. The matrix A determines the stability of
the associated pure ecological model. By assumption A has eigenvalues
satisfying $|\lambda_i + 1| < 1$ for any $(p_1, \ldots, p_s)$ and therefore at $(\hat{p}_1, \ldots, \hat{p}_s)$
in particular.

1)   Suppose the equilibrium gene frequency in each species is polymor-
phic. Then at equilibrium p_i satisfies $\partial W_i/\partial p_i = 0$ for $i = 1, \ldots, S$.
This condition makes $d_{ii}) = 0$; the block D is then a block of zeroes.
Therefore eigenvalues of the matrix are

$$\lambda_i = \partial(\Delta p_i)/\partial p_i \quad \text{for } i = 1, \ldots, S \tag{9}$$

together with the S eigenvalues of the matrix A. Explicitly

$$\lambda_i = (1/2)\hat{p}_i(1-\hat{p}_i)\partial^2 W_i/\partial p_i^2. \tag{10}$$

Then it is easy to verify that λ_i is between -2 and 0 (actually between
-1 and 0) if and only if $\partial^2 W_i/\partial p_i^2 < 0$, *i.e.* W_i is maximized with respect to p_i.

2)   Suppose the equilibrium gene frequency in each species is 0 or 1.
Then all the elements in the matrix C equal zero,

$$c_{ij} = \hat{p}_i(1-\hat{p}_i)\partial/\partial N_j[1/W_i \ \partial W_i/\partial p_i] = 0.$$

With C a block of zeroes we again have the eigenvalues given by (9) and the S eigenvalues of A. Explicitly

$$\lambda_i = \tfrac{1}{2}\, \partial W_i / \partial p_i \text{ if } \hat{p}_i = 0 \text{ and } \lambda_i = -\tfrac{1}{2}\, \partial W_i / \partial p_i \text{ if } \hat{p}_i = 1 \qquad (11)$$

It is then easy to verify that the boundary $\hat{p}_i = 0$ is unstable if $\partial W_i / \partial p_i > 0$ and stable if $\partial W_i / \partial p_i < 0$; the signs are reversed at $\hat{p}_i = 1$. Thus a boundary is stable if W_i decreases as p_i is varied away from the boundary. The case of complete dominance where $\partial W_i / \partial p_i = 0$ at the boundary requires subtler methods.

3) Suppose the equilibrium gene frequencies represent polymorphism in some species and fixation in others. The characteristic equation is given by $\Delta(\lambda) = 0$ where

$$\Delta(\lambda) = \begin{vmatrix} (B - \lambda I) & C \\ D & (A - \lambda I) \end{vmatrix}.$$

As $(B - \lambda I)$ and D are diagonal matrices they commute. Hence $\Delta(\lambda) = |(B - \lambda I)(A - \lambda I) - DC|$. The matrix DC is always zero. To see this, calculate the (i,j)'th element:

$$(DC)_{ij} = \sum_k d_{ik}\, C_{kj} = d_{ii}\, C_{ij},$$

because D is a diagonal matrix. From 1) and 2) we get that either $d_{ii} = 0$ or $C_{ij} = 0$. Thus $\Delta(\lambda) = |A - \lambda I|\,|B - \lambda I|$ and the characteristic equation factors into that of A and that of B. The eigenvalues of A satisfy $|\lambda_i + 1| < 1$ by assumption and the eigenvalues of B are given by (9). These are each between -1 and 0 if and only if p_i maximizes W_i as discussed above.

Lemma 2 is the central population genetic result about purely density-dependent coevolution. The subsequent theorems develop the implications of this lemma in ecological terms.

With the lemma we can identify which of the evolutionarily possible communities actually results from coevolution. We term an equilibrium community which represents a stable equilibrium in the coevolution model a *coevolutionarily stable community*. We derive two different but equivalent ways to identify a coevolutionarily stable community. The first is based on examining, for each species, whether the equilibrium gene frequency within the species maximizes or minimizes the population size of that species.

Result 1. Condition for a coevolutionarily stable community from the effect of p_i *on* $\hat{N}_i$ *for each i:*

    Assume that the fitnesses are purely density-dependent and that the associated pure ecological model has a unique locally stable equilibrium $(\hat{N}_1, \ldots, \hat{N}_s)$ *for any set of fixed gene frequencies* $(p_1, \ldots, p_s)$.

    A. An equilibrium community, $(\hat{p}_1, \ldots, \hat{p}_s, \hat{N}_1, \ldots, \hat{N}_s)$, *is coevolutionarily stable if and only if, for each species-i,* $\hat{N}_i$ *is maximized or minimized at* $\hat{p}_i$ *according to the following criterion:*

    1. If the sign of the feedback in the subcommunity obtained by deleting species-i is negative [i.e. $F_i(\hat{p}_i, \ldots, \hat{p}_s) < 0$] *then* $\hat{N}_i$ *is a local maximum with respect to* p_i.

    2. If the sign of the feedback in the subcommunity obtained by deleting species-i is positive [i.e. $F_i(\hat{p}_1, \ldots, \hat{p}_s) > 0$] *then* $\hat{N}_i$ *is a local minimum with respect to* p_i.

    B. If the feedback in the subcommunity obtained by deleting species-i is identically zero [i.e. $F_i(p_1, \ldots, p_s) \equiv 0$] *then evolution within species-i has no effect on the equilibrium abundance realized by species-i.*

    C. At least one species in a community must satisfy condition A1 above. That is, at least one species is such that pure density-dependent coevolution leads to the maximization of its population size.

Result 1 is an immediate consequence of Lemma 1 and 2. Part C restates the fact that $\Sigma F_i < 0$ in a stable community. Therefore at least one F_i must be negative and hence at least one population satisfies condition A1 above.

The criteria involving F_i follow from the expressions relating $\partial \hat{N}_i / \partial p_i$ and $\partial^2 \hat{N}_i / \partial p_i^2$ to $\partial W_i / \partial p_i$ and $\partial^2 W_i / \partial p_i^2$ respectively. The $\hat{N}_i$ as functions of the gene frequencies are defined implicitly by the equation system (7). Differentiating every equation with respect to p_i yields a system of linear equations for $\partial \hat{N}_j / \partial p_i$. Solving by Cramer's rule for $\partial \hat{N}_i / \partial p_i$ and applying the definitions of feedback (2) and (3) we obtain

$$\partial \hat{N}_i / \partial p_i = (\partial W_i / \partial p_i) \hat{N}_i \, F_i(\hat{p}_1, \ldots, \hat{p}_s) / F(\hat{p}_1, \ldots, \hat{p}_s) \qquad (12)$$

In Lemma 1 we observed that $F < 0$. Hence a boundary equilibrium in species-i maximizes W_i if and only if it also maximizes $\hat{N}_i$ provided $F_i < 0$ or minimizes $\hat{N}_i$ provided $F_i > 0$. To prove the result for polymorphic

equilibria we need to inspect the signs of the second derivations. Differentiating the system of equations again with respect to p_i and taking
account that $\partial N_j / \partial p_i = 0$ if p_i is a polymorphic equilibrium (see below)
yields a system of equations for the second derivatives. Solving by
Cramer's rule and applying the definitions of feedback yields

$$\partial^2 \hat{N}_i / \partial p_i^2 = (\partial^2 W_i / \partial p_i^2) \hat{N}_i \ F_i(\hat{p}_1, \ldots, \hat{p}_s) / F(\hat{p}_1, \ldots, \hat{p}_s) \qquad (13)$$

Therefore a polymorphic equilibrium in species-i maximizes W_i if and
only if it also follows the criterion above. Finally, note that if $F_i =$
0 then any evolution within species-i does not affect the equilibrium
abundance achieved by species-i.

Condition A2 is especially important because when it is met evolution within a species reduces its own equilibrium population size. This
is true even though the mean fitness is being maximized. Also, this result is not to be confused with phenomena like the evolution of predation rates which overexploit the prey, *etc.* Such phenomena are a reflection of the effects of *interspecific frequency-dependence*. It can be
shown that interspecific frequency dependence generally causes p_i to
assume a value which leads to an $\hat{N}_i$ which is neither a local maximum
nor minimum. Instead, $\hat{N}_i$ at the coevolutionary equilibrium lies some-
where between the minimum and maximum. In contrast, the result above
shows that even without interspecific frequency-dependence the result
of evolution within a species may be to *minimize* its abundance. This
possibility arises solely from the ecological network in which the species is embedded and does not involve schemes of frequency-dependent
selection.

The next result presents a criterion for identifying a coevolutionarily stable community based on examining the effect of every species
upon a given species.

*Result 2.   Condition for a coevolutionarily stable community from
the effect of every gene frequency p_j on the population size of a given
population $\hat{N}_i$:*

*·Assume that the fitnesses are purely density-dependent and that the
associated pure ecological model has a unique locally stable equilibri-
um $(\hat{N}_1, \ldots, \hat{N}_s)$ for any set of fixed gene frequencies $(p_1, \ldots, p_s)$.*

*A.   An equilibrium community $(\hat{p}_1, \ldots, \hat{p}_s, \hat{N}_1, \ldots, \hat{N}_s)$ is coevolutio-
narily stable, if and only if, for any chosen species $\hat{N}_i$, every p_j influences $\hat{N}_i$ as follows:*

1. *for* $j = i$ *then* $\hat{N}_i$ *is maximized or minimized wrt to* p_i *according as* $F_i < 0$ *or* $F_i > 0$ *as discussed above in Result 1; and*
      2. *for all* $j \neq i$ *then*
         a. $\hat{N}_i$ *is minimized wrt* p_j *if the feedback of the connecting
web from j to i is negative (i.e.* $F_{ij} < 0$*) or*
         b. $\hat{N}_i$ *is maximized wrt to* p_j *if the feedback of the connecting
web from j to i is positive (i.e.* $F_{ij} > 0$*).*
   B. *If the feedback in the connecting web from j to i is identically
zero [i.e.* $F_{ij}(p_1,\ldots,p_s) \equiv 0$*] then evolution within species-j has no
effect on the equilibrium abundance realized by species-i.*

Part A.1 of Result 2 is contained in Result 1 and part A.2 of Result 2 is proved by a similar method, namely by examining the expres-
sions relating $\partial \hat{N}_i/\partial p_j$ and $\partial^2 \hat{N}_i/\partial p_j^2$ to $\partial W_j/\partial p_j$ and $\partial^2 W_j/\partial p_j^2$ respectively.

Both results may be used together to give a complete picture of how
evolution within all of the populations shapes the abundance of one
another and their interactions. We now illustrate these results with
three examples.

Examples.

Three examples are presented to illustrate the results above. In
each example there are three variables, N_1, p_1, and N_2. Thus only spe-
cies-1 is evolving but its evolution may effect the equilibrium abundan-
ces realized by both species.

Example 1. Two competitors. Consider a community with the following
graph

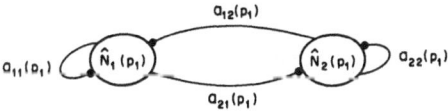

From this graph we can infer the configuration of a coevolutionarily
stable community. To apply Result 1 we inspect the sign of the feedback
in the subcommunity obtained by deleting species-1. G_1 is simply

The feedback in this subcommunity is negative ($a_{22} < 0$). Therefore, by
Result 1 the coevolutionarily stable community is that associated with
a value of p_1 which *maximizes* $\hat{N}_1$. Alternatively, we may focus attention
on species-2. To apply Result 2, we inspect the sign of the feedback of
the connecting web from 1 to 2. G_{21} is simply

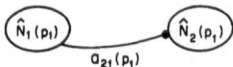

The feedback from this connecting web is negative. Therefore by Result
2 the coevolutionarily stable community is that associated with a value
of p_1 which minimizes $\hat{N}_2$. Figure 1 illustrates the coevolutionary process leading to the community characterized above. The W's are of the
forms $W_{1,ij} = 1 + r_1 - (r_1/K_{1,ij})N_1 - r_1(\alpha_{1,ij,2}/K_{1,ij})N_2$ and $W_2 =$
$1 + r_2 - (r_2/K_2)/N_2 - r_2(\alpha_{2,1}/K_2)N_1$. Notice that p_1 evolves to a value
which simultaneously maximizes $\hat{N}_1$ and minimizes $\hat{N}_2$.

   Example 2. Predator-prey system with prey evolving. Consider the
standard Volterra predator-prey model with the additional assumption of
negative density dependence in the prey. This additional assumption is
needed for stability in a discrete-time model. There is no density-de-
pendence in the predator, however. The community graph is (with $a_{22}(p_1) =$
0)

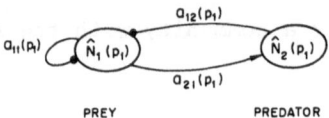

The appropriate subgraphs are

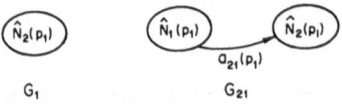

By Result 1 evolution within species-1 does *not* affect the equilibrium
abundance of species-1 because the feedback in G_1 is zero for all p. By
Result 2 evolution within species-1 maximizes $\hat{N}_2$ because the feedback
of the connecting web, G_{21}, is positive. Figure 2 illustrates the coevo-

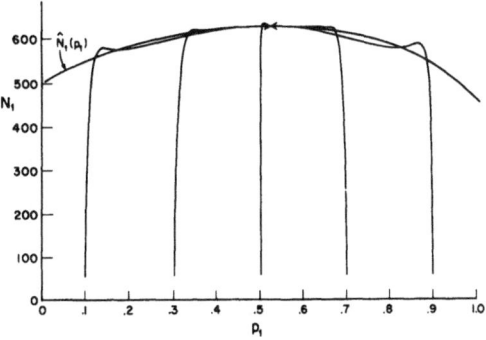

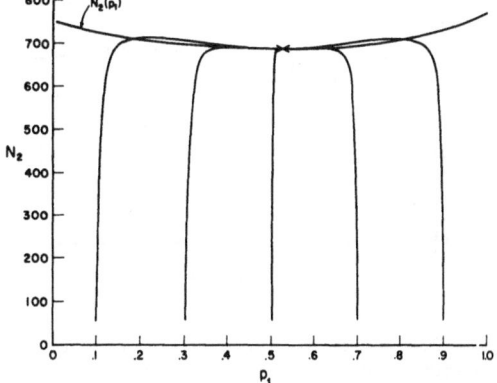

Figure 1.   The coevolutionary process for a community of two competitors. For the three genotypes A_1A_1, A_1A_2 and A_2A_2 of species 1 the competition coefficients α_{12} are .7, .05 and .4, and the carrying capacities K_1 are 1,000, 750 and 800. The corresponding parameters of species 2 are α_{21} = .5 and K_2 = 1,000, respectively. For both species the growth rate r_i are set to unity.

lutionary process leading to the coevolutionarily stable community. The W's in the illustration are $W_{1,ij}$ = 1 + r_1 - r_1N_1/K_{1ij} - a N_2 and W_2 = 1 + ba N_1 - d. In the illustration the heterozygous genotype in the prey was assigned the highest K thereby producing a polymorphism. But notice that as p_1 evolves to its final value there is no effect on $\hat{N}_1$. However, p_1 does maximize the predator's abundance. As Michael Rosenzweig has pointed out, increasing the K of the prey has a destabilizing efffect on system. Indeed if K evolved to a sufficiently high level the system would no longer have a stable equilibrium point.

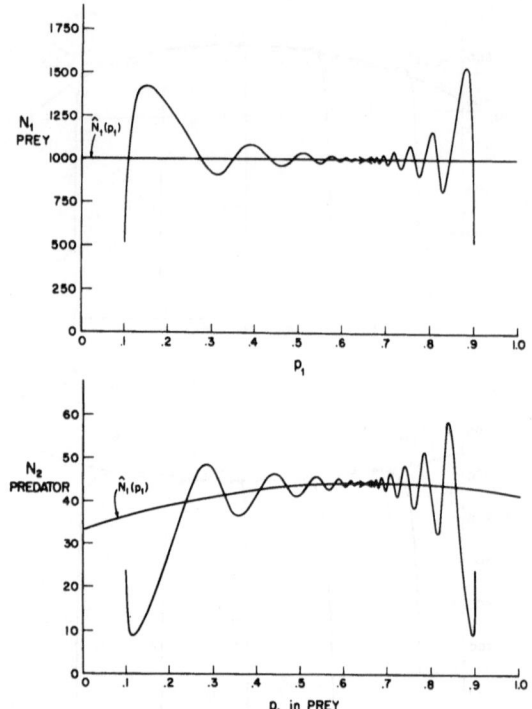

Figure 2. The coevolutionary process for a predator-prey system with the prey evolving. The carrying capacities K_1 of the genotypes A_1A_1, A_1A_2 and A_2A_2 of the prey are 1,700, 2,000 and 1,500, respectively. The other parameters are $r_1 = 1$, $a = .01$, $b = .1$, and $d = 1$.

It is well known that evolution of high predation rates may destabilize a predator-prey system by causing over exploitation of the prey.
But here a new effect is involved: evolution of increased productivity
by the prey is destabilizing the system.

   Example 3. A pioneer and a competitor. Consider a community graph

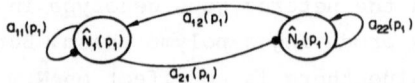

The pioneer species exerts a positive effect on itself and on the other
species, while the competitor exerts a negative effect on itself and on
the pioneer. The pioneer species might be viewed as a species which sta-
bilizes the soil and causes the build-up of organic content in the soil

whereas the competitor is the second to arrive. The appropriate subgraphs
are

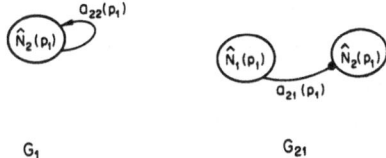

G_1 G_{21}

By Result 1 evolution within species-1 *minimizes* its own abundance because $F_1 > 0$. By Result 2 evolution within species-1 also minimizes the
abundance of species-2 because $G_{21} < 0$. Figure 3 illustrates the coevolutionary process. The W's are $W_{1,ij} = 1 + r - r N_1/K + \alpha_{ij,2} r N_2/K$,
$W_2 = 1 + r + r N_2/K - \alpha_{21} r N_1/K$. In the illustration the heterozygote

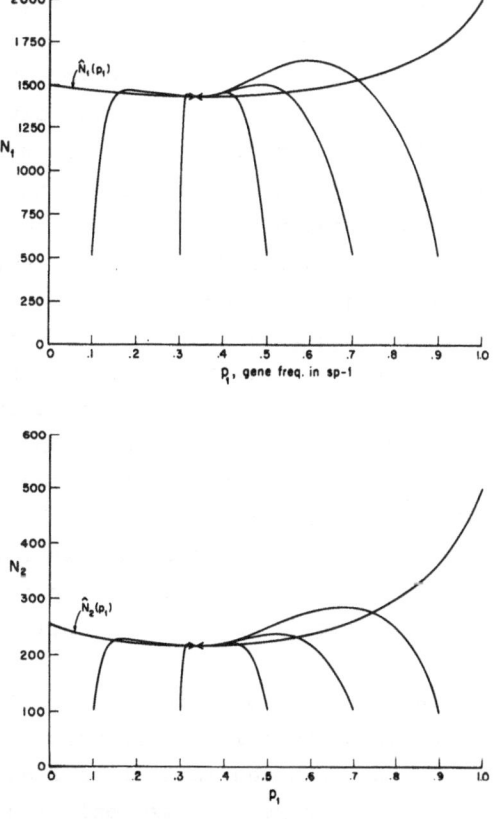

Figure 3. The coevolutionary process in a community of a pioneer and a competitor. The
values of α_{12} for A_1A_1, A_1A_2 and A_2A_2 are 3, 5 and 4, respectively. The other parameters
are $r = .25$, $K = 500$ and $\alpha_{21} = .5$.

was assigned the highest α_{12} thereby producing a polymorphism. Notice
that coevolution leads to the lowest population size for both species.

This example highlights the possibility of coevolution causing the
extinction of a species. Evolution within a species may reduce *its own*
equilibrium abundance below the threshold level which causes extinction.
Of course evolution may also cause the extinction of populations through
the effect of one species causing another species to decline to a level
which brings extinction. But in this example a species may evolve itself
to extinction.

CONCLUSION

I wish to conclude with some general remarks about how coevolution influences community structure. The traditional view of coevolution is
that it leads to mutually accommodated and co-adapted species. It is
presumed that a community of coevolved species satisfies one or more of
the following descriptions: it is maximally resistant to invasion by a
new species, maximally resistant to destruction from environmental perturbations, and maximally likely to persist through time. These presump-
tions have resurfaced frequently in the discussion of complexity and
stability in ecosystems. Robert May and others have shown that complex
ecosystems are less likely to be stable than simple ones. This result
contradicts the previous supposition. However, the complexity-stability
issue has generally been discussed in terms of purely ecological models
with no possibility of coevolution among the interacting populations.
Critics of these models on the complexity-stability issue have often
dismissed the conclusions on the basis that they apply only to "random
assemblages" and not to a coevolved biological system. So the question
is, when coevolution *is* taken explicitly into account does coevolution
generally improve community stability in any of the senses mentioned
above?

On the basis of this paper the answer is no. Coevolution certainly
restructures and molds a community but its end result is not generally
more stable than the initial configuration. This message is especially
clear in this paper because only the best behaved coevolution has been
considered. It is *a priori* evident that coevolution with complicated interspecific frequency dependence can greatly destabilize a community.

Indeed examples are easily made where the mean fitness is minimized at
a stable polymorphic equilibrium and the mean fitness maximized at an
unstable polymorphic equilibrium all because of interspecific frequency-
dependence. But in purely density-dependent coevolution the mean fitness
is always maximized at a stable equilibria, In this case our intuitions
about the optimum strategies of evolution *within* each species are correct. Nonetheless the optimum strategy within each species for maximizing
fitness, even when natural selection will in fact maximize fitness, does
not necessarily improve the abundance of that species --- whether it does
or not depends on the ecological network in which it is embedded.

   In two of the examples coevolution tends to destabilize the community. In the predator-prey example the evolution of increased productivity
by the prey had no effect on the realized prey abundance but maximized
the predator's abundance. Increasing the K of the prey destabilizes the
system as termed the "paradox of enrichment" by Rosenzweig.

   But in the third example coevolution destabilizes the system in a
much stronger sense. In that example species-1 is the species which sta-
bilizes the system in the first place. If species-1 were removed the subcommunity which remains could not exist because of its inherent positive
feedback. However, evolution within species-1 lowers its own abundance.
It may evolve itself to extinction and if so the remaining community is
inherently unstable. By Result 1, *whenever* a species is connected to an
unstable subcommunity (*i.e.* $F_i > 0$) then evolution within that species
will lower its own population size. Thus the coevolutionary fate of a
community composed of unstable subcommunities connected to an occasional
stabilizing species is bleak indeed.

   Of course we may imagine that some combined process of population
turnover together with coevolution will eventually produce communities
composed only of interconnected stable subcommunities. Such a community
would indeed be very stable. But is this carrying optimism to an extreme?

REFERENCES

Chen, W. 1971.
    Applied Graph Theory, North Holland Pub. Co. Pp. 484.

Levins, R. 1974. The qualitative analysis of partially specified systems in mathe-
    matical analysis of fundamental biological phenomena.
    Annals of the New York Acad. of Sci. 231: 123-138.

Chapter 5. Human Evolution

Selection and Non-Mendelian Variability

Marcus W. Feldman and Luigi L. Cavalli-Sforza

"It is clear that attitude toward a product or combination of the three products is not significantly related genetically to any other variable. However, the amount consumed of cigarettes is genetically related to anxiety. The amount consumed of cigarettes and coffee, and cigarettes and alcohol has a significant genetic component as well. The analysis definitely reveals that the consumption of cigarettes has a significant genetic component both by itself and in relation to one personality trait (anxiety) and the consumption of two other products (coffee and alcohol)".

"A substantial portion of marketing effort is directed toward the development of selective demand rather than primary demand for a product category. The tentative conclusions established in this study, that consumption has a genetic component apply to the primary demand for the product category and not for the selective demand of a brand. Management considerations aside what this study suggests is that heredity is a variable worth examining in connection with attempts to understand and explain consumer behavior." (Perry, 1973).

The quotation is taken from a study made by Perry of product consumption behavior in mono- and di-zygotic twin pairs.

Magni (1952), Cavalcanti and Falcao (1954), and Malagalowkin (1958) describe a cytoplasmic condition in *Drosophila bifasciata, Drosophila*

prosaltans, Drosophila willistoni, and *Drosophila pauliotorum* which results in the death of male embryos in mothers with the condition. This
phenotype, called "sex-ratio" may be transferred to normal females by
injection of infected tissue (Malagalowkin and Paulson, 1957). If the
genotype of an infected mother is appropriate, the sex-ratio in her
offspring may be normal (Malagalowkin, 1958).

The heritability of cigarette, alcohol and coffee consumption, and
the problem of cytoplasmic sex-ratio might be worlds apart as far as
their influence on corporate profits is concerned. But there are some
conceptual points of similarity between them which may be of general
evolutionary importance. For the former problem it would be extremely
difficult to ignore a familial effect differentially transmitted to MZ
and DZ twins; no data on family habits was presented in that study.
Such transmission of behavior is non-Mendelian. "Consumption" of the
various products would be regarded by geneticists as a quantitative
character. If it were transmitted by imitation, learning, or any other
cultural medium it would still, in general, be non-Mendelian.

The sex-ratio condition on the other hand is definitely biologically determined. The phenotype, as mentioned above, is under some sort
of genetic control, but its transmission is certainly non-Mendelian.
It is reminiscent of the transmission of an infectious disease. There
are many examples of traits like cytoplasmic sex-ratio. They include a
spectrum of modes of infection from the purely maternal to the completely horizontal. These have been extensively reviewed by Fine (1974).

The mathematical theory of the evolution of traits transmitted from
phenotype to phenotype is not highly developed, especially if any genetics is involved. Cytoplasmic sex-ratio was modelled by Watson (1960)
and there have been a few treatments of epidemic theory with genetic
input (see Dietz, 1976 and references). For continuously varying traits
there have been few evolutionary treatments which go beyond classical
biometrical genetics as far as the transmission from parent to offspring is concerned. Slatkin (1970) introduced a class of models for
the evolution of continuously varying phenotypes which we have modified to include the possibility of both genotypic ang phenotypic transmission. It is our purpose here to discuss the evolutionary properties
of two such models which superimpose cultural transmission of the phenotype over well-known genetic structures. The first is a model for
stabilizing selection on a continuously varying trait which may be
transmitted via the phenotype of the parents as well as genetically.
The second is a simple model of a discontinuous phenotype which is un-

der selection, and which is transmitted in a manner which may depend
on the genotype of the offspring as well as the phenotypes of the parents.

GAUSSIAN SELECTION AND PHENOTYPIC TRANSMISSION

The theory of quantitative inheritance, namely the statistical relationships of measures of a continuous variable among relatives and
within and between populations, almost always employs the genetic model suggested by Fisher (1918). Usually a somewhat imprecisely defined
multiple locus analog to Fisher's formulation of three genotypes A_1A_1,
A_1A_2, A_2A_2, with genotypic values -a, d, a, underlies the statistical
analysis. The results are in terms of quantities such as genetic and
phenotypic variance, and heritability, which are inferred from correlations between relatives. These methods have been taken over by social scientists, such as those interested in improving the marketing
of cigarettes, alcohol and coffee (Perry, *loc. cit.*) or those interested in racial differences in educability (Jensen 1969, 1973). A characteristic of the reasoning used by these investigators is identification of the relative magnitude of the "genetic variance", inferred from
correlations among relatives, with the degree to which the character
has a genetic cause (Lewontin, 1974).

The underlying theory here is not evolutionary and is not generally assumed to apply to natural populations. The selection schemes are
artificial and there has been some analysis of gene frequency change
in simple models for this case (Falconer, 1960 ch. 11). One and two locus models with both artificial and natural selection were discussed
in an evolutionary framework by Karlin and Carmelli (1975) and Carmelli
and Karlin (1975). There have been a few recent studies which have
dealt with natural selection on truly continuous variation. Most notable among these is that of Slatkin (1970) whose formulation of the
problem is as follows. The frequency of phenotypes with value x at time
t is p(x,t). The probability that individuals of phenotypic values y
and z produce an offspring of phenotype x is L(x;y,z). The function
S(x) is the proportion of offspring of phenotype x which survive to reproduce in the next generation. Then the phenotypic frequencies in the
next generation are

$$p(x,t+1) = \frac{S(x) \iint p(y,t)p(z,t)L(x;y,z) \, dy \, dz}{\iiint p(y,t)p(z,t)L(x;y,z)S(x) \, dx \, dy \, dz} \qquad (1)$$

This formulation produced some interesting properties of the phenotypic variance as a function of $S(.)$ and $L(.;.,.)$. Slatkin and Lande (1976) have employed this formulation in an ecological context and Lande (1975) has used it in an attempt to construct a multilocus theory of quantitative inheritance with an extension to include random drift.

Kimura (1965) used a quadratic deviations fitness model in a diffusion theoretic treatment of the same problem. Latter (1970) makes a very direct argument using a single locus and, after some approximation, is able to obtain results analogous to those of Kimura.

We have taken an approach which to some extent unifies the above theory while allowing its extension to the situation of cultural transmission and the inclusion of departures from the random mating implicit in (1). Our approach (Cavalli-Sforza and Feldman, 1976; Feldman and Cavalli-Sforza, 1977) is to define each individual in the population by a pair of numbers (g_t, f_t). The number g_t is the genotypic value at time t after selection, and f_t is the phenotypic value at time t after selection. The process of producing the population distribution of (g_{t+1}, f_{t+1}) at time t+1 has two stages. In the first, mating and transmission of the genotype and phenotype to offspring take place, after which the genotype and phenotype of the offspring are g^*_{t+1}, f^*_{t+1} respectively. Then selection takes place on the phenotypic value to produce the values in the next generation (g_{t+1}, f_{t+1}). If $\varphi_t(g,f)$ is the joint distribution of (g_t, f_t) then the above description can be written

$$\varphi_{t+1}(g,f) = \bar{W}^{-1} W(f) \iiiint \Psi_t(h,k,h',k')T(g,f;h,k,h',k') \, dh \, dk \, dh' \, dk' \qquad (2)$$

where $W(f)$ is the selection function and $\bar{W}$ is the normalizer obtained by integrating the rest of (2) with respect to f and g. The function Ψ is a mating rule which in the case of random mating would be

$$\Psi_t(h,k,h',k') = \varphi_t(h,k)\varphi(h',k') \qquad (3)$$

$T(g,f;h,k,h',k')$ is the joint transmission rule for genotypic and phenotypic values and generalizes the function L in Slatkin's formulation
(1). It is obvious that some specialization of these functions is necessary before (2) is iterable.

We have chosen to consider the case where $\varphi_t(g,f)$ is a bivariate
normal density function and W(f) is a single variable normal density
with mean μ and variance S. This is commonly called stabilizing selection about the optimum μ. S measures the strength of selection; large
S is weak selection, small S is strong selection. Two mating rules have
been considered, random mating (Cavalli-Sforza and Feldman, 1976), and
assortative mating based on phenotype (Feldman and Cavalli-Sforza,
1977). It is the transmission rule, T, which presents the greatest
problem of definition. It has to contain two rules; one which describes the offspring's genotypic value in terms of those of the parents taking into account the effect of mutation, and one for the transmission of the phenotype. If there were no selection or mutation, and
the trait in question were controlled by a single locus with additive
effects then the average genotypic variance of the offspring would be
one half of the total genotypic variance in the parental population
(Falconer, 1960). We have made the assumption, knowing full well that
for one locus under selection it is an approximation, that the variance of the progeny from a mating between parents with genotypic values h and h' is $G_t/2$, where G_t is the genotypic variance in the parental population, and that this is independent of h and h'.

For the phenotypic value in the offspring before selection a class
of linear transmission rules can be used, which are similar in spirit
to that devised by Cavalli-Sforza and Feldman (1973). This class en-
compasses the simple "genetic" model we introduced recently (Cavalli-
Sforza and Feldman, 1976) which can be expressed as

$$g_{t+1}^* = (g_t + g_t')/2 + x. \qquad (a)$$
$$\qquad\qquad\qquad\qquad\qquad\qquad (4)$$
$$f_{t+1}^* = g_{t+1}^* + y \qquad (b)$$

Here (g_t,f_t) and (g_t',f_t') are copies of the parental random variables
and, under the assumptions above, x and y are Gaussian random variables with $x \sim N(m_g, G_t/2 + M)$ and $y \sim N(0,E)$. M and m_g are the effects

of mutation on the genotypic variance and mean, respectively. E is an
environmental variance which in reality would probably change in time;
we assumed it to be constant over time. Under random mating f_t and f_t'
would be independent. Assortative mating can be built in by a correlation between them.

At this stage, after completion of transmission the distribution
$\phi_t(g,f)$ has been transformed to $\phi_{t+1}(g,f)$ say. Since all distributions
are Gaussian, this transformation can be expressed in terms of changes
from the parental means $\mu_{g,t}$, $\mu_{f,t}$, their genotypic and phenotypic
variances G_t, F_t and the covariance C_t to corresponding values $\mu^*_{g,t}$,
$\mu^*_{f,t}$, G^*_t, F^*_t and C^*_t in the offspring before selection.

Selection then occurs according to the function $W(f) =$
$\exp[-(f^*_{t+1} - \mu)^2/2S]$. Here μ is the optimum value of the phenotype
and S is the measure of the strength of selection, large S is weak selection, low S is strong selection. After this operation a full cycle
has been completed, and recursions for the moments of the geno-pheno-
typic distribution can be written. We showed (Cavalli-Sforza and Feldman, 1976) that these can be iterated and that the equilibrium genotypic variance after selection is

$$\hat{G} = M\left\{ [1 + 4(E + S)/M]^{1/2} - 1\right\}/2 \qquad (5)$$

with the corresponding phenotypic variance

$$\hat{F} = S(\hat{G} + E + M) / (\hat{G} + E + M + S). \qquad (6)$$

It is obvious that this scheme cannot be literally representative
of all selection schemes even at a single locus. It has in it the assumption that the average variance among sibships is realized in each
sibship. This may be a reasonable premise if the number of genes is
large and effects are small. Obviously when there is dominance the departure of x from $N(m_g, G_t/2 + M)$ will be a function of the gene frequencies and the degree of dominance. A partial compensation for this
might be made by assuming that $x \sim N(m_g, G_t(1 + D)/2 + M)$, so that there
is a constant fraction of G_t ascribable to these other forces, namely
D. The same iteration process produces

$$G = \frac{M}{2+D}\left\{ \frac{D(E+S)}{2M} - 1 + \left[\left(\frac{D(E+S)}{2M} + 1 \right)^2 + \frac{4(E+S)}{M} \right]^{1/2} \right\} \qquad (7)$$

The equilibrium variance (5) is the same as that obtained using totally different methods by Latter (1970) and Kimura (1965).

Suppose instead of (4b) we postulated that

$$f^*_{t+1} = (f_t + f_t')/2 + y \qquad (8)$$

with y as in (4), then we have the case of purely phenotypic transmission - in fact, blending inheritance. In the absence of the environmental contribution, E, this would result in the disappearance of all phenotypic variability, a classical fact. When E ≠ 0, it is easy to prove that

$$F_t \to \hat{F} = (S/2 + E) \left\{ \left[1 + 2SE/(S/2 + E)^2 \right]^{1/2} - 1 \right\} \qquad (9)$$

Between (4b) and (8) is a continuum of models allowing varying degrees of "genetic" and parental phenotypic contribution to the offspring phenotype. We have recently completed a study of this class of models (Feldman and Cavalli-Sforza, 1977) in which the phenotype of an offspring is the result of a proportion γ due to its genotypic value, and a complementary proportion δ = 1 - γ due to the mid-parental phenotypic value. In terms of cultural transmission the latter represents an environmental contribution to the phenotype of an offspring, since the parents' phenotypes can legitimately be considered part of the environment of the progeny. The details of the analysis of this case are presented in the above-mentioned paper. Karlin (1977) will present a multi-dimensional version with a different perspective from the genetic one we have attempted to give. Perhaps the most important conclusion of this more general analysis concerns the role of the proportions γ and δ in determining the phenotypic and genotypic variances in the population. It is quite natural to conceive of γ as the degree of genetic determination of the phenotype before selection. Yet when γ is small (and δ = 1 - γ large), the genotypic and phenotypic values evolve almost independently. The mutation variance M becomes the major determinant of G_t and the latter increases. The function G_t has become uncoupled from the restraining influence of selection which limits the growth of F_t. Hence the relative magnitudes of G_t and F_t may not reflect the relative contribution of genotype and mid-parental phenotype to the offspring phenotype after selection. A numerical example of this is presented in Table I. Feldman and Cavalli-Sforza (1977) also present an analysis of a class of assortative mating models which can be derived from the above formulation by incorporation of some classi-

Table 1.   Equilibrium Phenotypic and Genotypic Variances for E = 1

		s = 100, M = 0.001;		s = 100, M = 0.01;		s = 100, M = 0.1	
γ	δ	$\hat{F}^*$	$\hat{G}^*$	$\hat{F}^*$	$\hat{G}^*$	$\hat{F}^*$	$\hat{G}^*$
1.0	0.0	1.301	0.317	1.970	1.000	4.057	3.128
0.75	0.25	1.292	0.371	1.873	1.171	3.682	3.679
0.50	0.50	1.349	0.479	1.822	1.513	3.296	4.777
0.25	0.75	1.511	0.804	1.824	2.504	2.801	8.054
0.00	1.0	1.924	(a)	1.924	(a)	1.924	(a)

(a): $\gamma = 0$, $\delta = 1$ entails $G^* \to \infty$.

cal arguments by Fisher on correlations between relatives.

It is obvious that the theory cannot be viewed as precisely representing either genotypic or phenotypic mechanisms. On the other hand there are features of the single locus genetic theory (see, e.g., Latter, 1970) which our theory handles in a qualitatively adequate manner. For this reason it is of some interest to investigate the extensions to include different selection functions, different transmission rules, and an independently transmitted environment.

SELECTION ON A SIMPLE DISCONTINUOUS NON-MENDELIAN TRAIT

In the previous section we have regarded the "cultural transmission" of the phenotype through the phenotypic transmission from parents to offspring. When $\gamma \neq 0$, phenotypic and genotypic values are correlated. The phenotypic transmission, therefore, has a tinge of genotype about it, depending of course on the size of γ and S.

The next model we introduce is an interactive one between genotype and phenotype at a much more naive level.

Suppose that there are three genotypes in the population A_1A_1, A_1A_2, A_2A_2 and that each genotype may exist in one of two phenotypic forms. One form we might call "infected" (using an analogy from the theory of epidemics). The other we call "native". These states are represented as $\overline{A_1A_1}$, $\overline{A_1A_2}$, $\overline{A_2A_2}$ for the infected and A_1A_1, A_1A_2, A_2A_2 for the native. We then have six "pheno-genotypes". In terms of the example in the introduction the infected would have cytoplasmic sex-ratio condition, although the sex-limitation of this example needs further detail. Our treatment will not really take account of sex differences except in a rather trivial way.

The two phenotypes have different viabilities, which are not genotype dependent. Thus the infected have viability $1 + s$ relative to 1 for the native. It is not necessary to specialize the sign of s yet, since although "infected" is a loaded word, it could refer to the acquisition of something advantageous such as a skill. The frequencies of A_1 and A_2 are p and $q = 1-p$ respectively. (It requires proof that one genetic variable is sufficient but this is easy). The proportions of A_1A_1, A_1A_2 and A_2A_2 that are infected are k_1, k_2, k_3 at a given generation, so that k_1 is the number of $\overline{A_1A_1}$ divided by $(\overline{A_1A_1} + A_1A_1)$, etc.

At any mating, one parent is arbitrarily defined as the transmitting parent, to be denoted T-parent. If the T-parent is infected the probabilities that its offspring of genotypes A_1A_1, A_1A_2, A_2A_2 are native are N_1, N_2, N_3 respectively. If the T-parent is native then these probabilities are n_1, n_2, n_3. The model is then completely specified by the assumption that mating is at random.

Write $\alpha_i = 1-n_i$ and $\beta_i = (1+s)(1-N_i) - \alpha_i$ ($i = 1,2,3$). Then a recursion system can be written expressing the frequencies p', k_1', k_2' and k_3' in terms of those in the previous generation:

$$p' = p(1+sK_1)/(1+sK) \tag{10}$$

$$k_1' = (\alpha_1 + \beta_1K_1)/(1+sK_1) \tag{11}$$

$$k_2' = (1/2)\left\{(\alpha_2 + \beta_2K_1)/(1+sK_1) + (\alpha_2 + \beta_2K_2)/(1+sK_2)\right\} \tag{12}$$

$$k_3' = (\alpha_3 + \beta_3K_2)/(1+sK_2) \tag{13}$$

where $K_1 = pk_1 + qk_2$, $K_2 = pk_2 + qk_3$. The variables K_1 and K_2 are in a sense the marginal effects of the infection. $K = pK_1 + qK_2$ is the mean infection in the population.

The detailed equilibrium properties of (10) - (13) have been presented for the case $\alpha_i = 0$ (all i) by Feldman and Cavalli-Sforza (1976).

Among the main results are the following. When $s > 0$ a polymorphic gene frequency equilibrium exists and is stable provided

$$(\beta_2 - 1)^2 > (\beta_1 - 1) \qquad\qquad (14a)$$

and

$$\beta_2 > \beta_1, \; \beta_2 > \beta_3. \qquad\qquad (14b)$$

In other words when the heterozygote has the advantage in becoming infected, and the infection is advantageous then the polymorphism will be stable provided that the advantage to being infected is sufficiently large.

In the case where $\alpha_i \neq 0$ (all i) we have recently shown that a necessary and sufficient condition for the gene A_1 to be protected (i.e., prevented from fixing at $p = 0$) is that $\hat{k}_2 > \hat{k}_1$, $\hat{k}_3$ when $s < 0$ where $(\hat{k}_1, \hat{k}_2, \hat{k}_3)$ is the unique equilibrium of (10) - (13) obtained by setting $p = 0$. A similar condition holds at $\hat{p} = 1$. It is not yet known whether this automatically ensures that there is a unique stable polymorphism.

When $N_i = 0$, $n_i = 1$ (all i), that is all offspring of infected T-parents are infected and all offspring of native T-parents are native, it is possible to iterate (10) - (13) explicitly. Obviously there are no genetic differences in the infection parameters here. Thus any gene frequency equilibrium should depend on the initial frequencies. In fact it can be shown that the equilibrium p_∞ for p in this case can be written as

$$p_\infty = p_0 + [p_0 q_0 (K_{10} - K_{20})/K_0]\left(1 - \sum_{j=1}^{\infty} \left\{ 2^j [(1-K_0) + (1+s)^j K_0] \right\}^{-1} \right) \quad (15)$$

where K_{10}, K_{20}, K_0 are the initial values of K_1, K_2, K and p_0 is the initial gene frequency. Of interest here is that substantial change in gene frequency may occur so long as there are initial differences in K_{10} and K_{20}. This occurs despite the fact that there are no genetically based selection or transmission differences.

A variant of this class of models can be obtained by defining transmission probabilities which depend on the genotypes of neither offspring nor parents. Instead the probability of a child becoming infected is b_3, β or b_0 depending on whether both, one or none of the parents is infected. With the same selection as before (1+s:1) the frequency of infected individuals in the whole population can be iterated alone. It is possible that the infected phenotype fixes with s < o as a result of high infection probability.

The mathematical structure of (10) - (13) includes a delayed frequency dependence which seems worthwhile pursuing in more detail. Other authors (e.g., Bartholomew, 1973) have pointed out the numerous analogies between social interaction processes and epidemiology. It is probable that many evolutionary sequences have been strongly influenced by the transmission of information either within families or across groups in a "contagious" way. It may turn out that certain sociobiological phenomena often reduced by some investigators to a genetic etiology are more economically explained in this fashion.

Summary

Models are described for the transmission of phenotypes and genotypes jointly from parents to offspring. Both continuous and discrete, two valued, traits are discussed. The equilibrium properties are presented for some simple examples, and the result interpreted in the context of evolution of behavior.

Acknowledgements

The research was supported in part by grants NIH 10450-12, NSF GB 37835 and USAEC AT(04-3) - 326.

REFERENCES

Bartholomew, D.J. 1973. *Stochastic Models for Social Processes*. 2nd Ed. Wiley, New York.

Cavalcanti, A.G.L. and Falcao, D.N. 1954. A new type of sex-ratio in *Drosophila prosaltans* Duda.
Proc. IX Int. Congr. Genet., Bellagio, *Caryologia*, 6 (Suppl. II): 1233-1325

Carmelli, D. and Karlin, S. 1975. Some population genetic models combining artificial and natural selection pressures I. One locus theory.
Theor. Pop. Biol. 7: 94-122.

Cavalli-Sforza, L.L. and Feldman, M.W. 1973b. Cultural versus biological inheritance: Phenotypic transmission from Parents to Children.
Amer. J. Hum. Genet. 25:618-637.

Cavalli-Sforza, L.L. and Feldman, M.W. 1976. Evolution of continuous variation: Direct approach through joint distribution of genotypes and phenotypes.
Proc. Nat. Acad. Sci. U.S.A. 73 (in press).

Dietz, K. 1976. The effect of immigration on genetic control.
Theor. Popul. Biol. 9: 58-67.

Falconer, D.S. 1960. *Quantitative Genetics*. Oliver and Boyd, London.

Feldman, M.W. and Cavalli-Sforza, L.L. 1976. Cultural and biological evolutionary
processes. Selection for a trait under complex transmission.
Theor. Popul. Biol. 9 (in press).

Feldman, M.W. and Cavalli-Sforza, L.L. 1977. Evolution of continuous variation II.
On the effects of complex transmission and assortative mating.
Theor. Popul. Biol. (to appear).

Fine, P.E.M. 1974. The epidemiological implications of vertical transmission.
Thesis. Univ. of London, Faculty of Medicine.

Fisher. R.A. 1918. The correlations between relatives on the supposition of Mendelian inheritance.
Proc. Roy. Soc. Edinb. 52: 399-433.

Jensen, A.R. 1969. How much can we boost IQ and scholastic achievement.
Harvard Educ. Rev. 39: 1-123.

Jensen, A.R. 1973. The differences are real.
Psychology Today 7: 80-84.

Karlin, S. 1977. Manuscript in preparation.

Karlin, S. and Carmelli, D. 1975. Some population genetic models combining artificial and natural selection pressures II. Two locus theory.
Theor. Popul. Biol. 7: 123-148.

Kimura, M. 1965. A stochastic model concerning the maintenance of genetic variability in quantitative characters.
Proc. Nat. Acad. Sci. U.S.A. 54: 731-736.

Lande, R. 1976. The maintenance of genetic variability by mutation in a polygenic
character with linked loci.
Genet. Res. Camb. 26: 221-236.

Latter, B.D.H. 1970. Selection in finite populations with multiple alleles. II.
Centripetal selection, mutation and isoallelic variation.
Genetics 66: 165-186.

Magni, G.E. 1952. Sex-ratio in *Drosophila bifasciata*.
Rend. Inst. Lomb. di Sci. e Lett. 85: 391-411.

Malagalowkin, C. 1958. Maternally inherited "sex-ratio" conditions in *Drosophila
willistoni* and *Drosophila paulistorum*.
Genetics 43: 274-285.

Malagalowkin, C. and Paulson, D.F. 1957. Infective transfer of maternally inherited abnormal sex-ratio in *Drosophila willistoni*.
Science 126: 32.

Perry, A. 1973. Heredity, personality traits, product attitude and product consumption - an exploratory study.
J. Marketing Res. X: 376-379.

Slatkin, M. 1970. Selection and polygenic characters.
Proc. Nat. Acad. Sci. U.S.A. 66: 87-93.

Slatkin, M. and Lande, Russell. 1976. Niche width in a fluctuating environment -
density dependent model.
Amer. Natur. 110: 31-55.

Watson, G.S. 1960. The cytoplasmic "sex-ratio" condition in *Drosophila*.
Evolution 4: 256-265.

4. HUMAN EVOLUTION

ESTIMATION OF THE CHARACTERISTICS OF RARE VARIANTS

E. A. Thompson

This paper arises in connection with a study of the data on rare protein variants in American Indian populations. Several American Indian tribes have been extensively sampled (Neel 1973), and the variants found range from those present only in a single individual or nuclear family, through those that have spread through a village or to several villages, to one which has achieved polymorphic frequencies in a single tribe. It is perhaps doubtful whether this last should be considered a "rare variant", but we include it since it is fairly certain that it has arisen since tribal differentiation, and in spite of inter-tribal migration it has not spread to neighbouring tribes. For variants apparently localised in areas where sampling has been intensive, accurate estimates of the total number of individuals carrying the variant allele may be made.

There are many questions we would like to be able to answer about these rare variants. Two fundamental characteristics are the age and the selective effect, and we shall consider an approach towards estimating these parameters. Previous writers have considered the age of an allele through a diffusion analysis of changes in allele frequency (see for example Kimura and Ohta 1974 and Li 1975). However, most of the variants we consider will have arisen from a single mutation, spread only in a very localised area, and will never achieve polymorphic frequencies in the population at large; instead we shall consider the actual numbers

in which the variant is replicated. Maruyama (1974) considers the numbers
rather than the frequencies of alleles, but approximates the process by
a diffusion equation in continuous space. We shall consider a discrete
branching process.

Such a model has the advantage that the spread of a variant depends
only on the distribution of the number of variant offspring of a single
variant gene. Such considerations as population size do not enter the
problem. The only assumptions are that the variant arose from a single
mutation, that the parameters of the offspring distributions remain constant in time, and that the variant is sufficiently rare for all the individuals carrying it to be heterozygotes. This last requirement is in
order that variant genes reproduce independently. A likelihood approach
will be taken to the estimation problem. Given the parameters of the variant offspring distribution and the length of time since the variant
arose we have a probability distribution for current numbers. Conversely
currently observed numbers provide a likelihood for the parameters.

TOTAL NUMBER OF REPLICATES OF A RARE VARIANT

We consider first the information about the characteristics of a single
rare variant that is contained in the numbers alone. Clearly a single
observation can allow us to estimate only a single function of the parameters. One case where an explicit form for the probability distribution
of the number of replicates at any given generation can be found is for
a generalised geometric offspring distribution for the variant. Thompson
(1976) has considered the estimation of the age of a rare variant on
the basis of this model, and details of the derivation of the following
may be found in that paper. It is found to be convenient to express the
model in terms of the mean of the offspring distribution (m), and a second parameter (h) which allows us to choose also the variance of this
distribution. A large value of m may indicate a selective advantage for
the variant, but may also simply reflect an increasing population. Nevertheless, m is the parameter of interest in considering the problem of
selection. Conditional upon non-extinction of the variant, the probability that after t generations the variant will have k replicates, given
that it arose as a single copy, is

$$p_t(k|k{\neq}0) = M(1-M)^{k-1} \quad \text{for } k = 1,2,\ldots \tag{1}$$

where

$$M = (m-1)h/(m^t-1 + (m-1)h). \tag{2}$$

Given a necessarily non-zero observation k of current numbers, (1) may then be considered as a likelihood function for M. M is a function of the age, t, and of the parameters m and h, and any one of the three may be estimated if the other two are known.

In the analysis of data, since we have a discrete generation model, the first problem is which variant individuals should be counted. It is found to be most reliable to count numbers in the generations of current-ly reproducing adults. The relevant offspring distribution is then that of the number of offspring, heterozygous for the variant allele and sur-viving to adulthood, of a given adult heterozygote. New mutant individu-als not surviving to adulthood in their initial generation are thus dis-regarded. Using data from American Indian family-size distributions (Neel and Chagnon 1968) it can be shown that the assumed distribution with h = 1.5 provides a good fit to the data for a variety of mean fami-ly sizes (Thompson 1976). Indeed the fit is better than can be obtained with a negative binomial distribution, which is the standard distribu-tion used in this context (Kojima and Kelleher 1962).

Thompson (1976) considered the likelihood (1) as a function of m and t for fixed h (equal to 1.5), for several different American Indian variants. Considered as a function of t with m known, it was found that although in general likelihood confidence intervals for the age are ex-tremely wide, and more serious problems arise for m<1 (see also Stigler 1970), there are situations in which total numbers alone can contain useful information.

Although (1) may also be considered as a likelihood for m for known t and h, it would normally require very strong selection indeed for a neutral null hypothesis to be rejected. (We refer to m = 1 as the neutral hypothesis, which in absolute terms it is, although as commented above it may not be so relative to the general population if this is increas-ing or decreasing). The major problem is, however, not the weakness of information available, which must necessarily be a problem due to the very high variance intrinsic in the process of gene replication, but

the fact that m and t cannot, even in principle, be simultaneously estimated. In order to consider both unknowns we must consider what further
information we have in addition to total numbers. There are basically
two possibilities. We can consider either the distribution of numbers
of different variants, or of the distribution of numbers of a single variant in different subunits of the population. Each of these will be considered in the following sections.

THE DISTRIBUTION OF RARE VARIANTS

If we have a population in which variants arise and thereafter reproduce
according to specified processes there will be a resulting distribution
of rare variant numbers in the population. Subject to various ascertainment problems, it is possible that the observed distribution of rare variant numbers may convey information about the possible selective effects
of variants. Ewens (1972, 1973) has previously taken this approach to
allele frequencies or numbers at a single locus on the basis of a Wright-
Fisher model. Of more direct relevance to the present approach is the
paper of Karlin and MacGregor (1966), who consider alleles arising according to some "input process" and thereafter reproducing according to
a continuous-time Markov process with transition probability matrix
$Q_{ij}(t)$. If the input process is non-homogeneous Poisson, with intensity
function $u(t)$, and $N_k(t)$ denotes the number of different mutants represented by k replicates at time t, then the random processes $N_k(t)$, k =
1,2,... are independent non-homogeneous Poisson processes with expectations

$$E(N_k(t)) = \int_0^t Q_{1k}(t-s) \, u(s) \, ds. \tag{3}$$

This general result may be readily adapted to the particular case
of the present model. The direct continuous-time analogue of the generalised geometric offspring distribution is a simple birth and death process. Further for a constant mutation rate in a stable population, we
have a Poisson input of constant rate, or in a population with growth
rate m* we have $u(t)$ proportional to $(m*)^t$. In the stable-population

case we obtain

$$E(N_k(t)) \propto (1/k)\ (1 - M)^k \qquad\qquad (4)$$

(Recall that M is a function of m,h and t, or equivalently of t and the
birth and death rates). In theory therefore the observed distribution
of variant numbers contains information about selective effects. We
could take t to be some estimate of the time since the American Indians
entered South America, and consider the distribution of variants assumed
to have arisen since then. In particular, with sufficient data, we could
distinguish the case $m \geq 1$ from $m \leq 1$. However, the processes $N_k(t)$ are
Poisson, with variances equal to the mean, and accurate data are difficult to obtain. Also it is doubtful whether the model is applicable over
such long periods of time. Further the power of the method is small;
even if variants are in general disadvantageous, there will be a large
number of variants in the population, and some will be both numerous
and old.

Perhaps the most fundamental objection is, however, that in considering the distribution of variants the basic assumption must be that
all variants have the same selective characteristics. This is unlikely
to be so, and we are interested in estimating the characteristics of
particular variants, not variants in general. For all these reasons we
shall not pursue this approach further in this paper.

THE SPREAD OF A RARE VARIANT

One way in which we can get more information about a single variant is
by considering its spread. If it is confined to a single village, this
is an indication, regardless of its current number, that it is young.
A variant with say 20 representatives located in a single village would
thus be evidence for a young variant with a very rapid growth rate, and
hence possible selective advantage. A second variant, also with a current
total of 20 adult representatives, but spread over widely dispersed villages would be evidence for a much older variant. Of course initial
growth may be the result of chance social influences rather than of genetic advantage. MacCluer et al. (1971) noted in the Yanomama population
a significant father-son correlation in fertility, presumably caused by

social factors, and the existence of a few males with exceptionally large
numbers of grandchildren. Nevertheless the possibility of estimating such
a rapid growth is of interest, regardless of its underlying cause.

Suppose therefore that the population consists of r villages labelled
$i = 1, \ldots, r$, and that individuals migrate independently. The probability that an adult individual in village i migrates to village j before
reproducing is denoted by b_{ij}, and

$$b_{ii} = 1 - (\sum_{j \neq i} b_{ij})$$

is the probability of remaining in the village of birth. Let $B = (b_{ij})$
denote this migration matrix. We retain the model by which individuals
of a given variant type reproduce according to a generalised geometric
offspring distribution. We consider a single variant allele, and a gene
of this allelic type located in village i is said to be of "type" i. The
process may then be viewed as a multitype branching process. The generating function for the total number of adult variant replicates of a variant gene in an adult of the proceeding generation is, as given by
Thompson (1976),

$$g(z) = 1 - mh/(1+h) + m(h/(1+h))^2 z/(1-z/(1+h)). \tag{5}$$

These individuals then migrate according to the migration matrix B.
Thus the generation function

$$\phi^{(i)}(z) = E(\prod_{j=1}^{r} (z_j)^{N_j})$$

for the number of replicates N_j in village j produced by a single gene
in village i, after migration but before subsequent reproduction, is
$\phi^{(i)}(z) = g((Bz)_i)$ or $\phi(z) = g(Bz)$ where we define g by $(g(Bz))_i =$
$g((Bz)_i)$. Hence the t th. generation generating function is

$$\phi_t(z) = (gB)^t(z). \tag{6}$$

Note that, since $B0 = 0$ and $B1 = 1$, where 0 and 1 denote column vectors
or zeros and ones respectively, we have as required $\phi_t(0) = g_t(0)1$ and
$\phi_t(1) = 1$. Unfortunately (6) does not provide an explicit form for the
distribution. However, we shall derive the mean and variance.

The mean of the offspring distribution is the matrix (or set of vectors)

$$A = \{\delta\phi^{(i)}/\delta z_j\}_{z=1} = \{g'(1)\ b_{ij}\} = mB.$$

Thus if $N(t)$ denotes the vector of variant numbers at generation t $E(N(t+1)|N(t)) = m\ N(t)'B$ and hence $E(N(t)) = m^t N(0)'B^t$. The probability of extinction of the variant allele before it reaches age t is

$$P(N(t) = 0) = P(N_j(t) = 0,\ j=1,\ \ldots,\ r) = g_t(0) = 1 - m^t M,$$

where M is given by (2).

Thus conditional upon non-extinction we have

$$E(N(t)|N(t) \neq 0) = E(N(t))/(1-g_t(0)) = N(0)'B^t/M. \qquad (7)$$

Harris (1963) gives the recurrence formula for the variance-covariance matrix $C(t)$ of the vector $N(t)$;

$$C(t+1) = A'\ C(t)A + \sum_{i=1}^{r}\ (V_i\ E(N_i(t)), \qquad (8)$$

where A is the mean offspring matrix, or in our case mB, and V_i is the variance-covariance matrix for the offspring vector of a single gene in village i. The components of the matrix V_i for our case are easily derived;

$$(V_i)_{jj} = (2m/h - m^2)\ b_{ij}^2 + m\ b_{ij},\ i=1,\ \ldots,\ r,\ j=1,\ \ldots,\ r,$$

and

$$(V_i)_{jk} = (2m/h - m^2)b_{ij}b_{ik} \qquad \text{for } j \neq k$$

Thus

$$\sum_{i=1}^{r}\ (V_i)\ E(N_i(t)) = m^{t+1}((2/h - m)\ B'\Delta_t B + \Delta_{t+1}),$$

where Δ_t is the diagonal matrix whose diagonal components are those of of the vector $(N(0)\ B^t)$, and from (8) we obtain

$$C(t) = \Sigma_{s=1}^{t} \ m^{2t-s} \ \{(2/h - m) \ B'^{(t-s+1)} \ \Delta_{s-1} B^{(t-s+1)} + B'^{(t-s)} \Delta_s B^{t-s}\}.$$

$$(9)$$

It may be readily checked that in the case of no migration (B=I, the identity matrix) (9) reduces to

$$C(t) = m^t((m^t-1)/(m-1)) \ (2/h - m + 1) \ diag(N(0)) = W_t \ diag(N(0)),$$

$$(10)$$

where $diag(N(0))$ is the diagonal matrix with diagonal components $N(0)$ and W_t is the value of the t th generation variance of the total number of descendents of a single individual.

Conditional upon non-extinction we have, as for the mean (7),

$$E(N_j(t)N_k(t) \,|\, N(t) \neq 0) = E(N_j(t)N_k(t))/(1 - g_t(0)), \quad or$$

$$\Big(C(t|N(t) \neq 0) + \{E(N(t)\,|\,N(t) \neq 0)\}\{E(N(t)\,|\,N(t) \neq 0)\}'\Big) =$$

$$\Big(C(t) + \{E(N(t))\}\{E(N(t))\}'\Big)/(1 - g_t(0)).$$

Thus substituting for $g_t(0)$ and for $E(N(t)\,|\,N(t) \neq 0)$ from (7) we have the covariance matrix conditional upon non-extinction

$$C(t|N(t) \neq 0) = \{C(t)/m^t M\} - ((1-m^t M)/(m^t M)^2 \ \{E(N(t))\}\{E(N(t))\}'. \quad (11)$$

Formulae (9) and (11) are unfortunately too complex to be used analytically, but the variance-covariance matrix may be numerically evaluated in any particular case, in order to investigate the reliability of any estimation procedure.

Returning to the mean vector (7), the part $N(0)'B^t$ involves only the age of the variant and the migration pattern, while M incorporates the parameters of the offspring distribution. Thus we can consider the effect of migration on this mean separately from that of the rate of increase of the variant. The information contained in the spread of the variant is summarized by the moments of the mean spread. As an example consider the case of uniform symmetric linear migration at rate $\frac{1}{2}q$ to each of the two neighbouring villages, and suppose that the villages in the infinite sequence are labelled ..., $-r$, -1, 0, 1, ..., r, ... Assuming that the variant arose in village 0, $N(0) = (..., 0, 0, 1, 0, 0, ...)$. Also $(Bz)_i = \frac{1}{2}qz_{i-1} + (1-q)z_i + \frac{1}{2}qz_{i+1}$. We shall denote by $w(k)$ the vector of k th powers of the integers;

$$w(k) = (...,(-r)^k,(-r+1)^k,...,(-1)^k,0^k,1^k,2^k,...,r^k,...),k=0,1,2,3,...$$

and define $D_k(t) = (N(0)'B^t)w(k)$. $D_k(t)$ is thus a form of normalized k th moment of the mean spread of the variant. Now

$$B^t 1 = 1 \text{ for all } t, \text{ or } D_0(t) = 1 \text{ for all } t. \tag{12}$$

Further, by symmetry, the odd moments of the mean vector are all zero;

$$D_k(t) = 0 \text{ for all } t, \text{ for } k = 1,3,5,7,... . \tag{13}$$

The even moments may be determined recursively, for

$$D_{2k}(t) = (N(0)'B^t) \, w(2k) = (N(0)'B^{t-1}) \, (Bw(2k))$$

or, since

$$(r+1)^{2k} + (r-1)^{2k} = 2\sum_{s=0}^{k} \binom{2k}{2s} r^{2s},$$

$$D_{2k}(t) = (1-q) \, D_{2k}(t-1) + q\sum_{s=0}^{k} \binom{2k}{2s} D_{2s}(t-1). \tag{14}$$

Also $D_{2k}(0) = 1$, and hence $D_2(t) = D_2(t-1) + q \, D_0(t-1)$ or $D_2(t) = qt$, and $D_4(t) = qt(1 + 3q(t-1))$, and the higher moments may be established similarly.

The linear migration model has an obvious parallel in the ladder model for electrophoretic profiles (Moran 1975), and the expected moments of the variant spread may be further analyzed by the methods of that paper, adapted to the branching process case. In this paper, however, our aim is to consider whether the spread of a variant will contain useful information, rather than to pursue a particular migration pattern. We see that, provided the migration rate is known, the total numbers $(\Sigma_j \, N_j(t) = N(t)'1)$ and the second moments of the observed distribution $(N(t)'w(2))$ alone provide a method of estimating both the age and the rate of increase of the variant, for we have

$$E(N(t)'1 \mid N(t) \neq 0) = 1/M \quad \text{and}$$

(15)

$$E(N(t)'w(2) \mid N(t) \neq 0) = qt/M.$$

The normalized second moments of the observed distribution could thus be
used as an estimate of qt, or to estimate t if q is known, while the total observed numbers is an estimate of 1/M and hence provides an estimate
of m if t is known. (The latter is in fact the maximum likelihood estimate
(Thompson 1976)).

A similar analysis may be carried out for the case of two-dimensional
migration, or the pattern of moments resulting from any specific migration pattern arising in practice may be numerically evaluated. The variance of these statistics may be numerically computed from equations (7),
(9) and (11) since

$$\text{var}(N(t)'w(k) \mid N(t) \neq 0) = w(k)'C(t \mid N(t) \neq 0) \, w(k). \tag{16}$$

Although the theoretical justification for the estimates based on (15)
is not clear, accurate estimates of the age of an allele may be made from
normalized moments, provided the age is not too large. The method is most
appropriate for variants of sufficient age to have spread to several
villages, but not sufficient for the distribution to have become too
diffuse. For older variants not only are the variances of the moments
very much greater, but it is also unlikely that the migration pattern
has remained constant over the period since the variant arose. In fact,
although current migration patterns may be quite accurately estimated,
it is unlikely that these, or even the villages themselves, are of many
generations standing. The effect of the assumption of constant migration
on practical results is difficult to assess. Another practical problem
is that the moments (15) are about the position of origin of the allele.
For a young variant the village of origin may be clear, but if not a
correction may be made to allow for taking moments about some "median"
village.

The aim in the analysis of the spread of a rare variant has been to
show how estimates of both the age and of the growth rate of the variant
could be made. We have seen that there is information on the age of an
allele contained in the moments of its spread, and that the age may thus
be estimated without knowledge of the growth rate, provided the latter
is constant. Thus in theory the aim is achieved, but it is clear that

any estimate of m will be very unreliable. We saw in section II that even for known t an estimate of m has very wide confidence bounds, and using an estimated t they must be wider still. Unfortunately, there seems to be very little information on m contained in the distribution of numbers, beyond what is contained in total numbers. Substituting the observed for the expected values of the moments there is indeed no such information, and although in the distribution of $N(t)$ the migration and growth parts do not separate, indicating that the normalized distribution is not independent of m, we have no explicit form for this distribution. However, even if we cannot yet estimate m, the spread does provide a very much more satisfactory estimate of the age of a variant than does the total number of replicates. The analysis of rare variant data on the basis of the model of this section can provide further understanding of these data.

Added note.

Since this paper was presented a paper of Crump and Gillespie (1976) has been published. Their paper also analyses the spread of a mutant allele using a branching process model. Their aim was to look at the theoretical values of parameters of the total dispersion of a mutant before its extinction, rather than at the analysis of data, but some parts of their discussion parallel that independently given here.

Summary.

     Previous approaches to the problem of estimating the parameters, such as those of selection and age, involved in the probability distribution for the frequency of an allele have used diffusion equations. Such an approach is not suitable for the particular problem of this paper, for we wish to consider rare variants in South American Indian populations, which are assumed to have arisen from a single mutant gene and have spread only in a very localised area. A branching process model for the actual numbers in which a variant is replicated is therefore considered.

     Assuming a generalised geometric offspring distribution, and using a likelihood approach, it is shown that the total numbers in which a variant is replicated can be used to provide an estimate of a single function of the age and offspring distribution parameters, and hence of any one of these provided the others are known. However, in order to obtain estimates both of selection effects and of age, further information is required.

     One type of possible information is provided by the distribution of numbers of different variants, and estimation on the basis of this is considered. However, the practical applicability of this method is doubtful since it involves the assumption that all the variants under consideration have offspring distributions with the same parameter values.

A more useful line of inquiry is estimation on the basis of the spread of a single variant. The distribution of variant numbers amongst villages can be viewed as a multitype branching process, where the "type" of a variant gene is the village in which it is located. The t th generation mean vector and variance-covariance matrix may be found, and these used to obtain formulae for the mean and variance of the moments of the spatial distribution of the variant; the special case of linear migration is considered as an example. Using this approach it is possible to obtain simultaneous estimates both of the age and the rate of increase of the variant, provided the migration pattern is known. It is found that potentially useful estimates of the age of a variant may be made, provided that this is not too large, but with regard to the rate of increase the confidence limits for the estimate will be very wide; there is little information about this rate contained in the spread of the variant, other than is contained in the total numbers alone.

Acknowledgment.

     This work was begun while the author was visiting the Department of Human Genetics, University of Michigan, financed by NSF grant BMS-74-11823.

REFERENCES

Crump, K. S. and Gillespie, J.H.   1976.   The dispersion of a neutral allele considered
     as a branching process.
     J. Appl. Prob. 13: 208-218.

Ewens, W.J.   1972.   Sampling theory of selectively neutral alleles.
     Theor. Pop. Biol. 3: 87-112.

Ewens, W.J.   1973.   Testing for increased mutation rate for neutral alleles.
     Theor. Pop. Biol. 4: 251-258.

Harris, T.E.   1963.   The theory of branching processes.
     Springer Verlag, Berlin.

Karlin, K. and MacGregor, J.   1966.   The number of mutant forms maintained in a popu-
     lation.
     Proc. V Berk. Symp. 4: 415-438.

Kimura, M. and Ohta, T.   1974.   The age of a neutral mutant persisting in a finite
     population.
     Genetics 75: 199-212.

Kojima, K. and Kelleher, T.M.   1962.   Survival of mutant genes.
     Am. Nat. 96: 329-346.

Li, W.H.   1975.   The first arrival time and mean age of a deleterious mutant in a fi-
     nite population.
     Am. J. Hum. Genet. 27: 274-286.

MacCluer, J.W., Neel, J.V. and Chagnon, N.   1971.   Demographic structure of a primitive
     population; a simulation.
     Am. J. Phys. Anthrop. 35: 193-208.

Maruyama, T.   1974.   The age of a rare mutant gene in a finite population.
     Am. J. Hum. Genet. 26: 669-673.

Moran, P.A.P. 1975. Wandering distributions and the electrophoretic profile.
   Theor. Pop. Biol. 8: 318-330.

Neel, J.V. 1973. "Private" genetic variants and the frequency of mutation.
   Proc. Nat. Acad. Sci. USA 70: 3311-3315.

Neel, J.V. and Chagnon, N. 1968. Demography of two tribes of primitive relatively
   unacculturated American Indians.
   Proc. Nat. Acad. Sci. USA 59: 680-689.

Stigler, S. 1970. Estimating the age of a Galton-Watson process.
   Biometrika 57: 505-512.

Thompson, E.A. 1976. Estimation of the age and rate of increase of rare variants.
   Am. J. Hum. Genet. 28: 442-452.

5. HUMAN EVOLUTION

THE GENETICS OF HLA AND DISEASE ASSOCIATIONS

Glenys Thomson and Walter F. Bodmer

During the past few years a number of studies have shown an association between human histocompatibility (HLA) antigens and certain diseases. The search for associations between HLA and specific diseases
in man was stimulated by the discovery, in several animal species, of
specific immune response (Ir) genes closely linked to the species
histocompatibility system. Although early studies in man showed only
weak associations with Hodgkin's and some other diseases, subsequent
studies have shown very striking associations with a number of diseases. Many of these diseases have a presumptive or suspected auto-
immune aetiology, including ankylosing spondylitis, psoriasis, coelica
disease, myasthenia gravis and multiple sclerosis. Weak associations
have been found with some cancers and with some diseases which are
possibly due to infectious agents (McDevitt and Bodmer 1974; Svejgaard *et al.* 1975).

It is important when discussing disease association to remember
that the HLA system is a linked complex of many genes, perhaps hundreds
or even a few thousands. Genes so far identified control a variety
of functions connected with the cell surface and the immune system.
The original serologically detected antigens of the HLA system, present in almost all cell types, behave as if they were controlled by
multiple alleles at three loci. These loci are defined as HLA-A, HLA-B
and HLA-C (previously LA, FOUR and AJ). Another well defined locus is

HLA-D which controls mixed lymphocyte culture (MLC) response. On the
basis of observed recombinants the map order of these four loci seems
to be D, B, D, A. The recombination fraction between the D and B loci
is .008, between the B and C loci .002, and between the B and A loci
.008. Loci controlling immune response (Ir) and probably immune associated (Ia) cell surface determinants are also found within the HLA
region, although their map positions are as yet unknown.

Almost all the cases of increased frequency of HLA antigens have
involved pronounced increased incidences of a B locus antigen (see
Svejgaard et al. 1975), a notable exception being the association of
HLA-A3 with multiple sclerosis. MLC typing has been done in only a
limited number of diseases. However, it turns out that in some cases
the D locus MLC determinants show a stronger association with a disease than do the B locus antigens, as for example, DW2 with multiple
sclerosis and, probably, DW3 with coeliac disease.

McDevitt and Bodmer (1972,1974) and others suggested that the associations between HLA and disease are not due to the direct effects
of the serologically determined antigens or MLC determinant but to
the effects of closely linked loci, Immune response (Ir) genes. This,
at the present time, is the most popular candidate among the factors
which are thought to be responsible for associations between HLA
and disease though at the moment the evidence favouring the involve-
ment of Ir genes is largely circumstantial. Other mechanisms involving
the direct effects of the antigens have been discussed.

However, it certainly seems most reasonable to make the assumption
that there are loci within the HLA region which control susceptibility
to disease. Throughout the following we will assume that such loci
exist and will refer to them simply as "disease" loci. The observed
associations of diseases with the HLA loci would then imply that
there is significant linkage disequilibrium between the alleles at the
"disease" loci and those at the HLA loci. The stronger associations of
the B and D loci with diseases implies that the "disease" loci are
most likely closer to these two loci than to the A and C loci. Associations due to linkage disequilibrium are likely to be much more
complex than those resulting from say a simple direct antigenic effect, and may vary substantially from one population to another.

The diseases under consideration are ones which do not show simple
Mendelian segregation, but nevertheless in many cases have an obvious
inherited component. The evidence that heredity plays some part comes
from the usual observation that the incidence of the disorder is high-

er among the relatives of affected individuals than it is in the general population. Several different modes of transmission have been
proposed to explain the inheritance patterns of these types of diseases. These range from single locus two-allele models with variable
penetrance in one or all genotypes to multifactorial models in which
many genes of small effect and environmental sources of variance play
a role. Attempts to distinguish between the two modes of transmission
using family data have generally not been very successful (Smith 1971,
Kidd and Cavalli-Sforza 1973). Despite this, we will in this paper
consider only disease susceptibility models where there is a single
"disease" locus with a major detectable effect. (The model does not
exclude the possibility that other genes are involved in the disease,
but assumes that there is a single locus with a major effect). There
is in fact a strong rationale for believing that these may be the appropriate models to consider in the analysis of diseases showing
an association with HLA. If we exclude the possibility that it is the
HLA antigens themselves which are the factors predisposing an individual to the disease, and there is strong evidence that we can do this,
then the most likely explanation of an observed association is that
there is a major "disease" gene, or complex of tightly linked genes,
close to or essentially a part of the HLA region, and in linkage disequilibrium with certain alleles of the A, B, C or D loci of the HLA
system. It is much more difficult to explain the associations on
a multifactorial basis with no detectable major effect of an HLA
linked gene. So throughout the following we will consider models with
single "disease" loci. In the following two methods are outlined which
will make it possible to discriminate between dominant and recessive
modes of inheritance of the "disease" genes. The first method involves
looking at HLA genotype frequencies amongst diseased. The second method
looks at the HLA similarity of affected sib pairs.

ANTIGEN GENOTYPE FREQUENCIES AMONGST DISEASED

A simple two locus disease association model is considered. The first
locus, denoted A, is the antigen locus. The two alleles A and a denote
the presence or absence of a particular antigen. The second locus, denoted D, with alleles D and d, is the disease susceptibility locus
(henceforth we will often refer to this simply as the "disease" locus).

The use of the letter A and D here should not be confused with the loci HLA-A and HLA-D of the HLA system mentioned in the introduction. There are four gametic or chromosome types possible (also referred to as haplotypes) namely AD, Ad, aD and ad. Their frequencies are denoted by x_1, x_2, x_3 and x_4 respectively. The gametic frequencies can be written in two forms. Both notations will be given and used interchangeably throughout the analyses. This may seem unnecessarily complicated but it turns out that it is much simpler to express some quantities in one or other of the notations.

The first notation is the well-known formulation in terms of gene frequencies and linkage disequilibrium, namely

$$x_1 = p_A p_D + D, \ x_2 = p_A p_d - D, \ x_3 = p_a p_D - D, \ x_4 = p_a p_d + D \quad (1)$$

where p_A, p_D are the frequencies of the alleles A and D respectively, etc. and $D = x_1 x_4 - x_2 x_3$ is the coefficient of linkage disequilibrium.

The second notation is in terms of gene frequencies and the ratio of antigen presence to absence ($k : 1-k$) in gametes containing the "disease" allele D, namely

$$x_1 = k p_D, \ x_2 = p_A - k p_D, \ x_3 = (1-k) p_D, \ x_4 = p_a - (1-k) p_D. \quad (2)$$

The two notations are completely interchangeable since

$$k = p_A + D/p_D \quad (3)$$

The limits on the value of k are

$$\max (0, \ 1 - p_a/p_D) < k < \min (1, \ p_A/p_D) \quad (4)$$

Two models will be considered in the sequel.

For the dominant model individuals with "disease" genotypes DD and Dd are susceptibles and dd individuals are non-susceptibles. For the recessive model only DD individuals are susceptibles, Dd and dd individuals are non-susceptibles. For both dominant and recessive models a 2 × 2 table denoting presence or absence of disease susceptibility versus presence or absence of antigen can be set up. The genotypes making up the four phenotypic classes and the frequencies of the classes can be written in the forms given in Tables 1 and 2 for the dominant and recessive cases respectively. The frequencies of the classes

Table 1.   2 × 2 table for the dominant disease model. The four classes
denote presence or absence of disease susceptibility versus
presence or absence of the antigen. The genotypes making up
these classes and their frequencies e, f, g and h, are given
in terms of the antigen and "disease" gene frequencies and
$\Delta = D(D + 2p_a p_d)$ (see equation (1) and Bodmer and Bodmer 1974)

Disease sus-	Antigen		Σ
ceptibility	+	−	
+	$AD/AD,\ AD/Ad,\ AD/aD,$ $AD/ad,\ Ad/aD$ $e_{dom} = (1-p_a^2)(1-p_d^2) + \Delta$	$aD/aD,\ aD/ad$ $f_{dom} = p_a^2(1-p_d^2) - \Delta$	$1-p_d^2$
−	$Ad/Ad\quad Ad/ad$ $g_{dom} = p_d^2(1-p_a^2) - \Delta$	ad/ad $h_{dom} = p_a^2 p_d^2 + \Delta$	p_d^2
Σ	$1-p_a^2$	p_a^2	

e, f, g and h in a population cannot themselves be measured since it is
not possible to determine the disease susceptibility status of an individual. However, by making the assumption that a proportion x of susceptibles get the disease then a 2 × 2 table of disease status versus
presence or absence of antigen can be set up, as shown in Table 3. For
both the dominant and recessive models, the frequencies of the four
classes in Table 3 are given by

$$r = xe,\ s = xf,\ t = g + (1-x)e,\ u = h + (1-x)f, \qquad (5)$$

where the appropriate values of e, f, g and h for the dominant and
recessive models are obtained from Tables 1 and 2 respectively. The
frequencies of the classes r, s, t and u in a population are observable, with a certain degree of accuracry.

Now in these two models of disease association we have a number of

Table 2. As for Table 1 except that for the recessive disease model
$\delta = D(2p_a p_D - D)$

Disease sus-ceptibility	Antigen		Σ
	+	−	
+	$AD/AD, AD/aD$	aD/aD	
	$e_{rec} = p_D^2(1-p_a^2) + \delta$	$f_{rec} = p_D^2 p_a^2 - \delta$	p_D^2
−	$AD/Ad, AD/ad, Ad/Ad$ $Ad/aD, Ad/ad$	$aD/ad, ad/ad$	
	$g_{rec} = (1-p_D^2)(1-p_a^2) - \delta$	$h_{rec} = (1-p_D^2)p_a^2 + \delta$	$1-p_D^2$
Σ	$1-p_a^2$	p_a^2	

Table 3. 2 × 2 table denoting classes with presence or absence of disease versus presence or absence of antigens. $r = xe$, $s = xf$, $t = g + (1-x)e$, $u = h + (1-x)f$, see equation (5) and Tables 1 and 2

Disease	Antigen	
	+	−
+	r	s
−	t	u

Table 4. Variables in the "disease" gene models

---

p_A	antigen gene frequency. Known from population data
p_D	"disease" gene frequency
D (or k)	association of antigen and "disease" loci
x	proportion of susceptibles who get disease
R	recombination fraction between the two loci

Dominant or recessive "disease" gene

Table 5. The expected values of the frequency of the disease in the
population (FDP), the frequency of the antigen amongst the
diseased (FAD), and the antigen genotype frequencies amongst
the diseased, for the dominant and recessive models. (See
equations (6) and (7) for definitions)

	Dominant	*Recessive*
1. Frequency of disease in population (FDP)	$x(1-p_d^2)$	xp_D^2
2. Frequency of antigen amongst diseased (FAD)	$1-p_a^2 + \Delta/(1-p_d^2)$	$1-(1-k)^2$
3. Antigen genotype frequencies amongst diseased	$AA:\ p_A^2 - \Delta/(1-p_d^2) +$ $2p_d D/(1-p_d^2)$	$AA:\ k^2$
	$Aa:\ 2p_A p_a + 2\Delta/(1-p_d^2) -$ $2p_d D/(1-p_d^2)$	$AA:\ 2k(1-k)$
	$aa:\ p_a^2 - \Delta/(1-p_d^2)^2$	$aa:\ (1-k)^2$

---

variables, only one of which, the antigen gene frequency, is directly
known. These variables are listed in Table 4. In the population various
quantities are observable. The expected values of these quantities, in
terms of the unknown variables, under the two models of dominant and
recessive "Disease" genes, can be calculated. Three quantities which
can be observed in the population and turn out to be useful in discriminating between the two modes of inheritance are the frequency of the
disease in the population (FDP),

$$FDP = r + s \qquad\qquad (6)$$

(see Table 3, Tables 1 and 2, and equation (5)), the frequency of the
antigen amongst the diseased (FAD),

$$FAD = r/(r+s), \qquad\qquad (7)$$

and the antigen genotype frequencies amongst the diseased denoted
$f_d(AA)$, $f_d(Aa)$ and $f_d(aa)$.

The appropriate values of these quantities for the dominant and recessive cases are given in Table 5.

The expected antigen genotype frequencies for the recessive case
(see Table 5) suggest a very simple test that can be used to determine
if a set of observed genotype frequencies are compatible with a recessive mode of inheritance of the "disease" gene. An estimate of k,
for the recessive case, is given by

$$k_{rec} = 1 - (1 - FAD)(1 - FAD)^{1/2} \qquad\qquad (8)$$

(see Table 5 and equation (7)). Using this value of k the expected antigen genotype frequencies amongst diseased under a recessive model can
then be determined (see Table 5) and a x^2 goodness-of-fit used to test
if the observed genotype frequencies are compatible with those expected. If the true mode of inheritance of the "disease" gene is dominant,
and the frequency of the antigen amongst diseased (FAD) is large compared with the frequency in the general population, this test will allow one to determine that the mode of inheritance is not compatible
with a recessive "disease" gene. Very simply the basis of the test is
as follows. If a "disease" gene is recessive then a high association
of the disease with a particular antigen must imply a high frequency
of individuals homozygous for the antigen amongst the diseased, where-

as a dominant "disease" gene does not imply this.

No such direct test is available for the dominant case, since estimates of the parameters involved cannot be made with only population incidence data. However, if one makes the assumption, in the dominant case, that the "disease" gene frequency p_D is small, further progress can be made. If p_D is small the expressions for the frequency of the antigen amongst diseased (FAD) and the antigen genotype frequencies amongst diseased simplify and are given by the following.

$$FAD_{dom} \approx 1 - (1-k)p_a \text{ (if } p_D \text{ small)} \tag{9}$$

(see Bodmer and Bodmer 1974). The antigen genotype frequencies amongst diseased (for p_D small and dominant "disease" gene) are

$$f_d(AA) = k(1-p_a)$$

$$f_d(Aa) = kp_a + (1-k)p_A \tag{10}$$

$$f_d(aa) = (1-k)p_a$$

where p_A is the antigen gene frequency in the general population. An estimate of k, for the dominant case, is then given by

$$k_{dom} \approx 1 - (1-FAD)/p_a \tag{11}$$

The expected antigen genotype frequencies amongst diseased, under a dominant model (with small "disease" gene frequency) can then be calculated from (10) and a χ^2 goodness-of-fit performed. The logical order of testing then would be as follows. First, see if the observations are compatible or not with a recessive mode of inheritance (test 1), then, test for a dominant mode of inheritance with small "disease" gene frequency (test 2). These tests would not allow one to discriminate between a recessive "disease" gene and a dominant "disease" gene with a high frequency and very low probability x of a susceptible getting the disease.

These tests have been applied to data on ankylosing spondylitis patients from the Oxford file (Hill, Hill and Bodmer 1976). Twenty-three patients with a definite diagnosis of ankylosing spondylitis were HLA typed. The antigen HLA-B27 has a very strong assocaition with ankylosing spondylitis (McDevitt and Bodmer 1974; Svejgaard et al. 1975)

Of the twenty-three patients only one lacked the antigen HLA-B27
twenty-one were confirmed heterozygotes and one was a suspected homozygote for HLA-B27. The test for a recessive mode of inheritance of
the "disease" gene was applied to this data (test 1). If the mode of
inheritance is recessive then, from (10), the estimate of k is

$$\hat{k}_{rec} = .7914 \tag{12}$$

(in this example FAD = 22/23 = .9565). With this estimate of k it
follows, from Table 5, that the expected number of individuals in the
three antigen genotype classes (denoted AA, Aa and aa where A denotes
the antigen HLA-B27) amongst the diseased, for the recessive model, are
14.41, 7.60 and 1.0 respectively. A χ^2 goodness-of-fit of the observed
values 1, 21 and 1 to these expected gives $x_1^2 = 36.13$ (P < .001). So
these observations are certainly not compatible with a recessive mode
of inheritance. The next step is to test if the data is compatible with
a dominant mode of inheritance of a "disease" gene with small frequency
(test 2). The appropriate estimate of k in this case is, from (13) and
taking p_A = .0357 (from McDevitt and Bodmer 1974),

$$\hat{k}_{dom} = .9549 \tag{13}$$

The expected antigen genotype frequencies amongst diseased are then,
from (12), 0.78, 21.22 and 1.0 respectively for the classes AA, Aa
and aa, where A denotes HLA-B27. A χ^2 goodness-of-fit of the observed
values 1, 21 and 1 to these expected gives $\chi_1^2 = .0642$. Thus, the observations are certainly compatible with a dominant "disease" gene
with low frequency. It follows from (4) that for the estimate of k giv-
en by (15) the upper limit for the dominant "disease" gene frequency
p_D is .037. (If the presumed HLA-B27 homozygous individual is in fact
heterozygous for HLA-B27 and a blank antigen the above results still
obtain. For test 1 $\chi_1^2 = 41.72$ (p < .001) and for test 2 $\chi_1^2 = .8086$).
    As pointed out above these tests will only have strong discrimina-
tory powers in cases where the frequency of the antigen amongst diseased (FAD) is large compared to the frequency of the antigen in the general population. The exact limits on the variables have not been determined. It remains to be seen if the tests are useful in many cases.
The problem is that although a number of associations of diseases with
HLA have been found the majority of the associations are probably not
strong enough to allow the above tests to discriminate the appropriate

mode of inheritance. This follows from the fact that the goodness of
fit tests being used have very weak power. However, as more loci are
determined within the HLA region, and in particular as "typing" methods
for Ir and Ia genes are found, it is possible that stronger associations
for certain diseases will be found so that the above tests would then
become applicable.

HLA SIMILARITY OF AFFECTED SIB PAIRS

Day and Simons (1976) have recently developed an elegant technique for
studying the inheritance pattern of HLA haplotypes in sibs who are both
affected by a disorder. Their models are more general than those considered in the previous section but the basic assumption is the same,
namely that there is a disease susceptibility locus close to the HLA
region. The method used by Day and Simons considers the HLA region as
a single entity by means of multiple case family studies. By looking
at inheritance patterns within families we know that the disease susceptibility gene will be inherited with the HLA haplotype, discounting
the possibility of recombination, which is rare within the HLA region
and can often be detected and so eliminated from the analyses. Thus
the basis of the method involves the use of HLA haplotypes as markers
to trace the inheritance pattern of closely linked disease susceptibility genes. The existence of a disease susceptibility gene is demonstrated by showing that related cases have greater HLA similarity
than would be expected by chance.

Day and Simons assume that all genotypes are susceptible to getting
the disease but those carrying the "disease" gene D carry a greater
risk, given by R, of contracting the disease, relative to a risk of 1
otherwise. So for the recessive model the relative risks of disease
for the three "disease" genotypes are (Day and Simons model)

$$DD \quad Dd \quad dd$$
$$R \quad\quad 1 \quad\quad 1 \tag{14}$$

and for the dominant model

$$
\begin{array}{ccc}
DD & Dd & dd \\
R & R & 1
\end{array}
\qquad (15)
$$

The basis feature of the method is that only diseased individuals are
used in the calculations, since for individuals without the disease one
cannot determine whether the disease will develop later or not.

For various values of p_D (the disease susceptibility gene frequency)
and R (the increased risk carried by those with the "disease" genotype)
Day and Simons calculated the probability of affected sibs and cousins
having two, one and no HLA haplotypes in common, under recessive and
dominant models. From observed data their tables can be used to estimate a minimum value of R for the recessive and dominant models, although the method does not distinguish whether the "disease" gene is
recessive or dominant.

In the following the method proposed by Day and Simons will be applied to the disease susceptibility models presented in the last section. As pointed out above these models are simpler than those proposed
by Day and Simons. The rationale for looking at these models, as distinct from the more complicated ones of Day and Simons, is that logically the best method of determining the mode of inheritance of a disease is to assume, as a first step, the simplest possible model. If
the observed data turns out to be compatible with the predictions of
this simple model, then this is probably the best explanation of the
underlying process. If the observations are not compatible with a
simple model then more complex models must be considered, but initially it seems appropriate to consider the simplest models. So, in
the simplest case, for the recessive model, the relative risks for the
three "disease" genotype are (compare with (14))

$$
\begin{array}{ccc}
DD & Dd & dd \\
x & 0 & 0
\end{array}
\qquad (16)
$$

and for the dominant model (compare with (15))

$$
\begin{array}{ccc}
DD & Dd & dd \\
x & x & 0
\end{array}
\qquad (17)
$$

(this effectively assumes that $R (\approx x) \gg 1$).

The method follows exactly that given by Day and Simons and the reader is referred to their paper for details of the analysis. A simple outline of the steps involved is as follows. The possible parental disease chromosome types are determined together with their respective prior probabilities. For each chromosomal combination of the parents the relative likelihood of getting a sib pair with the disease is calculated. Using Bayes' theorem these relative likelihoods can be combined with the prior probabilities of particular parental combinations to give the probability of each parental chromosome arrangement given the occurrence of a sib pair with the disease. For each arrangement the probabilities of the diseased sibs having both or one HLA haplotype in common can be calculated. These probabilities are then multiplied by the posterior probability of the particular arrangement to give the overall probability of an affected sib pair having both, one or no haplotypes in common.

It turns out for the recessive case that the probabilities of an affected sib pair sharing both, one or no HLA haplotypes, denoted by X, Y and Z respectively, are given by

$$X_{rec} = 1/(1+p_D)^2, \quad Y_{rec} = 2p_D/(1+p_D)^2, \quad Z_{rec} = p_D^2/(1+p_D)^2 \quad (18)$$

Note that by using data only from affected sib pairs the parameter x is eliminated and the probabilities of HLA similarity of affected sib pairs given by (18), are dependent only on the (unknown) disease susceptibility gene frequencies, p_D. These probabilities, for particular values of p_D, are given in Table 6.

For the dominant case the respective probabilities of affected sib pairs having both, one or no haplotypes in common are

$$X_{dom} = (1 + p_d)/[4(1 + p_D p_d) + p_D p_d^2] \qquad (19)(i)$$

$$Y_{dom} = 2(1 + p_D p_d)/[4(1 + p_D p_d) + p_D p_d^2] \qquad (19)(ii)$$

$$Z_{dom} = (1 + p_D p_d - p_d^3)/[4(1 + p_D p_d) + p_D p_d^2] \qquad (19)(iii)$$

Again these probabilities are independent of x and dependent only on the (unknown) disease susceptibility gene frequency p_D. Various values of these probabilities are tabulated in Table 7.

The question now is to determine if a given set of observed frequencies of sib pairs who share both, one or no HLA haplotypes can

Table 6. The probabilities of an affected sib pair sharing both, one
or no HLA haplotypes (X, Y and Z respectively) for various
values of the disease susceptibility gene frequency p_D,
for the recessive case

p_D	X	Y	Z
.0	1.0	.0	.0
.05	.907	.091	.002
.1	.827	.165	.008
.15	.756	.227	.017
.2	.694	.278	.028
.25	.640	.320	.040
.3	.592	.355	.053
.35	.549	.384	.067
.4	.510	.408	.082
.45	.476	.428	.096
.5	.445	.444	.111
.55	.416	.458	.126
.6	.391	.469	.140
.65	.367	.478	.155
.7	.346	.484	.170
.75	.326	.490	.184
.8	.309	.494	.198
.85	.292	.497	.211
.9	.277	.499	.224
.95	.263	.5	.237
1.0	.25	.5	.25

Table 7. As for Table 6 for the dominant case

p_D	X	Y	Z
.0	.5	.5	.0
.05	.460	.495	.045
.1	.428	.491	.081
.15	.401	.488	.111
.2	.377	.487	.136
.25	.358	.486	.156
.3	.341	.485	.174
.35	.326	.485	.189
.4	.313	.486	.201
.45	.302	.487	.211
.5	.293	.488	.219
.55	.284	.489	.227
.6	.277	.490	.323
.65	.271	.492	.237
.7	.265	.494	.241
.75	.261	.495	.244
.8	.257	.497	.246
.85	.254	.498	.248
.9	.252	.499	.249
.95	.25	.5	.25
1.0	.25	.5	.25

be used to distinguish whether a disease susceptibility gene is dominant or recessive. Unfortulately, the maximum likelihood estimate of
p_D for the dominant case has no simple analytical solution so that a
calculation of the relative likelihood of the two models would have to
be done numerically. This involves solution of a 6th degree polynomial. This exact approach will not be used here, instead a very simple
graphical method will be used. This is discussed later.

Day and Simons in their calculations used an observed frequency of
the occurrence of both HLA haplotypes in sib pairs to obtain a confi-
dance interval for p_D and R values. Similarly in the present case an
observed frequency of sharing both HLA haplotypes in sib pairs could
be used to obtain a confidence interval for p_D, in both the dominant and
recessive models. However, we have more information available than this.
The tabulations in Tables 6 and 7 give us information on the expected
frequencies of the three classes of affected sibs,namely those sharing
both, one or no HLA haplotypes. What we need is a test to use all this
information in helping to discriminate between the two possible modes
of inheritance. As pointed out above the exact test would be to calculate the maximum likelihood estimates of p_D for the dominant and recessive models and calculate the relative likelihoods of the two models.
This is numerically complicated. Instead, a graphical method will be
used. The graphical method has the advantage of giving a clear demonstration of the basis of the method of discrimination involved.

The graphical method (suggested by Nick Weed) involves plotting
the relative values of the three classes of affected sibs on a trian-
gular diagram (the di Finetti representation). The dominant and recessive models give quite different sets of possible values of X, Y and
Z (the respective probabilities of an affected sib pair sharing both,
one or no haplotypes). The possible values of X, Y and Z for the two
models, are plotted in Figure 1.

The two curves have two points in common. One is (1/4, 1/2, 1/4)
consistent with p_D = 1 in both the recessive and dominant cases. The
other is (.31, .49, .20) consistent with p_D = .386 in the dominant case
and with p_D = .804 in the recessive case. For these two points the
dominant and recessive models are indistinguishable. This is not a
problem in the present analysis since we are primarily concerned with
diseases whose gene frequency is assumed to be small.

In any particular case the mode of inheritance of a disease is determined by plotting the observed set of (X,Y,Z) values on the composition triangle and seeing which curve, that is dominant or recessive,

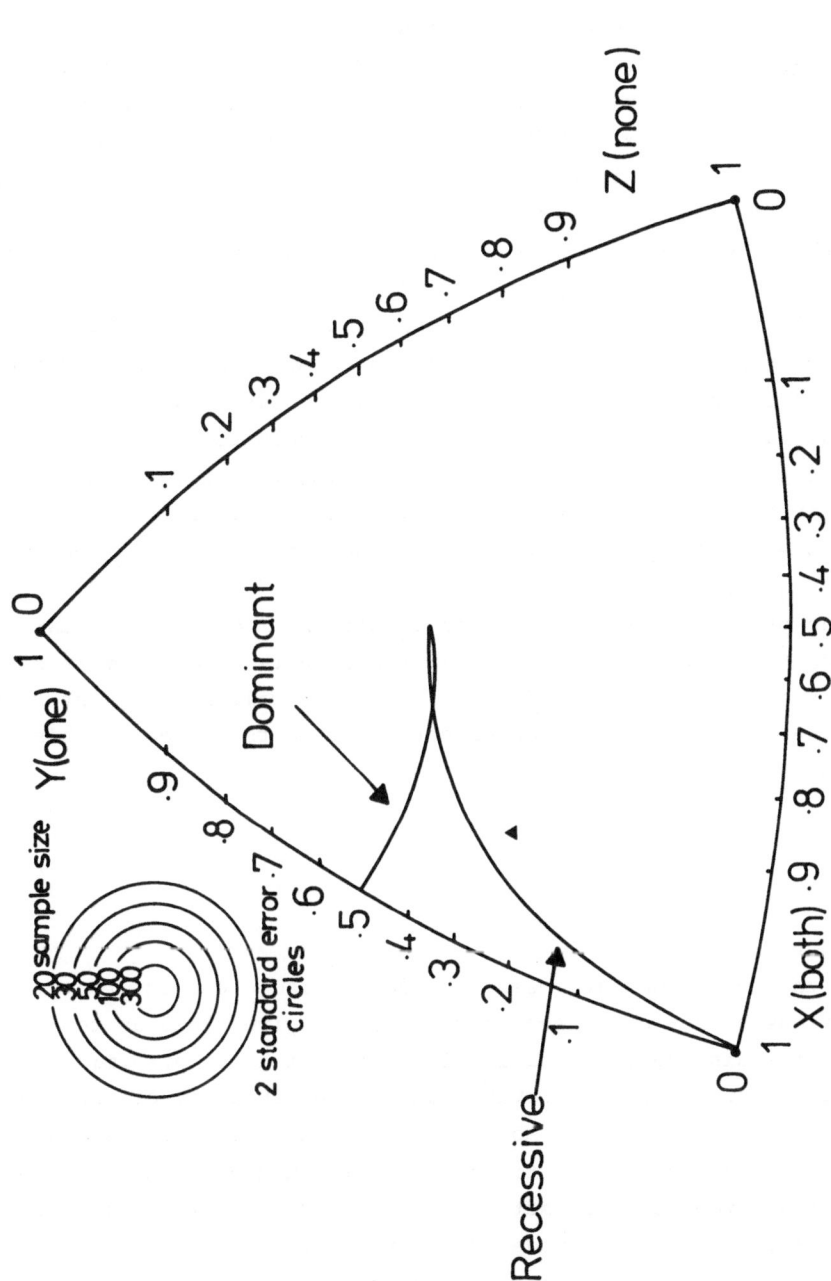

Figure 1. The relative values of the three classes of affected sibs *AA*, *Aa* and *aa* under dominant and recessive disease models. See text for details. The figure represents a mapping of the ordinary composition triangle with the property that the error is linear (supplied by A.W.F. Edwards). The point ▲ represents data on fifteen sib pairs affected with juvenile diabetes mellitus (Cudworth and Woodrow 1975).

the data is most compatible with.

There is an effect of the method of ascertainment of the affected
sib pairs on the expected values of X, Y and Z. If we are only dealing
with families of size 2 or if we just choose a sib pair at random from
families and discard those without both sibs affected there is no problem and the above analysis is exact. However, if we are dealing with
cases where we search for affected sib pairs then family size enters
into the calculations. (Nick Wedd, personal communication). These results will be published elsewhere. However, the main point is that the
general method outlined above remains valid.

Cudworth and Woodrow (1975) have presented family data on juvenile
diabetes mellitus which is appropriate for the above analysis. They
considered 15 affected sib pairs and found that 10 of these shared
both HLA haplotypes, 4 had one haplotype in common and 1 sib pair had
no haplotypes in common. Day and Simons used this observation to estimate minimum values for R of 9 and 19 in the recessive and dominant
models respectively. From the composition triangle, Figure 1, and Table
6 we see that these observed values of X, Y, Z (.667, .267, .066) are
quite consistent with a recessive mode of inheritance of a "disease"
gene with frequency p_D = .22.

With the present small sample size a dominant mode of inheritance
cannot be completely ruled out although the results are much more compatible with a recessive mode of inheritance. It will be very interesting to obtain further HLA data on affected sib pairs to see if this result can be confirmed.

DISCUSSION

The aim of the models we have been considering is to use information
on associations of various diseases with HLA to give clues as to the
genetics of the diseases. The basis of the models is to determine the
expected values of various observable quantities in terms of the unknown variables, under the two models of dominant and recessive "disease" genes. Using these predictions it should then be possible to
determine if a given observed set of data is more likely compatible
with a dominant or recessive mode of inheritance. It is important to
make use of all possible information when trying to determine the mode
of inheritance of a disease. If one test indicates that say a recessive

mode of inheritance is the most probable explanation of the data then
it is necessary to test if all the other tests and observations are
compatible with a recessive model.

It is of course difficult to get extremely accurate estimates for
many of the quantities of interest in studies of disease inheritance.
Complications such as variable age of onset of a disease, difficulties in obtaining accurate diagnoses in pedigree work, etc. make the
estimation of parameters difficult. However, the tests outlined above
should enable a much clearer demarkation of results expected under the
two modes of inheritance than has previously been possible. So the
fact that estimates of some quantities are not that accurate need not
be crucial. For instance, in cases where the frequency of an antigen
amongst diseased is high as compared to the population frequency of
the antigen the differences in expected antigen genotype frequencies
amongst diseased for the dominant and recessive models are quite striking. At the moment there are a limited number of diseases which fall
into this category. However, as more loci are determined within the
HLA region it is possible that stronger associations will be found for
some diseases. The test using expected antigen genotype frequencies
would then provide a great deal of information regarding the inheritance of the disease gene. The method developed by Day and Simons
(1975) has the advantage that it enables one to get close to the actual disease susceptibility loci. This is done by using the HLA haplotypes as markers to trace the inheritance pattern of closely linked
disease susceptibility genes. This method is useful even in cases
where the association of the disease with HLA is weak.

So far very little has been done to apply the above tests to diseases known to have an association with HLA. The relevant data is limited at present, particularly data on affected sib pairs. However, the
two preliminary results indicating a dominant gene for ankylosing spondylitis and a recessive gene for juvenile diabetes mellitus show that
the analyses may be very useful in determining modes of inheritance of
"disease" genes. Further information and analyses on other diseases
will be of interest.

Summary.

A number of studies in recent years have shown an association between histocompatibility (HLA) antigens and certain diseases. It seems reasonable to make the assumption that there are loci within the HLA region which control susceptibility to
disease. The observed associations of diseases with the HLA loci would then imply
that there is significant linkage disequilibrium between the alleles at the "disease" loci and those at the HLA loci. Assuming single "disease" loci two methods
are outlined which will make it possible to discriminate between dominant and recessive modes of inheritance of "disease" genes and possibly estimate their population frequency. The first method involves looking at HLA genotype frequencies amongst
diseased. The second method looks at the HLA similarity of affected sib pairs. Preliminary data suggests that susceptibility to ankylosing spondylitis is determined
by a dominant gene and to juvenile diabetes mellitus by a recessive gene.

REFERENCES

Bodmer, W.F., and Bodmer, J.G. 1974. The HL-A Histocompatibility Antigens and Disease.
Tenth Symposium in Advanced Medicine, pp. 157-174. Pitman Medical Press, London.

Carter, C.O. 1974. Multifactorial Genetic Disease.
In *Medical Genetics* 1974, pp. 191-208. Edited by V.A. McKusick and R. Claiborne.
H.P. Publishing Co., New York.

Cudworth, A.G., and Woodrow, J.C. 1975. Evidence for HL-A linked genes in "juvenile"
diabetes mellitus.
Br. Med. J. 3: 133-135.

Day, N.E., and Simons, M.J. 1975. Disease Susceptibility Genes - their identification by multiple case family studies.
Tissue Antigens 8: 109-119.

Hill, F.A., Hill, A.G.S., and Bodmer, J.G. 1976. The Clinical Diagnosis of Ankylosing Spondylitis in Women and Relation to Presence of HLA-B27.

Kidd, K.K., and Cavalli-Sforza, L.L. 1973. An analysis of the genetics of schizo-
phrenia.
Soc. Biol. 20: 254-265.

McDevitt, H.O., and Bodmer, W.F. 1972. Histocompatibility antigens immune respon-
siveness and susceptibility to disease.
Amer. J. Med. 52.: 1-8.

McDevitt, H.O., and Bodmer, W.F. 1974. HL-A, Immune Response Genes and Disease.
Lancet, pp. 1269-1275.

Smith, C. 1971. Discriminating between different modes of inheritance in genetic
disease.
Clin. Genet. 2: 303-314.

Svejgaard, A., Platz, P., Ryder, L.P. Nielsen, L.S., and Thomsen, M. 1975. HL-A
and disease associations - A survey.
Transplant. Rev. 22: 3-43.

Wilson, S.R. 1971. Fitting of models of incomplete penetrance to family data.
Ann. Hum. Genet. 35: 99-108.

Wilson, S.R. 1974. Fitting of multifactorial models to family data.
Ann. Hum. Genet. 38: 231-241.

Editors: K. Krickeberg;
S. Levin; R. C. Lewontin;
J. Neyman; M. Schreiber

Biomathematics

Vol. 1:   **Mathematical Topics in Population Genetics**
Edited by K. Kojima
55 figures. IX, 400 pages. 1970
ISBN 3-540-05054-X

This book is unique in bringing together in one volume many,
if not most, of the mathematical theories of population
genetics presented in the past which are still valid and some
of the current mathematical investigations.

Vol. 2:   E. Batschelet
Introduction to Mathematics for Life Scientists
200 figures. XIV, 495 pages. 1971
ISBN 3-540-05522-3

This book introduces the student of biology and medicine to
such topics as sets, real and complex numbers, elementary
functions, differential and integral calculus, differential equations, probability, matrices and vectors.

M. Iosifescu; P. Tautu
Stochastic Processes and Applications in Biology and Medicine

Vol. 3:   Part 1: Theory
331 pages. 1973

ISBN 3-540-06270-X

Vol. 4:   Part 2: Models
337 pages. 1973
ISBN 3-540-06271-8

Distribution Rights for the Socialist Countries: Romlibri,
Bucharest
This two-volume treatise is intended as an introduction for
mathematicians and biologists with a mathematical background
to the study of stochastic processes and their applications in
medicine and biology. It is both a textbook and a survey of the
most recent developments in this field.

Vol. 5:   A. Jacquard
The Genetic Structure of Populations
Translated by B. Charlesworth; D. Charlesworth
92 figures. Approx. 580 pages. 1974
ISBN 3-540-06329-3

Population genetics involves the application of genetic information
to the problems of evolution. Since genetics models based on
probability theory are not too remote from reality, the results
of such modeling are relatively reliable and can make important
contributions to research. This textbook was first published
in French; the English edition has been revised with respect
to its scientific content and instructional method.

Springer-Verlag
Berlin
Heidelberg
New York